CIVILIZATION AND THE CULTURE OF SCIENCE

Stephen Gaukroger, who was educated at the University of London and the University of Cambridge, is Emeritus Professor of History of Philosophy and History of Science at the University of Sydney. He is the author of fourteen books and the editor of nine collections of essays. His recent publications include *The Emergence of a Scientific Culture: Science and the Shaping of Modernity, 1210–1685* (Oxford 2006), *The Collapse of Mechanism and the Rise of Sensibility: Science and the Shaping of Modernity, 1680–1760* (Oxford 2010), and *The Natural and the Human: Science and the Shaping of Modernity, 1739–1841* (Oxford 2016). His work has been translated into Arabic, Chinese, French, German, Italian, Portuguese, and Russian.

How did science come to have such a central place in Western culture? How do cognitive values—and subsequently moral, political, and social ones—come to be modelled around scientific values? In *Civilization and the Culture of Science*, Stephen Gaukroger explores how these values were shaped and how they began, in turn, to shape those of society. The core nineteenth- and twentieth-century development is that in which science comes to take centre stage in determining ideas of civilization, displacing Christianity in this role. Christianity had provided a unifying thread in the study of the world, however, and science had to match this, which it did through the project of the unity of the sciences. The standing of science came to rest or fall on this question, which the book sets out to show in detail is essentially ideological, not something that arose from developments within the sciences, which remained pluralistic and modular. A crucial ingredient in this process was a fundamental rethinking of the relations between science and ethics, economics, philosophy, and engineering. In his engaging description of this transition to a scientific modernity, Gaukroger examines five of the issues which underpinned this shift in detail: changes in the understanding of civilization; the push to unify the sciences; the rise of the idea of the limits of scientific understanding; the concepts of 'applied' and 'popular' science; and the way in which the public was shaped in a scientific image.

T0177848

Civilization and the Culture of Science

Science and the Shaping of Modernity, 1795–1935

STEPHEN GAUKROGER

OXFORD
UNIVERSITY PRESS

OXFORD
UNIVERSITY PRESS

Great Clarendon Street, Oxford, OX2 6DP,
United Kingdom

Oxford University Press is a department of the University of Oxford.
It furthers the University's objective of excellence in research, scholarship,
and education by publishing worldwide. Oxford is a registered trade mark of
Oxford University Press in the UK and in certain other countries

© Stephen Gaukroger 2020

The moral rights of the author have been asserted

First published 2020
First published in paperback 2022

Impression: 3

All rights reserved. No part of this publication may be reproduced, stored in
a retrieval system, or transmitted, in any form or by any means, without the
prior permission in writing of Oxford University Press, or as expressly permitted
by law, by licence or under terms agreed with the appropriate reprographics
rights organization. Enquiries concerning reproduction outside the scope of the
above should be sent to the Rights Department, Oxford University Press, at the
address above

You must not circulate this work in any other form
and you must impose this same condition on any acquirer

Published in the United States of America by Oxford University Press
198 Madison Avenue, New York, NY 10016, United States of America

British Library Cataloguing in Publication Data
Data available

Library of Congress Cataloging in Publication Data
Data available

ISBN 978–0–19–884907–0 (Hbk.)
ISBN 978–0–19–286628–8 (Pbk.)

DOI: 10.1093/oso/9780198849070.001.0001

Printed and bound in Great Britain by
Clays Ltd, Elcograf S.p.A.

Links to third party websites are provided by Oxford in good faith and
for information only. Oxford disclaims any responsibility for the materials
contained in any third party website referenced in this work.

Preface

This is a book about the relation between science and 'Western' culture in the nineteenth and early twentieth centuries. I have taken Condorcet's 1795 essay on the 'progress of the mind' as offering the first fully fledged statement of the view that scientific progress is distinctive of Western civilization. As my *terminus ad quem* I've taken the notional date of 1935, standing in for the period in which there were a number of high-profile efforts to deploy science in an attempt to revive the cosmopolitan Western culture lost in the Great War.

The present volume is the fourth and final in a series which I began thinking about in the mid 1990s. Dissatisfied with triumphalist accounts of the development of modern science on the one hand, and social and cultural reductionist ones on the other, I started to look at the historiographical question of whether there was a more fruitful way of studying the development of the sciences, one that explored the relation between science and culture in the West. I had originally thought in terms of a comparative history of science—comparing 'Western' science with the culturally very different Chinese, Arab-Islamic, and Iberian-Catholic scientific programmes—and although I learned much from this, it ultimately proved unworkable. After a number of false starts, I concluded that what was needed was a long-term investigation into science and the shaping of modernity. A theme that has gradually come to the fore in these volumes has been the way in which questions that were once treated as purely conceptual problems were transformed into empirical ones, where, among the empirical resources drawn upon, history figured prominently. This has been a feature not only of the subject matter of the volumes, however, but also of the approach that I have followed. History provides a lever with which to prize open questions that would otherwise remain hidden and intractable, and I have set out to show that questions of the nature and standing of science, which philosophers and others have tended to treat as purely conceptual questions, have a profound, formative historical content.

Probing this historical content is the only way to deepen our understanding of how science has shaped, and in turn been shaped by, the values of our own society. The project as a whole has of necessity a long time-span and a broad coverage, and one of the challenges has been to refine and tailor the varied resources that I have drawn on so as to yield something with a clear focus and with genuine explanatory power. In particular, *longue durée* history requires close and constant attention to historiographical questions, and I have set out to write conceptually informed history.

To the extent to which I have been successful in generating a coherent and fruitful strategy in the present volume, I have benefitted from conversations and correspondence with: Thomas Ahnert, Max Bennett, Ada Bronowski, Stephanie Buchenau, Sharon Carleton, Alan Chalmers, Peter Cryle, Peter Harrison, Helen Irving, Frank James, Roy MacLeod, Jennifer Mensch, Jennifer Milam, Charlotte Morel, Dalia Nassar, Michael Olson, Lydia Patton, Bill Reddy, Dean Rickles, Margaret Schabas, Elizabeth Stephens, Udo Thiel, and Grahame Thompson. I am particularly grateful to Laura Kotevska, Donald Sassoon, Anik Waldow, Ryan Walter, Ian Wills, and Charles Wolfe for their comments on particular chapters; to Floris Cohen, whose work has overlapped with mine in numerous ways since the 1990s; and above all to Ursula Klein for her detailed and incisive comments on early drafts of the chapters making up Part II. Material from the book has been presented at talks in Aarhus, Berlin, Brisbane, Florence, London, Melbourne, Munich, Paris, Prague, St. Andrews, Sydney, Turin, and Venice.

Finally, a personal note. We shall encounter in passing in what follows, among the many promoters of science and learning for the working classes, the first Mechanics' Institute, founded in London in 1823. This thrived and subsequently became one of the colleges of the University of London, Birkbeck, where in the early 1970s, as an evening student who had done poorly in the orthodox educational system up to that point, I discovered the challenges and joys of a properly organized intellectual life. This is an appropriate place to record my inestimable debt to that institution, and to the nineteenth-century culture that, in the face of significant resistance and in many cases in spite of itself, initiated and promoted learning and scholarship for all. We have much to learn from it.

Contents

List of Illustrations

List of Plates

Introduction

In his 1906 essay, 'The Place of Science in Modern Civilization', Thorstein Veblen set out a widely held conception of the role of science in contemporary culture:

The making of states and dynasties, the founding of families, the prosecution of feuds, the propagation of creeds and the creation of sects, the accumulation of fortunes, the consumption of superfluities—these have all in their time been felt to justify themselves as an end of endeavour; but in the eyes of modern civilized men all these things seem futile in comparison with the achievements of science. They dwindle in men's esteem as time passes, while the achievements of science are held higher as time passes. This is the one secure holding-ground of latterday conviction, that the 'increase and diffusion of knowledge among men' is indefeasibly right and good . . . and no other cultural ideal holds a similar unquestioned place in the convictions of civilized mankind.[1]

He continues:

The question here is: How has this cult of science arisen? What are its cultural antecedents? How far is it in consonance with hereditary human nature? and, What is the nature of its hold on the convictions of civilized men?[2]

Translated into a more modern idiom, these are the questions with which we shall be concerned. Specifically, we shall be asking how science, in the period from the end of the eighteenth to the middle of the twentieth century, came to have such a central place in Western culture. And in the background of this enquiry will be a deep and particularly intractable question, but one on which everything ultimately hinges: what exactly is it that we want out of science?

Science had not always had the high standing that it enjoyed at the time that Veblen was writing. In the seventeenth century—the century of the 'Scientific Revolution'—the public reception of science, natural philosophy as it was then, was decidedly hostile. Far from being regarded as a worthwhile and significant form of investigation, at the beginning of the century it was wholly marginalized. In the case of Britain, for example, the plans for the national reform of science of Francis Bacon, by the early decades of the seventeenth century one of the most

---

[1] Thorstein Veblen, 'The Place of Science in Modern Civilization', *American Journal of Sociology* 11 (1906), 585–609: 587.

[2] Ibid, 588.

Civilization and the Culture of Science: Science and the Shaping of Modernity, 1795–1935. Stephen Gaukroger, Oxford University Press (2020). © Stephen Gaukroger.
DOI: 10.1093/oso/9780198849070.001.0001

powerful figures in English political life, were completely ignored: science was simply not on any public agenda. By the middle of the century, natural philosophy was being ridiculed as a useless and pointless form of activity. The reaction to the Royal Society is indicative. Within three years of receiving its charter, it almost collapsed through lack of attendance at meetings and shortage of funds, and its patron, Charles II, made fun of the fellows, whom he referred to as 'my ferrets', at their doing nothing but 'weighing ayre'. Fellows such as Hooke, Evelyn, and Oldenburg were distraught at the public reactions to which the Royal Society was subject, which veered between complete neglect and ridicule. Shadwell's 1676 long-running comedy *The Virtuoso* cruelly lampooned Hooke. The audiences found it hilarious and Hooke complained in his diary about the 'dammed dogs' in the theatre who turned around to look at him and 'almost pointed'. At the same time, parodies of the society's reports were circulated, emphasizing what was seen as their triviality, and the journalist Ned Ward was having much fun describing the 'Philosophical Toys' (scientific instruments and experiments) at 'Maggot-Mongers Hall' (the Royal Society). It was not just the court and the literati who were unappreciative. Traditionalists objected on different grounds, with Robert South, the public orator of the University of Oxford, referring to the Fellows of the Royal Society as 'the profane, atheistical, epicurean rabble, whom the nation so rings of, and who have lived so much to the defiance of God'. Nor was the hostility confined to England. In 1740 Linnaeus was complaining about people protesting that science was of no use. 'Such people', he writes, 'think that natural philosophy is just about the gratification of curiosity, just an amusement to pass the time for lazy and thoughtless people.'[3]

By the middle of the eighteenth century, however, the public perception of science was beginning to change. Electrical demonstrations, which were more spectacular than any of those in displays in mechanics, chemistry, or optics, played a significant role here. A report from the *Gentleman's Magazine* for 1745, for example, notes that

from the year 1743 they discover'd phenomena, so surprising as to awaken the indolent curiosity of the public, the ladies and people of quality, who never regard natural philosophy but when it works miracles. Electricity became all the subject in vogue, princes were willing to see this new fire which a man produced from himself, and which did not descend from heaven. Could one believe that a lady's finger, that her whale-bone petticoat, should send forth flashes of true lightening, and that such charming lips could set on fire a house? The ladies were sensible of this new privilege of kindling fires without any poetical figure, or hyperbole, and resorted from all parts to the publick lectures of natural philosophy, which by that means became brilliant assemblies.[4]

[3] See Stephen Gaukroger, *The Emergence of a Scientific Culture: Science and the Shaping of Modernity, 1210–1685* (Oxford, 2006), 36–9 for details of the examples given here.
[4] *Gentleman's Magazine* 15 (1745), 194. On the spectacle of electricity, see Iwan Rhys Morus, *Shocking Bodies: Life, Death and Electricity in Victorian England* (London, 2011).

The electrical displays helped to project natural philosophy as a whole into the public realm. Although spectacle undeniably played a role, it was not just a matter of spectacle. As an indication of the shift in public perception, consider two paintings of scientists (in this case alchemists/chemists), one from 1661 (Plate 1) and one from 1771 (Plate 2). The first depicts an isolated figure sharing a dilapidated room with his wife mending clothes in the background, and his infant child eating scraps. This 'sooty empiric', as sixteenth- and seventeenth-century writers would have described him, depresses his bellows to mundanely domestic effect. The latter portrays a magus, dignified and Moses-like, with his smartly dressed young disciples in the background. By contrast with the effect of bellows in the first scene, the scientist in the second produces a spectacular illumination. Both paintings represent the same kind of activity, but the standing of this activity has risen appreciably.

This does not mean that there was no significant resistance to science even in the early decades of the nineteenth century. In 1806, the police reformer Patrick Colquhoun was opposing science education on the grounds that 'science and learning, if universally diffused, would speedily overturn the best constituted government on earth',[5] and the anonymous author of the 1826 *Consequences of a Scientific Education to the Working Classes of This Country Pointed out* writes that attempting to replace distinctions based on 'wealth and rank' by those based on 'talent or science' would 'plunge the country into irretrievable ruin and despair'.[6] At the same time, there was considerable resistance to universal scientific education even on the part of scientists, as is clear from the opposition to the Parochial Schools Bill of 1807.[7] In 1817, it was noted in the Parliamentary Debates that the Seditious Meetings Act of the same year, which required all 'assemblies of divers Persons' to apply for licences, was allowing country magistrates to resist 'the application of a Mineralogical Society on the presumption that the investigation of such subjects led to blasphemy'.[8] Henry Cole, in a published 1832 Cambridge sermon, *Popular Geology Subversive of Divine Revelation!*, was urging public resistance to science.[9] Two years later the Oxford Movement cleric and philosopher William Sewell, while telling us how much he rejoiced in the flourishing of physics, warns nevertheless that: 'it is its defects, its abuse, its false reasonings, its corruptions by the human heart, the atmosphere of conceit that surrounds it, the

[5] Patrick Colquhoun, *A Treatise on Indigence* (London, 1806), 148–9.

[6] 'A Country Gentleman', *The Consequences of a Scientific Education to the Working Classes of This Country Pointed out; and the Theories of Mr Brougham on That Subject, Confuted; in a Letter to the Marquess of Lansdown* (London, 1826), 52–3.

[7] See D. S. L. Cardwell, *The Organisation of Science in England* (London, 1972), 38.

[8] Quoted in Ian Inkster, 'London Science and the Seditious Meetings Act of 1817', *The British Journal for the History of Science* 12 (1979), 192–6: 195.

[9] Henry Cole, *Popular Geology Subversive of Divine Revelation! A Letter to the Reverend Adam Sedgwick . . . being a Scriptural Refutation of the Geological Positions and Doctrines Promulgated in his Lately Published Commencement Sermon, Preached in the University of Cambridge, 1832* (London, 1834), 44–5.

idolatry with which it is worshipped, the miserable coldness it engenders, its selfishness, its dreams of vanity, its alienation from God, this is what we ought to dread'.[10]

But the 1830s had marked a turning point in England, and Mary Buckland, well-known scientific illustrator and wife of the geologist William Buckland, registered the shift in correspondence with William Whewell in May 1833, writing that: 'My poor husband—could he be carried back half a century, fire and faggot would have been his fate.'[11] Writing in 1842, William Grove, pioneer in the improvement of electrical cells, asks:

Why is England so great a nation? Is it because her sons are brave? No, for so are the savage denizens of Polynesia: She is great because their bravery is fortified by discipline, and discipline is the offshoot of Science. Why is England a great nation? She is great because she excels in Agriculture, in Manufactures, in Commerce. What is Agriculture without Chemistry? What Manufactures without Mechanics? What Commerce without Navigation? What Navigation without Astronomy?[12]

By the second half of the century, the transformation was complete, and science had become firmly established as the pre-eminent form of human understanding. In a banquet speech in Glasgow in 1873—reported approvingly in the new science journal *Nature*, founded four years earlier—Benjamin Disraeli proclaimed:

How much has happened in these fifty years—a period more remarkable than any, I will venture to say, in the annals of mankind. I am not thinking of the rise and fall of Empires, the change of dynasties, the establishment of Governments. I am thinking of those revolutions of science which have had much more effect than any political causes, which have changed the position and prospects of mankind more than all the conquests and all the codes and all the legislators that ever lived.[13]

This assessment is seconded, thirty-one years later, by another British prime minister, Arthur J. Balfour, who was telling the British Society for the Advancement of Science, of which he was president, of the way in which developments in science in the nineteenth century had completely transformed our understanding

---

[10] William Sewell, *A Second Letter to a Dissenter on the Opposition of the University of Oxford to the Charter of the London College* (Oxford, 1834), 5–6.

[11] Cited in Jack Morrell and Arnold Thackray, *Gentlemen of Science: Early Years of the British Association for the Advancement of Science* (Oxford, 1981), 235.

[12] William Robert Grove, *On the Progress of the Physical Sciences* (London, 1842), 37.

[13] Quoted in *Nature* 9 (27 November 1873), 71. Similar claims by politicians abound. Typical is the speech of the American congressman Daniel Webster on opening the Northern Railroad in 1847: 'It is an extraordinary era in which we live. It is altogether new. The world has seen nothing like it before. I will not pretend, no one can pretend, to discern the end; but every body knows that the age is remarkable for scientific research into the heavens, the earth and what is beneath the earth; and perhaps more remarkable still for the application of this scientific research to the pursuits of life.' Quoted in Leo Marx, *The Machine in the Garden: Technology and the Pastoral Ideal in America* (New York, 1964), 214–15.

of the world and, through this, our lives.[14] The British Society for the Advancement of Science, founded in 1831, had in fact included in its ranks not only prime ministers, such as Sir Robert Peel and Lord Palmerston, but also earls, marquises, viscounts, and the Prince Consort himself (president in 1859), along with clerics (the largest group), physicians, military officers retired on half-pay at the end of the Napoleonic wars, manufacturers, engineers, poets, and others.[15] The topic of the central role of science in the affairs of the nation was a common one from the middle of the nineteenth century onwards, and by 1934 Ritchie Calder, the scientific correspondent of the *Daily Herald*, was proposing that the House of Lords be replaced by what he called a 'Senate of Scientists'.

The political significance of science was matched by its cultural significance. In the nineteenth century, in the rapidly expanding urban centres of the industrial north of England, for example, scientific reading and enquiry took over from literary, political, or classical studies as the leading form of entertainment among the literate,[16] and from the 1830s there was a significant increase in reviews and discussion of scientific books in journals such as the *Edinburgh Review*, the *North British Review*, the *Quarterly Review*, *The Foreign Quarterly Review*, and the *Monthly Review*.[17] As Richard Yeo has pointed out:

The major quarterlies and magazines treated science as part of a broad intellectual framework which included theology, literature, philosophy, and political economy. Reviewers often moved freely across these disciplines, and science was presented in a manner which assumed an informed readership. The great debates on geology, evolution, and their religious implications were conducted in these periodicals, and this meant that the meaning and status of science were matters of public discussion.[18]

The change affected institutions equally. In the eighteenth century, there was only one institution in Britain devoted exclusively to the study of natural philosophy, the Royal Society. This changed in the early decades of the new century, and by the time of the Great Exhibition in 1851, which millions attended, the attitude to science had been thoroughly transformed, as every kind of science education institution flourished, from mechanics institutes to the newly established London universities, University College and King's College. Moreover, in the course of the

---

[14] Arthur J. Balfour, 'Address: Reflections Suggested by the New Theory of Matter', *Report of the Seventy-Fourth Meeting of the British Association for the Advancement of Science* (London, 1905), 3–14.

[15] See Morrell and Thackray, *Gentlemen of Science*, xix. Disraeli, Gladstone, Peel, and Prince Albert had also been among the driving forces behind the establishment of the Royal College of Chemistry in 1845. The Royal College later became the first constituent college of Imperial College.

[16] See, for example, Arnold Thackray, 'Natural Knowledge in Cultural Context: The Manchester Model', *American Historical Review* 79 (1974), 672–709.

[17] On the reviews of scientific books, see Richard Yeo, *Defining Science: William Whewell, Natural Knowledge and the Public Debate in Early Victorian Britain* (Cambridge, 1993), 77–87.

[18] Richard Yeo, 'Science and Intellectual Authority in Mid-Nineteenth Century Britain: Robert Chambers and *Vestiges of the Natural History of Creation*', *Victorian Studies* 28 (1984), 5–31: 9.

nineteenth century science was extended outside the confines of serious research, and was installed at the centre of leisure, for both adults and children. In the case of adults, natural history and machines predominated, whereas in the case of children, as well as pedagogic material there was a new focus on such children's daytime activities as rock collecting, shell collecting, and the identification of plants, and the night-time activity of star-gazing.[19] The turn to science is reflected in the first comprehensive history of nineteenth-century thought to appear at the turn of the new century, that of John Theodore Merz, who, summing up its achievements, writes that:

It will be generally admitted that the scientific spirit is the prominent feature of the thought of our [nineteenth] century as compared with other ages. Some may indeed be inclined to look upon science as the main characteristic of this age. The century may be called with some propriety the scientific century, as the last was called the philosophical century.[20]

If the nineteenth century was 'the age of science', however, it was not just in virtue of its technical achievements. Science had begun to shape standards of culture and civilization early in the century. Parallel with this, and even more consequential for the standing of science, was the fact that, by the beginning of the century, the West's sense of its superiority was shifting from its religion, Christianity, and its achievements in the arts and humanities, to its science. The novelist Thomas Love Peacock had written as early as 1820 that 'as the sciences of morals and of mind advance towards perfection, as they become more enlarged and comprehensive in their views, as reason gains the ascendancy in them over imagination and feeling, poetry can no longer accompany them in their progress, but drops into the back ground, and leaves them to advance alone'.[21] Ten years later, Coleridge was countering with an opposing view, writing that 'the science of theology was the root and trunk of the knowledge that civilized man, because it gave unity and the circulating sap of life to all the other sciences, by virtue of which alone they could be contemplated as forming, collectively, the living tree of knowledge'.[22] But when historians and others subsequently began to reflect on the shift, a strong consensus emerged that it was science, rather than Christianity or literature, that

[19] See for example Melanie Keene, *Science in Wonderland: The Scientific Fairy Tales of Victorian Britain* (Oxford, 2015).
[20] John Theodore Merz, *A History of European Thought in the Nineteenth Century* (4 vols, Edinburgh and London, 1904–12), i. 89. Merz was a Newcastle chemist and industrialist, a key figure in making Newcastle the site of the largest integrated supplier of electricity in Europe: see David Edgerton, *The Rise and Fall of the British Nation: A Twentieth Century History* (London, 2018), 137.
[21] Thomas Love Peacock, 'Four Ages of Poetry', in F. H. B. Brett-Smith, ed., *Peacock's Four Ages of Poetry, Shelley's Defence of Poetry, Browning's Essay on Shelley* (Boston, 1921), 1–19: 9. Originally published in *Olniers Literary Miscellany in Prose and Verse* (1820), 183–200.
[22] Samuel Taylor Coleridge, *On the Constitution of the Church and State, According to the Idea of Each* (3rd edition, London, 1839), 51.

was the bearer of the values of civilization, and in the decades that followed there were moves in France, Germany, and England to replace Christianity's role as the guardian of a moral code with science. In France, Comte had called, in the 1850s, for a secular religion of humanity to replace that of Christianity in his *Système de politique positive*.[23] In 1889 the zoologist Alfred Giard was celebrating the displacement of religion by science, and the philosopher Alfonse Darlu was writing in 1895 that the 'task of our century' had been 'to secularize religious ideas'.[24] In Germany, in 1858 Virchow was one of a number of physiologists and others calling for ethics to be guided by science,[25] and in an 1873 talk on 'The Sciences and their Importance for the Ethical Education of Humanity', he argues that, for progress and prosperity, natural scientists must participate in applying the scientific method 'to problems of the spirit, and of the conscience; we demand that each from his own standpoint help to develop morality as an empirical science according to the rules which general natural science has constituted'.[26] In Britain, Francis Galton was calling for a 'scientific priesthood' to tend to the health and welfare of the nation (1875), Thomas Huxley was arguing that scientific progress had laid 'the foundations for a new morality',[27] and Herbert Spencer was urging that ethical principles be derived not from religious precepts but from scientific ones in his *The Principles of Ethics* (1892).[28] By the early decades of the twentieth century there was if anything an increase in the number of calls for science to act as the basis for morality,[29] summed up later in the century in E. O. Wilson's claim that the time might have come for 'ethics to be removed temporarily from the hands of philosophers and biologicized.'[30]

*

What is at issue in the consolidation of a scientific culture in the West is the ascendancy of a culture in which all cognitive values, and subsequently many

---

[23] August Comte, *Système de politique positive, ou Traité de sociologie, Instituant la Religion de l'Humanité* (4 vols, Paris, 1851–4).

[24] Alfonse Darlu, 'Réflexions d'un philosophe sur la question du jour. Science, morale et religion', *Revue de métaphysique et de morale* 3 (1895), 239–51: 244.

[25] Rudolph Virchow, 'On the Mechanistic Interpretation of Life', in Rudolph Virchow, *Disease, Life, and Man: Selected Essays of Rudolph Virchow*, ed. and trans. L. J. Rather (Stanford, 1958), 102–19. On the more general movement in Germany see Keith Anderton, 'The Limits of Science: A Social, Political, and Moral Agenda for Epistemology in Nineteenth Century Germany', unpublished PhD dissertation, Harvard University, 1993.

[26] Rudolph Virchow, 'Die Naturwissenschaften in ihrer Bedeutung für die sittliche Erziehung der Menschheit', in Karl Sudhoff, *Rudolf Virchow und die deutschen Naturforscherversammlungen* (Leipzig, 1922), 122–47: 132. Quoted in Anderton, 'The Limits of Science', 332.

[27] Thomas Henry Huxley, *Lay Sermons, Addresses and Reviews* (London, 1877), 10–11.

[28] See Gaukroger, *The Emergence of a Scientific Culture*, 24–5.

[29] See Peter J. Bowler, *Reconciling Science and Religion: The Debate in Early Twentieth-Century Britain* (Chicago, 2001).

[30] E. O. Wilson, *Sociobiology: The New Synthesis* (Cambridge, MA, 1975), 562. For an example of a biological reading of ethics, see Richard Joyce, *The Myth of Morality* (Cambridge, 2001).

moral, political, and social ones, come to be modelled around scientific values, and we shall be exploring how these values were shaped and how they began in turn to shape those of society. The argument that I shall be presenting continues the trajectory of three earlier volumes, which traced key aspects of the legitimation of science and the establishment of a scientific culture up to the early decades of the nineteenth century.[31] The core nineteenth- and twentieth-century development is that in which science comes to take centre stage in shaping ideas of civilization. This understanding changes significantly in the wake of the First World War, with its poison gas and indiscriminate mechanized slaughter. Following the war, fundamental questions began to be raised about whether science, far from fostering or even defending civilization, was in fact the single greatest cause of its destruction, or whether it should instead be argued—in an attempt to isolate science from the sensitive issues of the day—that it was the application of science, by contrast with science itself, that was at fault.[32] The context in which this occurred was crucial: this was at a time when the ability of the educated general public in understanding new scientific developments was severely diminished, for the complexity and technical nature of science now meant, as a writer in *The Nation* put it as early as 1906, that 'one may say not that the average cultivated man has given up science, but that science has deserted him'.[33]

I have set out, in what follows, to explore the perceived legitimacy of modern scientific culture. Our starting point is provided by Condorcet's 1795 *Esquisse d'un tableau historique des progrès de l'esprit humain*, which offered what was to become the canonical statement of the idea of the inevitability of progress, and of the idea that what drives the move towards a new, modern form of society is science. But we are not dealing with a linear or one-dimensional development here, and I have identified five issues that have underpinned the transition to a scientific modernity of the kind with which we are familiar.

In Part I, I explore the way in which the understanding of civilization shifted away from religious, political, social, and artistic achievements in the direction of

[31] Gaukroger, *The Emergence of a Scientific Culture*, which covers the period from 1210–1685; *The Collapse of Mechanism and the Rise of Sensibility: Science and the Shaping of Modernity, 1680–1760* (Oxford, 2010); *The Natural and the Human: Science and the Shaping of Modernity, 1739–1841* (Oxford, 2016).

[32] As David Edgerton notes, on the view that science and war are 'radically different enterprises': 'These stories, which were intended to be taken literally, were fairy stories, but ones that bewitched many students of the relations of science, technology, and war. The divergence between this picture and the most straightforward empirical analysis of the relations of science and war... will astonish even the most hard-bitten cynic.... [T]he divergence cannot be explained by secrecy. There was always sufficient information in the public domain to yield a very different picture.' David Edgerton, 'British Scientific Intellectuals and the Relations of Science, Technology, and War', in P. Forman and J. M. Sánchez, eds, *National Military Establishments and the Advancement of Science and Technology: Studies in Twentieth Century History* (Dordrecht, 1996), 1–35: 2.

[33] Cited in Daniel J. Kevles, *The Physicists: The History of a Scientific Community in Modern America* (Cambridge, MA, 1971), 94. Cf. Theodore M. Porter, 'How Science Became Technical', *Isis* 100 (2009), 292–309.

science in the course of the modern era. In Part II, we see how the question of the unity of science was a pivotal issue from the beginning of the nineteenth to the middle of the twentieth century, and how in crucial respects it represents a failed attempt to recapture a notion of unity that collapsed with the demise of a pre-modern idea of Christian unity of understanding. In particular, I reject the idea that the sciences can ultimately be reduced to physics, or that, failing this, they can be thought of in terms of layers of emergent properties: as far as explanation is concerned, pluralism and modularity pervade all the sciences, and attempting to arrange them into a hierarchy is fruitless and counterproductive. The transform-ation of the understanding of science from the early decades of the nineteenth century onwards is the subject of Part III, where we look at the attempts to change the nature of philosophy in a fundamental way, making it simply a form of reflection on science, and indeed in some respects incorporating it into science. In Part IV, I explore the need to replace our understanding of science as a theoretical exercise with something that more exactly matches how the scientific enterprise was pursued from the nineteenth century onwards. While 'applied science' and 'popular science' have traditionally been treated as dependent subsidiary exercises, there is a compelling case to be made that they are actually integral to science. On the one hand, we cannot think of science simply in terms of 'scientific theory', and must take seriously its non-discursive products, such as machines. On the other, we need to realize that it was not science as a model of truth that placed it at the centre of modern culture, but science as a model for the future. In Part V we examine the application of the idea of the unity of scientific understanding to human behaviour, focusing on what was arguably the most important aspect of how science shaped modern culture: how changing conceptions of civilization, increasingly moulded by science, bear on the question of the attributes expected of, and to be fostered within, the individual citizens of the new order.

In a little more detail, the argument of the book is as follows. Civilization, I argue in Part I, was not associated with science before the late eighteenth century, evident from the literature on China, considered to be a civilized culture and one from which the West might even have something to learn, but believed to be lacking in scientific development. Voltaire's use of China as a contrast case prompted a questioning of the Christocentric and Eurocentric conception of history that had predominated up to that point. This conception had provided a sense of what the assumed uniqueness of the West consisted in, and with its demise a new source of this uniqueness began to be explored, one that resulted not directly from a concern with the East, but with an account of how learning in Europe had stagnated in the Middle Ages. China was generally agreed to be a static society, and this was now taken to be a feature of medieval culture as well, the corrective to which had been the emergence of scientific enquiry. Together with the growing belief that science was the only effective weapon against the superstition that had held Europe back in the Middle Ages, this promoted the view that science was the key to the progress that was now associated with civilization. While the relation between science and civilization had been more than merely descriptive in its eighteenth-century advocates, by the nineteenth century it took on a trenchantly prescriptive tone. In Comte's influential *Cours de Philosophie Positive*, society was to be reformed on a resolutely scientific basis, and we can distinguish two kinds of development of a Comtean programme, evident in Buckle's explicitly Comtean attempt to place the historical understanding of civilization on a scientific basis, and in Spencer's attempt to account for the evolution of civilization along biological lines.

The nineteenth-century cultural elevation of science puts it in some respects in an analogous position to that previously occupied by religion in marking out Western civilization, and there is one particular respect in which it takes over a task from religion, although this is rarely appreciated as such. The doctrine of the unity of science is typically taken to be a feature of the sciences themselves. In Part II, we see that this is not the case. The idea of the unity of the sciences is driven by varying factors, which can be mistaken for one another: some are political and social, some ideological, some based on a questionable gloss on attempts to reconcile different theories, and some unreflectively extrapolated from genuine but localized forms of reduction or amalgamation. Accordingly, in asking why the unity of understanding was such a significant issue, and why this took the form of the search for a unity of science in the nineteenth and twentieth centuries, we need to explore how seemingly extrinsic considerations bear on science in these respects. On the question of significance, the issues go back to the rise of Aristotelian scholasticism in the thirteenth century, with its rejection of the Christian-Platonist conception of a universe in which what were subsequently distinguished as the natural and supernatural realms were integrated. The supernatural had provided a unifying thread, and with its removal, the natural realm lost its unifying principle. At the end of the eighteenth century, science was promoted as occupying this unifying role. This is why, beginning in the nineteenth century in Germany, France, and Britain, there developed a comprehensive cultural investment in the idea of the unity of science. And this is what we need to make sense of: we need to ask what possible connection there could be between the unity of science and political and moral unification programmes of the time, for example. In the course of exploring the sources of these demands for unity, we can begin to question the various devices that have been introduced to save the unity of science, most notably the doctrine of emergent properties. Accordingly, one of the primary aims of Part II is to probe the history of claims for large-scale reduction and unification, and to bring to light the modularity and explanatory pluralism of the natural and the life sciences.[34]

---

[34] I use the term 'natural sciences' to refer to physics and chemistry, by contrast with the life sciences. This is a little awkward, and a more usual nomenclature would be to refer to all three as the natural sciences, with a division between the life sciences and the physical sciences (physics and chemistry). But in what follows there is a crucial conceptual and material distinction between those

Chapter 3 examines the attempts of the British Association for the Advancement of Science to regiment the sciences, that is, to decide what to include and what to exclude from the rubric of science, and to order and rank those that it included. In Chapter 4, I go on to consider whether the relations between the physical sciences could actually have followed the reductionist, hierarchical approach. The aim is to show in some detail the intractable problems that dogged attempts to unify physics in the nineteenth and early twentieth centuries. More generally, I show that unification aspirations are in many cases the result of mapping political notions of unification on to scientific ones. In Chapter 5, we explore the material sciences, and I show the impossibility of unifying physics and chemistry. Examining the case of the establishment of atomism in the first decade of the twentieth century, we see that it is a mistake to consider this to be a case of unification of physics and chemistry.

In Chapter 6, I turn to the development of physiology and the reductionist aspirations of nineteenth-century physiology. We see that the scientific standing of physiology is not dependent on the extent to which it can be reduced to the physical and material sciences, and that the distinctiveness of the life sciences, what marks them out as autonomous with respect to the physical and material sciences, does not depend on their ability to secure the existence of some non-physical force that distinctively shapes and guides biological processes. I argue that the notion of emergent properties, rather than being an alternative to reductionism, is actually designed to save it, and that, from an explanatory point of view, it is empty. At the same time, we see that trying to formulate questions of emergent properties in terms of ontology is a disastrous way of thinking about scientific explanation. In Chapter 7, which deals with the historical path in the life sciences, I explore attempts to unify evolutionary theory with other developments in the life sciences, and argue that the life sciences remain irreducibly pluralistic, and in this capacity can act as a model for the sciences more generally.

In Part III, we explore how science was transformed in the nineteenth century in terms of what it included, and how it was legitimated at the philosophical level. Each of these transformations bears on, and deepens our understanding of, the questions of civilization and the unity of science discussed in Parts I and II.

On the first question, for the idea of the unity of the sciences to work, not all claims to scientific standing can be accommodated. Inclusion in the sciences, and exclusion from them, needed to be policed so as to maintain the character of science, in particular its sense of unity and coherence. For advocates of the general programme of unification, what was at issue was no longer so much unification as

---

disciplines that make up physics (the physical sciences) and those that make up chemistry (the material sciences), and this is an important substantial point, not a semantic one, in which questions of reduction and emergence are in play, so physics and chemistry cannot be referred to collectively as the 'physical sciences'. But they need to be marked out nevertheless from the life sciences, so have been designated 'natural sciences'. This is, of course, not to deny that the life sciences are part of the natural sciences in the usual sense.

controlled unification, where the skill came in knowing where the line was to be drawn, and being able to justify this as a natural line, as it were, not something decided by fiat or ideology. In Chapter 8, we examine claims to scientific standing that were highly contested, and provoked a new kind of metascientific enquiry, in which disciplines were accredited, ranked, and ordered in terms of priority. These accreditations, rankings, and priorities were rationalized in terms of the internal structure of science, but they were predominantly extra-scientific in origin, and were more than anything else an elaborate exercise in legitimation. The issues centred on accounts of human behaviour that had traditionally been the preserve of religious and metaphysical teaching. These included ethics, where efforts were now afoot to put it on a scientific standing, as well as areas that had the character of a loose combination of moral, political, and economic views which, with the advent of Ricardo's political economy, could now be claimed to have been put on a scientific footing. Such new 'scientific' disciplines were fiercely opposed, and the most prominent dispute in this respect, that between Whewell and Mill, became transformed into a philosophical project of understanding the nature of science. Whewell was engaging in one of the last attempts to provide a comprehensive Christian understanding of the natural and moral realms, whereas Mill was offering one of the first comprehensive, wholly secular understandings of the natural and moral realms, one which played a significant role in shaping the public understanding of science.

On the second question, we turn, in Chapter 9, to the redefinition of philosophy in Germany in the second half of the nineteenth century and early decades of the twentieth. The issues arose initially due to claims by German scientists and writers of popularizations of science that science exhausted everything that could be known, and consequently philosophy could no longer continue to play any role in our understanding of the world. In finding a new role for philosophy, the influential Neo-Kantian Hermann Cohen construed it as a form of clarification and systematization of science, of which it now effectively formed an essential part. The legitimacy of science as the pre-eminent form of understanding the world, and consequently as the basis for culture, for civilization, depends on its unity, but it cannot establish this relying on its own resources: for that, philosophy, qua theory of the foundations of science, is needed. Philosophy *completes* science, by establishing its unity on an a priori basis. As in Mill's construal, philosophy becomes transformed into a new kind of para-scientific endeavour. Cohen's version of Neo-Kantianism did not directly engage the human sciences, however. When attempts, such as that of Windelband, to address this question within the context of Neo-Kantianism were made, the centrality of the human sciences to the elaboration of a viable understanding of science once again became evident. Neo-Kantian attempts to 'ground' science in philosophy, whether those of Cohen or Windelband, raise the question whether philosophy is best placed to provide such grounding. These questions come to a head in the wake of the Great War, as science came to be implicated in the horrors of its scientifically generated technology of war, and the chaos that it engendered and which continued in its wake. As we shall see, for many, the aspirations of a scientifically driven Enlightenment turned out to be ineffectual and empty.

Science has traditionally been taken to be complete in itself and autonomous, but with two wholly dependent subsidiary exercises: 'applied science', offering applications to practical questions, and 'popular science', offering simplifications for a wider public. In Part IV, we see that technology, broadly conceived, is not something deriving from science, but in many respects, at least from the 1850s, is integral to its success, and that cognate considerations hold for 'popular science', which also in crucial respects forms an integral part of the success of the scientific enterprise. The way in which the values of a 'pure' theoretical science bear on the understanding of modern civilization is not, and could not be, the same as that of the relation between civilization and a complex mix of technology, scientific theory, and popularization. Yet it can only be the last of these that drives any sense of what modern civilization owes to science.

In Chapter 10 we examine how science was transformed by its interaction with technology, and how it was promoted in the public sphere. When the idea of the role and benefits of science in the development of civilization are being proposed, it is rarely 'pure science', that is, a technology-free form of science, that is being mooted. Science enters the equation because it is being assumed that technology is applied science. But this construal of technology, as we shall see, cannot be made to work. In this connection, I explore two questions central to understanding the nature and role of technology in the nineteenth and the early decades of the twentieth century. First, there is the problem of how technology engages with science. Here I shall be arguing that, to the extent to which science and technology can be integrated, what might once have been thought of as scientific developments should in fact be conceived in terms of a mixture of theory, experiment, and theory-free invention. This unstable mixture is what confers on 'science' its unruly character. Second, I want to look at the different values of science and engineering: at the relative standing of scientists and engineers in the nineteenth and early twentieth centuries, and differences in approach to problem solving. In particular, I shall look at a case where science and engineering have a problematic fit: physical and engineering approaches to aerodynamics in the early decades of the twentieth century. Here we shall see how the separation of scientific and engineering concerns was unable to stem this unruliness, despite the claims of a rigorous foundational approach by those in a tradition of mathematical physics. The association of science and engineering means that we must take seriously the non-discursive products of science, particularly machines. Once we consider not just the function and construction of machinery, but also the operation of machinery, we encounter questions very different from those that concern us in the study of 'pure' science, but to which we need to be attentive.

In Chapter 11, we explore how science was able to interact with a culture outside that of professional scientists and philosophers: how those outside the community of scientists and philosophers would have encountered science. As scientific values became embedded in those of civilization, a new set of demands was made on science, and it needed to establish its legitimacy in such a role. The association of science and civilization, crucial to the standing of science in the modern era, would simply not be possible without popular science—popular natural history books, children's science books, museums, science fiction— because science doesn't have the resources to effect this association in its own right. In order to meet the expectations of the vastly expanded role that it had assumed from the end of the eighteenth century, one in which it was displacing religion as the key to understanding our place in the world for example, it was crucial that science be extended into the non-propositional realm, engaging with the world in terms of desires, expectations, anxieties, fears, hopes, goals, raw beliefs, etc. Only popular science could do this, but it could never have just been a promotional exercise, however much scientists may have wished for this. It raises both expectations and anxieties about the standing of science in the modern era.

Having looked in earlier parts of the book at how the public encountered the sciences, in Part V we move to consider how the sciences encountered the public, and in particular how science attempted to shape its public in its own image. Just as the project of the unity of the sciences was concerned to include certain disciplines and exclude others, and to tier and rank those that it included, so the resources that the shaping of the population drew upon followed this model in crucial respects. Continuing the tradition of distinguishing 'civilized' from 'un-civilized', a number of disciplines emerged in the late nineteenth and early twentieth centuries that distinguished different types of societies and personalities, and set out to explain characteristic features of modernity. In Chapter 12, we examine how the 'civilizing process', a feature of Western society since the Renaissance, came to be articulated in scientific terms in the nineteenth and twentieth centuries. In particular, the tiering and ranking of different societies came to be extended to differences between individuals, an exercise in which the newly developing discipline of psychology played a crucial role. Central to this exercise was the establishment of norms for everything from bodily proportions to social behaviour. These developments were linked to widespread worries about how natural selection could lead to the degeneration of the population in a social system with welfare structures. What resulted was the advocacy of what its proponents considered a 'new science', eugenics, which identified what were considered to be favourable and unfavourable inherited traits, promoting the former and introducing measures to inhibit reproduction of the latter. As part of this process, what emerged was the invention of the modern notion of 'intelligence', which now becomes a criterion of social standing, notionally replacing those of class, race, and birth. Finally, we examine a shift of mentality inherent in these developments, in the concern to shape the population into the kinds of person who can occupy a scientifically modelled form of civilization.

At the core of this lies the shift from reason to rationality, and we explore some of the dire consequences of this shift.

This is the final volume in a set of four which has set out to explore the emergence of a scientific culture from the early thirteenth century to the early decades of the twentieth. The direction that the arguments have taken may not always have been a familiar one, particularly since I have been concerned with a history of scientific culture rather than the history of science as such. Accordingly, the final chapter is a conclusion to the series as a whole, where I have developed some of the most important threads in the general argument.

PART I

CIVILIZATION

1

Science and the Origins of Civilization

> As long as what is meant by culture is essentially the promotion of science, culture will pass by the great suffering of the human being with pitiless coldness, because science only sees problems of knowledge, and because within the world of the sciences suffering is really something improper and incomprehensible.
>
> Friedrich Nietzsche, 'Schopenhauer als Erzieher'[1]

From the middle of the eighteenth century, the ideas of science and civilization have been transformed in such a way that they have come to be inextricably linked. In understanding these ideas—as well as cognate notions such as progress and technology—we need to be able to register changes in the way in which they are conceived over two centuries, while at the same time keeping a firm sense of what it is that we are tracing the development of. The terms themselves undergo significant changes, and they are multi-layered. As well as having a network of meanings, they indicate goals, in the case of science, and standards, in the case of civilization. Complications arise because the two ideas shift to reflect forms of mutual accommodation that vary over time. 'Civilization', for example, shifts in the late eighteenth century from something atemporal to something that, like science, progresses over time by stages; and, for reasons that have to do with its standing as a bearer of civilization, the understanding of 'science' comes to incorporate technology in the early decades of the nineteenth century, only for attempts to be made to dissociate technology from science in the wake of the First World War. Nevertheless, to begin with a tracking of the terms and the ideas they convey, in a way that separates them to some extent, allows us to impose some order on our enquiry, and it enables us to build up connections gradually as we develop a firmer grasp of what it is that is being connected. Accordingly, our primary task in this chapter is to map out the development of the notion of civilization. This focus will mean that, in sketching how civilization comes to be connected to science, we will of necessity be operating with a somewhat rarefied notion of science in the first instance, something that will be corrected in following chapters, where we can start to build up a more comprehensive and integrated picture.

---

[1] Friedrich Nietzsche, *Unzeitgemässe Betrachtungen* (2 vols, Leipzig, 1899), ii. 69–70.

Civilization and the Culture of Science: Science and the Shaping of Modernity, 1795–1935. Stephen Gaukroger, Oxford University Press (2020). © Stephen Gaukroger.
DOI: 10.1093/oso/9780198849070.001.0001

In an influential series of lectures at the Sorbonne in 1820, the historian François Guizot set out a rationale for a history of world civilization:

We can ask ourselves whether it is universally true that there is a universal civilization of the human species, a destiny of humanity, whether the various peoples have handed down from century to century something that has never been lost, which must increase, forming an ever larger mass, and is thus passed on to the end of time. For my own part, I am convinced that there is in fact a general destiny of humanity, a transmission of the aggregate of civilization, and consequently that there is a universal history of civilization to be written.[2]

What Guizot was offering was a history of European civilization that would, at the same time, be a history of world civilization. That European civilization was the same thing as world civilization was a universally shared assumption among European thinkers.[3] As Jürgen Osterhammel, in an important comparative history of the nineteenth century, has noted:

The nineteenth century stands out from the sequences of ages by the fact that never before, and never again after the First World War, were the political and educational elites of Europe so sure of marching at the head of progress and embodying a global standard of civilization. Or, to put it the other way around: Europe's success in creating material wealth, in mastering nature through science and technology, and in spreading its rule and influence by military and economic means, brought about a sense of superiority that found symbolic expression in talk of Europe's 'universal' civilization. Toward the end of the century, a new term for this made its appearance: modernity. The word had no plural; only in the final years of the twentieth century would scholars begin to speak of 'multiple modernities'. The concept of modernity has to this day remained enigmatic; there has never been agreement as to what it means and when the corresponding phenomenon emerged in historical reality.[4]

There has been a significant literature on modernity since the 1960s,[5] and the ideas that come to the fore in this literature include stable political societies, industrialization, individual political rights, the move from large extended households to nuclear families, secularism, free markets, the rise of the huge impersonal

[2] François Guizot, *Histoire de la civilization en Europe: depuis la chute de l'empire romain jusqu'à la révolution française* (Paris, 1828), 9.

[3] On the common identification of European history with world history, see Georg G. Iggers, 'The Idea of Progress in Historiography and Social Thought since the Enlightenment', in G. Almond, M. Chodorow, and R. Pearce, eds, *Progress and its Discontents* (Berkeley, 1982), 41–66.

[4] Jürgen Osterhammel, *The Transformation of the World: A Global History of the Nineteenth Century* (Princeton, NJ, 2014), 836. Cf. Christopher A. Bayly, *The Birth of the Modern World, 1780–1914: Global Connections and Comparisons* (Oxford, 2004), 9–12.

[5] Beginning effectively with Shmuel N. Eisenstadt, *Modernisation, Protest and Change* (Englewood Cliffs, NJ, 1966), though the theme is ultimately a Weberian one. See also Shmuel N. Eisenstadt, 'The Civilizational Dimension of Modernity: Modernity as a Distinct Civilization', *International Sociology* 16 (2001), 320–40.

metropolis, and the decline in religious mentality. But exactly what modernity committed one to varied. For example, one commentator has noted that there had been an assumption that there was a common direction to modern change—with Western societies in the vanguard—and that societies that failed to move along must be suffering from some fatal flaw. He goes on to argue that this was a very common approach in the 1960s and 1970s, when the United States was trying to promote economic development in many countries as an antidote to communism.[6] But at the same time, many in the third world in this period associated modernity with a socialist economy.

Needless to say, the notion of modernity is a contentious one, though the idea of setting out to modernize societies sounds a little more benign than the idea of setting out to civilize them, given the implicit understanding that what one civilizes are 'savages', and the fact that a 'civilizing mission' was frequently little more than a post hoc rationale for the often inhumane behaviour that accompanied the imperial conquests of the nineteenth century. Yet if the term 'modernity' were to be substituted for 'civilization' in an effort to rid the latter of its negative connotations, it is difficult to see how such a sanitizing move could be successful.[7] It would just shift the problem. After all, it is difficult to separate twentieth-century 'modernity' from its mechanized warfare, concentration camps, indiscriminate bombing of civilians, and its chemical, nuclear, and biological weapons. The 'modern world' is not obviously superior to the cruelties of the Age of Empire.

Our immediate concern is not with the moral issues, however, still less with semantic ones. There is a different reason why we should not take 'modern' societies to be just a new name for what used to be called civilized ones. In exploring the relations between science, civilization, and modernity there is a basic distinction that we need to capture. It is that civilization was not traditionally thought of in terms of scientific achievement, whereas the idea of modernity is intimately associated with that of science. As Marwa Elshakry points out:

It was not really until the nineteenth century, after all, that this 'world order of knowledge' actually went global. Perhaps the chief spur to the global interest in science at the end of the nineteenth century was the idea that it was the secret behind Europe's rise and its industrial and imperial success. How else to explain that by the end of the nineteenth century, Europeans traded with and held dominion over more than three-quarters of the globe? It is no coincidence that it was around this time, too, that the very notion of 'Western science' was itself invented. It was thanks to the new empires of the late

[6] Peter N. Stearns, *The Industrial Turn in World History* (London, 2017), x–xi.

[7] As well as 'modernity', the idea of 'culture' has replaced 'civilization'. As one commentator has noted: 'Sociologists, anthropologists and historians have learned to avoid civilization, and instead, analyze everything with culture. Culture is "in" and civilization is "out", because civilization must be "contrasted with *savagery* or *barbarism*", as Raymond Williams declared in *Keywords*.' Wolf Schäfer, 'Global Civilization and Local Cultures', *International Sociology* 16 (2001), 301–19: 302.

nineteenth century that 'science' acquired a new sense of its place in the world: at once Western, modern, and universal.[8]

To the extent to which civilization and modernity can be equated, civilization has been transformed into something in which science plays a defining role. It is the standing of modernity as a scientific culture that we shall be primarily concerned with, and by 'modernity' I shall refer to a culture in which cognitive values generally become subordinated to scientific ones, and where, correspondingly, science is expected to provide an archetypical path to cognitive success not just for scientific disciplines but for any discipline with cognitive aspirations. There are of course other understandings of modernity, and I have used the term simply as a convenient one under which to bring a complex phenomenon whose salient features I have set out to capture and explore. At the heart of the project is an exploration of the ways in which science, and the culture in which it has flourished, have mutually influenced one another.

CIVILIZATION AND BARBARISM

In the second half of the eighteenth century, the theme of the role of science in shaping a new triumphant Western culture began to emerge, as the progress of science became intimately connected with the progress of civilization.[9] The link between science and civilization was never an organic association in which the two naturally converged, but was rather something that needed to be created and promoted. As Herren et al. note, nineteenth-century historiography differs sharply from its predecessors. In the eighteenth century, they point out, 'historians, who were merely interested in the mode of unintentional discovery expressed by the concept of "serendipity", regarded Asia with curiosity and admiration. In the nineteenth century, European historians invented the Orient as underdeveloped, thus confirming Western progress.'[10]

---

[8] Marwa Elshakry, *Reading Darwin in Arabic, 1860–1950* (Chicago, 2013), 10.

[9] The idea of 'Western' culture, conceived as a culture in interrupted continuity with ancient Greece and Rome and independent of developments in the 'East', is deeply problematic. See, for example, Peter Frankopan, *The Silk Roads: A New History of the World* (London, 2015). Nevertheless, what is important for the account that I am presenting is the West's self-image, primarily as it affects its pursuit of science, so there is some justification for bracketing off such questions. It should be noted, moreover, that many accounts that (rightly) stress the importance of non-Western developments in shaping Western culture, in their enthusiasm misunderstand or exaggerate technical scientific and mathematical developments outside the West. More seriously, they fail to take into account the differences in science as a cultural project, one that includes its distinctive role and aims, thereby effectively assimilating technical developments outside the West to a thoroughly Western model. See the discussion of the distinctiveness of the Western understanding of science in ch. 1 of Gaukroger, *The Emergence of a Scientific Culture*.

[10] Madeleine Herren, Martin Rüesch, and Christiane Sibille, *Transcultural History: Theories, Methods, Sources* (Berlin, 2012), 12. See the exemplary account in Jürgen Osterhammel, *Die Entzauberung Asiens: Europa und die asiatischen Reiche im 18. Jahrhundert* (Munich, 1988), 11–38.

Civilization itself, in anything resembling a modern understanding, is a relatively recent idea, and it appears at much the same time in the commercial societies of France and Britain. There had been no word for 'civilization' in Latin (Leibniz uses the term *cultus* as an antonym for barbarism but it hardly corresponds in meaning),[11] or in the European vernaculars before the second half of the eighteenth century.[12] Moreover, what had traditionally been the contrast class for civilization, namely barbarism, was thought of in a rather different way prior to then. How to save oneself from, and distinguish oneself from, barbarism was a common theme in sixteenth- and seventeenth-century thought, beginning with Erasmus' *Antibarbarorum* (1520), which called for a secular educational programme based on Greek and Latin authors, in order to instil the values of classical learning in a Christian society. The contrast was between the demands of Christianity and barbarism, and the role of classical education was to bolster the former. The choice remains that between Christianity and barbarism: classical learning in itself is an adjunct to Christianity, not an alternative to it.

The association with Christianity is retained in an early appearance of the term 'civilisation' in French, in 1756 in Mirabeau's *L'Ami des hommes*.[13] Mirabeau writes that: 'Religion is indubitably the first and most useful restraint on humanity: it is the original source of civilization; and it counsels and reminds us constantly of sociability.'[14] Yet as Catherine Larrère has noted, *L'Ami des hommes* is a treatise on commerce, agriculture, and population, and its context is that of mid-eighteenth-century discussion on the values of commercial society, for example in Montesquieu and Rousseau, where the term 'civilisation' is not yet in currency but where the subject of discussion is what Mirabeau is concerned with when he subsequently uses the word.[15] The strong association between

---

[11] See François Zourabichvili, 'Leibniz et la barbarie', in Bertrand Binoche, ed., *Les Équivoques de la civilisation* (Seyssel, 2005), 33–53.

[12] The first English dictionary definition appears in 1775: 'Civilization' is defined as 'the state of being civilized', and 'civilized' as 'reclaimed from savageness': John Ash, *A New and Complete Dictionary of the English Language* (2 vols, London, 1775). Johnson did not include the word in the fourth 1772 edition of his dictionary. On the question of vernaculars, note that England was exceptional in having a near-universal vernacular. In France at the time of the Revolution, the French language was dominant 'in only 15 of the country's 89 departments, sharing the stage with German, Basque, Breton, Occitan, Provençal, and other patois. One of the great crusades of the early Revolution was, in fact, to make French universal *in France*': Michael D. Gordin, *Scientific Babel: How Science was Done Before and After Global English* (Chicago, 2015), 18.

[13] See Bertrand Binoche, 'Civilisation: Le mot, le schème et le maître-mot', in Bertrand Binoche, ed., *Les Équivoques de la civilisation* (Seyssel, 2005), 9–30. The two canonical works on the history of the French term are Lucien Febvre, 'Civilisation: Evolution d'un mot et d'un groupe d'idées', in L. Febvre et al., *Civilisation: Le mot et l'idée* (Paris, 1930), 1–59; and Joachim Moras, *Ursprung und Entwicklung des Begriffs der Zivilisation in Frankreich (1756–1830)* (Hamburg, 1930).

[14] Victor Riqueti, marquis de Mirabeau, *L'Ami des hommes, ou Traité de la population* (2 vols, Avignon, 1756–8), i, Part I, 136.

[15] Catherine Larrère, 'Mirabeau et les physiocrates: l'origine agrarienne de la civilisation', in Bertrand Binoche, ed., *Les Équivoques de la civilisation* (Seyssel, 2005), 83–105. Note however that Voltaire had used the term as early as 1731 in his *L'Histoire de Charles XII*.

civilization and a newly emerging commercial society here is reflected in the works in which the English word 'civilization' first appears, John Gordon's *A New Estimate of Manners and Principles* (1760–1), and Adam Ferguson's *Essay on the History of Civil Society* (1767). Germany is a different case, and indeed is anomalous in this respect. The French term was identified with the French Revolution in Gödicke's 1796 *Neues französisches Wörterbuch*, and is translated very awkwardly as 'Sittenverfeinerung oder Sittenverbesserung': refined or improved customs/morality. Even though there was subsequently a German word, *Zivilisation*, it was the very different concept of *Kultur* that played an analogous role to the French and English terms.[16] We will return to this issue later in the chapter. For the moment, our focus will be on the concept of civilization.

From the mid-eighteenth century onwards, there is a new element in the way in which civilization was contrasted with barbarism and savagery. What is at issue is no longer just what marks out civilization from barbarism, but what marks it out, maintains it, and protects it from barbarism. The answer lies in science. By contrast with the thesis of Gibbon's *Decline and Fall of the Roman Empire*, the first volume of which appeared in 1776, that the barbarians ultimately overran the Roman Empire largely because Roman citizens themselves had lost their civic virtues, above all their discipline and industry, the contrast with barbarism is now presented in different terms. Linnaeus, for example, as early as 1759 was remarking that 'only the Sciences distinguish Wild people, Barbarians and Hottentots, from us',[17] and a century later, in 1866, Thomas Huxley writes of science saving civilization from barbarism, talking of a 'new nature' created by science and manifested 'in every mechanical artifice, every chemically pure substance employed by manufacture, every abnormally fertile race of plants, or rapidly growing and fattening breed of animals'. This new nature, we are told, is

the foundation of our wealth and the condition of our safety from submergence by another flood of barbarous hordes; it is the bond which unites into a solid political whole, regions larger than any empire of antiquity; it secures us from the recurrence of pestilences and famines of former times; it is the source of endless comforts and conveniences, which are not mere luxuries, but conduce to physical and moral well-being.[18]

In 1877, the German physiologist Emil du Bois-Reymond was suggesting that the fall of the Roman Empire was due not to social, political, or economic conditions

---

[16] See Emmanuel Renault, 'Le concept hégélien de civilisation et ses signifiants: *Bildung, Kulture* (et *Zivilisation*?)', in Bertrand Binoche, ed., *Les Équivoques de la civilisation* (Seyssel, 2005), 225–39. As an indication of the linguistic difference, note that the original German title of Freud's *Civilization and its Discontents* is *Das Unbehagen in der Kultur*, not *Das Unbehagen in die Zivilisation*. 'Culture and its discontents' would have meant something quite different in English.

[17] Quoted in Lisbet Koerner, *Linnaeus: Nature and Nation* (Harvard, MA, 1999), 94.

[18] Thomas Huxley, *Collected Essays* (9 vols, New York, 1893–4), i. 51.

but to their lack of science: their culture rested 'on the quicksand of speculation'.[19] Historical progress was due to science alone—'the history of natural
science is the actual history of mankind'—and it was science 'that made mankind
mankind'.[20] This estimation of the civilizing role of science was similarly shared
by engineers, despite some degree of antipathy between engineers and scientists at
this time.[21] J. J. Carty, for example, chief engineer of American Telegraph and
Telephone, in his 1916 presidential address to the American Institute of Electrical
Engineers, announced that: 'By every means in our power, therefore, let us show
our appreciation of pure science and let us forward the work of the pure scientists,
for they are the advance guard of civilization. They point the way which we must
follow.'[22]

The theme of science and civilization was pursued by many around the turn of
the century, and in a 1923 essay on 'The Beginnings of Science', included in a
collection devoted to the connection between science and civilization, the author
sees the progress of civilization accompanying the progress of science in particularly revealing terms. He rejects the common way of thinking about civilization,
in terms of cultural achievements such as the arts and monumental architecture.
Civilization, and civilized behaviour, is associated with a scientific mentality, and
explicitly dissociated from the arts and crafts, for example, which he treats as being
completely independent of civilized values:

We speak of primitive 'cultures'; and, therein, of arts or crafts, and other expressions of
primitive personality, amazingly advanced in technique and aesthetic value; yet alongside
of these modes of self-expression, so rationally responsive to daily needs of humanity, such
'cultures' exhibit instances of an almost incredible stupidity—as it seems to us—and of
intolerable, almost inhuman self-repression, all the more surprising from their association
with such intimate interplay of hand and tongue with brain, such orderly, intelligible—in
a word, reasonable—accommodation of effort to need.[23]

[19] Emil du Bois-Reymond, 'Kulturgeschichte und Naturwissenschaft', in Emil du Bois-
Reymond, *Reden* (2 vols, Leipzig, 1912), i. 567–629: 584.

[20] Ibid., i. 596.

[21] To take just one example, physicists and other professional scientists were excluded from the
US Naval Consulting Board, set up following the sinking of the Lusitania. The reason for this, as
Edison's chief engineer explained, was that it was Edison's 'desire to have this board composed of
practical men who are accustomed to *doing* things, and not *talking* about it': quoted in Daniel
J. Kevles, *The Physicists: The History of a Scientific Community in Modern America* (Cambridge, MA,
1971), 109. These were the not only reasons for excluding scientists however: there were also
commercial considerations. When the Naval Consulting Board had to outsource experimental
work on anti-submarine defence, academic physicists were kept out on the grounds that their
inclusion would 'complicate the patent situation', ibid., 120.

[22] John J. Carty, 'The Relation of Pure Science to Industrial Research', *Science* 44 (1916),
511–18: 518.

[23] J. L. Myers, 'The Beginnings of Science', in F. S. Marvin, ed., *Science and Civilization* (Oxford,
1923), 7–42: 7. Cf. Thorstein Veblen, 'The Place of Science in Modern Civilization', *American
Journal of Sociology* 11 (1906), 585–609: 'To modern civilized men, especially in their intervals of
sober reflection, all these things that distinguish the barbarian civilizations seem of dubious value and
are required to show cause why they should not be slighted' (587).

What is at issue here goes well beyond the benefits of science to society. Rather, it is a question of the way in which specific benefits define a particular kind of society. It is not so much that science aids social programmes, or even that a particular kind of society requires a particular kind of scientific input, but rather, it will turn out, that what science can do shapes what society can do. This is a distinctive feature of modernity for our purposes, and it relies on science and civilization mutually shaping one another. The theme is the focus of the report of Vannevar Bush, commissioned by President Roosevelt at the end of 1945, in which proposals are called for turning science from warfare to curing disease, to the development of scientific talent in American youth, fuller and more fruitful employment, and a more fulfilling life. In the first chapter, 'Scientific Progress is Essential', he writes:

Advances in science when put to practical use mean more jobs, higher wages, shorter hours, more abundant crops, more leisure for recreation, for study, for learning how to live without the deadening drudgery which has been the burden of the common man for ages past. Advances in science will also bring higher standards of living, will lead to the prevention or cure of diseases, will promote conservation of our limited national resources, and will assure means of defense against aggression.[24]

What exactly is the connection between science and civilization? How did they come to be associated in the West? What is at issue, I shall be arguing, should be seen as an element in the legitimation of science. It can be traced back to developments between the seventeenth and twentieth centuries in European scientific culture. It is ultimately a question of the ability of science to establish for itself a standing as a legitimate form of activity. But it takes on a new significance in the wake of the Enlightenment, as a conception of civilization is formed around the ideas of science and technology, and ideas of science and technology are in turn shaped around conceptions of civilization.

This reciprocity becomes possible with the emergence of a notion of scientific progress which is mapped onto the understanding of civilization. Although a question must arise right from the outset as to whether the kind of progress that occurs in science is the same as that invoked in the idea of social and political progress, nevertheless the one is imposed on the other. The most striking example of this lies in the idea of the progress of society towards civilization. At stake is the question of modernity, with its notion of progressive, directed social change. There was no essential temporal dimension to civilization as traditionally conceived, in the way that there is to modernity. Before the end of the seventeenth century, it was not thought that civilized societies were the outcome of a particular process of development. As Osterhammel notes,

---

[24] Vannevar Bush, *Science, the Endless Frontier: A Report to the President on a Program for Postwar Scientific Research* (Washington, DC, 1946), 10.

Nothing like the triad of antiquity, Middle Ages, and modern times—which Europe had gradually accepted since the 1680s—came into use in any other civilization that could look back at a continuous and comparably documented past. There were periods of renewal and rebirth, but before contacts with Europe it rarely occurred to anyone that they were living in an age superior to the past.[25]

While Athens and Rome remained models of civilization for many writers in eighteenth- and nineteenth-century Europe—especially if one included architecture, sculpture, and literature among the principal marks of civilization—the association of science and civilization brought with it a transfer of the notion of scientific progress to the understanding of civilization, transforming it into modernity.

There is a fundamental shift in understanding here. The shift lies at the core of the issues with which we shall be engaging. In particular, the question how science and civilization came to be associated in the West in the wake of the Enlightenment is intimately bound up with that of legitimation of science. Given this trajectory, one of our most pressing tasks will be to investigate how science, from the mid-eighteenth century onwards, took on the features that transformed it into the kind of thing that could act as a bearer of the values of civilization. Accordingly, our central concern in this chapter will be with the question of how civilization—traditionally associated with the degree to which a state has attained successful political institutions, a cultural and religious life, literacy levels, social cohesion, prosperity, and a system of laws, for example—could be tied so intimately to science, often previously regarded either as something merely technical or as a useless form of speculative activity. How was it possible for the one to come to be seen as the natural complement to the other? Or, to put it in other terms, what was the nature of the transformation—of both the understanding of science and the understanding of civilization—that enabled science to be treated as the criterion for civilization?

THE UNIQUENESS OF THE WEST

In 1954, the first volume of Joseph Needham's *Science and Civilization in China* appeared, initiating a large-scale project designed to show that China had a long history of science which, for various reasons, had gone unrecognized. Although 'civilization' appears in the title, *Science and Civilization in China* hardly mentions the history of China's political, economic, cultural, or legal achievements. Needham's volumes were almost exclusively a history of Chinese science. This prompts the question why a history of science would be called a history of 'science and civilization'. Could a history of science be *ipso facto* a history of civilization? In the

---

[25] Osterhammel, *The Transformation of the World*, 51.

Preface to his first volume, in setting out the general project, Needham in effect argues that for China to have a place in 'the history of human civilization', the history of its science must be established.[26] The assumption is that, in establishing that China did indeed have a science, we can no longer be in any doubt about its claims to have had a civilization. Science and civilization each need to be conceived in particular ways if this kind of association is to be possible, and the association we are concerned with is a very specific one where science is in effect treated as a necessary condition of civilization. We need to know why, from among a range of possible options, this is the one that emerged triumphant in the West, and we need to know how the association was developed in the first place.

Our starting point in uncovering what lies behind this phenomenon will be to identify its origins, and to explore how the notions of science and civilization that initially nurtured it differed from those that preceded it. The association of science and civilization is so well-entrenched in our own culture, however, that it may look natural, so the distancing provided by a contrast class is valuable. China is a particularly revealing case in this respect, for here the mutual misunderstandings evident in the early modern encounter between the Jesuits and their Chinese patrons, for example, allows us to highlight some of the peculiarities of the European understanding of science and its relation to broader social and cultural issues. An appropriate starting point is 1673, when *La Science des chinois*, by the Sicilian Jesuit missionary Prospero Intorcetta, was published in Paris.[27] Despite the title, this was not an account of Chinese science as we now understand the term. It was a French translation of Intorcetta's Latin translation of a Chinese text, *Chum Yùm: Sinarum scientia politico-moralis*, published in two parts, in Canton in 1667 and Goa in 1669. As the Latin title indicates, the science in question is not the natural or medical sciences, which are hardly mentioned, and the discussion is taken up almost exclusively with the political and moral philosophy of China, largely reformulated in terms of Aristotelian moral philosophy. What China was thought to be able to offer was not astronomical or medical science but the basis for a system of stable government.

Intorcetta's treatise was typical of reports on China of the time. Nevertheless, there was a small and somewhat marginal interest in China and the natural sciences. At the beginning of the seventeenth century, Francis Bacon had re-marked on Chinese inventions, such as gunpowder, magnetic compass needles, printing, and paper, that predated those in the West by centuries, but he concluded that these had been discovered by accident, not in a systematic way, and he had hoped that Western science, by contrast, would be able to introduce systematic enquiry and yield equally significant inventions more reliably. At the end of the century Leibniz, in a letter to the Jesuit procurator of missions, Antoine Verjus, on 2 December 1697, was enthusiastic about an exchange of information

---

[26] Joseph Needham, *Science and Civilisation in China* (7 vols, Cambridge, 1954 onwards), i. 3.
[27] Prospero Intorcetta, *La Science des chinois* (Paris, 1673).

between China and Europe, writing that such an exchange would provide Europeans with the knowledge that the Chinese had built up over thousands of years.[28] But even Leibniz considered the attainments of Chinese science to be limited and, generally speaking, there was a very low estimation of China's achievements in the physical and medical sciences in the first half of the eighteenth century. In a series of letters written in the 1730s, for example, Jean-Baptiste Dortous de Mairan, director of the Paris Académie des Sciences, set out a number of questions and observations on China to the Jesuit Dominique Parrenin. Parrenin had spent a considerable time in China as a missionary, and one question that Mairan put to him was how a people who had achieved such a significant level of civilization could have made so little progress in science, despite the thousands of years that they had devoted to astronomy and medicine.[29] A decade later, in 1742, David Hume was drawing attention to China as a well-organized state which had failed to establish the kind of successful scientific culture that he believed had contributed so much to the greatness of the West.[30]

None of these writers expressed any doubt that China was a civilized nation. One of the earliest writers on China, Matteo Ricci, had singled out the cultivation of moral virtue, the integrity of the family, and the promotion of good governance as qualities of Confucianism from which the West could learn,[31] and China clearly manifested the cultural achievements that Voltaire, for example, had identified in France since the death of Louis XIV in the areas of language, conduct, and thought.[32] By contrast with the West, its scientific achievements did not match its cultural ones, but this did not diminish or disqualify the latter: quite the contrary, they showed that whether or not a society was civilized was independent of any scientific achievements. In the pre-industrial era in Europe, the standards by which Europeans judged and compared non-Western cultures with their own typically focused on such factors as religion, physical appearance, and social patterns. As Michael Adas notes, on those occasions on which science and technology were discussed, they were 'generally treated as part of a larger

---

[28] Gottfried Wilhelm Leibniz, *Leibniz Korrespondiert mit China*, ed. R. Widmaier (Frankfurt, 1990), 55. See Franklin Perkins, *Leibniz and China: A Commerce of Light* (Cambridge, 2004), and Michael C. Carhart, *Leibniz Discovers Asia: Social Networking in the Republic of Letters* (Baltimore, 2019), ch. 9.

[29] The letters were subsequently published as Jean-Baptiste Dortous de Mairan, *Lettres de M. de Mairan, au R. P. Parrenin, Missionaire de la Compagnie de Jesus, à Pekin. Contenant diverses Questions sur la Chine* (Paris, 1769).

[30] David Hume, *Essays and Treatises on Several Subjects* (2 vols, Edinburgh, 1793), i. 123.

[31] Matteo Ricci and Nicolas Trigault, *China in the Sixteenth Century: The Journals of Matthew Ricci, 1583–1610* (New York, 1953), 337. The seventeenth-century Spanish Dominican Domingo Navarrete even suggested that Confucianism provided a remedy for the political crisis afflicting his native Spain: *The Travels and Controversies of Friar Domingo Navarrete, 1618–1686*, ed. J. S. Cummins (2 vols, Cambridge, 1962) i. 6.

[32] François Marie Arouet de Voltaire, *Essai sur les mœurs, et l'esprit des nations et sur les principaux faits de l'histoire depuis Charlemagne jusqu'à Louis XIII* (Geneva, 1756); *Le Siècle de Louis XIV* (Berlin, 1751). Generally on China and the eighteenth century, see Part II of Jürgen Osterhammel, *China und die Weltgesellschaft. Vom 18. Jahrhundert bis in unsere Zeit* (Munich, 1989).

configuration of material culture. Within this configuration, monumental architecture, sailing vessels, and even housing were often more critical than tools or
astronomical concepts in determining European attitudes toward different non-
Western peoples.'[33]

Intorcetta, Mairan, Hume, and their contemporaries all believed that China
had a civilization, but that it had no significantly developed science. Leibniz saw
this as a failing of China, but more importantly he also saw it as a significant
opportunity, believing that a demonstration of Western scientific superiority,
particularly in mathematics and astronomy, might be enough to carry with it a
more general sense of Western supremacy, one which included the superiority of
its religion, Christianity. In this respect, he mirrored the hopes of the contemporary Jesuit missionaries with whom he was in correspondence.

If we focus not on the Jesuit perspective on their interchange with the Chinese
scholars, but rather on the Chinese perspective on the interchange, a very different
picture emerges from that reported by the Jesuits. The first point to note was that
the Jesuit missionaries did not interact with 'the Chinese', but with a small group
of Chinese scholar-patrons. This latter group—comprising literati, that is, land-
owning cultural elites distinguished by their knowledge of classical scholarship
and lineage ritual—saw themselves very much as patrons, and portrayed the
Jesuits as tributary officials who had come to China to serve the emperor. The
literati had a particular agenda, namely securing their access to, and standing in,
the Ming court. This determined how they dealt with the Jesuits' offerings, and
their estimation of the importance of the Jesuit mission bore little relation to the
Jesuits' own assessment. As Elman notes:

Each side sought to efface the other by simple reduction of the other to themselves. Their
actual common ground was a hybrid that assumed each side had the same agenda, but each
aimed to achieve diametrically opposite results. Ricci and the Jesuits tried to efface the
classical content of the investigation of things with western European natural studies,
which would then enable the Chinese to know heaven and accept the Church. Chinese
effaced Western learning with native traditions of investigating things and extending
knowledge, which would allow them to assert that European learning originated from
China and thus was assimilable.[34]

Taking advantage of the interest in science at court to improve their standing, the
Chinese scholars presented themselves as the bearers of the vastly superior science
which they had taken from the Jesuits. Interestingly, like the Jesuits, they
associated the superiority of Western science with the superiority of Western
religion and politics: but the latter superiority was not something that had
originated in the West on their reckoning. Quite the contrary, on their view it

[33] Michael Adas, *Machines as the Measure of Man: Science, Technology, and Ideologies of Western Dominance* (Ithaca, NY, 1989), 6.
[34] Benjamin Elman, *On Their Own Terms: Science in China, 1550–1900* (Cambridge, MA, 2005), 113.

Fig. 1.1. Matteo Ricci and Xu Guangqi, frontispiece to Chinese edition of Euclid's *Elements* (1670)

had originated in the East. This was encouraged by the fact that the Jesuits, in order to accomplish their task of converting the Chinese to Christianity, had had to contextualize selected doctrines in Chinese terms. As Florence Hsia notes, the 'ideal Jesuit was thus a shape-shifter, putting on different clothes and different roles, speaking in as many tongues and modulating his message in as many different forms of diction as his audiences required'.[35] The upshot of this in the case of China was that the writings that the Jesuits brought with them, including religious writings, had, to their intended Chinese recipients, an unmistakable Chinese flavour about them, and were viewed by them as recovered ancient Chinese writings[36] which had been lost two thousand years earlier in the great burning of the books (213 BCE) by the Emperor Shih Huangdi, but preserved in the West in the doctrines of heaven, hell, and the great Lord (*Tianzhu*).[37] The situation was further complicated by the fact that Ricci and his followers, who defended the idea of an original Chinese monotheism in the 'Chinese rites' controversy,[38] believed that their task was to 'reawaken' the old faith, which tended to assimilate an archaic Chinese religion to Christianity.

In sum, what was at stake for most seventeenth- and early eighteenth-century writers on China was whether it offered a model of long-term social and political stability, and whether there was a continuity between its religious beliefs and those of Christianity. There was another respect in which China was compared with the West, however, that was to prove even more significant. This was the antiquity of its culture, a question that had significant consequences for the understanding of Christianity, and for the Christocentric model of history on which notions of civilization were effectively premised.

The Chinese view of the antiquity of their religious doctrines was doubtless encouraged by the fact that Chinese chronology extended much further back than the Christian chronology current in the West, as both European and Chinese writers were aware.[39] European writers rejected the claims of the great antiquity of

---

[35] Florence Hsia, *Sojourners in a Strange Land: Jesuits and Their Scientific Missions in Late Imperial China* (Chicago, 2009), 1.

[36] Much the same kind of misunderstanding occurred in Japan, where 'God' was translated by the character for Sun-Buddha and 'Christianity' by that for Buddhist law, with the result that the Jesuit missionaries were taken to be sectarian reformers from India. See Urs App, *The Birth of Orientalism* (Philadelphia, 2010), 16–19.

[37] See Sangkeun Kim, *Strange Names of God: The Missionary Translation of the Divine Name and the Chinese Responses in Late Ming China, 1583–1644* (New York, 2004).

[38] The Chinese rites controversy was a dispute over whether Chinese ritual practices were religious, and thus in conflict with Christianity, or secular. See Michela Fontana, *Matteo Ricci: A Jesuit in the Ming Court* (Lanham, MD, 2011). There is a collection of primary sources in Donald F. St. Sure, Ray Robert Noll, and Edward Malatesta, *100 Roman Documents Concerning the Chinese Rites Controversy (1645–1941)* (San Francisco, 1992).

[39] The Jesuit missionary Martino Martini, in his *Sinicae historiae decas prima* (Amsterdam, 1658), first brought the existence of Chinese records dating from antediluvian times to European readers. Chronology problems had been recognized independently of the Chinese calendar however. Edward Stillingfleet, in his *Origines Sacrae* (London, 1662), warns that one of the 'most popular pretences of the Atheists of our Age, have been the irreconcileableness of the account of Times in Scripture with

Chinese civilization, particularly the extent to which these contradicted biblical calculations not only of the creation of the earth, but more to the point, of the time when Noah was supposed to have repopulated the earth.[40] Moreover, some of the Chinese datings could be discounted on astronomical grounds: Jean Dominique Cassini calculated that the conjunction of five planets in the reign of 'the fifth Emperor of China' did not occur in 2513 BCE as claimed, but five centuries later, in 2012 BCE.[41] But the Chinese themselves had no doubts at all about the general authenticity of their own records, and there was compelling evidence that the biblical Flood (dated by Christian writers to 2348 BCE) occurred some time later than the dates of the early Chinese sage-kings.[42] In the light of this, it was natural for them to think of Western civilization as something comparatively recent, and as the beneficiary of an earlier Chinese civilization.

One thing that the Chinese chronology controversy highlights is that the question of civilization turned on more than just cultural and political achievements, but also on that of the antiquity of one's culture. Questions of antiquity were questions of credentials, and accordingly closely bound up with those of legitimacy, and they bore directly on the issue of whether a Christocentric and Eurocentric model of history, so crucial for establishing the primacy of Christian and European values at the fountainhead of civilization, had the legitimacy it

---

that of the learned and ancient Heathen Nations' (Preface, unpaginated). Stillingfleet seems to have been unaware of the controversies over Chinese chronology, and has as his targets Egypt, Phoenicia, Chaldea, and Greece.

[40] See, for example, Thomas Burnet, *Archaelogiae philosophiae: sive doctrina antiqua de rerum originibus* (London, 1692).

[41] Jean Dominique Cassini, 'Reflexions sur la Chronologie chinoise par Monsieur Cassini', in Simon de la Loubere, *Description du royaume de Siam* (2 vols, Amsterdam, 1691), ii. 304–21. For a summary of the objections to the accuracy of Chinese astronomy at the time, see George Costard, 'A Letter from the Rev G. Costard to the Rev. Thomas Shaw, D.D. F.R.S. and Principal of St. Edmund-Hall concerning the Chinese Chronology and Astronomy', *Philosophical Transactions* 44 (1747), 475–92.

[42] Not all Christian writers accepted that Chinese chronology was inconsistent with the story of Noah. As Elman notes: 'Martinus Martini's 1654 history of China challenged European chronology by accepting the Fu Xi reign [2952–2838 BCE]. Hence, Martini also preferred using the Septuagint chronology to place the Great Deluge in China before 3000 B.C., although he was sceptical that the Chinese flood in Yao's reign was the same as Noah's flood. John Webb, on the other hand, drew from Martini's account to argue that Emperor Yao was in fact Noah. Others chose Noah's son Sem as the first to get to China, while Anthanasius Kircher (1601–1680) chose Cham.' *On Their Own Terms*, 140. Note also that the claims of century-long reigns of the Chinese sages, which might have attracted criticism as implausible, were in fact accepted by Christian writers, who assimilated them to similarly long-lived Old Testament Patriarchs. The Benedictine chronologist Paul Pezron, for example, writes that: '*Fohi* [Fu Xi] was born in the time of the Patriarchs, having been a contemporary of *Heber*, *Phaleg*, and *Rehu*, the great grandfathers of Abraham, and men then still lived two, three, or four centuries, as is evident from Sacred History. It follows from this that there exist such long-term reigns, i.e. 115 years, as the Chinese aver, because their length of life approached that of the Patriarchs.' *L'Antiquité des tems rétablié et defenduë contre les Juifs & les nouveaux chronologists* (Paris, 1687), 246. On these questions see Edwin J. Van Kley, 'Europe's "Discovery" of China and the Writing of World History', *The American Historical Review* 76 (1971), 358–85.

assumed itself to have. The issues came to a head in the second half of the eighteenth century, and they centred on the question of 'universal history'.[43]

UNIVERSAL HISTORY

From the late seventeenth century, a detailed periodization which divided European secular history into antiquity, the Middle Ages, and the modern period, came into general use, and with it the notion that the present was superior to the past in virtue of having occurred at a later stage of a process of development.[44] What emerged, and what became distinctive of the West's sense of itself, was what Osterhammel has called a 'future-orientated rhetoric of new beginnings', fuelled by rebirths and renewals.[45] But secular history had been ineluctably entangled with a Christian 'universal history' which offered a rather different story, one couched in terms of perfection, fall, and redemption. Although it had eschatological elements, this was very much a past-orientated schema and, problematically, it was one that went to the core of the understanding of European civilization,[46] for at least before the middle of the eighteenth century there was agreement that Christianity—particularly through its moral dictates—had played the crucial role in effecting the transition from primitive society to civilization. Even if not quite everybody considered civilization to be something essentially Christian, there is no doubt that, before the middle of the century, Christian Europe provided a universal model for understanding the values of, and transition to, civilization. Even those *philosophes*, such as Diderot, who questioned Christian origins adhered, perhaps unconsciously, to the biblical narrative, seeking not only a single origin for the human race, for example, but also for human civilization.[47] A Eurocentric account of the emergence of civilization was in essentials a Christocentric one, and the civilized versus non-civilized division effectively mirrored that between the Christian and the non-Christian. Yet within the category of the non-Christian there was clearly a difference between primitive groups, on the one hand, and large-scale societies such as those of China and the Ottoman Empire. China and the Ottoman Empire could hardly be counted as primitive, and, if thought of in purely secular terms, could indeed be treated as non-Christian forms of civilization, which is effectively how Voltaire, for example, treated China.

[43] As John Gray remarks, 'just as the category of *civilization* is a central element in the Enlightenment project, so the idea of *a universal history of the species* is integral to it'. John Gray, *Enlightenment's Wake: Politics and Culture at the Close of the Modern Age* (London, 1995), 125.

[44] See John G. A. Pocock, 'Perceptions of Modernity in Early Modern Historical Thinking', *Intellectual History Review* 17 (2007), 55–63.

[45] Osterhammel, *The Transformation of the World*, 51–2.

[46] Moras notes that the term 'la civilisation européenne' first appears in 1766 in a work on the French colonies in North America probably by the physiocrat Nicolas Baudeau (who also introduced the term 'économiste'): Moras, *Ursprung und Entwicklung*, 47.

[47] See René Hubert, *Les Sciences sociales dans l'Encyclopédie* (Paris, 1923).

The alternative was to Christianize the apparent civilizations of the East, a strategy deployed in the case of China, where, as we have seen, there were attempts to accommodate the Jesuit reports, for example by tying China to a Judeo-Christian heritage, seeking to find traces of a Noachian religion in ancient Chinese texts. Such a procedure was not without precedents. Classical Greek culture had been treated as proto-Christian by the Church Fathers, with claims of Plato having familiarity with the writings of Moses.[48] But even putting to one side the credibility of such accounts by the eighteenth century, the Ottoman Empire, for example, would seem to present insuperable difficulties for this kind of approach. It post-dated Christianity and its religious base showed no indebtedness to it, so it was difficult to trace origins in the way that had been attempted for China.

Whatever the accommodation strategies, there was a manifest conflict between a Christian model of civilization and the future-orientated model that had emerged by the nineteenth century. Historians and theologians were aware of this, and abandoning the Christian model gradually became the preferred option, although it inevitably left traces in the new conception. But how was the transition from a Christian historiography to a future-orientated one effected? The key move in initiating the transition occurred in what had been the essentially Christian genre of universal history. The standard work on universal history was Bossuet's 1681 *Discours sur l'histoire universelle*, which set out the model for reconciling Christian and secular history.[49] For Bossuet, the histories and religions of all peoples were rooted in the events described in the Old Testament. His account is unapologetically Christocentric in that everything is subordinated to the history of the Church, which provides the organizing thread. It is also unapologetically Eurocentric in that there is no mention of China, India, or Japan. Developments outside Christian Europe were irrelevant for Bossuet's 'universal history', which confined itself to the 'succession of empires' only insofar as they have a 'necessary connection with the history of the people of God'.[50]

Fewer than half of the universal histories written during the last half of the seventeenth century included China, but most eighteenth-century ones did so. Nevertheless, many of these eighteenth-century histories cannot be said to have incorporated it into their conception of world history in any comprehensive way: it was simply accommodated to the Eurocentric-Christocentric model. Some

---

[48] See Gaukroger, *The Emergence of a Scientific Culture*, 49–59

[49] Christian universal history had a genealogy going back to the *Chronici canones* of Eusebius of Caesarea in the early fourth century. Eusebius began his chronology with the creation of the world, but included Greek, Roman, Jewish, Assyrian, Persian, and Egyptian rulers.

[50] Jacques-Bénigne Bossuet, *Discours sur l'histoire universelle a monseigneur le dauphin pour expliquer la fuite de la religion et les changemens des empires* (Paris, 1681), 430. Other writers saw the issue in terms of proximity either to the Garden of Eden or Mount Ararat: Turgot, for example, suggests that the Chaldeans are at the fountainhead of culture because 'they are closest to the source of the first traditions' (*Oeuvres*, ii. 599).

writers simply identified the early Chinese emperors with the Hebrew patri-archs,[51] while others followed the biblical model in every other respect but made China the first empire after the Flood, displacing Babylon.[52] Yet others, while rejecting China's claims to great antiquity, nevertheless included long descriptions of China. Augustine Calmet, for example, in his *Histoire universelle* of 1735,[53] did cover China, despite otherwise taking Bossuet as his model. By the middle of the century, China was beginning to be treated as an essential part of universal histories, and the sixty-eight-volume *Universal History* published in London between 1747 and 1768 offers a wide coverage of Chinese material.[54] Equally extensive was the treatment in Lambert's *Histoire générale*, which adver-tised itself in its subtitle as covering virtually every aspect of social, political, scientific, and artistic life in Europe, Asia, Africa, and America.[55]

There was, however, one mid-eighteenth-century work that stood out. Vol-taire's *Essai sur les mœurs* is a watershed in comparative history. The *Essai* has a twofold significance. First, it uses information about China in a way that no other work had done, as a contrast case to identify features of self-conceptions of Europe, and the historiographical assumptions of its religion, Christianity, that had otherwise escaped attention. Second, the long history of Voltaire's reworking of *Les Mœurs*, from the first fragments dating from 1745 to the third edition of the work in 1769, reveals a shift from a concern to offer a contrast case, China, to a reconstruction of the non-Christian origins of civilization in ancient India.

Voltaire's interest in China was motivated in large part by a concern to establish a form of deism as the original religion, displacing Christianity (and Judaism) in this role.[56] Religious questions therefore play a significant role in his discussion. But because of the Christocentric basis of 'universal histories', there are also profound historiographical issues in play. Voltaire not only refuses to deploy a Eurocentric model to examine China, but uses China to raise questions about Europe.

[51] E.g. Pezron, *L'Antiquité des tems*.

[52] Johann Matthias Hase, *Historiae universalis politicae quantum ad eius partemn i. ac ii. idea plane nova et legitima tractationem summorumn imperiorum etc.* (Nuremberg, 1743); Etienne André Philippe de Prétot, *Analyse chronologique de l'histoire universelle, depuis le commencement du monde, jusqu'a l'empire de Charlemagne inclusivement* (Paris, 1753).

[53] Augustine Calmet, *Histoire universelle sacrée et profane, depuis le commencement du monde jusqu'à nos jours* (Strasbourg, 1735).

[54] George Sale et al., *An Universal History, from the Earliest Account of Time* (68 vols, London, 1747–68).

[55] Claude-François Lambert, *Histoire générale, civile, naturelle, politique et religieuse de tous les peuples du monde, Avec des observations sur les mœurs, les coutumes, les usages, les caracteres, les differentes langues, le gouvernement . . . les arts & les sciences des différents peuples de l'Europe, de l'Asie, de l'Afrique & de l'Amerique* (15 vols, Paris, 1750).

[56] On Voltaire's interest in China, see Walter Engemann, *Voltaire und China* (Leipzig, 1933); Raymond Schwab, *The Oriental Renaissance: Europe's Rediscovery of India and the East, 1680–1880* (New York, 1984), ch. 6; and App, *The Birth of Orientalism*, ch. 1.

A key theme in his earlier *Le Siècle de Louis XIV*, published in 1751, five years before the publication of the *Mœurs*, is that of persecution and tolerance.[57] The book concludes with a final chapter on the Chinese rites controversy, which he sees as manifesting an important feature of Christianity:

It is not enough for the disquiet of our minds, that we have disputed at the end of seventeen hundred years upon the articles of our own religion, but we must likewise introduce into our quarrels those of the Chinese. This dispute did not produce any great disturbance; but, more than any other, it goes to characterize that restless, quarrelsome, and contentious spirit that prevails with us.[58]

Here, what was the crucial point of interest in China for Christian writers is dismissed as little more than argument for argument's sake. The significance of China must lie elsewhere, not as a lesson in the extent to which China can be assimilated to the West, but in the extent to which the West can provide a legitimate framework for understanding China. *Les Mœurs* begins not only with an account of ancient China but compares it favourably with the Christian West, on antiquity, its moral and social virtues, and what Voltaire considers its mono-theism, construed very much in the form of deism. Monotheism is no longer the property of Judaism and Christianity but is rather, for Voltaire, a form of natural religion common to all cultures.

Note however that Chinese religion, which he identifies with Confucianism, only provides an alternative form of monotheism, not its primordial form, even though it is one with an earlier provenance than Judaism and Christianity. As *Les Mœurs* comes to be revised by Voltaire in the later editions, his attention starts to be focused less on China and more on India, particularly after the discovery of ancient texts of Indian origin significantly pre-dating Chinese ones. A case could now be made that India, whose religion he takes to be monotheistic,[59] was the cradle of civilization. In his *La philosophie de l'histoire* of 1765, Voltaire traces a genealogy of the human race, which he considers began with savages roaming the forests, followed, much later, by the emergence of the idea of a creating and punishing God, an idea that arose in temperate, densely populated regions such as India, China, and Mesopotamia, with the people around the Ganges river forming the earliest of these human settlements.

The first European secular institution for the study of oriental languages, the École Spéciale des Langues Orientales Vivantes in Paris, was founded in 1795, and its first director, Louis-Mathieu Langlès, was influenced by Voltaire's ideas of

---

[57] See John G. A. Pocock, *Barbarism and Religion* (6 vols, Cambridge, 1999–2015), ii. 83–96.

[58] Voltaire, *Oeuvres complètes de Voltaire* (72 vols, Gotha, 1784–90), xxi. 399.

[59] Voltaire is following Mathurin Veyssière de la Croze, *Histoire du Christianisme des Indes* (The Hague, 1724), who argued that Noah's religion made its way to the Indies soon after the deluge and was preserved there; and Johann Lukas Niekamp, *Kurtzgefasste Mißions-Geschichte, oder historische Auszug der evangelischen Mißions-Berichte aus Ost-Indien von dem Jahr 1705 bis zu Ende des Jahres 1736* (Halle, 1740), translated into French in 1745.

the Indian origin of all civilization, as well as the work of British scholars of India such as William Jones. He writes that he supports Voltaire's view that the Chinese and the Egyptians learned what they knew from the Indians,[60] and that the Bible was an imitation of the far older Veda. Urs App sums up the situation well, when he writes:

In the course of the eighteenth century, Europe's dominant ideological matrix experienced a deepening crisis, and its hitherto unassailable biblical foundation showed ever more threatening fissures. The loss of biblical authority, which was due to many factors, occurred at a time when Judaism and Christianity themselves began to be increasingly viewed as local phenomena on a dramatically expanded, worldwide canvas of religions and mythologies. At the end of the eighteenth century, Volney . . . portrayed Christianity as a relatively insignificant and young local religion based on local varieties of solar myth.[61]

In one sense it is clear that these developments were, despite themselves, shaped by a biblical perspective. The question of origins, for example, was crucial. Before the last decades of the second century, the attraction of the Pauline religion had lain in its novelty and its radical difference from everything else. But at the same time there had developed a view among its critics that something barely a century and a half old could not make claims to being the true religion. Theophilus, the bishop of Antioch, realized that it needed a longer genealogy, and, believing that the sacred Jewish books had foretold the coming of Christ, he maintained that the history of Christianity was continuous with that of Judaism, and on this basis concluded that the origins of Christianity predated those of any other religion. The antiquity of Christianity, its standing as the oldest religion, became a crucial element in defending its legitimacy. But it was not just the antiquity of the practices and beliefs that mattered. Equally important was the antiquity of its book, the Bible. Consequently, when the existence of earlier, seemingly mono-theistic, religious texts became evident, biblical claims to authority as the prim-ordial religious text were no longer supportable.

To the extent to which the history of civilization was the history of Christian-ity, and to the extent to which this standing was dependent on the idea that Christianity—in the form of Judeo-Christianity—lay at the historical origins of the human world, the undermining of its relative antiquity meant that the association of the history of Christianity and the history of civilization lost legitimacy. But, with a few possible exceptions,[62] it was not being proposed that some Indian Ur-religion replace Christianity in this role. The undermining of the association between Christianity and civilization did not favour Indian religion over Christianity as the basis of civilization, but rather had the effect of dissociating the development of civilization from religion generally speaking.

[60] Louis Mathieu Langlès, *Fables et contes indiens: Nouvelles traduits, avec un discours préliminaire et des notes sur la religion, la littérature, les moeurs, &c. des Hindoux* (Paris, 1790).
[61] App, *The Birth of Orientalism*, xiii–xiv.
[62] See Schwab, *The Oriental Renaissance*, on the complex range of responses.

This did not mean that Western 'civilization' was not marked out from other cultures, or that Enlightenment thinkers abandoned any sense of the superiority of their own culture, but rather that, increasingly, they did not see religion as the source of this superiority. Moreover, there was no longer any reason why one should follow the Christian model in tracing the superiority back to origins. In particular, one would not be seeking long-term continuity if one thought, as *philosophes* generally did, that the period roughly between classical antiquity and the sixteenth century—the period in which, in their view, the Catholic Church held back and stunted intellectual enquiry—had regressed from the achievements of earlier times. What mattered was the distinctiveness of the present. What was it, then, that marked out Enlightenment Western societies from others for the *philosophes*?

Voltaire gives a hint of where the answer might lie. In his writing on China, he promoted the standing of China as an advanced civilization, and tells us that despite being indifferently proficient in the sciences, they were 'the first people in the world in morals and policy'.[63] There was however one significant difference between China and the West to which he drew attention: the absence of 'progress' in China. By contrast with Enlightenment France and Britain, for example, China was a static society. The sheer number and variety of proposed explanations for this that Voltaire adduces—from the racial characteristics of the Chinese, the difficulties of language and the Chinese script, the backwardness of the educational system, the predominance of ancestor worship, and the Tartar conquest[64]—is an indication of the intractability of the phenomenon. Although he touches on the question on a number of occasions, he nevertheless does not see it as a decisive one. This changes, and as the century progressed there was significantly increased interest in the question. In particular, Volney, in his very popular *Les Ruines, ou méditation sur les révolutions des empires* (1791), argued in detail for one of the most popular solutions: that the backwardness of the East could be put down to its despotism.[65]

But how was despotism to be combatted? For some, the answer lay in science, and it derived not from reflection on the East but on their own European past. Contemporaneous with Voltaire's historical writings, there emerged in Paris a number of spirited defences of an Enlightenment culture which identified science as the remedy for the kind of social stagnation that Voltaire had identified in other societies such as China, but which they identified in their own medieval past. The idea arose that it was science that had enabled their predecessors to

---

[63] Voltaire, *Oeuvres complètes*, xxi. 399.

[64] See J. H. Brumfitt, *Voltaire Historian* (Oxford, 1958), 80.

[65] Constantin-François de Volney, *Les Ruines, ou méditation sur les révolutions des empires* (Paris, 1791); on Volney see App, *The Birth of Orientalism*, ch. 8. In Mary Shelley's *Frankenstein* (1818), the monster learns of the rights of man, the origins of social inequality, and the disaster of organized religion from Volney's *Ruines*.

overcome the despotism of the Church, and the stagnation it induced. Before we turn to these accounts, however, we need to note an understanding of civilization that was in some respects at odds with that whose development we have been following.

CIVILIZATION VERSUS *KULTUR*

In 1868, Matthew Arnold wrote:

What I admire in Germany is, that while there, too, Industrialism, that great modern power, is making at Berlin and Leipzig and Elberfield most successful and rapid progress, the idea of Culture, Culture of a true sort, is in Germany a living power also. Petty towns have a university whose teaching is famous throughout Europe; and the King of Prussia and Count Bismarck resist the loss of a great savant from Prussia as they would resist a political check.[66]

Between the middle of the nineteenth century and the outbreak of the Great War, Germany was at the forefront of scientific and technological research. As Nye notes, 'from 1840 to World War I, nearly eight hundred British and Americans earned doctoral degrees in chemistry at one of twenty German universities.... Englishmen studying in Germany sometimes directed research, including the work of other Englishmen in German university laboratories'.[67] At the same time, she points out, the German system produced a higher level of literacy and culture among a broader spectrum of its population than was true anywhere else in the West.[68]

In 1914, the English view of Germany suddenly changed. In the First World War, the British, French, and Americans considered the fight against Germany as a fight for civilization. The British Victory medal was inscribed 'The Great War for Civilisation, 1914–1919' (Fig. 1.2). The Germans, by contrast, considered themselves as defenders of *Kultur*, and German values were ridiculed in Allied propaganda. A British manufacturer, for example, produced cast-iron 'iron crosses' with 'For Kultur' on one side and the names of cultural sites destroyed by the Germans on the other (Fig. 1.3), while a US recruitment poster had a drooling German gorilla carrying the limp half-naked body of a woman in one arm and a club inscribed with the word '*Kultur*' in the other (Plate 3). The contraposition of civilization and *Kultur* was not merely an instrument of Western propaganda, however: the Marburg Neo-Kantian Paul Natorp, for example,

---

[66] Matthew Arnold, *Schools and Universities on the Continent* (London, 1868), 256.
[67] Mary Jo Nye, *Before Big Science: The Pursuit of Modern Chemistry and Physics, 1800–1910* (Cambridge, MA, 1996), 3.
[68] Ibid., 8.

Fig. 1.2. First World War Victory Medal (reverse)

contrasted the two, associating the former with Western materialism and defend-ing Germany's special cultural vocation.[69]

'*Kultur*' does not directly translate as 'culture'. On the contrary, it marks out a mentality that in some ways stands as an alternative to that of civilization. As Norbert Elias points out, 'civilization'

does not mean the same thing to different Western nations. Above all, there is a great difference between the English and French use of the word, on the one hand, and the German use of it, on the other. For the former, the concept sums up in a single term their pride in the significance of their own nations for the progress of the West and of mankind. But in German usage, *Zivilisation* means something which is indeed useful, but neverthe-less only a value of the second rank, the surface of human existence. The word through which Germans interpret themselves, which more than any other expresses their pride in their own achievement and their own being, is *Kultur*.[70]

---

[69] See Paul Natorp, *Der Tage des Deutschen: Vier Kriegsaufsätze* (Hagen, 1915). See the discussion in Thomas E. Willey, *Back to Kant: The Revival of Kantianism in German Social and Historical Thought, 1860–1914* (Detroit, 1978), 116–24.

[70] Norbert Elias, *The Civilizing Process: Sociogenic and Phylogenetic Investigations* (Oxford, 2000), 4. Elias traces the antithesis between *Zivilisation* and *Kultur* back to Kant: ibid., 8.

Fig. 1.3. British First World War fake Iron Cross

He continues:

The French and German concept of civilization can refer to political or economic, religious or technical, moral or social facts. The German concept of *Kultur* refers essentially to intellectual, artistic, and religious facts, and has a tendency to draw a sharp dividing line between facts of this sort, on the one side, and political, economic, and social facts, on the other. The French and English concept of civilization can refer to accomplishments, but it refers equally to the attitudes or 'behaviour' of people, irrespective of whether or not they have accomplished anything. In the German concept of *Kultur*, by contrast, the reference to 'behaviour', to the value which a person has by virtue of his mere existence and conduct, is very minor.[71]

This conception of *Kultur* was particularly marked in academic circles. In an 1877 speech, du Bois-Reymond warns against the potential 'Americanization' of German science, with its utilitarian devotion to technology and capitalist material-ism,[72] and in his history of German universities, Ringer notes that 'the German idea of academic freedom was always at least partly informed by the conviction that *Geist* must not be asked to descend from the realm of theory in order to

---

[71] Ibid., 4. Many examples of the distinction can be found. Einstein, for example, presents himself as a man of *Kultur* rather than a man of civilization when, in his 'Autobiography', he writes: 'The essential in the being of a man of my type lies in precisely *what* he thinks and *how* he thinks, not what he does or suffers'. Einstein, 'Autobiographical Notes', in P. A. Schilpp, ed., *Albert Einstein, Philosopher-Scientist* (2 vols, New York, 1959), i. 1–96: 33.

[72] Du Bois-Reymond, 'Kulturgeschichte und Naturwissenschaft', i. 624.

involve itself in practice'.[73] Wolf Lepenies has described the German preoccupation with *Kultur* as something that its advocates see as a 'noble substitute for politics'.[74]

The idea of *Kultur* has often been associated with an anti-science agenda, for example with Spengler's widely read *Decline of the West*, the first volume of which was published in 1918,[75] offering a popular version of the idea of the decline of civilization in the modern era, a theme originally developed in Nietzsche, and in Max Nordau's widely discussed *Entartung* ('degeneration') of 1892.[76] Spengler railed against the idea of an analytic, cause-seeking science, and saw the pervasiveness of science as evidence that Western civilization was in the final stages of decline.[77] He was not alone in this, and Paul Forman has characterized the dominant intellectual tendency of Germany in the late 1920s and early 1930s as 'a neo-romantic, existentialist "philosophy of life", revelling in crises and characterized by antagonism towards analytical rationality generally and toward the exact sciences and the technical applications particularly. Implicitly or explicitly, the scientist was the whipping boy of the incessant exhortations to spiritual renewal, while the concept—or the mere word—"causality" symbolized all that was odious in the scientific enterprise.'[78] But the specifically German idea of *Kultur* is intimately associated with the no less distinctively German notion of *Bildung*—the idea of cultivating a particular kind of educated, cultured, and enlightened persona for oneself[79]—and there is no inevitable commitment to an anti-science agenda. In itself, the idea of *Bildung* is neither pro- nor

[73] Fritz Ringer, *The Decline of the German Mandarins: The German Academic Community, 1890–1933* (Cambridge, MA, 1969), 111. There were dissenting voices however. Rudolph Virchow is cutting in his 1847 lecture 'Standpoints in Scientific Medicine': 'From the period of philosophical confusion we have clung to a conception which is nowhere further developed than in Germany and which has nowhere produced more harm than in medicine—I refer to the idea of 'Science for its own sake', a detached science with its own goals, a science for the sake of mere knowing.' Rudolph Virchow, *Disease, Life, and Man: Selected Essays by Rudolph Virchow*, ed. and trans. L. J. Rather (Stanford, 1958), 29.

[74] At the same time he notes that this is not a uniquely German phenomenon: specific instances can be found in France in the eighteenth century, in Spain in the nineteenth, in Russia at the end of the nineteenth century, and in Ireland at the same time. Wolf Lepenies, *The Seduction of Culture in German History* (Princeton, NJ, 2006), 5–11.

[75] Oswald Spengler, *Der Untergang des Abendlands* (2 vols, Munich, 1918–28).

[76] Max Nordau, *Entartung* (2 vols, Berlin, 1892–3).

[77] The book had a receptive audience: note Siegfried Kracauer's comment on his youth in the Weimar republic, that it was marked by 'a "hatred of science" rampant among the best of today's youth.' *The Mass Ornament: Weimar Essays* (Cambridge, MA, 1995), 213–14.

[78] Paul Forman, 'Weimar Culture, Causality, and Quantum Theory, 1918-1927: Adaptation by German Physicists and Mathematicians to a Hostile Intellectual Environment', *Historical Studies in the Physical Sciences* 3 (1971), 1–116: 4.

[79] I discuss *Bildung* in Gaukroger, *The Natural and the Human*, 342–50. For a more exhaustive treatments see Hans Weil, *Die Entstehung des deutsches Bildungsprinzips* (Bonn, 1930); Rudolph Vierhaus, 'Bildung', in O. Brunner, W. Conze, and R. Koselleck, eds, *Geschichtliche Grundbegriffe* (5 vols, Stuttgart, 1972–89), i. 508–51; and Walter H. Bruford, *The German Tradition of Self Cultivation: Bildung from Humboldt to Thomas Mann* (Cambridge, 1975).

anti-science. One of its greatest advocates, and one of the greatest advocates of *Kultur*, Lessing, writes of the search for truth, for example, that 'the true value of a man is not determined by whether, in reality or not, he possesses the truth, but rather by his sincere striving to reach the truth. It is not possession of the truth, but rather the pursuit of truth that allows him to extend his powers, and in which his constantly increasing capacity for perfection is to be found.'[80] Although the idea of *Bildung* finds its paradigm application in the realm of the arts, this idea that it is the search for truth, not the truth itself, that is of greatest value, fits the work of the scientist perfectly well. It offers a very different understanding of the value of scientific enquiry from what is in effect the results-orientated conception of science that accompanies the eighteenth- and nineteenth-century notions of civilization with which we have been concerned, and which Weimar intellectuals rejected.

The *Bildung-Kultur* conception, by focusing on the practice of science rather than its results, allows for the elevation of science to the highest level of cultural achievement: the practice of science becomes a way of giving one's life a direction. But the situation is complex. In one respect, this conception can enhance the prospects of science as a worthwhile form of activity. The idea of *Bildung* was a response to a set of older values, which contemporaries characterized in terms of the idea of the *Gelehrten*, academic scholars. It was an alternative to academic scholarship for its own sake, which was becoming widely ridiculed in Germany from the middle of the eighteenth century onwards, beginning with Lessing's 1748 play *Der junge Gelehrte*.[81] But the move has a number of dimensions. R. Steven Turner has argued that there is a fundamental shift in German intellectual culture with the concentration of scientific enquiry in the state-funded universities from the beginning of the nineteenth century onwards. This shift, which he terms the 'Great Transition', is one of the demise of traditional ideals of learnedness: erudition, eloquence, Latinity, and a polymathic command of a common tradition of learning.[82] The shift was consolidated in the early decades of the nineteenth century,[83] moving German intellectual culture from

---

[80] 'Anti-Goetze: Eine Duplik' (1778) in: Gotthold Ephraim Lessing, *Werke*, ed. H. G. Gölpert et al. (8 vols, Berlin, 1978), viii. 32.

[81] See John H. Zammito, *Kant, Herder, and the Birth of Anthropology* (Chicago, 2002), ch. 1.

[82] See R. Steven Turner, 'The Growth of Professorial Research in Prussia, 1818 to 1848', *Historical Studies in the Physical Sciences* 3 (1971), 137–82; R. Steven Turner, 'University Reformers and Professorial Scholarship in Germany, 1760–1806', in Lawrence Stone, ed., *The University in Society, volume 2* (Oxford, 1974), 495–531; R. Steven Turner, 'The *Bildungsbürgertum* and the Learned Professions in Prussia, 1770-1830', *Histoire Social—Social History* 13 (1980), 105–35; R. Steven Turner, 'The Great Transition and the Social Patterns of German Science', *Minerva* 25 (1987), 56–76.

[83] There is some dispute over just how long the 'learned world' remained viable after 1800: Turner sees 1800 as the turning point, but Denise Phillips, in *Acolytes of Nature: Defining Natural Science in Germany 1770–1850* (Chicago, 2012), 14–15, argues that it persisted into the nineteenth century. What is clear is that the shift was consolidated by the 1830s at the latest: cf. Moritz Wilhelm Drobisch: 'Mathematics and the natural sciences have worked their way up to unsuspected heights,

erudition, now beginning to be considered introverted and pointless, to functional expertise. Functional expertise is of course supportive of a scientific culture, and there can be little doubt that between the middle of the century and the outbreak of the First World War it was Germany, rather than Britain, that was at the forefront of scientific and technological research.[84] The problem with the rise of functional expertise is that the idea of *Kultur* is being used to promote an association of science with politics and other 'second-rank' pursuits.

In any case, a link between science and civilization would not necessarily be to the benefit of either. Which of the routes was followed depended on complex contextual circumstances, and these questions become of particular significance in the wake of the First World War, where the context in which enquiry was pursued was very different from that of the second half of the nineteenth century. The complexity of the post-war situation as it affected the standing of science in Germany was shaped by the commandeering of its patents and manufacturing processes in its chemical industries by the Americans,[85] the boycott of German and Austrian scientists from international conferences (where much of science was conducted in this period) from 1919 to 1931,[86] and a concomitant collapse of German as the 'language of science'.[87] In the circumstances, we might expect any association between science and civilization in the germanophone world in the interwar years to be far from straightforward, and somewhat different from that in other European countries. But in the period up to 1914 there is, by comparison, more common ground with British and French notions of civilization, and, on the part of many British and French intellectuals, a sympathy with a cosmopolitan understanding of *Kultur*,[88] by contrast with the triumphalist and parochial understandings of civilization that had become prevalent in imperial powers such as Britain.

and . . . acquired a classic standing that is almost in competition with the aesthetic classicism of ancient literature': *Philologie und Mathematik als Gegenstände des Gymnasialunterrichts betrachtet, mit besoner Beziehung auf Sachsens Gelehrtenschulen* (Leipzig, 1832), 27.

[84] The 'age of science' had been publicly announced by the industrialist and scientist Werner von Siemens in his 1886 address to the Convention of German scientists and physicians: 'Das Naturwissenschaftliche Zeitalter', *Tageblatt der 59. Versammlung deutscher Naturforscher uns Ärzte zu Berlin* (Berlin, 1886), 92–6.

[85] See Gavin Weightman, *The Industrial Revolutionaries: The Creation of the Modern World 1776–1914* (London, 2007), 381.

[86] See Brigitte Schroeder-Gudehus, 'Challenges to Transnational Loyalties: International Scientific Organizations after the First World War', *Science Studies* 3 (1973), 93–118; Brigitte Schroeder-Gudehus, *Deutsche Wissenschaft und internationale Zusammenarbeit: Ein Beitrag zum Studium kultureller Beziehungen in politischen Krisenzeiten* (Geneva, 1966); Roy MacLeod, 'Der wissenschaftliche Internationalismus in der Krise: Die Akademien der Allierten und ihre Reaktion auf den Ersten Weltkreig', in Notker Hammerstein et al., *Die Preussische Akademie der Wissenschaften zu Berlin, 1914–1945* (Berlin, 2000), 317–49. More generally, see Anne Rasmussen, 'Science and Technology', in John Home, ed., *A Companion to World War I* (Chichester, 2012), 307–22.

[87] See Gordin, *Scientific Babel*, ch. 6.

[88] On this cosmopolitan reading of *Kultur*, see for example Edward Skidelsky, *Ernst Cassirer: The Last Philosopher of Culture* (Princeton, NJ, 2008).

2

The Evolution of Civilization

How could the association between civilization and science have arisen? If there is an obvious candidate for the forms of activity that not only accompany but mark out civilization (e.g. in the cases of ancient Athens and Rome, which still served as models), it is not science but the arts. There can be no doubt that artistic achievements would have had a more traditional claim to being a sign of the development of civilization. Moreover, it is not as if progress in particular sciences, such as mechanics or astronomy or optics, could be associated with the march of civilization: they were technical practices pursued by a miniscule group, and even to the extent that their researches reached a wider reading public, they would seem to be well insulated from matters of politics and culture.

The association depends on the 'science' in question. Not only is it not some particular science that is at issue, nor is it the collection of the particular sciences. It is science abstractly conceived: a metalevel conception, something that goes beyond technical achievements and provides a distinctive way of viewing and thinking about the world. This conception of science is what enabled its association with civilization, and it has two features that are of importance: its claims to universality, and its embodiment of progress.

STADIAL THEORIES OF SCIENCE AND CIVILIZATION

Science's claim to universality is different from its claim to progress in a crucial respect. Progress is a distinctively Enlightenment notion. Christianity had made no claims to progress, and the notion was alien to it.[1] By contrast, universality had

[1] This is particularly true of the Catholic Church. Pope Pius IX, for example, in his 1864 Encyclical, identified in his list of Errors—Error number 79 of a world sadly awash with errors (including the error that held that Protestantism is an alternative form of Christianity)—the ridiculous notion that 'the Pope can and must reconcile himself... with progress, with liberalism, and with modern civilization': A. Roger and F. Chernoviz, *Lettres Apostoliques de Pie IX, Grégoire XVI, Pie VIII* (Paris, 1901), 35. In 1839 his predecessor, Gregory XVI, had issued a decree forbidding scientists of the Vatican state to attend the first meeting of Italian scientists in Pisa in 1839, and instructed them not to correspond with attendees: see Luigi Pepe, 'Volta, the Istituto Nazionale and Scientific Communication in Early Nineteenth-Century Italy', in *Nuova Voltiana: Studies on Volta and his Time* 4 (Milan, 2002), 101–16: 102. On the traumas afflicting the nineteenth-century papacy on the question of modernization, see David I. Kertzer, *The Pope Who Would be King: The Exile of Pius IX*

Civilization and the Culture of Science: Science and the Shaping of Modernity, 1795–1935. Stephen Gaukroger, Oxford University Press (2020). © Stephen Gaukroger.
DOI: 10.1093/oso/9780198849070.001.0001

been a cornerstone of medieval Christianity, and a claim to universality had been central to the standing of Catholicism from the Reformation onwards. By the eighteenth century, it had become central to the standing and legitimacy of science. Unlike in the arts, where one can appreciate wide diversity, and correlatively where one can grasp and see the point of different kinds of aesthetic judgement, the value of scientific laws is unvarying and absolute. But to locate the universality of science in this way, simply in its theoretical characteristics such as its laws, removes it from any association with the culture which successfully nurtures this science. It has nothing to tell us about the role or success of science in some cultures, and its failure to emerge, or (as was considered to be the case in China) failure to develop, in others.

The crucial thing about universality was that it was something that needed to be nurtured and promoted. This nurturing and promotion of a distinctive scientific universality took a material form: measurement. Precise measurement had been the *sine qua non* of scientific instrument making, which received a huge boost in the seventeenth century,[2] and if the instruments were to function as required, they depended on standards of calibration. But the concerted move to precision measurement is something distinctive to the second half of the eighteenth century. One of the most significant eighteenth-century manifestations of precision was Lavoisier's identification of chemical elements, for example, which was possible only on the basis of his extremely elaborate measuring apparatus, constructed specially for him by over seventy carefully chosen manufacturers of laboratory instruments,[3] enabling him to measure minute losses and gains in weight as a result of chemical reactions.[4] At the same time, there developed what might be termed an ideology of measurement. As Norton Wise notes, the rhetoric of precision acquires the power to carry conviction 'in virtually any domain in which it was applied; and only then was it applied in every imaginable domain'.[5]

---

and the Emergence of Modern Europe (Oxford, 2018). The Catholic rejection of science continued into the twentieth century in the writings of some of its apologists. See for example Hilaire Belloc's essay 'Science as the Enemy of Truth', in his *Essays of a Catholic Layman in England* (London, 1933), 195–236.

[2] See Anita McConnell, 'Instruments and Instrument-Makers, 1700-1850', in Jed Z. Buchwald and Robert Fox, eds, *The Oxford Handbook of the History of Physics* (Oxford, 2013), 326–57.

[3] See Marco Beretta, 'Between the Workshop and the Laboratory: Lavoisier's Network of Instrument Makers', *Osiris* 29 (2014), 197–214. An inventory of Lavoisier's laboratory carried out after his execution revealed that he had (at present rates) over a million dollars worth of glassware and scientific instruments: see Sharon Bertsch McGrayne, *Prometheans in the Lab: Chemistry and the Making of the Modern World* (New York, 2001), 10.

[4] This is not to say that this way of pursuing chemistry was universally accepted by Lavoisier's contemporaries: see Jan Golinski, '"The Nicety of Experiment": Precision of Measurement and Precision of Reasoning in Late Eighteenth-Century Chemistry', in M. Norton Wise, ed., *The Values of Precision* (Princeton, NJ, 1995), 72–91.

[5] Matthew Norton Wise, 'Precision: Agent of Unity and Product of Agreement', in M. Norton Wise, ed., *The Values of Precision* (Princeton, NJ, 1995), 92–100: 92.

In the late eighteenth and nineteenth centuries, projects for universal standards of calibration moved out of the sphere of scientific experiments and into the public realm. The late nineteenth-century standardization of time zones was probably that with the most immediate impact as far as the public were concerned, but the introduction of the French metric system in the wake of the Revolution was of equal moment.[6] The idea of a standardized system of weights and measures had been proposed by the astronomer Jérôme Lalande. In 1771, using data from the observations of the transit of Venus in 1761 and 1769, he calculated the 'astronomical unit'—the mean distance from the centre of the earth to the centre of the sun—to a far greater degree of accuracy than earlier attempts. In April 1789, in the immediate wake of the Revolution, he denounced the 'unconscionable and multiple abuses of the diversity of measures',[7] a not uncommon complaint at the time,[8] and proclaimed that Paris measures should be the national standard. This proposal was subsequently rejected on the grounds that it did not fully meet the aspirations to universality of the 'enlightened', and in its place there was proposed a fundamental measure from nature itself, one that, by contrast with its predecessors and competitors, would be eternal and the common property of everyone: it would be truly universal in a way that no other kind of measure could be.[9] In March 1790 the Republican government legislated for a new standard, equivalent to one ten-millionth of the length of a meridian of longitude between the equator and the North Pole, and a platinum bar was cut to provide the new 'metre' length. At the same time, other units of measurement, such as weight, were to be correlated with the metre, and all such units were to be decimalized.[10]

One of the great defenders of the metric system was Condorcet, for whom, as we shall see below, there was a great deal of intertranslatability between science and politics. On Condorcet's account, reforming measurement systems on the

[6] See Ken Alder, *The Measure of All Things: The Seven-Year Odyssey that Transformed the World* (London, 2002).
[7] Jérôme Lalande, *Article pour les cahiers dont les 36 rédacteurs sont prier instament et requis expressément de faire usage* ([Paris], 1789). Cited in Alder, *The Measure of All Things*, 81.
[8] The problem was to persist. In 1816 Biot noted that: 'Everyone who has had occasion to carry out extensive researches has noted with regret the scattered state of the materials of this fine science, and the extent to which it still labours under uncertainty. One result is allowed in one country, and another in another. Here one numerical value is constantly employed, while in another place it is regarded as doubtful or inaccurate. Even the general principles are far from being universally adopted.' Jean Baptiste Biot, *Traité de physique expérimentale et mathématique* (4 vols, Paris, 1816) i. p. ii.
[9] Enlightenment thinkers did not have the monopoly on 'universal' measures. On the question of the standardization of time zones, the Catholic Church objected to the meridian being based in England, at Greenwich, and argued that it should be set in Jerusalem, the 'universal city'. See Richard J. Evans, *The Pursuit of Power: Europe 1815–1914* (London, 2016), 393.
[10] When an international bureau of metric weights and measures was established in 1875, no one cared about the 'natural' basis for the measure (which, since the meridian of longitude passed through Paris, was not in fact wholly 'natural'). What mattered was the convenience of a decimal system. See Peter Galison, *Einstein's Clocks, Poincaré's Maps* (London, 2004), 84–6.

basis of universal principles has its correlate in the realigning of different national laws to yield a single set of legal and political principles,[11] a key element in his conception of the civilizing process. Such connections are far from idiosyncratic. In recent literature, for example, the interrelation between scientific and political conceptions has been explored in the context of a conception of modernity whose distinctive feature is that of 'empty' space and time, that is, one in which space is empty of sacred spaces, and non-human agents such as gods, and time is free from providence, destiny, end times, and apocalypse.[12] As William Reddy has put it: 'Modernity, it is said, arose, first, as the temporality of modern science, in which mathematically expressed laws apply uniformly across time and space, and, second, as the temporality of the nation-state, in which laws apply with perfect uniformity throughout its territory, in which politics are national in scope, citizenship a unifying status, and history the story of the nation's self-liberation.'[13]

With regard to the question of progress, the critical characteristic of science is that it undergoes a type of development that is quite different from that in the arts, one that comes to be central to the understanding of civilization. It exhibits cumulative progress,[14] whereas the arts do not: the achievements of classical literature and sculpture are not, simply in virtue of being earlier, less adequate or less satisfactory than later forms.[15] We can identify three factors that contribute to this association of civilization and science through the notion of progress. First, from the sixteenth century, science began to be thought of as a progressive historical process in which there was a gradual increase in scientific knowledge. Second, beginning with Francis Bacon, this progressive process was also considered to proceed in discrete stages that were shaped by features of the societies in which the scientific developments occurred, or failed to occur, and these

[11] See Louis Maquet, 'Condorcet et la création du système métrique décimal', in Pierre Crépel and Christian Gilian, eds, *Condorcet, mathématicien, économiste, philosophe, homme politique* (Paris, 1989), 52–62.

[12] See Richard Wolin, '"Modernity": The Peregrinations of a Contested Historiographical Concept', *American Historical Review* 111 (2011), 741–51; and Michael Saler, 'Modernity and Enchantment: A Historiographical Review', *American Historical Review* 111 (2006), 692–716.

[13] William M. Reddy, 'The Eurasian Origins of Empty Time and Space: Modernity as Temporality Reconsidered', *History and Theory* 55 (2016), 325–56: 326. Reddy argues that, in fact, notions of empty space and time arose from background assumptions in wide use across Eurasia in the early modern period.

[14] This idea became constitutive of the understanding of science right up to the middle of the twentieth century, epitomized in George Sarton, who writes that 'scientific activity is the only one which is obviously and undoubtedly cumulative and progressive': *The History of Science and the New Humanism* (New York, 1956), 10.

[15] There were nevertheless disputes that did question the superiority of ancient art forms, although not on the grounds that they occurred earlier. The most famous of them, that initiated by Perrault in the Académie Française in 1687, hinged on the superiority of modern literature, poetry, and drama over that of the ancients, where the key contested figure was Homer. Charles Perrault, *Paralelle des anciens et des modernes, en ce qui regarde les arts et les sciences. Dialogves. Avec le poëme du Siecle de Louis le Grand, et une epistre en vers sur le genie* (Paris, 1688). See Noémi Hepp, *Homère en France au XVIIe siècle* (Paris, 1968).

socio-political features were given an explanatory role. Third, from the middle of the eighteenth century, French and Scottish Enlightenment writers began to think of civilization as the outcome of a progressive historical process, and by the end of the century, at least for French writers, this was explicitly on the model of science. Underlying this, there was postulated a connection between reason and civilization, where the progress of reason was considered to be manifested in the progress of science.

But just as the universality of science was manifest in a material form, measurement, so too was the progress of science manifested in a material form, 'technology' (although, as we shall see in Chapter 10, the relation between the two is far from straightforward). Evidence of the progressive enhancement of civilization by the progress of science, in the mid-nineteenth century for example, lay in such things as its railways, street lighting, piped water, sewerage systems, and steam-powered shipping. Without progress of this kind, the progress of science per se would not be of any public concern, nor would it have the standing or value it has had.

The key to earlier debates over the superiority of particular ages lay in the rejection of the idea of a Golden Age in classical antiquity. Matters were complicated, however, by the existence of a second, parallel notion of a Golden Age, a biblical myth of the Fall. The two were inevitably intertwined, and both of them raise questions about scientific understanding. The Greek ages were divided into the gold, silver, copper, and earthen, while the biblical myth offers a stadial understanding of history along the lines of perfection, fall, and redemption.[16] The concern with the Adamic Fall, a concern central to much early modern thought, takes as one of its key themes the image of Adam as intellectual exemplar, and it was still alive and well in the mid-eighteenth century.[17] Some thinkers, most famously Pascal, took the Fall to mean that we were doomed to a life of uncertainty. Bacon, like his contemporaries, had not thought a return to the Adamic state was possible, but he had believed that the deficiencies of understanding that had resulted from the Fall could be remedied to some extent. If the kind of unmediated understanding ascribed to Adam was not something that could be reproduced, there were nevertheless instrumental and experimental vehicles that allowed us to uncover natural processes that would otherwise remain hidden to us. Adamic science may not have been achievable any longer, but there

[16] This periodization was subject to much fine-tuning of course. The distinguished orientalist William Jones, for example, in lining up the biblical world ages with the Indian and Greek ones, divided them into the Diluvian or purest, the Patriarchial or pure, the Mosiack or less pure, and the prophetical or impure: 'On the Gods of Greece, Italy and India', *Asiatick Researches* 1 (1788), 221–75: 236–7. See Urs App, 'William Jones's Ancient Theology', *Sino-Platonic Papers* 191 (2009), 1–125: 12–13.

[17] See, for example, Andrew Michael Ramsay, *The Philosophical Principles of Natural and Revealed Religion Unfolded in a Geometrical Order* (2 vols, Glasgow, 1748–9), ii. 8.

could be significant improvements in knowledge that did not depend on a return to a prelapsarian state of understanding.

It is this achievability that was at issue in the debates over the idea of classical antiquity as the Golden Age. Golden Age notions were countered by sixteenth-century critics in a number of ways. Some, such as Cardano, listed the many modern practical achievements of a loosely technological nature unknown to the ancients, such as domestic furnaces, church bells, stirrups on saddles, and counterweights in clocks.[18] Others, such as Bodin and Le Roy,[19] went further, arguing that the modern age is superior in every respect to earlier ages. For such critics of Golden Age conceptions, living in a particular age is crucial to what one can achieve. This much they share with advocates of a Golden Age. But the historical model is now not one of decline but of progress. Later cultures supersede earlier ones because the process is modelled on the growth from infancy to maturity. It is Bacon who spells out this development most explicitly. In his *Redargutio Philosophiarum* of 1608, he tells us that the Greeks sat around chattering and arguing without producing anything, living in an age not that far removed from myths and fables, and with little historical or geographical knowledge.[20] Bacon is concerned with a description of the possibilities of the age here because he has a developmental model, that of infancy to maturity, which is progressive, and—like all thinkers before the end of the eighteenth century—he assumes his own age to be that of maturity.

Bacon had a significant influence after his death: Baconian themes were explicitly written into the founding of the Académie des Sciences in Paris, and the Accademia della Traccia in Italy;[21] Baconianism stood at the foundation of the Royal Society[22] (although his advocates tended to select Baconian doctrines to suit their convenience); and 170 years later William Harcourt, one of the founders of the British Association for the Advancement of Science, took the title for the new association from Bacon's *Advancement of Learning*, its methodology from his *Novum Organum*, and its programme from his *New Atlantis*.[23] Bacon linked the advance of science with the move to a more cosmopolitan culture which possessed greater general knowledge, and greater conversational and literary achievements.

---

[18] Girolamo Cardano, *De subtilitate rerum* (Nuremberg, 1550), book 7.

[19] Jean Bodin, *Methodus ad facilem historiarum cognitionem* (Paris, 1566); Loys Le Roy, *De la vicissitude ou variété des choses en l'univers, et concurrence des armes et des lettres par les premieres et plus illustres nations du monde, depuis le temps où a commencé la civilité, et memoire humain jusques à presente* (Paris, 1575).

[20] See Stephen Gaukroger, *Francis Bacon and the Transformation of Early-Modern Philosophy* (Cambridge, 2001), 110–14.

[21] See ibid., 2–3.

[22] See Charles Webster, *The Great Instauration: Science, Medicine and Reform (1626–1660)* (London, 1975).

[23] A. D. Orange, 'The Idols of the Theatre: The British Association and its Early Critics', *Annals of Science* 32 (1975), 277–94: 278. It should be said, however, that Harcourt's Baconian model was not that ultimately adopted, soon to be replaced by Whewell's Newtonian model, with its emphasis on mathematics and theory. See Morrell and Thackray, *Gentlemen of Science*, 267–76.

There was here the beginnings of a link between these developments and what he saw as the progress of scientific understanding, but this was not necessarily taken up by all those who admired Bacon. Certainly by the eighteenth century there began to emerge a strong association between the development of reason and the development of civilization, but in the German-speaking world, for example, reason was associated not with science but with metaphysics, so it was able to offer no bridge between science and civilization. In England, we can find such an association, but only implicitly. The language of Colin Maclaurin's *An Account of Sir Isaac Newton's Philosophical Discoveries* (1748), in describing the transition from savagery and barbarism to the science of Newton, for example, suggests an association of the progress of science and the progress of civilization, but the point is not followed up.

It was in France that the association of science and reason came into its own and, through this, to the explicit association of science and civilization. In mid-eighteenth-century Paris, a number of accounts appeared on the causes of the differential development of civilization at different times and places, of which the most notable were Montesquieu's *De l'esprit des lois* (1748),[24] Turgot's December 1750 lecture to the Sorbonne, 'Sur les progrès successifs de l'esprit humain',[25] the long introductory 1751 preliminary 'Discours' to Diderot and d'Alembert's *Encyclopédie*,[26] and Voltaire's *Essai sur les mœurs*. These accounts of civilization varied considerably. Montesquieu, for example, offers a climatic account, whereas Turgot defends blind passions and ambitions as being as important as reason in the development of civilization, contrasting the West with China, where he maintains that a rational avoidance of war in China has led to a lack of interaction between different peoples and to the progress that results from such interaction. By contrast, Voltaire and d'Alembert focused on the deleterious effects of Christianity on the progress of civilization, although it is only with d'Alembert that we find the progress of civilization mapped on to the progress of reason in an explicit and systematic way. And it is here that we find the connections between science and reason being drawn in such a fashion that the progress of civilization now comes to be associated with the progress of science.

In his general proposal for the reform of knowledge in the preliminary 'Discours' to the *Encyclopédie*, d'Alembert defends 'reason'—by contrast with religious teaching for example—as the sole ultimate criterion of judgement.[27] He

[24] Charles Montesquieu, *De l'esprit des lois: ou Du rapport que les loix doivent avoir avec la constitution de chaque gouvernement, les moeurs, le climat, la religion, le commerce, &c.* (Geneva, 1748).

[25] Turgot, *Oeuvres*, ii. 597–611.

[26] 'Discours préliminaire' to Denis Diderot and Jean le Rond d'Alembert, *Encyclopédie ou Dictionnaire raisonné des sciences, des arts et des métiers par une société des gens de Lettres, mis en ordre et publié par Diderot et quant à la Partie mathématique par d'Alembert* (2nd. edn., 40 vols, Geneva, 1777–9), i. pp.v–lxix.

[27] In what follows, I summarize some of the material in Gaukroger, *The Collapse of Mechanism*, ch. 12.

then makes the crucial move of associating reason with science. On his Lockean sensationalist programme, it is sensation alone that puts us in touch with the world and allows us to preserve our bodies and to provide them with their needs. We can achieve this either by our own observations and discoveries, or by those of others which have been communicated to us.[28] The ideas generated from these sources can then be combined and connected, and his ultimate model for these connections are the algebraic procedures established in the physical sciences. Once reason and our knowledge of the world have been associated with science, the main task for d'Alembert is the reconstruction of the history of the sciences, showing 'the steps by which we arrived at our present state'. What we need, he argues, is 'a historical explanation of the order in which the various parts of our knowledge succeed one another',[29] and this takes the form of a genealogy of reason showing how, in its historical forms, it converges on the project embodied in the *Encyclopédie*, which thereby represents the culmination of human cognitive endeavour and constitutes the starting point for further enquiry.

The journey to science and rationality is, on d'Alembert's account, a tortuous and circuitous one, and not a linear narrative: the antithesis of rationality was not to be found in antiquity, for example, but in the shift from reason to dogma in the Middle Ages. The development of science (natural philosophy), on d'Alembert's account, was a slow process obstructed by scholasticism. In the arts, poets and others had been allowed to celebrate pagan deities 'as a matter of innocent amusement', something that proved fertile ground for the imagination, and which was hardly a threat to Christianity, since no one was going to be led by this to revive the worship of Jupiter and Pluto. But things were different in science. Here, 'it was either understood, or claimed, that blind reason might wound Christianity'.[30] It was in this climate that religion, whose proper domain was restricted to faith and morals, began to take upon itself the teaching of natural philosophy, and the policing of these areas by the Spanish and Roman Inquisitions. But 'whilst ignorant or malevolent enemies thus made open war on science' it continued to be pursued in secret by some 'extraordinary men'.[31]

What d'Alembert provides in the 'Discours' is a vindication of the Enlightenment project of the *Encyclopédie* in the distinctively Baconian genre of a legitimating genealogy. Primarily at issue was the task of establishing a historical sequence in which one can follow a progression that starts with the origins of knowledge and traces a process of growth—while uncovering and analysing various false starts—which can be shown to culminate in the present. His genealogical programme was consolidated in Condorcet's *Esquisse d'un tableau historique des progrès de l'esprit humain*, where an account is given of proceeding through historical stages towards a goal defined in terms of rationality, manifested in a form of organization of society superior to earlier ones.

[28] *Encyclopédie*, i. ix.   [29] Ibid., i. xxxii.   [30] Ibid., i. xxxvii.
[31] Ibid., i. xxxviii–xxxix.

D'Alembert was instrumental in promoting the career of his protégé Condorcet to the position of permanent secretary of the Académie des Sciences, and, some literary achievement being considered advisable in securing a position like this, he began work on a history of the sciences, taking the Académie as the model for the development of the sciences. This eventually appeared as *Esquisse d'un tableau historique des progrès de l'esprit humain*, and it offered what was to be a canonical statement of the idea that the inevitability of progress is a basic social-historical law. Condorcet's approach follows that of a number of mid-eighteenth-century historians who had distinguished between the histories of individual actions and achievements, and large-scale cultural change.[32] Montesquieu for example noted that causes become less arbitrary to the extent that they have a more general effect, in that we know better what shapes the achievements of societies that have adopted a given way of life than we do what shapes the lives of individuals. And Hume argued that causal connections are more easily established when changes in human conditions produce changes in large-scale human behaviour, as in the case of the rise and progress of the arts and sciences, or that of the rise and progress of commerce, rather than on an individual level. This is Condorcet's approach: he is not concerned with individual achievements but with a large-scale social phenomenon which he seeks to subject to a general law.[33] The causality inherent in the process is manifested at the diachronic level, as it is the preceding moments that bring about the later ones,[34] and understanding this development has a genuine explanatory value for Condorcet, because following 'these observations on what man has hereforto been, and what he is at present, we shall be led to the means of securing and of accelerating the still further progress, of which, from his nature, we may indulge the hope'.[35]

The early chapters of the *Esquisse* offer a conjectural reconstruction of the early history of humanity, much along the lines of the stadial history of Turgot, and the more comprehensive comparative stadial histories of Scottish writers such as Ferguson and Stewart.[36] The first 'epoch' is that in which 'men united into hordes', the second comprises the formation of pastoral groups and the transition to agricultural society, and the third is identified with the shift from an

[32] See Gaukroger, *The Natural and the Human*, 267–70.
[33] Condorcet, *Outlines of an Historical View of the Progress of the Mind* (London, 1795), 2–3.
[34] Ibid., 3–4.
[35] Ibid., 4. Condorcet can be considered as standing at the origins of what Elizabeth Anderson terms 'institutional epistemology', which she characterizes as asking questions such as: 'do institutions of a particular type have the ability to gather and make effective use of the information they need to solve a particular problem? Given the epistemic powers of such institutions, what problems ought to be assigned to them? How can they be designed so as to improve their epistemic powers?' 'The Epistemology of Democracy', *Episteme* 3 (2006), 8–22: 8.
[36] See Gaukroger, *The Natural and the Human*, 258–66. Condorcet's account is marginally more cosmopolitan than that of d'Alembert, Turgot, or the Scottish writers in that he includes Arabic, Indian, and Chinese cases, although his treatment of these is perfunctory, filling out his story rather than using these cases to question the centrality of Europe, as Voltaire had done.

agricultural society brought about by the invention of the alphabet. With the fourth epoch, that of classical Greece, we encounter the beginnings of science in Pythagoras, whom Condorcet sees as a precursor of Descartes and Newton. Nevertheless, with the rise of science came the rise of the conflict between science and superstition, and the death of Socrates is identified as the 'first crime that the war between philosophy and superstition conceived and brought forth'.[37] From philosophy arose 'the first traces of that science, so comprehensive and useful, known at present by the name of political economy'.[38] The fifth epoch is that of the Aristotelian division of the sciences, by which the proper contents and limits of the individual sciences are distinguished. But the achievements of Greek and Hellenistic science were, we are told, alien to the Romans, whose eventual military despotism prevented 'the tranquil meditations of philosophy and science' from finding a place.[39] But the most serious decline came with the advent of Christianity:

Contempt for human sciences was one of the first features of Christianity. It had to avenge itself of the outrages of philosophy; it feared that spirit of investigation and doubt, that confidence of man in his own reason, and the scourge alike of all religious creeds. The light of the natural sciences was even odious to it, and was regarded with a suspicious eye, as being a dangerous enemy to the success of miracles: and there is no religion that does not oblige its sectaries to swallow some physical absurdities. The triumph of Christianity was thus the signal of the entire decline both of the sciences and of philosophy.[40]

In the sixth epoch, that of the Middle Ages proper, we witness 'the human mind rapidly descending from the height to which it had raised itself, while Ignorance marches in triumph, carrying with her, in one place, barbarian ferocity; . . . every where, corruption and perfidy'.[41] A general picture is becoming clearer in these passages. It is the sciences that offer the strongest defence against superstition, and with the decline of the sciences superstition is renewed, accompanied by an inevitable return to barbarism. Here we get a good sense of what the association between science and civilization is for Condorcet. The crucial connection is superstition. Superstition is the obstacle to civilization, and this is something that neither the arts, nor the development of good governance, nor the rise of commercial society, can offer any resistance to: only science, in a broad sense, can do that.

Condorcet's characterization of the seventh epoch connects the revival of learning in the West with a technological innovation: the birth of printing. In the first instance, however, printing did not in itself stimulate enquiry. Following d'Alembert, Condorcet notes that 'books were more studied than nature, and the opinions of antiquity obtained the preference over the phenomena of the universe'.[42] But the period was quickly followed by the eighth epoch, 'when the

---

[37] Condorcet, *Outlines*, 77.   [38] Ibid., 89.   [39] Ibid., 116.
[40] Ibid., 128. Translation emended.   [41] Ibid., 137.   [42] Ibid., 175.

Sciences and Philosophy threw off the Yoke of Authority'.[43] The printing press played a crucial role here, giving a broad mass of people access to information which they would not otherwise have had, whether through introductory books, dictionaries, and encyclopedias and the like, or through more abstruse publications. Fundamental questions of freedom are at issue here for Condorcet, for it is 'the press that has freed the instruction of the people from every political and religious chain', and 'that instruction which is to be acquired from books in silence and solitude, can never be universally corrupted'.[44] Nevertheless, while some sciences, notably the physical sciences, began to flourish, others, such as political economy, had still not emerged.[45] These omissions began to be corrected in the ninth epoch, Condorcet's own age, as reason triumphed in the works of Collins, Bolingbroke, Bayle, Fontenelle, Montesquieu, and others,[46] and as the various physical and mathematical sciences progressed at a phenomenal rate. Indeed the sciences and civilization are constantly linked, no more so than in the final chapter, the 'tenth epoch', which 'describes the future progress of mankind'. It is clear, we are told, 'from observation of the progress which the sciences and civilization have hitherto made', that 'nature has fixed no limits to our hopes'.[47]

It is science that secures an open-endedness to historical development here. This a novel idea which goes beyond d'Alembert's notion of the present as a culmination of the achievements of the past, for example. The very idea of continuity with the past, assumed as much in d'Alembert's genealogy as in Christian historians, is put in question in the revolutionary programme of the later chapters of the *Esquisse*. What drives the move towards the new form of society that Condorcet advocates is not a development in the arts or forms of government as such, but the sciences, including the new science of political economy. As the embodiment of reason, science stands above the events that history describes, providing a prescriptive guidance that enables us to secure not just a freedom from barbarism and superstition, but something unlimited in its potential.

THE SCIENCE OF CIVILIZATION

If d'Alembert's genealogy of reason sees progress as leading to the present, and Condorcet's account of the 'progress of the mind' construes this as a phenomenon that goes beyond the present, Comte's history of civilization completes this movement, reworking Condorcet so as to treat civilization in terms of a transition to a clearly identified future state. As with d'Alembert and Condorcet, Comte's idea of civilization has science at its core, but it is altogether more elaborate than

---

[43] Ibid., 178.   [44] Ibid., 183–4.   [45] Ibid., 203.
[46] Ibid., 247.   [47] Ibid., 319.

its two predecessors, offering a new kind of 'universal history', albeit one very much on a Eurocentric model. As we have seen, in the late eighteenth century there were two ways of providing a contrast class by which the present could be put in perspective: the Voltaire route via the East, and the medieval route followed by d'Alembert and Condorcet. It is the latter that Comte, who is in many ways radically reworking Condorcet's *Esquisse*, is following.

Comte's starting point was one of the pressing questions of the day: the standing and significance of the French Revolution. By contrast with many of his contemporaries, he does not treat the Revolution as an aberration, but as a phenomenon that must be located within a large-scale timeline defined by the separation of spiritual and temporal powers. The beginning of the decline and displacement of the former by secular power is traced to the end of the thirteenth century, and it is here we have the beginnings of 'modern history'.[48] The process finally culminates in the French Revolution. The period associated with the French Revolution, the Enlightenment, was part of what Comte identifies as a transitory period between that of theology and that of a 'positive' (that is, scientifically constructed) state. It is the effect, not of Enlightenment ideas as such, but of the advances of science and industry. The Enlightenment ideas associated with these advances, which he terms 'metaphysical', are effective in undermining the theological and monarchical edifice of pre-Revolutionary Europe, but they leave society unbalanced, with no moral authority, and are hence in themselves incapable of effecting the required transition to a new form of balanced society. Comte's conception of what this balanced society would look like changes radically between his *Cours de Philosophie Positive* (1830–42) and his *Système de politique positive* (1851–4), the former offering a secular conception of the positive society, the latter offering a 'religion of humanity'. The first was immensely influential in the nineteenth century, whereas the impact of the second was confined to a narrow circle of Comteans, and widely ridiculed elsewhere.[49] In both cases, however, there was a form of self-reflection that emerged in the final 'positive' stage: 'social physics', as it was initially termed, 'sociology' as it became later. Social physics was explicitly modelled on the other sciences: mathematics, astronomy, physics, chemistry, biology.[50] It emerges through detailed

---

[48] Comte, *Cours de Philosophie Positive* (6 vols, Paris, 1830–42) vi. 4–5.

[49] Buckle, for example, a staunch advocate of the *Cours*, remarks that the *Système* contains 'a scheme of polity so monstrously and obviously impracticable, that if it were translated into English, the plain men of our island would lift their eyes in astonishment, and would most likely suggest that the author should for his own sake be immediately confined.': Henry Thomas Buckle, 'Mill on Liberty', *Fraser's Magazine* 59 (May 1859), 509–42: 511. Comte's 'religion of humanity' did however attract a following in Brazil, with the first 'Temple of Humanity' erected in 1881. Its influence waned in the twentieth century, but evidently has not completely died out.

[50] Ethics became the seventh science in the *Système*. On the development of Comte's thought see Mary Pickering, *Auguste Comte: An Intellectual Biography* (3 vols, Cambridge, 1993–2009); and the summary account in Mike Gane, *Auguste Comte* (London, 2006). On the early writings see the editor's Introduction to Auguste Comte, *Early Political Writings*, ed. and trans. H. S. Jones

examination of the history and nature of the sciences in the first four books of the *Cours*, which are devoted in large part to the question of how each of these disciplines became sciences. From this, Comte derives an account of scientific method, which is then applied to the only area of investigation that has not been subjected to scientific treatment, social physics. The establishment of social physics as a science brings with it a corresponding form of practice, however: it is used to transform politics. Social physics, as an intellectual practice, has a directly transformative effect on the potentialities inherent in social and political life.

For Comte, society was to be reformed on a resolutely scientific basis, and here lay the attraction for a generation of those disaffected with the industrial and other developments of the early to mid-nineteenth century. As one commentator, writing on Comte's impact in Britain, has put it:

Comtism met the anxieties of classically-educated, amateur-orientated intellectuals by providing a scientifically sanctioned blueprint of the future which assured their continued relevance despite the increasing industrialization, specialization, and democratization of society. The nineteenth century produced no more explicit and comprehensive statement of intellectual populism. It could well be argued that the remarkable thing about Comtism was not that it attracted some support in England but that it did not attract much more.[51]

Crucial to Comte's project is the ranking of the sciences in order of their fundamental standing: mathematics, astronomy, physics, chemistry, physiology, and social physics.[52] In the final 'positive' stage of social development, social physics, the form of reflection that captures and accompanies the historical development of understanding, offers the one true perspective in which the absoluteness of claims to knowledge can be established. But his extensive account of social physics in volumes V and VI of the *Cours* is preceded by volume IV, on chemistry and biology. The biology section is the last of the series on the natural sciences, but it is also in some respects the first on the human sciences. The sequence of societies in volumes V and VI is explicitly grounded in the biological account set out in volume IV. Civilization, for example, is set out in terms that mix historical and biological questions, and comparisons are with animals as much as with primitive social states:

(Cambridge, 1998). On the *Système*, see Andrew Wernick, *Auguste Comte and the Religion of Humanity: The Post-Theistic Program of French Social Theory* (Cambridge, 2001).

[51] Christopher Kent, *Brains and Numbers: Elitism, Comtism, and Democracy in Mid-Victorian England* (Toronto, 1978), xiii. See also Terence R. Wright, *The Religion of Humanity: The Impact of Comtean Positivism on Victorian Britain* (Cambridge, 1986). Comte's impact in Britain was greater than it was in France: see Pickering, *Auguste Comte*, i. 494.

[52] See the diagrammatic summary, 'Tableau Synoptique', in *Cours de Philosophie Positive*, i. foldout facing page 17.

Civilisation develops, to an enormous degree, the action of man upon his environment: and thus it may seem, at first, to concentrate our attention upon the cares of material existence, the support and improvement of which appear to be the chief object of most social occupations. A closer examination will show, however, that this development gives the advantage to the highest human faculties, both by the security which sets free our attention from physical wants, and by the direct and steady excitement which it administers to the intellectual functions, and even the social feelings .... The influence of civilization in perpetually improving the intellectual faculties is even more unquestionable than in its effects on moral relations. The development of the individual exhibits to us in little, both as to time and degree, the chief phases of social development. In both cases, the end is to subordinate the satisfaction of the personal instincts to the habitual exercise of the social faculties, subjecting, at the same time, all our passions to rules imposed by an ever-strengthening intelligence, with the view of identifying the individual more and more with the species. In the anatomical view, we should say that the process is to give an influence by exercise to the organs of the cerebral systems, increasing in proportion to their distance from the vertebral column, and their nearness to the frontal region. Such is the ideal type which exhibits the course of human development, in the individual, and, in a higher degree, in the species. This view enables us to discriminate the natural from the artificial part of the process of development; that part being natural which raises the human to a superiority over the animal attributes; and that part being artificial by which any faculty is made to preponderate in proportion to its original weakness: and here we find the scientific explanation of that eternal struggle between our humanity and our animality.[53]

Comte's commitment to providing a scientific basis for his account of humanity was, however, highly selective. He dismissed a number of developments in science on the grounds that they were 'metaphysical'. Cellular theory in biology, for example, reminded him of Leibniz's metaphysical doctrine of monads, and so was rejected. Less obviously, he also found sidereal astronomy and probability theory 'metaphysical'. At the same time, in a very revealing move, he was hostile to experimental research because it could lead to awkward discoveries which could result in further uncertainties.[54] Nevertheless, a commitment to a scientific conception of civilization was the core of Comte's project. It manifested itself in two main ways, although his successors were not necessarily committed to both. The first was the commitment to the placing of history on a scientific basis, and I shall consider an explicitly Comtean project along these lines, that of Buckle. The second was the attempt to account for the evolution of civilization along biological lines, and here we shall be considering Spencer.

[53] I have quoted the contemporary English translation/condensation by Harriet Martineau (authorised by Comte himself): *The Positive Philosophy of Auguste Comte, Freely Translated and Condensed by Harriet Martineau* (2 vols, London, 1853), ii. 150–1. The original passages on which this reworking is based are in *Cours*, iv. 623ff.

[54] See Kent, *Brains and Numbers*, 61.

THE TRANSITION TO CIVILIZATION

Buckle considered Comte to have done more than anyone else to raise the standard of historical writing,[55] and he set out to transform history from a compilation into a science, to write a comprehensive account of the history of the world on a scientific footing, in which it could be seen to follow laws of historical development in the same way that the motions of the planets are subject to Newton's laws of motion.[56] In a letter of February 1853, in response to a request for progress on his enterprise, he writes:

I have long been convinced that the progress of every people is regulated by principles—or, as they are called, Laws—as regular and as certain as those which govern the physical world. To discover those laws is the object of my work. With a view to this, I propose to take a general survey of the moral, intellectual, and legislative peculiarities of the great countries of Europe, and I hope to point out the circumstances under which those peculiarities have arisen. This will lead to a perception of certain relations between the various stages through which each people have progressively passed. Of these general relations, I intend to make a particular application; and, by a careful analysis of England, show how they have regulated our civilization, and how the successive and apparently the arbitrary forms of our opinions, our literature, our laws, and our manners, have naturally grown out of their antecedents.[57]

The aim was to look beneath surface events in order to discern general underlying principles, for 'in the moral world, as in the physical world, nothing is anomalous; nothing is unnatural; nothing is strange. All is order, symmetry, and law.'[58] One of the techniques deployed by Buckle in this context was statistics, recently elevated to a prominent position in the analysis of social phenomena in Quetelet,[59] to whose work Buckle was indebted. Nevertheless, as Porter has noted, statistics was more critical to the justification of Buckle's project than to its execution and he invoked 'the results of statistics as compelling evidence for the existence of social laws, rather than as instances of the general principles governing history'.[60] The existence of such principles presupposed, in Buckle's view, the absence of providence and free will in human affairs, and it was his denial of these that catapulted him to notoriety with the publication of the first volume of the

---

[55] See Ian Hesketh, *The Science of History in Victorian Britain: Making the Past Speak* (London, 2011), 17.

[56] This kind of view has subsequently taken many forms, including Popper's claim that historical explanation must ultimately take the form of causal explanation, dependent on laws just as much as the physical sciences are, except that the interest of historical explanation in singular events means that such explanations take the form of general laws and singular initial conditions: Karl Popper, *The Poverty of Historicism* (London, 1975), 144.

[57] Quoted in Hesketh, *The Science of History*, 15.

[58] Henry Thomas Buckle, *History of Civilization in England* (2 vols, London, 1857–61), ii. 25.

[59] See Gaukroger, *The Natural and the Human*, 295–300.

[60] Theodore M. Porter, *The Rise of Statistical Thinking 1820–1900* (Princeton, NJ, 1986), 62.

History of Civilization in England, which had an impact equal to, if not greater than, the appearance two years later of Darwin's *Origin of Species*.[61]

In chapter 6 of the second volume of his *History*, Buckle devotes a large section to arguing that the emergence of modern science owes to poetry its ability to discover fundamental laws. Comte had argued that apprehension of bare facts was insufficient for scientific discovery, and must be accompanied by the work of the imagination, and Buckle rejected the Rankean notion of professionalized history in favour of one where the imagination was crucial. But a larger question was now at issue. Buckle, in a significant departure from Comte, was now in effect treating modern civilization as part of a cultural package in which science and the arts have an equal standing, and indeed it was their coming together had allowed science to progress on his account.[62]

History was one of the disciplines that fed into Comte's social physics/sociology, and as such the task of putting it on a secure scientific basis was crucial if the general project was to go through. Buckle's *History* contributes to this task, but at the same time it tempers the idea that science alone can adjudicate the direction of history. For it to proceed, a form of imagination more usually associated with the arts is needed, and this consideration applies as much to scientific historical enquiry as it does to other forms of science. This does not dislodge science from its key role in underpinning civilization, but rather sets out to show what is needed if science is to be successful in taking on this role.

A different kind of development of the idea of a scientific understanding of the progress towards civilization is one which sets out to show an evolution along biological lines. In the work of Spencer, civilization is the outcome of a biologically describable process. Spencer was not a Comtean, but his evolutionary project is in many ways a reworking of an approach that we find in Comte, and he tells us in his *Autobiography* that it was his 'pronounced opposition' to Comte's views that 'led me to develop some of my own views'.[63]

In his earlier writings, Spencer combined a form of utopian socialism with a staunch laissez-faire programme which denied the government any role in education, charity, religion, or even roads, considering such intervention a transgression of natural law. Robert Richards has suggested that Spencer's longer-term project has two crucial ingredients. The first is his utopianism, in which he envisioned an ideal society in which government, classes, and private wealth in land would have disappeared, with the result that individual freedom and social dependence would operate in harmony with one another. But, Richards argues—

---

[61] See Leslie Stephen, 'An Attempted Philosophy of History', *Fortnightly Review* 27 (1880), 672–95.

[62] See in particular his lecture 'On the Influence of Women on the Progress of Knowledge', *Fraser's Magazine* 57 (1858), 395–407, where he contrasts the inductive method of men, which starts from facts, with the deductive method of women, which starts with ideas, and shows how both come together in a thinker like Newton.

[63] Herbert Spencer, *An Autobiography* (2 vols, New York, 1904), i. 517–18.

and here is the second ingredient—he 'attempted to give scientific substance to this dream by projecting it as the inevitable outcome of the evolutionary process. But to cast evolution in that role required he formulate his evolutionary theory so that it could have such a consummation.'[64]

The first of Spencer's attempts to tie together social evolution and biological evolution is set out in *Social Statics* (1851), where he sees the process in terms of the evolution of civilization:

Concerning the present position of the human race, we must therefore say, that man needed one moral constitution to fit him for his original state; that he needs another to fit him for his present state; and that he has been, is, and will long continue to be, in process of adaptation. By the term civilization we signify the adaptation that has already taken place. The changes that constitute progress are the successive steps of the transition. And the belief in human perfectibility, merely amounts to the belief, that in virtue of this process, man will eventually become completely suited to his mode of life. If there be any conclusiveness in the foregoing arguments, such a faith is well founded. As commonly supported by evidence drawn from history, it cannot be considered indisputable. The inference that as advancement has been hitherto the rule, it will be the rule henceforth, may be called a plausible speculation. But when it is shown that this advancement is due to the working of a universal law; and that in virtue of that law it must continue until the state we call perfection is reached, then the advent of such a state is removed out of the region of probability into that of certainty.[65]

Civilization, and the processes that power it, are analogues of cognate evolutionary processes in the animal and plant realms, and are part of a necessary development:

Progress, therefore, is not an accident, but a necessity. Instead of civilization being artificial, it is a part of nature; all of a piece with the development of the embryo or the unfolding of a flower. The modifications mankind have undergone, and are still under-going, result from a law underlying the whole organic creation; and provided the human race continues, and the constitution of things remains the same, those modifications must end in completeness.[66]

Underlying Spencer's evolutionary story here is a broadly Lamarckian conception of evolution. His Lamarckism derives in the main from a reading of Lyell's *Principles of Geology* in 1840, and from Chambers' *Vestiges of Creation*, which he read when it appeared in 1843. Lyell had devoted significant space in the second volume of the *Principles* to refuting Lamarck, but Spencer by contrast was impressed by what he learned of Lamarck from Lyell's exposition. This was reinforced by his reading of Chambers' highly speculative and accordingly

[64] Robert J. Richards, *Darwin and the Emergence of Evolutionary Theories of Mind and Behavior* (Chicago, 1987), 246.

[65] Herbert Spencer, *Social Statics: Or, the Conditions Essential to Human Happiness Specified, and the First of them Developed* (London, 1851), 63–4. See also his 'A Theory of Population, Deduced from the General Law of Animal Fertility', *Westminster Review* 57 (1852), 468–501.

[66] Spencer, *Social Statics*, 65.

controversial account in the *Vestiges*.[67] According to the *Vestiges*, the sun and the planets had coalesced from an initial fire, and life had begun with a spark acting on a 'globule'. From this point on evolution was guided by 'vital statistics', following Quetelet (the English translations of whose works Chambers had published) and his construction of the 'average man'. All embryos start at the same point, and through a system of branching develop into various different kinds of organism, with human beings resulting from a process of averaging at their head. One important thing that Chambers and Spencer had in common was commitment to a discipline that covered animals and humans with significant explanatory pretensions, namely Gall's phrenology.[68] This gave direct support for a project that linked distinctively human social characteristics, revealed in phrenology, and zoological ones, also revealed in phrenology. The problem was that, as the question of progress was stated, there was little more than an analogy. We are none the wiser about the mechanism by which this process is supposed to occur, and are just told how natural it is. One obvious candidate for the mechanism would be providence, but that would not be a *natural* process.

It was at this point that Spencer's newfound interest in anatomy and physiology started to pay off.[69] Spencer was an unswerving advocate of the idea that what was distinctive about modern industrial society was the division of labour, and he had found what he considered to be a zoological version of the same thesis in T. R. Jones' *General Outline of the Animal Kingdom*, in Jones' account of how the differentiation of parts allowed the animal to withstand external forces acting on them, the parts mutually compensating for deficiencies in other parts. In *Social Statics*, he had written that:

In man we see the highest manifestation of this tendency. By virtue of his complexity of structure, he is furthest removed from the inorganic world in which there is least individuality. Again, his intelligence and adaptability commonly enable him to maintain life to old age—to complete the cycle of his existence; that is, to fill out the limits of this individuality to the full. Again, he is self-conscious; that is, he recognizes his own individuality. And, as lately shown, even the change observable in human affairs is still towards a greater development of individuality—may still be described as 'a tendency to individuation'.[70]

The parallels between the social division of labour characteristic of modern civilization and basic zoological principles were reinforced and given a more definite rationale as a result of his reading Henri Milne-Edwards' *Outlines of Anatomy and Physiology*, where he found an explicit use of the principle of the

[67] On Chambers, see James A. Secord, *Victorian Sensation: The Extraordinary Publication, Reception, and Secret Authorship of Vestiges of the Natural History of Creation* (Chicago, 2000).
[68] On the contemporary popularity of Gall's phrenology in England, see John Van Wyhe, *Phrenology and the Origins of Victorian Naturalism* (Aldershot, 2004).
[69] See Richards, *Darwin and the Emergence of Evolutionary Theories*, 268–74.
[70] Spencer, *Social Statics*, 400.

division of labour, and William Carpenter's *Principles of General and Comparative Physiology*, from which he picked up an account of parallels between embryonic developmental stages and the developmental stages 'exhibited by the permanent conditions of races occupying different parts of the ascending scale of creation'.[71]

   Spencer's project was not a narrowly biological one. As Hale notes, following the publication of Darwin's *Origin of Species* and the popularity it generated for evolutionary theories in general, Spencer 'quickly rose to be among England's foremost commentators on the subject. Pigeons were all well and good—Darwin had made them central to his first chapter—but people were more interested in the light that Darwin's work might throw on "the origins of man and his history", and this is what Spencer gave them and more.'[72] Darwin himself, in the sixth edition of *The Origin of Species*, sanctioned at least some of the Spencerian project of moving beyond purely biological matters:

In the future I see open fields for far more important researches. Psychology will be securely based on the foundation already well laid by Mr. Herbert Spencer, that of the necessary acquirement of each mental power and capacity by gradation. Much light will be thrown on the origin of man and his history.[73]

Comte and Spencer were among the most widely read and influential authors of the nineteenth century. Although their particular visions of the relation between a unified science and civilization did not survive into the next century,[74] there is a significant overlap between these visions and those of more circumspect physicists, biologists, and philosophers who have made the doctrine of the unity of science central to their understanding of science. In the latter case, the 'extra-scientific' context of the unity of science is often obscured or implicitly denied, and the impression is given that it is generated purely internally.[75] What the

---

[71] William Carpenter, *Principles of General and Comparative Physiology* (London, 1841), 195. The *Principles* was widely used in teaching physiology and a number of chapters in Chambers' *Vestiges* had also been inspired by it.

[72] Piers J. Hale, *Political Descent: Malthus, Mutualism, and the Politics of Evolution in Victorian England* (Chicago, 2014), 68. Not everyone had such a low view of pigeons. A publisher's reader for *The Origin of Species*, Whitwell Elwin, editor of *Quarterly Review*, disliked the book, which he considered lacked argument and substance. He recommended that Darwin should confine himself to pigeons, abandoning the rest of the work, since 'everyone is interested in pigeons', and such a book would be reviewed in every journal in the kingdom. See Janet Browne, *Charles Darwin: The Power of Place* (London, 2002), 74–6.

[73] Charles Darwin, *The Origin of Species by Means of Natural Selection, or the Preservation of Favoured Races* (6th edition, London, 1872), 428.

[74] There is, however, a case to be made that Sarton, the founder of the twentieth-century discipline of the history of science, was pursuing a Comtean programme in some respects. See Bert Theunissen, 'Unifying Science and Human Culture: The Promotion of the History of Science by George Sarton and Frans Verdoorn', in H. Kamminga and G. Somsen, eds, *Pursuing the Unity of Science: Ideology and Scientific Practice from the Great War to the Cold War* (London, 2016), 157–82.

[75] I have put 'extrascientific' is scare quotes because, as we shall see when we come to look at the question of technology and science, the boundary between the scientific and the extrascientific is greyer than it may at first seem.

projects for a science-based conception of civilization that we have looked at—those of d'Alembert, Condorcet, Comte, Buckle, and Spencer—bring to light is that it is not just a question of the triumph of civilization being equated with the triumph of science, but that science is caught up with civilization in deep and complex ways, not least in the manner in which it acts as a means of understanding civilization.

EAST AND WEST

Western notions of civilization and modernity were not simply exported and passively received internationally. Rather, they were selectively appropriated, in a complex process of recontextualization and embedding in cultures that were often very different from that in which they were generated. We can get a sense of the core issues by considering the reception of 'Darwinism' outside the West in the late nineteenth and early twentieth centuries. This 'Darwinism' is not the same as Darwin's evolutionary theory—very few proponents of evolutionary theory, even self-styled Darwinians, accepted all the details of Darwin's account[76]—and it was those versions, such as that of Spencer, that treated the evolution of species as part of a much larger process of cosmic evolution that had the greatest appeal.

It is clear that, in China, Japan, and in the Arabic world, for example, there was no doubt that the West was at the pinnacle of social evolution for those who thought in these terms,[77] and that this was manifest in its scientific and technological achievements.[78] What the evolutionary model of social development provided was a sense of a natural process by which societies evolved, a process which many of Darwin's contemporaries and successors linked to a very general form of cosmic evolution. Spencer's writings in particular were widely translated and had an immense international following. This is not surprising, for as Robert Richards has noted, Spencer's lifelong motivation was to understand how the natural process of evolution could produce a moral society.[79] His appeal, as Elshakry points out, lay in the fact that he 'was an evolutionist who cared more

[76] See Peter J. Bowler, *The Eclipse of Darwinism: Anti-Darwinian Evolution Theories in the Decades Around 1900* (Baltimore, 1983); David Hull, *Darwin and his Critics: The Reception of Darwin's Theory of Evolution by the Scientific Community* (Cambridge, MA, 1983).

[77] Note that the evolutionary terms and those of race were not the same. The traditional Chinese racial classification, for example, despite a recognition of the 'advanced' nature of Western scientifically sophisticated countries, had Chinese at the apex as far as race was concerned: see Frank Dikötter, *The Discourse of Race in Modern China* (London, 1992).

[78] On China, see Yan Haiyan, 'Knowledge Across Borders: The Early Communication of Evolution in China', in B. Lightman, G. McOuat, and L. Stewart, eds, *The Circulation of Knowledge Between Britain, India, and China* (Leiden, 2013), 181–208; on Japan, see G. Clinton Godart, 'Spencerism in Japan: Boom and Bust of a Theory', in B. Lightman, ed., *Global Spencerism: The Communication and Appropriation of a British Evolutionist* (Leiden, 2015), 56–77; on the Arab World, see Elshakry, *Reading Darwin in Arabic*.

[79] Robert J. Richards, *Darwin and the Emergence of Evolutionary Theories*, 241.

about a cosmic evolutionary process than the specifics of species change, and though he subscribed to a positivist philosophy of science, he also left room (or so many thought) for metaphysics; and, finally, despite his interest in the natural and physical sciences, it was as spokesman of a new social (and even ethical) science that he gained his reputation: that he saw the evolution of morality as both the cause and effect of human progression also appealed to many, and was recalled creatively, by many around the world'.[80]

The ranking of Western culture as the evolutionary pinnacle did not mean that everyone thought that the details of the ranking from primitive to civilized that particular stadial and evolutionary theories of civilization proposed was correct, and the ranking of some as primitive cultures was revised. Arab commentators understandably did not accept the politically convenient association of countries colonized by Western powers with primitiveness for example, and rejected Spencer's characterization of native American Indians as primitive.[81] Moreover, for those Arabic, Chinese, Indian, and Japanese literati who took up Spencer in the 1870s and 1880s, it was less a question of assimilating their own cultures to science, but rather a question of assimilating science to their own cultures. They typically traced lineages for the theory of evolution, usually in its grander 'cosmic' form. In China, Confucian notions of the perfectibility of the cosmic order were assimilated to evolutionary theory, and in India traditional Hindu cosmology was assimilated to evolution.[82] In the case of early Muslim readers of Darwin:

they embraced much of Darwin's message as their own. Supporters and critics alike pointed out that Muslim philosophers had long referred to the idea that species or 'kinds' (as the Arabic term *anwaʿ* suggests) could change over time. The notion of transmutation was also recalled in these discussions, and early Muslim philosophical and cosmological texts were cited whenever Darwin was discussed in Arabic, Farsi, or Urdu. Analogies were drawn with earlier notions of a hierarchy of beings, from matter and minerals to flora and fauna, and finally to humanity itself. That some medieval works also argued that apes were lower forms of humans provided more evidence for nineteenth-century Muslims that Darwin's theory was 'nothing new'.[83]

One significant feature of the reception of Darwin in the East was that the Western notion of a conflict between science and religion was not exported. It is true that in Egypt there was some familiarity, through extracts and popularizations, with Andrew Dickson White's *History of the Warfare of Science with Theology in Christendom* (1896), but it did not always meet with the reaction that one might expect: namely, that, just as pre-Enlightenment Christianity had

---

[80] Elshakry, *Reading Darwin in Arabic*, 39–40.

[81] Marwa Elshakry, 'Spencer's Arabic Readers', in B. Lightman, ed., *Global Spencerism: The Communication and Appropriation of a British Evolutionist* (Leiden, 2015), 35–55: 49 n.21.

[82] See C. Mackenzie Brown, 'Western Roots of Avataric Evolutionism in Colonial India', *Zygon* 42 (2007), 423–48.

[83] Elshakry, *Reading Darwin in Arabic*, 7–8. Such assimilations were not new in Arabic culture: see for example Christopher de Bellaigue, *The Islamic Enlightenment: The Modern Stuggle Between Faith and Reason* (London, 2017), 33–47.

done, so Islam had hindered social and intellectual development. A number of senior Islamic scholars believed that the opposition to Darwin was a peculiarly Christian phenomenon and that Darwinism could easily be accommodated to Islamic teaching, and White's *History* was used as evidence of the backwardness of Christianity.[84] In the case of India, where science has traditionally been seen as a secularizing force, one commentator has noted that, by the end of the nineteenth century, 'science serviced religion by effecting a wholesale transformation of practices and dispositions. Thus science became the mode of enchantment for an Indian modernity without banishing God. This was not, as orientalists had proclaimed, because India was "spiritual" rather than rationalist but, rather, because religion itself became disenchanted.'[85] Moreover, there are cases in which particular forms of the evolutionary story were replaced by others, revealing the underlying source of interest. In Japan, for example, in the 1870s and 1880s, during the high point of the Meiji reign, there was a 'Spencer boom', with the publication of a large number of translations of his works, together with commentaries. But Spencer was replaced in the 1890s by an alternative Hegelian evolutionary narrative—based on the work of a Hegelian critic of Spencer, Thomas Hill Green—which was more religiously orientated, and so met the prevailing religious sensibilities more effectively.[86]

As in the seventeenth- and eighteenth-century cross-cultural exchanges that we looked at earlier, the Chinese made sense of the Western developments to which they were introduced in their own terms. Moreover, what both Western travellers and missionaries on the one hand, and the Chinese on the other, saw as the distinguishing feature of the other culture was a stable and prosperous society, and both reflected on how it had been achieved. By the second half of the nineteenth century, with the uptake of 'evolutionary' narratives, the transition to modernity was increasingly articulated in terms of scientific and technological progress. Such progress was not considered to be some racially innate quality. It is instructive to note, given the popularity of Spencer, that he was not the only writer to provide the right kind of 'evolutionary story', that is, the right kind of large-scale developmental account of progress towards modernity. To understand what marked him out, we need to appreciate that it was his educational writings that were by far the most widely translated of his works, pointing to the importance of the fact that 'evolutionary' progress was largely a matter of education for Spencer. This was particularly important in China, where missionary translators were silent about Darwin and evolutionary theory, and did not translate the works of Darwin or the evolutionary writings of Spencer.[87] When a work of Spencer did appear, it was his educational writings. Spencer offered a programme which, rooted in the family, provided the means for progress. The path to progress rested on a moral as

[84] Ibid., 8–9.
[85] Shruti Kapila, 'The Enchantment of Science in India', *Isis* 101 (2010), 120–32: 131.
[86] Godart, 'Spencerism in Japan', 72.   [87] Elman, *On their Own Terms*, 345–51.

well as on an intellectual and social basis, and this is what its success lay in. Moreover, in common with many of his contemporaries, Spencer proposed particular educational reforms, notably the replacement of education in classical languages and literature by that in science. In Arabic and Chinese educational systems there seemed to be exact parallels with what Spencer had identified as the problematic choice of study material, and his solution appeared to be an appropriate one.

In a 1904 editorial article on Spencer in the Arabic popular science monthly *Al-Muqtataf*, Spencer's importance was identified as lying in his 'truly synthetic philosophy'. As Elshakry summarizes it, it offered a means of discovering the order of nature and the history of humanity in one:

What appealed to them, in particular, was that it provided the basis for uncovering the universal laws of nature and man alike. Indeed, for many Arabic readers, Spencer's philosophy was regarded as most valuable for providing the basic evolutionary principles of good governance. The editors emphasized how both our understanding of the cosmos and our role in it could be revealed through this synthetic and holistic yet still precise and positivist philosophy.[88]

A crucial feature of this 'synthetic philosophy' was its ability to unify our understanding of the world. It displayed the unity of nature, demonstrating 'how the universe is one complete picture—like the parts of a puzzle that fit together'. Elshakry notes that 'the roots of this desire for holism can surely be traced back in part to the natural-theological tradition that was so important for science popularizers everywhere; perhaps too in the case of these Arab intellectuals, ideas of the unity of God's creation dated back centuries and still formed part of a continuous tradition of Arabic translations and publications on the "Marvels of Creation"'.[89]

TECHNOLOGY AND CIVILIZATION

From the nineteenth century, in both the West and in those countries to which the Western association of science and civilization had been exported, the evidence for the progressive enhancement of civilization by the progress of science was considered to lie in such things as its railways, piped water, sewerage systems, steam-powered shipping, better food, warmer homes, softer clothing, and the massive transformation of domestic life and working conditions (not least working hours) brought about with the introduction of gas and electric lighting.[90]

---

[88] Elshakry, 'Spencer's Arabic Readers', 42.   [89] Ibid., 45.

[90] With respect to lighting, for example, note the comment of the chemist William Dibdin: 'The necessities of modern civilization having to so large an extent turned night into day both in the working world as well as in that of the world of pleasure and social intercourse when the day's work is done, a state of things has arisen in which artificial illumination holds the very first place, as without it

A few unworldly philosophers aside, when people thought about the benefits of modernity what they thought about were technological and medical achievements affecting the domestic and the working environment, not an increase in theoretical understanding of natural processes.

These signs of progress have traditionally been attributed to a science-driven technology, as examples of 'applied science'. But while science may have issued in advances in some cases, many of the most dramatic ones seem to have issued rather from 'invention', a form of enquiry independent of scientific theorizing. The question therefore arises how far it is invention that has driven modernity and how far it is something connected with science. We can distinguish developments before and after the middle of the nineteenth century. As Mokyr has noted, 'most of the devices invented between 1750 and 1830 tended to be a type in which mechanically talented amateurs could excel. In many cases British inventors appear simply to have been lucky . . . . When, after 1850, deeper scientific analysis was needed, German and French inventors gradually took the lead.'[91] But this deeper scientific analysis, rather than initiating the developments, was in large part devoted to following them up in ways determined by commercial imperatives.

Consider the case of oil. As Agar has remarked, although historians of science remember 1859 as the year of Darwin's *Origins of Species*, there is a good case for arguing that the discovery of oil in Pennsylvania had a comparable influence on the modern world.[92] In that year, three investors employed Edwin Drake to apply a drilling technique that had been used from antiquity for the extraction of salt to the search for oil, and in August 1859 he discovered a new light oil, from which kerosene was extracted. A crude version of kerosene, 'coal oil', had been known from the early eighteenth century as a by-product in the manufacture of coal gas and coal tar, although its smoky flame precluded domestic use until the Canadian geologist Abraham Gesner was able to produce a cleaner, comparatively smokeless version. His distillation process was far too costly to support the routine domestic use of kerosene, however, and outside cities where coal gas was in use, animal fats, particularly whale oil, were the preferred fuel, as they burned brighter and were cleaner than coal oil. The discovery of 'light' (low-density) oil discovered at 'Drake Well' changed that, and in a short space of time, kerosene for lighting, and soon afterwards for cooking, started to be produced from petroleum (Fig. 2.1).

the whole scheme of present day society would at once fall to the ground.' *Public Lighting by Gas and Electricity* (London, 1902), 18.

[91] Joel Mokyr, *The Lever of Riches: Technological Creativity and Economic Progress* (New York, 1990), 244. The traditional predominance of mechanically talented amateurs over university-educated scientists and engineers is noted elsewhere by Mokyr: 'Data show that out of 498 applied scientists and engineers born between 1700 and 1850, 329 had no university education at all. The proportion of notable engineers with no university education in the eighteenth century was 71 percent. Out of a sample of 244 inventors born before 1820, only 68 had enjoyed higher-level training.' Joel Mokyr, *A Culture of Growth: The Origins of the Modern Economy* (Princeton, NJ, 2017), 125.

[92] Jon Agar, *Science in the Twentieth Century and Beyond* (Cambridge, 2012), 174.

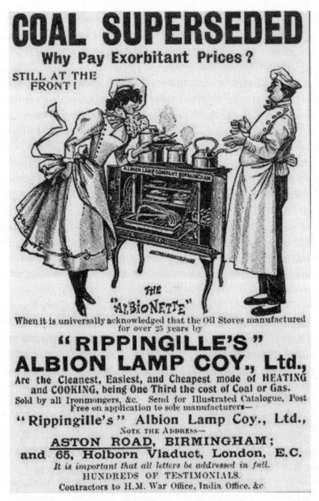

Fig. 2.1 Advertisement for oil stove, *c.*1900

By the 1870s and 1880s, American geologists' surveys were devising theories about the structure of oil reservoirs and the dynamics of subsurface fluid flow, and by the last decades of the century, as one commentator has noted, 'geologists had formulated a theoretical and practical science of petroleum, one of the chief intellectual contributions of nineteenth-century Americans'.[93] In both cases, the

---

[93] Paul Lucier, 'Geological Industries', in Peter Bowler and John Pickstone, eds, *The Cambridge History of Science, Volume 6: The Modern Biological and Earth Sciences* (Cambridge, 2009), 108–25: 121.

input from science does not initiate the process but comes last, aiding a set of developments that begins as an essentially commercial extra-scientific enterprise. As Paul Lucier notes in the case of American geological surveys, 'the principal reason for their establishment was economic. As in the case of the British survey, American ones were justified through the rhetoric of utility. In practice, American geologists, for the most part, put economic results before theoretical work.'[94] Chemistry was similarly placed in regard to petroleum research. Chemists played a crucial role in the refining of petroleum, and in the course of the twentieth century this allowed the creation of a huge number of products, from fuels, lubricants, tar, asphalt, and pharmaceuticals (in the form of organosulphur compounds) to plastics, that transformed living and working conditions. But again the chemical research was at the end of the process, its goals determined by commercial and pragmatic considerations. My point is not that the research programmes of geology and chemistry in general were dictated by such consider-ations. Rather, it is that in those areas where there is a significant contribution to changes in modern life—such as fuels, pharmaceuticals, and dyes—it is difficult to discern much that can be said to be science-led. Indeed, in a number of wholly transformative pharmaceutical discoveries—such as penicillin and cortisone—the discovery is purely accidental.

Nor is it simply a question of an absence of connection between technology and science. There is also the problem of a scientific approach posing obstacles to progress. Thomas Hughes has noted in the case of Edwin Armstrong, whose system of frequency modulation for radio was dismissed by mathematicians, that:

As Armstrong insisted, mathematicians often forced the messy world of invention and engineering into an overtly ordered and reductionist mold that precluded a thorough and usable explanation of how physical things worked. The same was true of scientists. Scientists, unfamiliar with the details of the new technology such as that being introduced by [independent inventors], often exasperated the inventors by insisting that they apply theory that the inventors knew was outmoded. Some scientists arrogantly ridiculed the empirical approach of the so-called Edison hunt-and-try method at the same time that they reasoned from anachronistic theory. Edison was impatient with stiff-necked, aca-demic scientists who argued that the theory of electrical circuitry, developed for arc lights, was valid for the newer incandescent lighting. Similarly, in the field of bridge building, Robert Maillart, the pioneer of reinforced-concrete construction, had to suffer unsolicited and erroneous suggestions from theoreticians who believed that the elegant theory worked out for older stone-and-iron construction was applicable.[95]

[94] Ibid., 114. Freeberg quotes a contemporary reaction to Edward Bellamy's utopian view of invention benefiting mankind: 'Inventors in actual life are generally distinguished by an insane desire for money [and] by the wildest overestimates of the wealth which their inventions will ultimately bring them'. *The Age of Edison: Electric Light and the Invention of Modern America* (New York, 2013), 153.
[95] Thomas P. Hughes, *American Genesis: A Century of Invention and Technological Enthusiasm, 1870–1970* (Chicago, 2004), 49. Cf. Vaclav Smil's comment: 'On March 29, 1879, just nine months before Thomas Edison demonstrated the world's first electrical lighting system, *American*

The case for science shaping modernity is, then, questionable at best. Accordingly, we might ask whether we would have a more plausible thesis if we were to substitute technology, broadly conceived, for science. Such a move would be radical, for it entails a rejection of the idea that what drives the modern world is something whose aim is the discovery of truth, and this idea is a necessary ingredient in the association of science with social and political progress. In its place, we would be left with a conception in which modernity is delivered by a pragmatic and often commercially driven form of enquiry in which the role of science is one of rationalizing and making sense of features of these developments, and perhaps improving them as a result, not of initiating them. This coheres with an aspect of modernity that is absent from (and in some cases antithetical to) traditional conceptions of civilization, but which is distinctive of the idea of technology playing a shaping role in a culture, namely urban industrialization.

The association between technology and modernity looks strong when compared with that of science and modernity, but once we begin to probe the extent to which technological change was the initiator of the developments that we associate with the shift to modernity, serious difficulties arise. The problem is that in the recent literature focusing on the Industrial Revolution, for example, there has been a compelling rethinking of the role of technology. Building on the work of Jan de Vries,[96] Christopher Bayly has promoted the idea of the 'industrious revolution'. The term refers to developments that de Vries identified in Northern Europe in the late eighteenth and early nineteenth centuries, in which family labour started to be used more efficiently by buying in goods and services from outside the household. Bayly writes that families acquired new 'packages' of consumer items which worked on each other to produce yet larger gains in productivity and social satisfaction:

For instance, the consumption of coffee and, later, tea went along with the purchase of sugar, fine breads, and easily replaceable plates off which to eat these items. The resulting package—let us call it 'breakfast'—gave people a higher calorific intake, a new time discipline, and a new pattern of sociability and emulation in the household. It gave a boost to the trade in specialist foods and, later, disposable crockery which replaced the old and heavy family possessions which had been passed down from generation to generation.[97]

Register concluded that "it is doubtful if electricity will ever be used where economy is an object." That same year, the Select Committee on Lighting of the British House of Commons heard an expert testimony that there is not "the slightest chance" that electricity could be "competing, in a general way, with gas."' *Creating the Twentieth Century: Technical Innovations of 1867–1914 and Their Lasting Impact* (Oxford, 2005), 11–12.

[96] Jan de Vries, 'The Industrial Revolution and the Industrious Revolution', *Journal of Economic History* 54 (1994), 240–70; Jan de Vries and Adriaan van de Woude, *The First Modern Economy* (Cambridge, 1997). See also Jan de Vries, *The Industrious Revolution: Consumer Behavior and the Household Economy, 1650 to the Present* (Cambridge, 2008).

[97] See Bayly, *The Birth of the Modern World*, 51.

The demand that arose from the 'invention of breakfast' and from new household crockery, cutlery, and furniture may have provided consumers for the early Industrial Revolution, but there is no reason to think that it led necessarily to early forms of industrialization. As Bayly notes, on most interpretations, the Industrial Revolution is assumed to be a 'supply side' revolution, resulting from the mechanization of production. But industrious revolutions could increase prosperity without relying on an increase in industrial production.[98] They created changes in patterns of consumer demand which was not a consequence of industrial production but could well be a cause of an increase in production, even if in a modest 'non-revolutionary' way. In this connection, it is worth noting that the development of middle-class commercial values in the late eighteenth and early nineteenth centuries was not associated with a revolution in industrial production, which was yet to come. Before the introduction of railways, modern industry was confined to small regions of northern and central England. Bayly points out that although here industry was acting to accelerate the expansion of commerce, elsewhere the rise of the commercial middle class owed more to a reorganization and globalization of consumption patterns and to the accumulation of changes in the artisan sectors of the economy.[99]

Bayly uses the idea of an industrious revolution in an imaginative and very productive way, as a better model for understanding the development of societies outside the West than that provided by the idea of the Industrial Revolution. Our aims are different. They are to discover, in broad terms, the extent to which distinctively modern living conditions were in fact a result of technological developments, as has generally been assumed. The work on the development of the industrious revolution indicates instead that technological developments can be, and in significant cases were, caused by patterns of consumer demand.[100]

Where does this leave us? The association of science and civilization, or rather that form of civilization we have been referring to as modernity, has been crucial for many thinkers: it has been considered to embody social and political progress and cosmopolitan values, and it has aspired to a unity of understanding to which Christianity, for example, could no longer lay claim in the face of the demands of the modern world. It was considered that science delivered two kinds of benefits (although whether these were actually benefits was of course contested): material benefits, delivered through the practical application of science, and spiritual or

[98] Ibid., 52.
[99] Ibid., 114. Among the many interesting parallel cases of social activities promoting technological and, in this case, scientific developments, that of late nineteenth-century systematics, is worth noting. By around 1900 collecting and inventory of plants and animals had been organized on a sufficiently grand scale to effectively complete Linnean classifications. A dominant role in this was played by survey collecting—supported by governments, universities, and museums—which had its roots in new forms of outdoor recreation, such as hiking, camping, and sports hunting: see Robert E. Kohler, *All Creatures: Naturalists, Collectors, and Biodiversity, 1850–1950* (Princeton, NJ, 2006).
[100] Cf. Robert Pool, *Beyond Engineering: How Society Shapes Technology* (Oxford, 1997) for late twentieth century examples.

intellectual ones, whereby science acted as a way of understanding the world that was free from prejudice, dogmatism, and the rule of arbitrary and inept authority. The claim was that the former brought radical changes in communications, domestic circumstances, and the productivity of industry, while the latter offered freedom from traditional religious and other intellectual constraints. Although it was important that these came as part of a package, what was explicitly promoted was 'science'. It was absolutely crucial that it was science that was given ultimate responsibility for the manifest benefits that accrued, not just because science had to be treated as responsible for the production of these benefits if it was to overcome the charge of uselessness that had dogged it in the seventeenth and eighteenth centuries, but more importantly because it was science, not technology, that was considered to exercise civilizing effects on the population. It was science, not technology, that claimed to offer a coherent ordered view of the world and our place in it.

Here we come to a central theme, one that we are about to pursue in detail in Part II. The question is whether science can provide a unity of understanding, as religious systems have traditionally aspired to do. I shall be arguing that the aspiration for a unified science is wholly misguided. It is an aspiration foisted upon science as a result of its taking a religious model of understanding. Christianity itself was increasingly perceived to have failed to achieve the required unity, and science, bolstered by political, ideological, and quasi-religious dreams of unification, took upon itself a unifying role in an exercise in legitimation. In fact, as we are now about to see, in their daily practice none of the physical, material, or life sciences succeeded as a result of a reliance on unification. Quite the contrary, so long as one does not confuse unification with a commitment to consistency, unification remained at the metascientific level, as a misguided form of rationalization and promotion of the success of science.

PART II

THE UNITY OF SCIENCE

3

The Promotion of Unification

With the demise, in the thirteenth century, of a Christian-Platonist conception of the world that had integrated the natural and the supernatural, the organizing thread was ripped out of the natural realm. It was with this epochal separation of the natural and the supernatural that modernity began to emerge in the Christian West. The aftershock was felt for centuries, and John Donne's 1611 lament for a pre-mechanical world, ''Tis all in pieces, all coherence gone', captures the continuing sense of loss of unity of understanding that marks out the modern era.[1] A distinctive feature of the modern era in its post-Scientific Revolution form is the attempt to re-establish a unity of understanding in which science fills the void left by religion, by attempting to provide an organizing thread intrinsic to the natural realm.[2] We need to be alert to science's mirroring of religious conceptions in this respect, and in particular to the way in which they shape its tasks.[3] Science's embodiment of progress enables it to replace Christianity as a civilizing force, its claims to universality contrasting sharply with a deeply divided church. But it is its insistence on its unity that is crucial for the claim to offer a genuinely comprehensive worldview, for without this, progress could only be a matter of localized phenomena. The stakes could not be higher, for 'science' takes on a highly contested cultural standing in the nineteenth century.

---

[1] The literature is vast, but there is no more informative and readable introduction to the issues than Lucien Febvre, *The Problem of Unbelief in the Sixteenth Century: The Religion of Rabelais* (Cambridge, MA, 1982). See also Mikhail Bakhtin, *Rabelais and his World* (Cambridge, MA, 1968).

[2] At the same time, Christianity lost its automatic hold on the natural realm, and attempted to compensate in various ways, for example through physico-theology. This would be a long and complex story, and presently a largely unexplored one: but see, for the early modern period, Peter Harrison, *The Bible, Protestantism, and the Rise of Natural Science* (Cambridge, 1998); Peter Harrison, *The Fall of Man and the Foundations of Science* (Cambridge, 2007).

[3] I am not concerned here with the question whether these are intrinsically religious conceptions. In his *The Legitimacy of the Modern Age* (Cambridge, MA, 1983), for example, Hans Blumenberg responds to Löwith's argument that central values of modernity are simply a secularized form of Christian conceptions, by pointing out that there is nothing intrinsically Christian about these values, which Christian thinkers in many cases took over from Stoic and Neoplatonist systems and simply Christianized: they were originally secular notions. In the present case, it makes no difference what the ultimate origins of the idea of an underlying unity are, because it is the Christian one that is being replaced.

Civilization and the Culture of Science: Science and the Shaping of Modernity, 1795–1935. Stephen Gaukroger, Oxford University Press (2020). © Stephen Gaukroger.
DOI: 10.1093/oso/9780198849070.001.0001

The project to unify science has been pursued in various ways. Among the more modest are those which set out simply to establish the unity of the physical and material (chemical) sciences. By contrast, the most ambitious, such as those of Comte and Spencer, set out to configure the life sciences, the human and social sciences, as well as the moral and political realms, in a hierarchy in which disciplines are ranked in terms of their proximity to a physical or mathematical base: the closer to the base the greater the explanatory power. In examinations of claims for unification, critical attention has generally been focused on the more ambitious projects, but while the problems are more evident there, they pervade all unification projects, whether modest or ambitious.

Putting to one side doubts about whether the original Christian-Platonist project was able to provide a unification of understanding only on the basis of intellectually impoverished resources, which collapsed once they were questioned,[4] the issue facing us is whether there could plausibly be a unity of understanding of the kind promoted in subsequent unification projects. I shall build up the case that the whole idea of the unity of understanding is misguided, both as a supposedly established fact about the nature of science, and as an idealized and aspirational objective that can legitimately guide scientific enquiry. It is not the practicality of the project for the unity of science that is at fault, it is the very idea of the unity of science, the attempt to realize in newly secular terms what is essentially a misguided project. There are a number of different ways in which these questions could be approached, and what is probably the most popular route—treating them as purely conceptual or philosophical issues—would, I believe, prevent us from delving sufficiently deeply or fruitfully into what is at stake. Questions about the unity and coherence of our understanding of the world are of necessity highly contextualized. They depend on what one wants the unity to achieve, and for this one needs to look in detail at the development of the schemes for the unity of science, what has motivated them, whether the motivation has been intrinsic or whether it has been shaped by social, political, and ideological concerns which have been mapped on to science from outside, and if so, what the mechanism is by which this mapping has been achieved.

In the nineteenth century, the question of the unity of science became a matter of urgency. The problem would have been a question of less pressing concern were it not for changes in the standing of science that occurred from the eighteenth century onwards. In French thought in the second half of the eighteenth century, for example, science and the Enlightenment were intimately linked. To be committed to the Enlightenment project was to take a stand on the cultural importance of science in relation to religion, medicine, pedagogy, and politics.[5] By the nineteenth century, science had become established as an

---

[4] See Gaukroger, *The Emergence of a Scientific Culture*, chs 2–4.
[5] See Gaukroger, *The Collapse of Mechanism*, chs 6 and 7; Gaukroger, *The Natural and the Human*, ch. 3.

irresistible cultural force, as the mark of civilization and modernity. In the case of Germany, as Denise Phillips notes, between 1770 and 1850 the term *Naturwissenschaft* or 'natural science' became 'invested with an emotional intensity it utterly lacked in the eighteenth century; it became a word that communicated fervor and inspired loyalty. In the eighteenth century, the word was a descriptive label for a (loosely specified) field of knowledge; in the nineteenth century, it referred to a powerful cultural force'.[6] And in the British case, as we are about to see, there were bitter disputes over the terms 'science' and 'scientist'. It is in this context that the particular questions of unity with which we are concerned became pressing. The worry was that, until its unity was secured, science's claim to be the sole form of understanding the world—and, on some accounts, the sole form of understanding of our place in the world—was not assured. This, more than anything else, is what has driven the assumption of the unity of science.

As a consequence, the unity of science takes on a new complexity. On the face of it, it looks as if unity could just be a matter of establishing particular kinds of relations between scientific theories, establishing a hierarchy and putting in place various forms of reduction for example. But this in effect assumes that unity is a context-independent matter, an activity to which scientists can turn their attention once the relevant theories have been developed. It ignores the reasons why unity is sought in the first place. If unification were a simple and straightforward technical matter, then this might pass unnoticed. But this is far from being the case. As often as not, unification projects have failed, have been abandoned, have met overwhelming resistance, have been compromised, have pointed enquiry in fruitless directions, have been in competition with other unification projects having quite different aims, or have been directed by political and ideological aspirations that have only a tenuous connection with the programmes being unified. In these cases, the question of what the attempt at unification is designed to achieve is of paramount importance. In examining the uncertain and chequered progress of unification projects—those which have led enquiry to dead ends as well as those where the case for unification is strongest—we are able to bring to the surface assumptions about the point of the exercise that might otherwise lie hidden.

Above all, it becomes evident that unification has never been defended by its proponents as, or considered as, a contingent fact about the relation between scientific theories; rather it is (implicitly) presented as an a priori feature of 'science', a metalevel abstraction which offers a rationalized image of selected scientific practices. It is not that there *is* a unity of science, it is that there *must be* a unity of science. However significant its explanatory failures, it would seem that nothing could displace it. Above all, as I hope to show, the commitment to the

---

[6] Phillips, *Acolytes of Nature*, 30. See also Rudolph Stichweh, *Zur Entstehung des modernen Systems wissenschaftlicher Disziplinen: Physik in Deutschland, 1740–1890* (Frankfurt, 1984), on the related question of the cultural aspects of the development of the discipline of 'physics'.

unity of science confuses ontological and explanatory questions, it assumes a teleological trajectory in the sciences that violates the most elementary safeguards of historical understanding, and by making the unity of the sciences in effect an a priori matter it removes it from the critical examination that it badly needs. Once we question the assumption that the unity of science has an a priori standing, and start to examine whether it has in fact been successfully established in the history of the sciences, it becomes clear that, while there are local forms of unification, there is nothing remotely resembling a general unification of even the natural (physical and material) sciences, let alone the natural and the life sciences.

THE ADVANCEMENT OF SCIENCE

In 1833, the members of the British Association for the Advancement of Science gathered, for their third annual meeting, in Cambridge, with William Whewell in the chair. Coleridge was present at the meeting, a sign of the breadth of the membership. He had earlier publicly proposed what he termed a 'clerisy', its aim being 'to secure and improve that civilization, without which the nation could be neither permanent nor progressive'.[7] The clerisy comprised an intellectual elite of which a select few 'were to remain at the fountain heads of the humanities, in cultivating and enlarging the knowledge already possessed, and in watching over the interests of physical and moral science', while instructing a larger group responsible for transmitting this guidance to the public in general.[8] In this vein, he objected to men of science—who were concerned with utilitarian pursuits, with practical questions and experiments—calling themselves natural philo- sophers since they were thereby usurping what he considered a higher calling. Coleridge was widely fêted at the meeting, and given his insistence that religion lay at the foundations of science and learning more generally, the scientifically aware clergymen of Trinity College Cambridge—of which Whewell was effect- ively the figurehead—in many respects provided just the kind of fund of talent on which the clerisy could be expected to build.[9] As Cambridge undergraduates, John Herschel, Charles Babbage, and George Peacock had founded the Analytical Society in 1815, with a view to revitalizing British mathematics and natural philosophy,[10] and they were soon joined by Whewell and others. By the mid- 1820s this group constituted the centre of power in the reform of British natural philosophy, playing major roles in the founding of the Cambridge Philosophical Society (1819) and the Astronomical Society of London (1820), as well as the

---

[7] Coleridge, *On the Constitution of the Church and State*, 47.　　　[8] Ibid., 46.

[9] See Silvan S. Schweber, 'Scientists as Intellectuals: The Early Victorians', in J. Paradis and T. Postlewait, eds, *Victorian Science and Victorian Values: Literary Perspectives* (New Brunswick, 1985), 1–38; Susan Faye Cannon, *Science in Culture: The Early Victorian Period* (New York, 1978).

[10] See Philip C. Enros, 'The Analytical Society (1812-1813): Precursor of the Renewal of Cambridge Mathematics', *Historia Mathematica* 19 (1983), 24–47.

British Association. There was competition with Cambridge for intellectual and political leadership, however, if not always exactly of the 'clerisy' mould: on the side of the clergy from what would become the Oxford movement, with Newman as its leader, and on the secular side from the group affiliated to the new University College London, namely Jeremy Bentham, James Mill, and Henry Brougham. The competing interests, and the fact that it was crucial for the future of the association that it harmonize the different interests of its very broad membership to the greatest degree, made it difficult for any one group to prevail over the others in its early years.

The British Association for the Advancement of Science, as its name announces, was formed to advance 'science', and there was debate at the early meetings on just what the practitioners of science should be called. Whewell reported on the debate in an article in 1834:

Formerly the 'learned' embraced in their wide grasp all the branches of the tree of knowledge; the Scaligers and Vossiuses of former days were mathematicians as well as philologers, physical as well as antiquarian speculators. But these days are past . . . . We are informed that this difficulty was felt very oppressively by the members of the British Association for the Advancement of Science, at their meetings at York, Oxford, and Cambridge in the last three summers. There was no general term by which these gentlemen could describe themselves with reference to their pursuits. *Philosophers* was felt to be too wide and lofty a term, and was very properly forbidden them by Mr. Coleridge, both in his capacity as philologer and metaphysician; *savans* was rather assuming and besides being French instead of English; some ingenious gentleman [Whewell himself] proposed that, by analogy with *artist*, they might form *scientist*, and added that there could be no scruple to making free with this termination when we have such words as *sciolist*, *economist*, and *atheist*—but this was not generally palatable.[11]

The term 'scientist' did slowly come into use, encouraged by Whewell, but it was not generally adopted until the early decades of the next century, being the subject of an inconclusive debate in *Nature* as late as 1924.[12] In the nineteenth century the resistance was significant, in part because, as Richard Yeo notes, 'some of the important men of science, such as Michael Faraday and T. H. Huxley, preferred to think of their work as part of broader philosophical, theological, and moral concerns'.[13]

---

[11] William Whewell, 'Mrs Somerville on the Connexion of the Sciences', *Quarterly Review* 51 (1834), 54–68: 59.

[12] In 1924, the physicist Norman Campbell wrote to the editor of *Nature*, noting that there was a prejudice against the word 'scientist' deriving from a time 'when scientists were in some trouble about their style', but that that time was past, and urging the editor to abandon other nomenclatures, such as 'man of science' and 'scientific worker' in favour of 'scientist'. See Melinda Baldwin, *Making Nature: The History of a Scientific Journal* (Chicago, 2015), 4. Baldwin goes on to note that, in 1925, the Royal Society, the British Association for the Advancement of Science, the Royal Institution, and Cambridge University Press were all still rejecting the exclusive use of the term: ibid., 5.

[13] Yeo, *Defining Science*, 5.

Why would Faraday and Huxley refuse the name 'scientists'? Despite their having some fun at the expense of the word—Huxley remarked that 'for anyone who respects the English language, I think "scientist" must be about as pleasing a word as "electrocution"'[14]—the answer lies less in the term itself than in the activity it describes: science. The British Association came to give the term 'science' a uniquely narrow and restricted meaning, one that was not only contrary to the long-standing notion of natural philosophy that it replaced, and to common usage in the English language (it took fifty years to become accepted into common usage in dictionaries), but also to that of the usage in the main scientific languages of the period: *science* (Fr.), *Wissenschaft* (Germ.), *scienza* (Ital.), *ciencia* (Span.), *nauka* (Russ.). The British Association's sense of its mission is crucial here, and it excluded literature from its deliberations, for example, by contrast with the many provincial societies that were connected with it, societies that had played a crucial part in its establishment: these regularly not only combined literary and scientific studies but included the two in their titles.[15] It also excluded areas such as the study of antiquities and philology, even though the former was a significant part of the activities of many British provincial societies,[16] and was important to the conception of scientific enquiry in France. Philology also played a significant role in German scientific societies, on which the British Association was otherwise modelled in key respects.[17]

The rationale behind the particular definition should be seen in terms of what Morrell and Thackray argue is a project to consolidate the role of science as the 'dominant mode of cognition of industrial society'. They write that the deliberate creation of 'boundaries between natural and religious or political knowledge, the conceptualization of science as a sharply edged and value-neutral domain of knowledge, the subordination of the biological and social to the physical sciences, the harnessing of a rhetoric of science, technology, and progress—these were some of the ways in which an ideology of science was constructed'.[18] This ideology was in many respects exclusive, as illustrated by the case of phrenology. Phrenology had been developed in continental Europe by Franz Joseph Gall and Johann Spurzheim at the beginning of the nineteenth century, and it was taken up in Britain by the Edinburgh lawyer George Combe, whose *The Constitution of Man*

---

[14] Quoted in James A. Secord, *Visions of Science: Books and Readers at the Dawn of the Victorian Age* (Oxford, 2014), 105.

[15] Note in particular that the subtitle of the *Athenaeum*, which is the journal that carried regular reports of the meetings of the individual sections of the association, and so came closest to being its house-journal, was *Journal of Literature, Science, and the Fine Arts*.

[16] Morrell and Thackray, *Gentlemen of Science*, 276.

[17] The British Association owed a considerable debt to German institutional arrangements: the Gesellschaft Deutscher Naturforscher und Ärzte, founded in 1822, acted as a model for the British Association in its annual change of location for meetings (in the German case presumably because there was no national capital), for example, and in the division into sections. The Royal Society finally established committees for special branches of science only in 1838.

[18] Morrell and Thackray, *Gentlemen of Science*, 32.

first appeared in 1828 and, as its popularity increased, subsequently went through a number of expanded editions. Combe argued that the brain was a composite of organs which specialized in different functions. These shaped the intellectual and emotional life of the subject, and the claim was that the skilled phrenologist could read their preponderance in relation to one another by registering their size through the contours of the shaven head of the subject.

Advocates for phrenology maintained that it was the first empirical science of the mind. As Combe put it:

Phrenology is not an *exact*, but an *estimative* science. It does not resemble mathematics, or even chemistry, in which measures of weight and number can be applied to facts; but, being a branch of physiology, it, like medical science, rests on evidence which can be observed and estimated only. We possess no means of ascertaining, in cubic inches, or in ounces, the exact quantity of cerebral matter which each organ contains, or of computing the precise degree of energy with which each faculty is manifested; we are able only to *estimate* through the eye and the hand the one, and by means of the intellectual the other.[19]

Supporters of Combe insisted that phrenology be included in a new section of the British Association, and indeed it was seen as the culmination of the unified sciences, a science of the mind standing at their apex, just as the physical and mathematical sciences stood at their base. In this respect, claims for phrenology in Britain mirrored claims for 'social physics' in Comte's schema. At the 1834 meeting of the British Association in Edinburgh, where Combe had founded the Edinburgh Phrenological Society twenty-four years earlier, the move to recognize phrenology was rejected. Ridiculed by the chair of the natural history section, there were moves by phrenologists to include it instead in the medical section, but with no success. In response, the phrenologists synchronized their annual meetings with those of the British Association, holding them at the end of the latter so that they had the appearance of being continuous with them, but the numbers of those attending collapsed and this ruse lasted only two years.[20] Nevertheless, phrenology, in seeking to complete, and occupy the pinnacle of, the unified sciences, raises a crucial question: whether projects for the unification of the sciences demand some kind of 'completion', ineluctably pointing to some final crowning discipline for example, or, in the case of reductionist conceptions, pointing to some final trophy, the last discipline to succumb to reduction. At the same time, phrenology raises the question of how we judge the empirical credentials of disciplines that clearly pursue a new empirical approach to issues such as the nature of the mind, and which claim a scientific standing on this basis.

Given the broad membership of the British Association, finding the appropriate characterization of science was no easy matter. As Morrell and Thackray put it:

---

[19] George Combe, *A System of Phrenology* (5th edition, 2 vols, Edinburgh, 1843), i, p. vii.
[20] See Morrell and Thackray, *Gentlemen of Science*, 276–81.

Only if science were separated from politics and theology could members of opposing social or religious groups unite for its advancement. Only if science were linked to progress and cut off from the controversial enquiries of statisticians and phrenologists would its appeal be authoritative and clear. Only if science were rendered attractive to various constituencies would it serve as an instrument of social expression and social integration.[21]

What, then, was the account that embodied this cautious but inclusive understanding of science? An 1834 report on the British Association gives a succinct definition: 'Science, in their new vocabulary, is restricted to the investigation of material substances.'[22] That is, material as opposed to spiritual substances. Certainly one thing motivating this kind of definition was to keep in place a strict separation between science and revealed religion, with geology—which offered estimates of the age of the earth at variance with the biblical account— being the area in which the two were in most danger of coming into conflict. Whewell's colleague John Herschel, in his *Preliminary Discourse on Natural Philosophy* of 1830,[23] had set out a programme that was at pains to show that the study of science presented no threat to revealed religion,[24] although it did offer a view of the benefits of science that accorded it a route to peace of mind that was independent of religious contemplation: 'The observation of the calm, energetic regularity of nature, the immense scale of her operations, and the certainty with which her ends are attained, tends, irresistibly, to tranquilize and re-assure the mind, and render it less accessible to repining, selfish, and turbulent emotions.'[25]

Herschel's programme was very much that which was taken up by the British Association, and it was cited extensively by British writers throughout the nine-teenth century.[26] In many respects the *Preliminary Discourse* mirrors d'Alembert's preliminary discourse to the *Encyclopédie*, and it is designed as a preliminary discourse to an encyclopedia, in this case the 'Cabinet Encyclopedia'. Like d'Alembert, Herschel offers a legitimatory account of science, but it differs from his in that, as well as not being politically radical, it focuses extensively on the questions that had come to the fore by 1830, above all the identification of

[21] Ibid., 224.

[22] Anon., 'The British Association for the Advancement of Science', *Oxford University Magazine* 1 (1834), 401–12: 401.

[23] John Herschel, *A Preliminary Discourse on the Study of Natural Philosophy* (London, 1830).

[24] What worried clergymen above all was the idea of an autonomous science. Scripturally motivated geologists such as Frederick Nolan were wholly opposed to science and religion being separated because an autonomous science was antithetical to the Mosaic account, while those clergymen that were to form the Oxford Movement were opposed to any group that could claim authority in intellectual affairs and those of state, potentially displacing spiritual authority. See Morrell and Thackray, *Gentlemen of Science*, 230–1.

[25] Herschel, *A Preliminary Discourse*, 16.

[26] Darwin refered to Herschel as 'one of our greatest philosophers' on the opening page of the *Origin of Species* for example. See Michael Ruse, 'Darwin's Debt to Philosophy: An Examination of the Influence of the Philosophical Ideas of John F. W. Herschel and William Whewell on the Development of Charles Darwin's Theory of Evolution', *Studies in History and Philosophy of Science* 6 (1975), 159–81.

scientific method and the classification of the physical sciences, which had made major advances since d'Alembert's day. It was also a response to Babbage's very polemical *Reflections on the Decline of Science in England* (1830), which had seen no future for the subject in England.[27] As Secord notes, Herschel was committed to establishing the claims of a scientific vocation as a way to truth, and as he realized in writing the *Preliminary Discourse*, 'the issue of conduct was central. If *Decline*, with its polemical tone, and political edge, was an example of how not to behave, the *Preliminary Discourse* could offer a model for the actions of the ideal seeker after truth.'[28]

The *Preliminary Discourse* is divided into three parts, the first of which, 'The Nature and Advantages of Physical Science', begins with an account of the two things that marked out human beings from animals, what made them civilized: science—a search for understanding of the empirical world—and a realization that there is something that transcends our understanding, namely God. The two come as inseparable parts of a package, something made even more explicit in Herschel's review of Somerville's *Mechanism of the Heavens* two years later, where, noting the simplicity and harmony that emerge from the intricacies of the planetary orbits, he adds a political message to the religious and natural-philosophical ones:

yet this intricacy has its laws, which distinguish it from confusion, and its limits, which preserve it from degenerating into anarchy. It is in this conservation of the principle of order in the midst of perplexity—in this ultimate compensation, brought about by the continued action of causes, which appear at first sight pregnant only with subversion and decay—that we trace the Master-workman with whom the darkness is even as the light.[29]

Nevertheless, to show that science is compatible with revelation is not to show its value to us. As Herschel makes clear in the *Preliminary Discourse*, that requires a different kind of demonstration, 'for if science may be vilified by representing it as opposed to religion, or trammelled by mistaken notions of the danger of free enquiry, there is yet another mode by which it may be degraded from its native dignity, and that is by placing it in the light of a mere appendage to and caterer for our pampered appetites'.[30] Accordingly, part II of the *Preliminary Discourse* deals with methodological questions, attempting to show how an abstract form of science, which occupies itself with higher-level generalizations, is compatible with inductive enquiry, in which theories are gradually built up from observation. Part III summarizes the results achieved in the various discrete sciences, taking up the question in the context of the history of science. Among other things, it presents a detailed attempt at a vindication of Newton's corpuscular theory of

---

[27] Charles Babbage, *Reflections on the Decline of Science in England and on Some of its Causes* (London, 1830).
[28] Secord, *Visions of Science*, 89.
[29] [John Herschel], 'Mechanism of the Heavens', *Quarterly Review* 47 (1832), 537–59: 541.
[30] Herschel, *A Preliminary Discourse*, 10–11.

light, which had recently been replaced by a revived version of Huygens' wave theory, reworked by Young and Fresnel. Herschel set out to show that Newton deployed an inductive method and argues that the corpuscular theory that resulted accounted for all optical phenomena known at the time, thereby vindicating inductivism as a reliable method of discovery. The final chapter of the book raises the question of the rapid progress of science since the sixteenth century. 'There is no more extraordinary contrast', Herschel writes, 'than that presented by the slow progress of the physical sciences, from the earliest ages of the world to the close of the sixteenth century, and the rapid development they have since experienced'.[31] One way to keep track of this new rapid development was to classify the various physical sciences, to bring them under some single rubric, setting out to capture what is distinctive about science as a form of enquiry in methodological terms. In his review of the book, Whewell praises it for being much more than a survey of the physical sciences. It contains an important element, namely 'his view of the philosophy of physical science, the principles on which its structure rests, the maxims by which its researches have been and must be successfully conducted . . . . It is, we believe, one of the first considerable attempts to expound in any detail the rules and doctrines of that method of research to which modern science has owed its long-continued steady advance and present flourishing condition.'[32]

Questions surrounding the understanding of science in the nineteenth century were not unique to Britain, and there were significant national variations. In comparing Britain with continental practices in the physical sciences in the first half of the nineteenth century, for example, Merz, in his history of nineteenth-century science, remarks:

On the Continent, both in France and in Germany, the sciences were rigidly marked off from one another, the connecting links were few and ill-defined, and speculations as to the general forces and agencies of nature were left to metaphysicians and treated with suspicion. In England alone the name of natural philosophy still obtained, and in the absence of separate schools of science, such as existed abroad, suggested, at least to the self taught amateur or to the practical man, the existence of a uniting bond between all natural studies.[33]

There is some truth in this observation—it is quite likely that Comte's emphasis on the division of labour as the key to success, for example, has its roots in the division of labour in the sciences, which he would have experienced directly as a student at the École Polytechnique—but it needs qualifying. French scientists and

[31] Ibid., 347.

[32] William Whewell, 'Review of *A Preliminary Discourse on the Study of Natural Philosophy* by J. F. W. Herschel', *Quarterly Review* 45 (1831), 374–407: 377.

[33] Merz, *A History of European Thought*, ii. 98. See also Crosbie Smith, '"Mechanical Philosophy" and the Emergence of Physics in Britain: 1800–1850', *Annals of Science* 33 (1976), 3–29.

philosophers referred to *sciences* in the plural, but the word covered all the natural sciences, and so was not that different from the English singular term in this respect. There was however a high degree of specialization in physics, chemistry, and natural history in France, and the subjects had a greater degree of autonomy than elsewhere, despite reductionist programmes such as that of Laplace.[34] In German, by contrast, the singular word *Naturwissenschaft* was a generic term for the physical and natural sciences more generally, and *Wissenschaft* designated a broad span of scholarly pursuits. There was significant overlap between these terms and the older British idea of 'natural philosophy'.

In sum, there is a greater degree of commonality than Merz's observation might suggest. Moreover, to the extent to which he distinguishes between philosophical (or 'metaphysical') and scientific treatments of links between the sciences, a difficulty arises. It is not so much that Whewell, for example, could be called both a philosopher and a scientist (he successively held the chairs of both mineralogy and moral philosophy at Cambridge), but rather that the kinds of metascientific problems he and his contemporaries and successors engaged with had an unmistakable philosophical element to them, whether explored by scientists or by philosophers. It is the nature of the undertaking, not who undertakes it, that is important. The chief danger that arises from ignoring this lies in concluding from the fact that it is predominantly professional scientists who are active in these debates that the questions of reduction and unity raised, for example, were generated purely internally, and consequently that the outcome was guaranteed by the credentials of the scientific discipline.

THE UNIFICATION OF THE SCIENCES

Whewell's comments on his proposal for the adoption of the word 'scientist' were contained in a review of Mary Somerville's *On The Connexion of the Physical Sciences*, which had appeared in 1834. *On The Connexion of the Physical Sciences* was a successor volume to Somerville's *Mechanism of the Heavens*, which was originally commissioned by the Society for the Diffusion of Useful Knowledge, founded in 1826 by Henry Brougham, the radical parliamentarian, anti-slavery campaigner, and co-founder of the influential *Edinburgh Review*. The aim of the Society was to publish new books at low cost for a very broad public on a range of scientific subjects, from hydraulics and botany to brewing, the habits of insects, and Egyptian antiquities. Brougham's own *Objects, Advantages, and Pleasures of Science*[35] had been a literary and commercial success, and in 1827 he announced his intention to prepare an accessible version of Laplace's mathematically

---

[34] Maurice Crosland and Crosbie Smith, 'The Transmission of Physics from France to Britain, 1800–1840', *Historical Studies in the Physical Sciences* 9 (1978), 1–61: 9.

[35] Henry Brougham, *A Discourse of Objects, Advantages, and Pleasures of Science* (London, 1827).

challenging five-volume *Mécanique céleste* for publication. Somerville, one of the few people in England competent to undertake the task, agreed to write a condensation of the first two volumes. But by the time she submitted the manuscript, sales of volumes in the society's series had fallen off drastically, and Broughton was no longer able to make good the offer of publication. The publisher John Murray was convinced to take over the project, and *Mechanism of the Heavens* appeared in 1831.[36]

The second volume, which Somerville had been working on, was not published, but the project did not come to a stop. There had been a long 'Preliminary Dissertation' to *Mechanism of the Heavens*, and she used this to serve as the basis for a new book. This new book was *On the Connexion of the Physical Sciences*. It was written for a more general audience than *Mechanism of the Heavens*, and while it was significantly more demanding than the popular science books of the period, it turned out to be one of the most successful introductions to the physical sciences in the nineteenth century, regularly updated and serving as the main guide to advances in the physical sciences. Moreover, it was no mere summary of results. James Clerk Maxwell wrote that it was among those 'suggestive books, which put into a definite, intelligible, and communicable form, the guiding ideas that are already working in the minds of men of science, so as to lead them to discoveries, but which they cannot yet shape into a definite statement'.[37] In his review of the book, Whewell writes:

The inconvenience of this division of the soil of science into infinitely small allotments have often been felt and complained of. It was one object, we believe, of the British Association, to remedy these inconveniences by bringing together the cultivators of different departments. To remove the evil in another way is one object of Mrs. Somerville's book. If we apprehend her purpose rightly, this is to be done by showing how detached branches have, in the history of science, united by the discovery of general principles.[38]

Somerville's aim was to establish fundamental connections between the various parts of the physical sciences. This turned out to be a comparatively straightforward exercise in the early sections on astronomy, which are centred around gravity and the nature of light. Indeed it is the thoroughly mathematical nature of astronomy that is presented as the key feature of the physical sciences, providing the reader with an intuitively imaginable and picturable universe. But mathematics is also what unifies the physical sciences, even if the programme is at best

[36] Mary Somerville, *Mechanism of the Heavens* (London, 1831). See Secord, *Visions of Science*, ch. 4, which offers an insightful account of Somerville's life and work, and to which I am indebted here.

[37] James Clerk Maxwell, 'Grove's "Correlation of Forces"', *Nature* 10 (20 August 1874), 302–4: 303.

[38] Whewell, 'Mrs Somerville on the Connexion of the Sciences', 60. Cf. Biot in 1831: 'A quick glance at science reveals to us the great extent of its riches. What it lacks is unity. It is the joining of the parts that makes a single body of it; it is the establishment of the data and the principles that gives the same direction to all efforts.' *Traité de physique expérimentale et mathématique*, i. p. vii.

underdeveloped when it comes to experimental disciplines such as magnetism and electricity.

The idea that the physical sciences were intrinsically mathematical disciplines offered a formal unification strategy in the nineteenth century, but this was not an entirely new view. With the introduction of a serious interest in mechanics and mathematical astronomy from the late sixteenth century, various mathematically based conceptions of physical understanding had been proposed. Galileo had famously maintained that the cosmos was written in the language of mathematics, and Kepler had treated the solar system as a material realization of constructions from solid geometry. But the most developed such approach was mechanism, which combined matter theory—in this case the view that all macroscopic physical behaviour could be reduced to the activity of micro-corpuscles or atoms—with mechanics, in the form of the view that this microscopic activity could be described comprehensively in terms of the size, shape, speed, and direction of motion of the micro-corpuscles.[39] By the beginning of the eighteenth century, however, mechanism faced overwhelming problems. In order to bring everything physical under its explanatory umbrella it had to discard a vast range of phenomena from the realm of the genuinely physical (primarily through the vehicle of the doctrine of primary and secondary qualities), and it was able to offer little more than promissory notes on how to reduce the behaviour of living things successfully to mechanically describable phenomena. At the same time, even within its much reduced domain of 'genuine' physical phenomena, it seemed to be able to explain very little, and what explanations it did propose were largely speculative.

Mechanism was really a form of matter theory with mathematical aspirations rather than a mathematically based theory. In the middle of the eighteenth century, there emerged a more genuine mathematizing project when the development known as rational mechanics (begun in the late seventeenth century with Huygens, Newton, and Varignon) took on new explanatory ambitions, particularly in the work of Euler. Rational mechanics had a secure and sophisticated mathematical apparatus as its starting point, and built up an account of physical kinds of matter from mass points, using these to develop descriptions of rigid bodies, flexible bodies, elastic bodies, and finally fluids. Euler in particular had hoped that the rest of physics, including the study of gravitation and electricity, could be constructed on this mechanical core.[40]

The idea that a mechanical description of the properties of matter could form a starting point for a more general description of the world, seemed to a number of scientists and philosophers to form the only viable starting point if there was to be a unified account of physical phenomena. And what secured mechanics in

---

[39] See Gaukroger, *The Emergence of a Scientific Culture*, chs 8 and 9.
[40] See Gaukroger, *The Collapse of Mechanism*, ch. 8.

this role was the fact that it could be formulated wholly in mathematical terms. The project received a fillip in the late eighteenth century from a development in the mathematical formulation of dynamics: Newton's formulation, which was vectorial in that it measured such things as force and velocity which are directional, was replaced with a much more streamlined formalism, analytical mechanics, where the fundamental qualities are scalar and dynamical relations obtained by a systematic process of differentiation of linear equations with time as the variable.[41] Such new ways of formulating the mathematics underlying the physical sciences helped powerful connections to be made between disciplines. But do these connections in fact point towards a general unification of the physical sciences, as is commonly assumed, or do they rather provide local points of unification which fall short of any general fusion of the physical sciences into a single body?

In the early decades of the nineteenth century, we witness the beginning of a number of attempts to bring together various physical disciplines under a single umbrella. In the 1810s, Fresnel had explored the possibility that light and heat were connected with the vibrations of a fluid, and by 1821 he was offering an account of optics in terms of the dynamics of a wave-propagating medium, a luminiferous ether,[42] a move that led him to consider a unified account of acoustics and optics. But he realized that if light behaved like sound, comprising longitudinal vibrations in the direction of the propagated wave, then the well-attested asymmetric behaviour of polarized light could not be explained. Consequently he abandoned unification along these lines, and opted instead for transverse waves, which turned out to enable him to encompass not only optical phenomena but a range of mechanical phenomena in the one account. Note however that there was no general move to unification here: it was rather a matter of replacing one model with another which focused on bringing a different set of phenomena together, sacrificing the unification of light and sound in the process. This consideration will be important below, as we see that there is no single overarching direction in unification projects in the physical sciences: historically speaking, they are often focused on different issues and lead in many different directions.

Just as Fresnel incorporated optics into a mechanical model, in 1822 Fourier demonstrated that heat could be incorporated into a framework previously occupied by purely mechanical problems, namely that of analytical mechanics.[43] The interesting thing here is that pursuing a unification strategy meant that mathematics comes to be substituted for physics. This is not surprising: as we

---

[41] See, in particular, Joseph Louis de Lagrange, *Mécanique analytique* (Paris, 1788).
[42] See P. M. Harman, *Energy, Force, and Matter: The Conceptual Development of Nineteenth-Century Physics* (Cambridge, 1982), ch. 2. I am particularly indebted to Harman's work in what follows.
[43] Joseph Fourier, *Théorie analytique de la chaleur* (Paris, 1822).

saw in the case of Somerville's *On the Connexion of the Physical Sciences*, mathematics works at a level of abstraction very conducive to grand unification programmes. In the present case, Fourier's account was purely mathematical, avoiding any mention of the physical nature of heat. The diffusion of heat was treated in terms of sets of differential equations, with no physical interpretation of the equations, and the explicit exemplar was Newton's account of gravitation: the cause was unknown, but the phenomenon itself could be observed and subjected to mathematical analysis. Fourier's approach was likewise to reduce physical problems to problems of mathematical analysis, working in terms of the effects of heat, not its hypothetical cause; it was concerned with the temperature distribution in bodies, for example, not with the way in which the repulsive power of heat determined the physical states of material substances.[44] Fourier's analysis acted as a guide for a number of subsequent attempts to incorporate physical phenomena into analytical mechanics. In 1842, William Thomson constructed an analogy between electrostatics and heat, using Fourier's mathematical account of thermal distribution as a model for electrostatic attraction. The analogy was again purely mathematical—no physical connection between the two phenomena could be established in this way—but it suggested that, at a deeper level, what was being uncovered was a basically mathematical structure of the world.

The attempt to achieve unification by minimizing the physics and dealing with questions at a level of mathematical abstraction was an issue that dogged unification models. In the early nineteenth century, the model for those who espoused a mathematical approach to physical questions was Laplace, and his most illustrious successor in England was William Rowan Hamilton, who developed an alternative formulation of analytical mechanics from the 1820s which allowed for mechanics and optics to be connected in a particularly fruitful way. Laplace had spoken in terms of 'atoms', but these were mathematical points, and as such, as Smith and Wise note, they could obey the inverse square law throughout its range of mathematical validity, even down to infinitesimal differences, 'whereas Newton's extended atoms would collide when their centres were separated by two radii. Clearly one could not justify point atoms by any physical analogy of nature.... He therefore appealed to the aesthetic criterion of mathematical simplicity and to hypothetico-deductive justification.'[45] Hamilton went further. He was a leading light in Section A (Mathematics and Physics) of the British Association for the Advancement of Science. In his view Section A was where the real physics was done, and it could be left to mathematics alone to open up the secrets of nature. Morrell and Thackray note that in 1837 he 'invaded Section B

---

[44] See Harman, *Energy, Force, and Matter*, 28.

[45] Crosbie Smith and M. Norton Wise, *Energy and Empire: A Biographical Study of Lord Kelvin* (2 vols, Cambridge, 1989), i. 156.

(chemistry and mineralogy) to quarrel with the theory of the real existence of atoms: he wanted to reduce the material world to mathematical points, because the nearer all the sciences approached Section A, the nearer they would be to perfection'.[46]

There are a number of examples in later nineteenth-century physics that could be mentioned where problems arise from fitting physics into a mathematical mould, and we shall return to this question in Chapter 10, but there is a particularly striking example from the end of the century in the work of Henri Poincaré, who thought of analytical mechanics, which he had done much to reshape, as the only way of pursuing physics. Poincaré was no conservative: he was the only significant figure at the turn of the century to consider the conflict between mechanics and electromagnetic theory to be absolutely critical,[47] and he was on the verge of solving the problems that Einstein was to resolve in terms of the theory of special relativity. But his own attempts, using his geometricized version of analytical mechanics, turned out to be unsuccessful, and he never accepted Einstein's solution. The physicist Maurice de Broglie reports that on one occasion, while Einstein was explaining his ideas, Poincaré asked him what mechanics he was using in his account of special relativity, to which Einstein answered that he was using no mechanics, which de Broglie observes 'surprised Poincaré'. As Galison remarks:

'Surprise' is perhaps an understatement. For Poincaré, whose conception of physics came down to mechanics, be it old or new, 'no mechanics' was an impossible response. Abstract mechanics, after all, was what Poincaré had held up to [his fellow members of the École Polytechnique] as constituting the very essence, the 'factory stamp' of that unique training for the world of Third Republic France.[48]

Poincaré's approach embodies a conception of the role of mechanics in the physical sciences as providing an indispensable foundation for the whole discipline, whereas Einstein successfully relied on his physical intuitions and constructed a somewhat ad hoc mathematical formalism to accommodate these.[49] Mathematics is not physics, even in the guise of analytical mechanics, and the danger was that mathematics was occasionally conceived to be able to underlie physics to such an extent that, in a Platonist fashion, the physical world came to

---

[46] Morrell and Thackray, *Gentlemen of Science*, 274.

[47] See Olivier Darrigol, 'The Electromagnetic Origins of Relativity Theory', *Historical Studies in the Physical and Biological Sciences* 26 (1996), 241–312; Olivier Darrigol, 'Henri Poincaré's Criticism of Fin-de-Siècle Electrodynamics', *Studies in History and Philosophy of Modern Physics* 26 (1995), 1–44.

[48] Galison, *Einstein's Clocks*, 297.

[49] Einstein's attitude to mathematics subsequently changed when he struggled with the difficulties of establishing a unified theory from the 1920s onwards—see Jeroen van Dongen, *Einstein's Unification* (Cambridge, 2010)—but he never got anywhere in this latter project, despite decades of work.

be seen as an—often imperfect—expression of a deeper mathematically structured reality.[50]

UNITY AND UNIVERSALITY

What lies behind the quest for the unity of the sciences? Unification programmes have become so entrenched that the original motivation can become lost, as the questions take on a life of their own, dissociated from any coherent rationale, and mistakenly treated as if they were *sui generis*. If the quest for unification goes beyond anything that could be generated internally to the programme of the sciences, how is it generated? Once we appreciate the need to go beyond questions of internal generation, we can begin to ask about the culture within which the search for the unity of the sciences is pursued. In particular, we can explore the extent to which demands on science are shaped by cultural and political imperatives, and why these take the form they do. I want to illustrate this by looking at four cases: the connections scientists such as du Bois-Reymond and Virchow made between a unified science and a unified state in nineteenth-century Germany; the way in which Wilhelm Ostwald conceived of a general science of energetics which would unify all forms of enquiry from mathematics to 'the science of culture'; the way in which the basis for claims of unification of nuclear physics shifted in Rutherford from an experimentally based reductionism to an appeal to the universality of instrumentation and measurement; and Neurath's programme for an 'encyclopedic' unity of science.

To get a sense of the issues, we can begin by briefly noting two later examples of the politics of the unity of the sciences, from the USA and the USSR. In the USA in the twentieth century there was a range of conflicting political investments in the unity of science. Dewey, in his 1938 'Unity of Science as a Social Problem', was talking of the unity of science as a bulwark against intolerance,[51] a view that led the conservative Catholic philosopher Mortimer Adler to remark that Nazism was not nearly so great a threat to democracy as John Dewey.[52] By the late 1940s and 1950s, at the height of the Cold War, the 'unity of science' was identified as a code word for communism by anti-communist critics of Logical Positivism such as Horace Kallen, as well as by the FBI, who were on the lookout for the term as

---

[50] This has become an especially pressing problem in recent decades, with the proponents of the Standard Model of particles and forces eschewing measurable predictions in favour the beauty of the mathematics. See Lee Smolin, *The Trouble with Physics: The Rise of String Theory, the Fall of Science, and What Comes Next* (New York, 2006); Sabine Hossenfelder, *Lost in Math: How Beauty Leads Physics Astray* (New York, 2018).

[51] John Dewey, 'Unity of Science as a Social Problem', in O. Neurath, R. Carnap, and C. Morris eds, *Foundations of the Unity of Science: Toward an International Encyclopedia of Unified Science* (2 vols, Chicago, 1955–70), i. 29–38.

[52] Quoted in David A. Hollinger, 'Science as a Weapon in *Kulturkämpfe* in the United States During and After World War II', *Isis* 86 (1995), 440–54: 442.

they sifted through the writings and private correspondence of Carnap and other émigré philosophers.[53] The case of the USSR is even more striking. It is perhaps no coincidence that the first atheist state in the world was, as Loren Graham has noted, the first state in the world to offer an official philosophy of science: its own version of the unity of science, namely 'dialectical materialism'. And lest one think that the connection between dialectical materialism and the doctrine of the unity of science is specious, it should be noted that before the mid-1930s, when it came to be interpreted in terms of the generation of contradictions and their resolutions, the doctrine was a classic statement of the doctrine of emergent properties. It held that nature could only be explained in terms of matter and energy ('materialism'), but at the same time it was staunchly anti-reductive ('dialectics'): chemistry couldn't be explained in terms of physics, biology in terms of chemistry, or society in terms of biology, because each level had its own laws, but the whole was anchored in physics.[54] This arrangement was one that Soviet scientists were evidently happy with because, even though at one level it was entirely vacuous, it meant that they could pursue their own specialities unhampered by an otherwise intrusive state.

These cases serve to alert us to just how overdetermined and culturally complex demands for the unity of science can be. We must also bear in mind that in the nineteenth and early twentieth centuries science was increasingly seen by many as an alternative to religion (i.e. to Christianity). For those who saw it as a legitimate alternative, science began to occupy a space vacated by religion, and it took on many of the aspirations of Christianity to provide an overall coherent picture of the world, one which not only made sense of the natural world but also of the place of human beings in the 'order of things'.[55] But the unity of science did not just provide a means of bringing a comprehensive ordering to scientific understanding for those who saw its replacement of religion as legitimate. It was also crucial for those who believed that such replacement was illegitimate. For such critics, it was important to establish the unity of science as a form of understanding that had strict limits, and to secure the boundaries of scientific understanding, preventing it from straying beyond these limits. In both cases, the unity of science lay at the core of the understanding of science in the nineteenth century, binding together cultural, political, religious, and technical issues in a distinctive fashion, one that bears directly on the question of how the relationship between science and civilization was envisaged.

A striking example of how political, cultural, and scientific notions of unification could be tied is provided by the reaction of some leading German scientists—

---

[53] George Reisch, *How the Cold War Transformed Philosophy of Science: To the Icy Slopes of Logic* (Cambridge, 2005), ch. 9.

[54] Loren R. Graham, *Science in Russia and the Soviet Union* (Cambridge, 1993), 100–3.

[55] This development was manifest in different forms from the seventeenth century onwards, and I have explored these different manifestations in earlier volumes in this series: *The Emergence of a Scientific Culture*, ch. 12; *The Collapse of Mechanism*, ch. 12; *The Natural and the Human*, passim.

scientists who were among the most committed to the unity of science and did the most to promote it—to the political events of 1848, and their support for the moves to unify the German states.[56] Du Bois-Reymond, for example, writes that: 'Physiology will fulfill her destiny...To chose an obvious comparison at the moment when I write this, it cannot fail to occur that one day physiology, giving up its peculiar interests, will enter completely into the great union of states of the theoretical sciences.'[57] For du Bois-Reymond, a unified science was the cultural embodiment of the political ideal of a national unification,[58] and later he wrote that 'the German unity which is finally nearing completion was first conceived in the German universities'.[59] In 1888 he named Darwin and Bismark as the great unifiers of the age.[60] Rudolph Virchow, the pathologist and reductionist physiologist, in a letter of 1848, writes that: 'As a natural republican, the realization of the demands which the laws of nature stipulate, and which are derived from the nature of the human being, can only be achieved in a republican state.'[61] Virchow's short-lived 1848 journal, *Die medicinische Reform*, constantly stressed the need for medical reform and political reform to accompany one another, and populist and nationalist themes are tied in with the reform of science.[62] And in 1855 the 'scientific materialist' Ludwig Büchner was writing that natural science 'has strong appeal these days. The public is demoralized by the recent defeat of national and liberal aspirations and is turning its preference to the powerfully unfolding researches of natural science, in which it sees a new kind of opposition against the triumphant Reaction.'[63] With the 1871 unification of Germany, Virchow, in a lecture in the same year at the National Convention of German Scientists, entitled 'On the Tasks of Science in the New National Life of Germany', announced:

The task of the future, now that the outer unity of the Reich has been established, is to establish...a true inner unity...a *unity of spirits*...If science seeks to achieve something especially for the life of our nation, then the first thing...is to fill the people with common understanding, and to give it the generally recognized principles of thought...

---

[56] On the connections between science and politics in Germany in the second half of the nineteenth century, see Kurt Bayertz, 'Darwinismus und Freiheit der Wissenschaft. Politische Aspekete der Darwinismus-Rezeption in Deutschland 1863–1878', *Scientia* 118 (1983), 267–81; and David Cahan, *An Institute for an Empire: The Physikalisch-Technische Reichsanstalt 1871–1918* (Cambridge, 1989), ch. 1.

[57] Du Bois-Reymond, 'Über dies Lebenskraft', in *Reden*, i. 21. Quoted in Anderton, 'The Limits of Science', 111. I am particularly indebted to Anderton's account here.

[58] Cited in Anderton, 'The Limits of Science', 133.

[59] 'Über Universitätseinrichtungen' (1869): *Reden* i. 368.

[60] See Gabriel Finkelstein, *Emil du Bois-Reymond: Neuroscience, Self, and Society in Nineteenth-Century Germany* (Cambridge, MA, 2013), 253.

[61] Quoted in Anderton, 'The Limits of Science', 114. See also Manfred Vasold, *Rudolf Virchow: Der große Arst und Politiker* (Stuttgart, 1988).

[62] Anderton, 'The Limits of Science', 127.

[63] Cited in Frederick Gregory, *Scientific Materialism in Nineteenth Century Germany* (Dordrecht, 1977), 105.

Only when we have succeeded in making our method the method of the whole nation-
... and gradually raising it to the original maxim of (all) thought and of moral behavior,
will the true unity of the nation be won.[64]

By the middle decades of the nineteenth century science was becoming a significant
participant in national cultures. This was not just the work of radicals such as
Büchner. Scientific and technological achievements increasingly become indicators
of national power and prestige, in such a way that science provided a model for
government administrators.[65] In the German context, the question of national
unity was persistently present in the thinking about the task of scientists. Virchow
remarked of the first meeting of the Convention of German Scientists and Doctors
in Leipzig in 1822, that it 'occurred not only in order to discuss with them matters
of science as science, but it also took place with the intention of awakening in the
scattered sons of the great fatherland the thought of an inner connection and thus in
fact of working together for a future unified nation'.[66] The association continued:
in his opening speech to the 1871 meeting of the convention, Benjamin Thierfelder
stated that the principles of national unity had for several decades survived only in
intellectual life, and the convention had been a leading agent of this.[67]

Just as Virchow and others were unifying physiology and mechanics, Ostwald
was arguing that mechanics was the wrong kind of foundational discipline, and
that 'energetics', a generalized form of thermodynamics, was what was needed.
Ostwald's starting point was physical chemistry—he co-founded the *Zeitschrift für
Physikalische Chemie* in 1887 and became its principal editor, receiving the Nobel
Prize for his work on chemical equilibria in 1909—and he decided that the
unificatory strategy that was involved was better served by energetics than mechanics, believing that he could derive the latter from the former.[68] Thermodynamics had presented significant problems for a general mechanics, and energetics was
designed to circumvent these problems. It offered a general science of energy, and

[64] Virchow, 'Ueber die Aufgaben der Naturwissenschaften im neuen nationalen Leben
Deutschlands', in Karl Sudhoff, *Rudolf Virchow und die deutschen Naturforscherversammlungen*
(Leipzig, 1922), 108–18; quoted in Anderton, 'The Limits of Science', 326. Virchow's stand is
very much a response to Pope Pius IX's 1864 list of 'errors'—in which he disparaged science and
claimed moral, social, and intellectual authority for the Church—and his subsequent announcement
of papal infallibility in such matters.
[65] As McCelland points out in his study of the development of the German university system, the
fundamentally modernizing forces of 'technical expertise and university *Wissenschaft* were accepted
because the rationally oriented principles of day-to-day administration favored the augmentation of
power and prosperity of the monarch's state.' Charles E. McClelland, *State, Society and University in
Germany, 1700–1914* (Cambridge, 1980), 153.
[66] Virchow, 'Über die Aufgaben der Naturwissenschaft im neuen nationalen Leben
Deutschlands' (20 Sept, 1871): quoted in Anderton, 'The Limits of Science', 174.
[67] Cited in Heinrich Schipperges, *Weltbild und Wissenschaft: Eröffnungsreden zu den
Naturforscherversammlungen, 1822–1972* (Hildesheim, 1976), 46–7.
[68] The problem, as Boltzmann pointed out, was that in his derivation of mechanics from
energetics, Ostwald had confused variation and differentiation: see Boltzmann, 'Ein Wort der
Mathematik an die Energetik', *Annalen der Physik* 57 (1896), 39–71.

Ostwald proposed two laws, based on the laws of thermodynamics. The first stated the interconvertibility of all forms of energy, energy being conserved in any such conversion. The second set out when a conversion of energy from one form to another had occurred.[69] But the process was not confined to physical and chemical processes, and in 1912, in the second volume of his *Monistische Sonntagspredigten*,[70] he set out a general schema for a unified account of natural knowledge (Fig. 3.1). Ostwald was not the only scientist of the time to see in the notion of energy something that could unify the whole of scientific understanding, and Haeckel, as we shall see in Chapter 7, took a similar approach in the life sciences. There is little in physics, chemistry, or the life sciences that might encourage the view that all natural phenomena can be accounted for in terms of the transformation of energy, still less that such transformations underlie human and cultural phenomena, but this is not where the rationale is to be found.[71] Rather, Ostwald's generalized energetics was directed against similarly ambitious attempts (in the most explicit case, those of the 'scientific materialists') to give mechanics a foundational standing. In connecting epistemology with psychology and physiology, and ethics with what he terms 'cultural science', the unity of scientific understanding is taken as given. The aim is to demonstrate how this unity is possible, and in particular how it is possible without falling into materialism. This

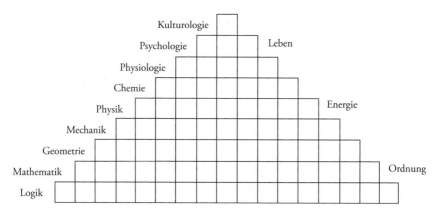

Fig. 3.1. Ostwald's pyramid of the pure sciences

[69] On Ostwald's energetics, see Robert Deltete, 'Wilhelm Ostwald's Energetics 1: Origins and Motivations', *Foundations of Chemistry* 9 (2007), 3–56; Robert Deltete, 'Wilhelm Ostwald's Energetics 2: Energetic Theory and Applications, Part I', *Foundations of Chemistry* 9 (2007), 265–316; Robert Deltete, 'Wilhelm Ostwald's Energetics 3: Energetic Theory and Applications, Part II', *Foundations of Chemistry* 10 (2008), 187–221.
[70] Wilhelm Ostwald, *Monistische Sonntagspredigten* (5 vols, Leipzig, 1911–16), ii. 346.
[71] See Casper Hakfoort, 'Science Deified: Wilhelm Ostwald's Energeticist World-View and the History of Scientism', *Annals of Science* 49 (1992), 525–44.

is to be done by showing how the replacement of mechanics by energetics enabled the move from physical to human phenomena to be made in a way that preserves a spiritual dimension. Even a phenomenon like happiness can be accounted for fully in terms of energetics, and Ostwald offers a formula for happiness:

$$G = (E + W)(E - W)$$

where G is the quantity of happiness, E is the energy applied in accord with the individual's will, and W the energy exerted unwillingly.[72]

A different kind of commitment to the unification of science can be found in the period after the Great War. As Kamminga and Somsen point out:

The world in which interwar pursuits of the unity of science emerged was one of deep division and splintering antagonism. Fragmentation not only seemed to characterize the academic realm and its ongoing specialization, but also, all too obviously, the world of international relations and social strife. Amidst the First World War, Bolshevik Revolution, nationalist agitation and the threat of another global conflict, the claim that any part of human endeavour was fundamentally united was bound to have broader import. It was precisely because science was seen to transcend these many oppositions that it carried a message to them. If the world was divided, science could unite it, and pursuing scientific unity inevitably had political and cultural as well as epistemological implications.[73]

Here it is not a question of national unification but one of internationalism that came to be at issue. The First World War had not only been the first international war in which science was comprehensively co-opted, it also signalled the collapse of a cosmopolitan Western culture. In response, attempts to use science as a model for international cooperation began to be promoted from the 1920s onwards. The case of nuclear physics in the 1920s, for example, seemingly hinges on purely technical matters of measurement, but what is of interest is the way in which these technical issues take on a cultural and political role in the decade after the Great War, and the way in which a culturally and politically neutral basis for the unity and universality of science is sought, unsuccessfully, in experimentation.

One of the most successful reductionist programmes after the Great War was Rutherford's resolutely experimental programme in nuclear physics at the Cavendish Laboratory. One commentator has put the attractions of the programme in these terms:

Against the mathematical and metaphysical speculations of theoreticians and philosophers in the new era of relativity in the early 1920s, the experimental elaboration of the basic building blocks of matter and their interrelations would provide the underpinnings of a fundamental and irreducibly material understanding of the properties and behaviour of

[72] Ostwald, 'Theorie des Glückes', *Annalen der Naturphilosophie* 4 (1905), 459–74: 461.
[73] Harmke Kamminga and Geert Somsen, 'Introduction', in Harmke Kamminga and Geert Somsen, eds, *Pursuing the Unity of Science: Ideology and Scientific Practice from the Great War to the Cold War* (London, 2016), 1–11: 4.

matter at micro and macro levels. Where the theorists disputed time, space and ether, airing troubling disputes about the philosophical foundations of science and employing the language of 'revolution' with the effect—intentional or otherwise—of undermining its historical stability, the experimentally grounded reductionist regime of research would provide the grounds for unifying agreement on the bedrock facts of matter.[74]

The idea here is that experiment provides a basis for a thoroughgoing reductionism, and hence a thoroughgoing unification of, at the very least, the physical and material sciences. This marks out the nuclear debates from the earlier atomic ones, in that the argument shifts onto exclusively experimental ground: scientific legitimacy now becomes a matter of experiment alone. Why was this? Hughes sets out the rationale:

After the Great War, when science in Britain increasingly came to be equated with industrial and military developments—exemplified not least by media hype about 'death rays' in 1924—Rutherford and his co-workers had to work hard to promote atomic and nuclear science as the most socially disengaged of 'pure' sciences, as epistemologically 'fundamental' and as capable of ultimately unifying the sciences both ideologically, through its implicit transnationalist character, and cognitively, through its revelation of the underlying and putatively universal constituents of matter.[75]

In the immediate post-war era, Rutherford was keen to move beyond the military scientific research in which he had been engaged during the war, and to play down the military applications of science, moving from a utilitarian to a purely experimental understanding of science. Experiment was considered to have a universality that transcended any differences that might be due to different mentalities, conceptual schemes, or ideologies, a pressing concern in the wake of the war. And if the case for reduction could be made purely on this basis, then the case for the unity of science—or at least for the unity of the natural sciences—can be made, for reduction is the ultimate form of unification.

Rutherford was initially interested in two phenomena, atomic structure and radioactivity, which he brought together in an ingenious way, pioneering the use of energetic particles emitted in radioactive decay as microscopic projectiles. When these projectiles were made to bombard, and therefore alter the structure of, substances, the results were used to make inferences about what the original structure of the substance must have been. Aided by the work of Bohr and Sommerfield,[76] Rutherford's model of the atom came to dominate the field by the 1920s, as an editorial in *Nature* at the time makes clear:

[74] Jeff Hughes, 'Unity through Experiment? Reductionism, Rhetoric and the Politics of Nuclear Science, 1918-40', in H. Kamminga and G. Somsen, eds, *Pursuing the Unity of Science: Ideology and Scientific Practice from the Great War to the Cold War* (London, 2016), 50–81: 50.

[75] Ibid., 51.

[76] The story is a complex one. See John L. Heilbron and Thomas S. Kuhn, 'The Genesis of the Bohr Atom', *Historical Studies in the Physical Sciences* 1 (1969), 211–90; Suman Seth, *Crafting the Quantum: Arnold Sommerfield and the Practice of Theory, 1890–1926* (Cambridge, MA, 2010); Helge

We think it no exaggeration to say that these experiments are some of the most fundamental which have ever been made. It is not often that a scientific discovery excites interest
outside the narrow circle of the laboratory or the scientific lecture-room . . . So fundamental are the consequences of this new discovery that the intellectual world at large must
follow with keenest interest the progress of the experiments associated with the name of
Rutherford.[77]

Nevertheless, Rutherford's reductionism was opposed by many scientists in the
1920s. Hughes notes that it needed to be actively promoted, and that 'numerous
popularisers and an uncritical media' gave it a broad bedrock of support, 'while
simultaneously marginalising critics and creating positive horizons of public
expectation for the outcomes of the work'.[78]

   For Rutherford and his supporters, the legitimacy of his account rested on
experimentally generated self-evidence. But is an experimental basis for reduction
secure? This is where the problems began with Rutherford's programme. It looks
as if it is simply a matter of producing experimental results in as open and neutral
a way as possible, so that they could not fail to gain assent. But the experiments
were very elaborate, and this impinged on the extent to which they could be
considered self-evident. When another group began experiments in nuclear
disintegration—at the Institut für Radiumforschung in Vienna—they yielded
different results from those found in Rutherford's Cavendish Laboratory, and
for over five years the Vienna scientists pursued research that seemed to contradict
Rutherford's conclusions, for example on the number of elements that could be
disintegrated. The problems derived from counting the scintillations, the mo-
mentary flashes that appear when particles impact on a zinc sulphide screen.
These were extremely tedious to count, and the protocols for counting them
varied between the Cambridge and Vienna laboratories. It was beginning to look
as if experiments were not up to providing the guarantees that Rutherford had
sought.[79] As Hughes notes, 'the legitimacy of the whole reductionist regime was
now potentially under threat, because the controversy with the Viennese threatened to expose the carefully concealed contingency of the Cambridge results and

Kragh, *Niels Bohr and the Quantum Atom: The Bohr Model of Atomic Structure, 1913–1925* (Oxford,
2012); Taddeus J. Trenn, *The Self-Splitting Atom: A History of the Rutherford–Soddy Collaboration*
(London, 1977).

   [77] Cited in Hughes, 'Unity through Experiment?', 50–81, 55. Nevertheless, Rutherford's was not
the only reductionist model, J. J. Thomson offering a different one: see Steven V. Sinclair,
'J. J. Thomson and the Chemical Atom: From Ether Vortex to Atomic Decay', *Ambix* 34 (1987),
89–116; and Michael Chayut, 'J. J. Thomson: The Discovery of the Electron and the Chemists',
Annals of Science 48 (1991), 527–44.

   [78] Hughes, 'Unity through Experiment?', 59.

   [79] On some of the problems with construing experimental results as definitive, see Harry
M. Collins, *Changing Order: Replication and Induction in Scientific Practice* (London, 1955); Peter
Galison, *How Experiments End* (Chicago, 1987).

thereby emphasize the fragility of the experimentally grounded reductionist programme'.[80] In fact the introduction of electronic particle counters by Geiger and Müller began to make scintillation-counting obsolete from 1928, and the controversy petered away. In its place, the idea of unity was now articulated not in terms of experimentation, but shifted to unity of instrumentation. Hughes sums up the situation by noting that 'epistemological unification—insofar as there was unification, given the continuing controversies in nuclear physics in the 1930s—depended on the material unification provided by particular kinds of electronic valves, plasticine and other commercially available products'.[81] Rutherford's own rhetoric shifted from the question of the experimental credentials of reductionism to an appeal to the universality of instrumentation and measurement, without reference to specific experimental results.

Unity of instrumentation is something we have already encountered in an eighteenth-century context, where precision measurement became a leading feature of Enlightenment thinking. It took on this role because it transcended national standards. It embodied the ideal of the universality of science, a universality that was a keystone of a world cosmopolitan culture, which was widely regarded as the essence of civilization prior to the Great War. This cosmopolitan culture, which many had considered to have been led by science, underwent a severe setback with the war, and there were significant efforts to revive it in the 1920s and early 1930s. The discourse of the universality of science now emerged to meet the new and unprecedented demands of the time, and the unity of science effectively became subsumed under a programme devoted to the defence of universal values. We shall return to these questions in Chapter 9. For present purposes, the crucial thing to recognize is that shifting the defence of the unity of science from theoretical developments to experimental ones did not work, and could not have worked. The reason is that commitment to the unity of science is not generated internally within the physical and material sciences, but is an outcome of extrinsic factors. This is evident from the way in which what was at stake shifted so radically on the part of its advocates such as Rutherford, from a defence of reductive unification to a focus on the universality of science and its leading role in the attempts to re-establish a cosmopolitan culture.

Finally, consider the most famous of the twentieth-century unity of science schemes. Neurath's *International Encyclopedia of Unified Science* is perhaps the most revealing of all on the question of what the significance of the unity of science is. 'The unity of science movement', he writes in the opening sentence of his introduction to the *Encyclopedia*, 'includes scientists and persons interested in science who are conscious of the importance of a universal scientific attitude.'[82]

---

[80] Hughes, 'Unity through Experiment?', 61–2.   [81] Ibid., 65.

[82] Otto Neurath, 'Unified Science as Encyclopedic Integration', in O. Neurath, R. Carnap, and C. Morris eds, *Foundations of the Unity of Science: Toward an International Encyclopedia of Unified Science* (2 vols, Chicago, 1955–70), 1–27: 1.

The contributions to the *Encyclopedia* in fact show no uniformity in approach, some, such as those of Hempel and Carnap, discussing 'empirical science' in a way that assumes basic shared precepts, others, such as that of Kuhn, seemingly making no assumptions on the question of unity. This is actually in keeping with Neurath's approach. His model is the *Encyclopédie*, which offered a significant degree of eclecticism,[83] and Neurath's conception of the unity of science is not one in which a system (he is thinking primarily of metaphysical systems) is imposed on the various sciences: 'An encyclopedic integration of scientific statements, with all the discrepancies and difficulties which appear, is the maximum of integration which we can achieve. It is against the principle of encyclopedism to imagine that one "could" eliminate all such difficulties.'[84]

Neurath's commitment to an encyclopedic approach nevertheless does not rule out a staunch commitment to physicalism. He writes, for example, that members of his Vienna Circle 'are in agreement that philosophy does not exist as a discipline: the body of scientific propositions exhausts the sum of meaningful statements'. But he continues: 'When reduced to unified science, the various sciences will be pursued in exactly the same way as when they were separate from one another.'[85] It appears from this that the unification programme has no effect on the individual sciences, but in the same article he writes that 'the proponents of "unified science" seek, with the help of laws, to formulate predictions in the "unified language of physicalism". This takes place in the sphere of empirical sociology through the development of "social behaviourism".'[86] Unlike other versions of the unity of science—but mirroring Rutherford's move from the universality of experimental results to the universality of the measuring instruments—what physicalism seems to provide for Neurath is not reductionist base, but a universal language. The need for a universal language was a central precept of the Vienna Circle (the principal point on which Wittgenstein disagreed with them). Language needed to be cleansed of metaphysics. Carnap, for example, devoted great effort to devising a formal logic in which to formulate all scientific statements,[87] but this was no good for everyday communication, which was just as important for him. Before he had ever considered formal logic, he was entranced by the idea of a truly international language whose structure would be transparent, and at the age of fourteen he began to learn Esperanto, later attending Esperanto conferences and staying in Eastern Europe with fellow Esperanto speakers. Among his motives, he writes, was 'the humanitarian ideal

[83] See Gaukroger, *The Emergence of a Scientific Culture*, 219.
[84] Neurath, 'Unified Science as Encyclopedic Integration', 20.
[85] Otto Neurath, 'Soziologie im Physikalismus', *Ekenntnis* 2 (1931/2), 393–431: 393.
[86] Ibid., 431.
[87] The logic was shaped by considerations of what was needed for physics: see Rudolph Carnap, 'Die physikalischer Sprache als Universalsprache der Wissenschaft', *Erkenntnis* 2 (1932), 12–26; and Rudolph Carnap, 'Psychologie in physikalischer Sprache', *Erkenntnis* 3 (1932), 107–42.

of improving the understanding between nations'.[88] Neurath went one step further, inventing a universal picture language, Isotype.[89]

The unity of science for Neurath and Carnap is a matter of a shared, scientifically constructed language. For Neurath in particular, the aim is to foster a 'universal scientific attitude', something that takes its bearings from science, but which has as its goal a form of world politics, and he praises Comte's and Spencer's 'endeavours to substitute an empirical scientific whole for philosophical speculations'.[90] In a 1946 paper, Neurath writes: 'I think that this gives a picture of the democratic attitude of the Unity of Science movement, which acknowledges from the start a multiplicity of possibilities. It is the problem of any democracy, which any actual scientific research organization has to solve: on the one hand, the nonconformists must have sufficient support; on the other, scientific research needs some cooperation.'[91] In Neurath the idea of the unity of science becomes almost completely subordinated to its utopian socio-political programme, now to be reconstituted on an empirical basis, and the democratic nature of the socio-political programme then becomes built into his conception of the kind of limited and open-ended unification to which empirical studies are susceptible. There is no clearer instance, after the nineteenth-century schemes of Comte and Spencer, of the socio-political imperatives behind the advocacy of the unity of science, than Neurath's programme. What is at stake in the defence of the unity of science is, and always had been, the cultural standing of science.

[88] Rudolph Carnap, 'Carnap's Intellectual Biography', in P. A. Schilpp, ed., *The Philosophy of Rudolph Carnap* (La Salle, IL, 1963), 3–84: 69.

[89] Otto Neurath, *International Picture Language: The First Rules of Isotype* (London, 1936).

[90] Neurath, 'Unified Science as Encyclopedic Integration', 8.

[91] Otto Neurath, 'Orchestration of the Sciences by the Encyclopedism of Logical Empiricism', *Philosophy and Phenomenological Research* 6 (1946), 496–508: 502. See Richard Creath, 'The Unity of Science: Carnap, Neurath, and Beyond', and Jordi Cat, Nancy Cartwright, and Hasok Chang, 'Otto Neurath: Politics and the Unity of Science', both in Peter Galison and David J. Stump, eds, *The Disunity of Science* (Stanford, 1996), 158–69, and 347–69 respectively.

4

The Unity of the Physical Sciences

Central to the attempt to reduce physics to mathematics is the idea that science is an essentially theoretical enterprise. Popper gives a succinct statement of this traditional view when he writes that science proceeds by devising theories first and then, in an entirely separate process (seemingly with separate personnel), uses experiments to test these theories:

The theoretician puts certain definite questions to the experimenter, and the latter, by his experiments, tries to elicit a decisive answer to these questions, and to no others. All other questions he tries hard to exclude . . . Theory dominates the experimental work from its initial planning up to the finishing touches in the laboratory.[1]

Compare this with David Kaiser's description of what happens in contemporary work by physicists setting out to explain the behaviour of subatomic particles:

Try as we might, we will never come across a 'theory' in the flotsam and jetsam of our sources—and thus we should be wary of letting the categories of 'theory construction and selection' direct our historical analysis. Instead, when we inspect the materials with which theoretical physicists have worked, night and day, we see tinkering and appropriation of paper tools—tools fashioned, calculations made, approximations clarified, results compared with data, interpretations advanced, analogies extended to other types of calculations or phenomena, and so on. 'Theories' do not appear, nor is it clear where they might even be found.[2]

In this chapter, we shall see that there is nothing new about these latter concerns. In particular, once we take seriously the roles of experimental practice, and the role of models—particularly physical, experimental models—it becomes evident that physical enquiry deploys a range of explanatory resources that fail to fit into the category of theory. At the same time, because reduction programmes tend to work in terms of the reduction of one kind of theory to a more 'fundamental' kind, the fact that theories do not remotely exhaust—and in some cases do not even capture—what physical enquiry and physical explanation consist in, means that reduction as traditionally conceived is simply not an appropriate or fruitful way to think about the relationship between the various forms of physical enquiry.

[1] Karl Popper, *The Logic of Scientific Discovery* (London, 1968), 107.
[2] David Kaiser, *Drawing Theories Apart: The Dispersion of Feynman Diagrams in Postwar Physics* (Chicago, 2005), 377.

Civilization and the Culture of Science: Science and the Shaping of Modernity, 1795–1935. Stephen Gaukroger, Oxford University Press (2020). © Stephen Gaukroger.
DOI: 10.1093/oso/9780198849070.001.0001

In what follows, we begin by considering experimental practice, showing how, from the seventeenth century onwards, there have always been forms of experimental investigation that have been independent of theoretical enquiry. I then turn to the role of experimental practices and experimental modelling in the development of the notion of energy, the most promoted candidate for a foundational concept in physics from the second half of the nineteenth century, and potentially the most powerful unifying notion in nineteenth- and early twentieth-century physics. We shall see that the problem that arises here is particularly intractable: namely, once one leaves the realm of purely mathematical axiomatic considerations, it is no longer clear that there are any plausible unification claims.

THE EXPERIMENTAL LEGACY

Experimental natural philosophy is a form of enquiry that can be traced back to Boyle's treatment of pneumatics, and Newton's treatment of the formation of spectral colours.[3] It is distinctive in that, unlike mechanism or mechanics, it is piecemeal, with localized procedures and results that often cannot be directly related to those in cognate enquiries. It deals with connections between the phenomena by relating these at the phenomenal level, rather than seeking to explain them in terms of something more fundamental. The most pressing task for defenders of experimental natural philosophy in the seventeenth and eighteenth centuries was establishing that experimental explanations could be standalone, because the problematic cases were those in which what was at issue was mechanist reduction. Many natural philosophers saw experimental explanations as being in competition with their own micro-corpuscularian ones, and concluded that the experimental ones were incomplete: they argued that, although the experiments may have identified new and interesting phenomena, they had not actually *explained* them because they had not identified the micro-corpuscular activity that generated them. Explanation for micro-corpuscularians was a matter of penetrating beyond the phenomena to another realm, so to speak. This had precedents in, and was indebted to, religious understanding,[4] not to mention philosophical understanding before Hume, but it was mapped on to science only at great cost, for it turned out to be a fruitless way of proceeding, as it became

---

[3] See Gaukroger, *The Emergence of a Scientific Culture*, ch. 10. Note that, contrary to much recent literature—see for example Peter Anstey, 'Experimental versus Speculative Natural Philosophy', in Peter Anstey and John Schuster, eds, *The Science of Nature in the Seventeenth Century: Patterns of Change in Early Modern Philosophy* (Dordrecht, 2005), 215–42—these developments cannot be traced back to Francis Bacon, who is doing something quite different and indeed antithetical to experimental natural philosophy: see Gaukroger, *The Emergence of a Scientific Culture*, 359–86. Cf. Lorraine Daston, 'The Empire of Observation, 1600-1800', in Lorraine Daston and Elizabeth Lunbeck, eds, *Histories of Scientific Observation* (Chicago, 2011), 81–113.

[4] One might even say that it was constitutive of a divine model of understanding. 'God sees not as man sees, for man looks at the outward appearance but God looks at the heart': 1 Samuel 16:7.

evident, in the eighteenth century, that the search for 'deeper' micro-corpuscularian explanations was an obstacle to investigation in a number of key areas, particularly electricity and chemistry.[5] The experimental philosophy approach is contrary to the idea that science proceeds by generating explanations theoretically and then testing them experimentally, and indeed is in some respects the opposite of this, for the direction of explanation is reversed, experimentation shaping and initiating the investigation.

Faraday was very much in the experimental natural philosophy tradition of Boyle and Newton, but particularly that of the early eighteenth-century electrical experimenter Stephen Gray. Gray's predecessor, the pioneer of electrical experiments, Hauksbee, had worked on the assumption that electricity was a property of bodies: some bodies were 'electrics' and their electrical power was released by rubbing, the friction generated imparting motion to particles making up a material 'effluvium' which was emitted from and returned to the electric. Accordingly, in his experiments Hauksbee had moved the electric around to test its effects on the various non-electrics, because it was the electric that was the causal agent, the non-electrics being on a par with passive recipients. The hierarchy was fixed in advance, as it were. Gray, by contrast, kept the electric fixed and devoted all his attention to the behaviour of the non-electrics. A network of effects was now the focus of attention, an 'electrical communication', not a centrally produced electrical 'vertue' which worked by means of attractions and repulsions. Gray had pared the discussion back to a very basic phenomenal level, not only discarding talk of causes, but also the Newtonian matter theory that had underlain the prevailing theory of electricity, and which required an effluvium because otherwise—since it was known since Boyle that electrical effects could occur in a vacuum—there would be action at a distance. Gray's experiments showed that electricity could not possibly be a matter of the emission of effluvia, and that it was a form of conduction.

Faraday's work on the interconvertibility of magnetic and electrical forces follows a similar, thoroughly experimental path. His interest was prompted by Ørsted's discovery in 1820 that, contrary to the prevailing view that electrical and magnetic forces were separate kinds of force, an electric current can cause a magnetic needle to move. But the way in which it moved was puzzling. It had been assumed that the magnetic effect would be along the direction of the axis of the current, whereas in fact the magnetic needle was deflected in opposite directions when placed above and below the wire carrying the electric current. Moreover, whereas in the case of static electricity like charges repel and unlike ones attract, Ørsted's experiments showed that, in the case of current electricity, like charges attract and unlike ones repel. Puzzled by this, Faraday designed a number of new experiments to vary features of the arrangement of the current-

⁵ See Gaukroger, *The Collapse of Mechanism*, chs 4 and 5.

carrying wire and the magnet. The earliest of these indicated that the wire had a tendency to rotate about the true pole of the magnet, independently of the influence of the pole at its other end, but the motion was hindered by the experimental apparatus, which he laboured to redesign so that the magnet was not in the path of the rotating wire, with the result that the wire and the pole rotated continuously about one another.

The converse effect—of an electric current on a magnet—was still obscure however. As Faraday put it, 'it appeared very extraordinary, that as every electric current was accompanied by a corresponding intensity of magnetic action at right angles to the current, good conductors of electricity, when placed within this sphere of action, should not have any current induced through them, or some sensible effect produced equivalent in force to such a current'.[6] He had hoped that, by placing magnets in various positions near current-carrying wires he could induce a current, but to no avail. In 1831, however, he finally had success. Based on his estimate of the most effective shape for a powerful electromagnet, Faraday had taken a thick iron split ring with wire wrapped around each of it halves, the coiling magnifying the magnetic field (Fig. 4.1). The wires at one end were attached to a battery (the primary circuit), those at the other to a galvanometer (the secondary circuit). Even though electric current could not flow directly between the two parts, when the battery was connected the ring became magnetized and the galvanometer registered a brief current, and when the circuit was broken, the ring demagnetized and another brief current was registered. His explanation for this behaviour was that the wire was put in what he considered to be a state of electrical 'tension'—what he called an 'electro-tonic' state—when in the magnetic field, and that the creation and dissolution of an electro-tonic state always resulted in a current.

But having established electromagnetic induction, Faraday's subsequent experiments led him to change his mind about the cause of the induction. Previously, he had maintained that only increase and decrease in the intensity of the electro-tonic state would cause induction, but a new experiment in which a copper disc is placed on a rotating magnet revealed that mere motion through an area of constant magnetic force could cause induction. It was in response to this that he introduced his theory of 'lines of force', arguing that the motion caused induction when the wire cut across lines of force, but if it merely moved along a line of force, no induction was produced. For Faraday electro-tonic states and lines of force had been alternative modes of representation of the same phenomenon: the rotating disc experiments indicated that the magnetism and the magnetic bar were independent of one another, so the electro-tonic state of the bar was replaced by the idea of 'magnetic curves' in the primary circuit. But by the 1850s, in accounting for the physical nature of the lines of force, he found

[6] Michael Faraday, *Experimental Researches in Electricity* (3 vols, London, 1839–55), i. 2.

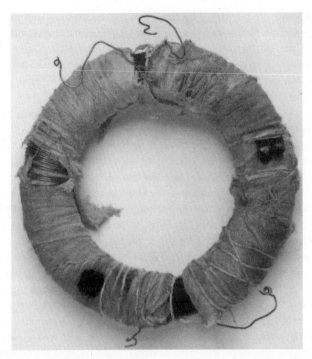

Fig. 4.1. Faraday's induction coil (1831)

himself having to reintroduce electro-tonic states: 'Again and again the idea of an *electro-tonic* state...has been forced on my mind: such a state would then constitute the physical lines of magnetic force.'[7] Combined with the revived electro-tonic conception, lines of force provided him with a model for a unified theory of static electricity, current electricity, and magnetism,[8] in which static electricity is a kind of tension, and current 'is a vibration when this tension is released'. He considered the mechanism of release of tension to be the same in both electrical and magnetic forces, though he was only able to establish it by analogy with static and dynamic electricities.

Faraday's experimental work remained at the centre of research in electromagnetism, but his attempts at rationalizing his experimental results in terms of electro-tonic states and lines of force were treated by his contemporaries as merely

---

[7] Faraday, *Experimental Researches*, iii. 420–1.

[8] See P. M. Harman, *The Natural Philosophy of James Clerk Maxwell* (Cambridge, 1998), ch. 4. A unified *mathematical* theory of static electricity, current electricity, and induction had been presented in 1846 by Wilhelm Weber in his *Elektrodynamische Maassbestimmungen* (Leipzig, 1846), on the assumption that electrical actions consisted in charged particles acting instantaneously through empty space.

speculative or illustrative devices. In his 1856 paper 'On Faraday's Lines of Force', Maxwell translated Faraday's ideas into a mathematical form, refining and transforming them in the process, but he also made clear that there was nothing 'mere' about illustrative devices.[9] Whereas Lagrange, in the Preface to the second edition of that bible of analytical mechanics, the *Mécanique analytique*, had boasted that no visual representations would be found in the work, 'but solely algebraic operations subject to a regular and uniform procedure',[10] Maxwell treats visual representations as crucial to the demonstration. He sets out a geometrical model of a field in which the direction of the forces is represented by lines of force filling space, and the intensity of the force by an incompressible fluid moving in tubes formed by lines of force. This geometrical model of fluid flow had no aspirations to be a physical model however: it was a mathematical analogy which—by contrast with analytical treatments—provided a visual representation of lines of force. His concern was with the spatial distribution of force in the field, and not with a physical representation of the field itself.[11]

Five years later, in his paper 'On Physical Lines of Force', Maxwell replaced lines of force by a comprehensive theory of the propagation of electrical and magnetic forces in terms of a stress in the electromagnetic medium.[12] The rotational momentum of force vortices is represented in strikingly visual terms as a honeycomb-like structure in which vortices rotate around parallel axes in the same direction (Fig. 4.2). The magnetic field itself is represented as a fluid filled with rotating vortex tubes, their geometrical arrangement corresponding to lines of force, and their angular velocities corresponding to the intensity of the field. Here the representation is no longer geometrical but physical: Maxwell considers it 'mechanically conceivable' but not a representation of something actually existing in nature. It is a heuristic model, and Maxwell uses it to present the discovery that light is a form of electromagnetic radiation.

In replacing the static lines of force, Maxwell had construed the electromagnetic medium as elastic, and this had allowed him to incorporate electrostatics into the model, the elastic stresses of the mechanical ether corresponding to the electrostatic field. When he calculated the velocity of the elastic waves corresponding to an electrical displacement in the electromagnetic medium, he found it to be the same as the velocity with which light waves were transmitted, which led him to conclude that the electromagnetic ether and the luminiferous ether were the same, although his way of representing the transmission made it difficult to

[9] James Clerk Maxwell, 'On Faraday's Lines of Force', *Transactions of the Cambridge Philosophical Society* 10 (1856), 27–83.

[10] Joseph Louis de Lagrange, *Mécanique analytique, nouvelle édition* (Paris, 1811), iii–iv.

[11] See Harman, *Energy, Force, and Matter*, 88.

[12] James Clerk Maxwell, 'On Physical Lines of Force', *Philosophical Magazine and Journal of Science* 21 (1861), 161–75, 281–91, 338–48; 22 (1861), 12–24, 85–95. There is an invaluable annotated guide to Maxwell's papers on electromagnetism in Thomas K. Simpson, *Maxwell on the Electromagnetic Field: A Guided Study* (New Brunswick, 1997).

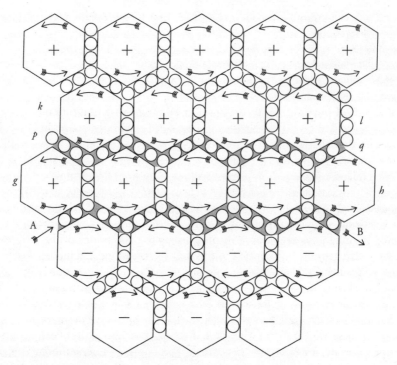

Fig. 4.2. Maxwell's representation of the electromagnetic medium

envisage, and his friend Monro wrote to him that 'a few such results are wanted before you can get people to think that, every time an electric current is produced, a little file of particles is squeezed along between a row of wheels'.[13]

Maxwell was by no means the first to notice that the speed of light and the speed of electrical propagation were the same. Gustav Kirchhoff, for example, in two 1857 papers on the conduction of electrical current through three-dimensional circuits,[14] had drawn attention to an analogy between the mathematics of electricity and that of wave propagation, noting that the wave moves 'with

[13] C. J. Monro to Maxwell, in Lewis Campbell and William Garnett, *The Life of James Clerk Maxwell. With a Selection from his Correspondence and Occasional Writings and a Sketch of his Contribution to Science* (London, 1882), 329–30. The problems were compounded by the fact that Maxwell's theory seemed to require that nothing flowed along conducting wires. This problem was largely resolved by Poynting in 1884, who, by means of what was to become known as the 'Poynting vector', was able to show how the energy of a battery was transmitted during conduction around an electrical circuit: John Henry Poynting, 'On the Transfer of Energy in the Electromagnetic Field', *Philosophical Transactions* 175 (1884), 343–61.

[14] Gustav Kirchhoff, 'Ueber die Bewegung der Elektricität in Drähen', and 'Ueber die Bewegung der Elektricität in Leitern', in Gustav Kirchhoff, *Gesammelte Abhandlungen* (Leipzig, 1882), 131–54, 154–68.

a speed very close to the speed of light in empty space'.[15] Wilhelm Weber, whose model of electrical currents Kirchhoff was developing, clearly found the prospect of a connection between light and electricity as developed by Kirchhoff interesting, but like Kirchhoff he did not take the coincidence of the measured value of the speed of light and calculated value of the speed of electricity to indicate such a connection.[16] Kirchhoff and Weber were no less concerned with the unification of sub-fields of physics than was Maxwell. But Kirchhoff's unification strategy, which was focused on the development of a unified electrodynamics, meant that the connections between light and electricity were not the crucial part of the unification exercise.

The projects of Maxwell and Kirchhoff were both genuine unification programmes in physics. The fact that they were directed towards different problems and accordingly led in different directions, and that Maxwell's came to marginalize Kirchhoff's achievements, should not lead us to think in terms of Kirchhoff 'failing' to unify physics. One of the problems with conceptions of the unity of science is that genuine unification projects, rather than acting as a model, tend to become subsumed under a wholly specious metascientific conception in which science is taken to be leading in a particular single direction. Unification projects that do not fit into this metascientific schema, which is guided by a teleological notion of convergence, are dismissed as unsuccessful and are forgotten, whereas in fact from these we can learn a great deal about how internal unification programmes operate and how they can actually lead in quite different directions at particular times. To the extent to which metascientific conceptions of the unification of science involve a selective identification of unification projects in physics as illustrations of their general thesis, they miss the opportunity to learn serious lessons from the development of the physical sciences.

Such lessons are evident in Maxwell, for his unification of optics and electromagnetism does not quite have the simple linear narrative it seems. Maxwell did not immediately abandon the hypothetical terms in which he discussed his ether, but with the rise of energy physics he rethought the nature of the ether that constituted the electromagnetic field in terms of a repository of energy, and this, by contrast with the mechanical model, he insisted must be construed literally. But this should not be taken as a commitment to a hierarchical notion of unification that we found in those concerned with mathematical reduction for example. Maxwell's attitude may be compared with that of his contemporary William Thomson (Lord Kelvin). Thomson captures a common nineteenth-century commitment to the unification of the natural sciences when he explains that 'a few simple, almost axiomatic principles, founded on our common

---

[15] Ibid., 146–7.
[16] Se the discussion in Kalil Swain Oldham, 'The Doctrine of Description: Gustav Kirchhoff, Classical Physics, and the "Purpose of All Science" in 19th-Century Germany', unpublished PhD dissertation, University of California Berkeley, 2008, 153.

experience of the effects of force, the general laws which regulate all phenomena, presented in any conceivable mechanical action, are established; and it is thus put within our power by a strict process of deductive reasoning to go back from these general laws to the actual results in particular cases of the operation of force'.[17] In a paper on the aims and progress of physical science of 1869, Helmholtz offers a similar assessment: 'The ultimate aim of physical science must be to determine the movements which are the real causes of all other phenomena and discover the motive powers upon which they depend; in other words, to merge itself into mechanics.'[18] But two years later, in 1871, Helmholtz is far more circumspect, declaring himself against 'the attempt to derive the foundations of theoretical physics from purely hypothetical assumptions concerning the atomic construction of physical bodies . . . . It has been realized that mathematical physics too is a purely empirical science; that it has no principles to follow other than those of experimental physics.'[19] Maxwell, who was equally at home with analytical models, geometrical lines of force models, and ingenious but contrived mechanical representations which strained the limits of constructability, was even more circumspect about the legitimacy of an axiomatic-deductive view of the physical sciences.[20] A remark in an 1856 paper is revealing in this respect:

Perhaps the 'book', as it has been called, of nature is regularly paged; if so, no doubt the introductory parts will explain those that follow, and the methods taught in the first chapters will be taken for granted and used as illustrations in the more advanced parts of the course; but if it is not a 'book' at all, but a magazine, nothing is more foolish to suppose than that one part can throw light on another.[21]

ENERGY AS A UNIFICATION STRATEGY

Maxwell's word of warning is apposite as we turn to the development of the notion of energy, the single most powerful unifying development in mid-nineteenth-century physics, with implications not only for the physical sciences

[17] Quoted in S. P. Thomson, *The Life of William Thomson, Baron Kelvin of Largs* (2 vols, London, 1919), i. 242.

[18] Hermann von Helmholtz, 'The Aim and Progress of Physical Science (1869)', in Hermann von Helmholtz, *Science and Culture* ed. David Cahan (Chicago, 1995), 204–25: 211.

[19] Herman von Helmholtz, 'Zum Gedächtniss an Gustav Magnus', *Vorträge und Reden* (3rd edition, 2 vols, Berlin, 1884), ii. 47.

[20] On his use of metaphors, analogies, and pictorial representations see Jordi Cat, 'On Understanding: Maxwell on the Methods of Illustration and Scientific Metaphor', *Studies in History and Philosophy of Modern Physics* 32 (2001), 395–441. See also the comparison of the extensive use of visual images in Faraday, Maxwell, Helmholtz, and Boltzmann in Arthur Miller, *Imagery in Scientific Thought* (Cambridge, MA, 1986).

[21] James Clerk Maxwell, 'Are There Real Analogies in Nature?', in Lewis Campbell and William Garnett, *The Life of James Clerk Maxwell, with a Selection from his Correspondence and Occasional Writings and a Sketch of his Contributions to Science* (London, 1882), 235–44: 243.

but also for the life sciences. In 1904, Merz makes the connection between the role of the concept of energy and the unification of the sciences in characteristically Whiggish terms:

The unification of scientific thought which was gained by any of these three views, the astronomical, the atomic, and the mechanical, was thus only partial. A more general term had to be found . . . which would give a still higher generalization, a more complete unification of knowledge. One of the principal performances of the second half of the nineteenth century has been to find this more general term, and to trace its all-pervading existence on a cosmical, a molar, and a molecular scale. It will be the object of this chapter to complete the survey of those sciences which deal with lifeless nature by tracing the growth and development of this greatest of all exact generalisations—the concept of energy.[22]

But rather than energy being the outcome of a concerted theoretical unifying programme, the emergence of the concept of energy took place, like that of electromagnetism, on a thoroughly experimental and somewhat more mundane basis.

Joule is an appropriate starting point. In 1843, his experiments establishing the equivalence of heat and mechanical work brought together mechanical and non-mechanical processes within the one schema.[23] These experiments were not the original form of Joule's thinking on these matters however. In the early 1840s, he had moved from the role of 'ingenious inventor' to that of 'experimental natural philosopher', and the transition illustrates the gap between a practical, and in an important respect extra-scientific, interest in machinery and a scientific one. His first publication, in 1838, was a proposal for an engine powered by electromagnetism, and appeared in a magazine edited by the popular electrical showman William Sturgeon.[24] Sturgeon was particularly interested in the economic advantages of an electromagnetic engine over a steam engine, and his approach to electromagnetism was at odds with that of scientists. As Crosbie Smith points out: 'Unlike Sturgeon's primary goals of improving display apparatus and electrical machines, elite London professors had built their credibility on a philosophical orientation towards experimental research: their concerns were primarily with the laws and causes of natural phenomena.'[25] Around 1838, Sturgeon moved from London to become superintendent of the Manchester Royal Victoria Gallery for the Encouragement and Illustration of Practical Science, which catered especially to electrical instrument and apparatus makers. He praised the machine proposed

[22] Merz, *A History of European Thought*, ii. 95–6.

[23] James Prescott Joule, 'On the Calorific Effects of Magneto-Electricity, and on the Mechanical Value of Heat', *Philosophical Magazine* 23 (1843), 263–76, 347–55, 435–43.

[24] James Prescott Joule, 'Description of an Electro-Magnetic Engine', *Annals of Electricity, Magnetism, and Chemistry; and Guardian of Experimental Science* 3 (1838), 122–3.

[25] Crosbie Smith, *The Science of Energy: A Cultural History of Energy Physics in Victorian Britain* (London, 1998), 55.

by Joule in terms of its 'ingenious arrangement of electro-magnets'.[26] What Sturgeon focuses on is the machinery as an ingenious invention, noting its potential as a source of motive power in locomotives and boats. But by 1839 Joule was beginning to dissociate himself from the role of inventor, and in subsequent communications to Sturgeon's *Annals* we find a shift, as Smith puts it, 'away from ingenious apparatus to experimental natural philosophy. Indeed, the whole mode of presentation had changed from that of personal letters to a series of consecutively numbered paragraphs . . . Faraday, not Sturgeon, had now become his role model.'[27] The difference was significant. As Morus notes, Sturgeon's 'mode of experimentation was to build new instruments and to progress by manipulating their components in order to maximize their efficiency. . . . This approach to experimentation was very different from that of Faraday, who went out of his way to present his results as disembodied facts, abstracted from the machines and the labor that had made them. He directed attention away from the material artifacts on his lecturing table because he argued that nature lay elsewhere.'[28]

The interest of this transition from 'ingenious inventor' to 'experimental natural philosopher' lies in the movement from an engagement with natural processes that might be described as that of an inventor to a theoretical one. The aim of the former is to produce an object—a working machine—not to produce a theory or a set of laws, or more generally to further general understanding of mechanical processes. Joule's attempts to transform the question of electromagnetic machines from one of invention to one of experimental science was something that involved not only the reformulation of the questions into a requisite form, but also the background and standing of the person making the experimental claims.[29] Joule devised and supported his theory that a portion of the heat contained in the steam expanding in the cylinder of a steam engine is converted into mechanical power on the basis of experiments on the heat produced by work done on compressing a gas,[30] where the crucial factor in these experiments is measurement, and the very exacting—and at the time

[26] William Sturgeon, 'Historical Sketch of the Rise and Progress of Electro-Magnetic Engines for Propelling Machinery', *Annals of Electricity, Magnetism, and Chemistry; and Guardian of Experimental Science* 3 (1838/9), 429–37.

[27] Smith, *The Science of Energy*, 58. Note that with Faraday as his model, Joule initially saw himself as a chemist, not a physicist. As John Forrester has noted: 'in the late 1840s Joule transferred his community allegiances in response to the physicists' excitement over his findings. Joule understood his work as chemistry, and chemists recognised its value for their science. But for physicists like Thomson, Joule's work was physics, and promised to create the foundations for a new science.' John Forrester, 'Chemistry and the Conservation of Energy: The Work of James Prescott Joule', *Studies in History and Philosophy of Science* 6 (1975), 273–313.

[28] Iwan Rhys Morus, *Frankenstein's Children: Electricity, Exhibition, and Experiment in Early-Nineteenth-Century London* (Princeton, 2014, 52.

[29] See ibid., ch. 4.

[30] James Prescott Joule, 'On the Changes of Temperature Produced by the Rarefaction and Condensation of Air', *Philosophical Magazine* 26 (1844), 369–83.

unique—thermometric skills that Joule deployed can be traced to procedures used in his family's brewing business. Indeed, unfamiliarity with these measuring techniques in the scientific community played a significant part in the delay in acceptance of Joule's experiments.[31]

Joule's is by no means an isolated case. A striking example of a similar reformulation is to be found in the transition from Carnot's original statement of the practical problem of improving heat engines to Thomson's formulation. Carnot's 1824 *Réflexions sur la puissance motrice du feu* were translated into a mathematically rigorous form ten years later by Émile Clapeyron.[32] It was through Clapeyron's version that William Thomson and his brother James came to develop Carnot's work, and in Clapeyron's version Carnot's references to the uses of heat engines to pump coal mines and to power steam ships are absent: the heat engine has become reduced to a set of questions of a scientific and mathematical nature.[33] Carnot asks: 'Can there be a limit to the improvement of the heat engine, which, by the very nature of things, cannot in any way be exceeded? Or, on the contrary, is it possible for improvement to proceed on indefinitely?'[34] Thomson reworks Carnot via Clapeyron, transforming these into two different questions: 'What is the precise nature of the thermal agency by means of which mechanical effect is to be produced, without effects of other kinds?' and 'How may the amount of this thermal agency necessary for performing a given quantity of work be estimated?'[35]

The question of the relation between invention and theoretical science is complex, and we will return to it in Chapter 10. Here my limited aim is twofold: to draw attention to the criteria for what was deemed to be science, by contrast with a process of invention; and to register the idea that rather than simply seeing technology as applied science, from the nineteenth century onwards there is a case to be made that science is in many respects best treated as theorized technology. Science does not generate its own problems but starts from technological problems and reworks them, attempting to give them a different kind of standing.

By 1843, once his transformation into an experimental natural philosopher was complete, Joule was attempting to establish the interconversion of electrical, chemical, and thermal phenomena, and this involved establishing their quantitative equivalence. To this end, he put his practical skills to use to construct a device (Fig. 4.3) in which mechanical work generated an electrical current, which in turn

[31] See Otto Sibum, 'Reworking the Mechanical Value of Heat: Instruments of Precision and Gestures of Accuracy in Early Victorian England', *Studies in History and Philosophy of Science* 26 (1994), 73–106.

[32] Emile Clapeyron, 'Mémoire sur la puissance motrice de la chaleur', *Journal de l'Ecole Royale Polytechnique* 22 (1834), 153–91.

[33] Smith, *The Science of Energy*, ch. 3.

[34] Sadi Carnot, *Réflexions sur la puissance motrice du feu et sur les machines propres à développer cette puissance* (Paris, 1824), 7.

[35] Talk to the Royal Society of Edinburgh, April 1849: William Thomson, *Mathematical and Physical Papers* (6 vols, Cambridge, 1882–1911), i. 114.

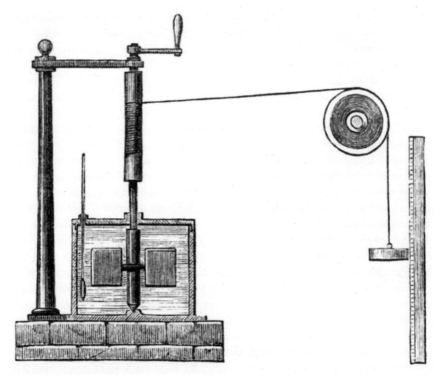

Fig. 4.3. Joule's apparatus to measure heat and mechanical work

generated heat, allowing the numerical relation between heat and mechanical work to be discerned. He inferred from the experiment that mechanical work could be transformed into an exactly equivalent amount of heat by friction, thereby giving a mechanical value of heat.

Joule concluded that heat was consumed in the generation of mechanical work, but Thomson worried that this seemed to contradict Carnot's theory of the heat engine, whereby work was generated by the passage of heat from a warm to a colder body, heat being conserved in the process (the warm body gets cooler but the cool body gets proportionately warmer). This led Thomson to question Joule's interconvertibility theory when he first encountered it.[36] In an 1850 paper Clausius had suggested how the two accounts could be reconciled, and this difficulty resolved, by abandoning the assumption that heat was conserved.[37] But the idea that heat could be lost absolutely seemed problematic to Thomson,

[36] See the account in Smith and Wise, *Energy and Empire*, i. 282–316.

[37] Rudolf Clausius, 'Über die bewegende Kraft der Wärme und die Gesetze, welche sich daraus für die Wärmelehre selbst ableiten lassen', *Poggendorffs Annalen* 79 (1850), 368–97, 500–24.

and in the following year he argued instead that it was dissipated. It was unrecoverable for the performance of work, but none could ever simply go out of existence.[38] The dissipation of heat was translated by Thomson into an instance of a general problem about the nature of irreversible processes. What occurred in these processes was that one form of energy was transformed into other forms, but, he argued, ultimately all forms of energy were forms of mechanical energy.

This thesis built on the pioneering work of Helmholtz, published four years earlier, in which he formulated a law in which light, electricity, heat, and magnetism were all subsumed under mechanical principles, specifically under the law of the conservation of energy.[39] A dozen or so researchers in Britain and on the continent were exploring the concept of energy at an advanced level mid-century,[40] and a number of scientists had suggested a principle of conservation of energy. The principle of conservation, as we now understand it, was initially devised in a physiological context, in Mayer and Helmholtz. Helmholtz's work began with an attempt to demonstrate that body heat and muscular action produced by animals could be derived from the physico-chemical process of oxidation of foodstuffs. In more general terms, what Helmholtz had done was to give a purely mechanical account of the energy expended in muscular actions, undermining the idea that there is a distinctively vital realm that operates via autonomous principles.[41] Yet Helmholtz's version, despite being proposed on a secure experimental basis and in a clear, mathematical form, initially met with resistance, and the leading physics research journal, *Annalen der Physik*, edited by Johann Poggendorff, had refused to publish the paper because of what the editor considered to be the philosophical style of argument of its preface.

Helmholtz's concept of energy provided an underlying basic physical concept that allowed the exploration of a number of physical questions that had previously been treated in discrete terms. It had been established, he writes, 'that all the forces of nature are measurable by the same mechanical standard, and that all pure motive forces are, as regards performance of work, equivalent. And thus one great step towards the solution of the comprehensive theoretical task of referring all natural phenomena to motion has been accomplished.'[42] This explicitly covers

---

[38] William Thomson, 'On the Dynamical Theory of Heat, with Numerical Results Deduced from Mr Joule's Equivalent of a Thermal Unit, and M. Regnault's Observations on Steam', *Transactions of the Royal Society of Edinburgh, 20 Part II (1851), 261–8, 289–98.*

[39] Published as *Über die Erhaltung der Kraft: eine physikalische Abhandlung* (Berlin, 1847).

[40] See Thomas Kuhn, 'Energy Conservation as an Example of Simultaneous Discovery', in M. Clagett, ed., *Critical Problems in the History of Science* (Madison, 1959), 321–56.

[41] Hermann von Helmholtz, 'Ueber den Stoffverbrauch in der Muskelaktion', *Archiv für Anatomie, Physiologie und Wissenschaftliche Medicin* (1845), 72–83.

[42] 'The Aims and Progress of Physical Science', in Helmholtz, *Science and Culture*, 214.

the living as well as the non-living: 'When we consider the work done by animals, we find the operation comparable in every respect with that of the steam-engine.'[43] Yet one gets the impression that Helmholtz hedges his bets on the question of reduction more than these quotes might indicate. Two years later, in a talk on the origins of the planetary system, he tells us that: 'The observer with a deaf ear only recognises the vibration of sound as long as it is visible and can be felt, bound up with heavy matter. Are our senses, in reference to life, like the deaf ear in this respect?'[44]

The central role of energy was explored in the context of the development of a systematic account of physics over a six-year period by Thomson and Peter Guthrie Tait, culminating in 1867 in their *Treatise on Natural Philosophy*,[45] subsequently dubbed the nineteenth-century *Principia*. It reworked analytical mechanics in such a way as to put a more physical interpretation on it, but above all made the principle of the conservation of energy the foundational concept of physics. The *Treatise* was organized comprehensively around the notion of energy, and everything else was either derived from it or at least offered as support for it. In his 1851 essay 'On the Dynamical Theory of Heat', Thomson had treated the development of the concept of energy as a collective and gradual one starting with Davy, proceeding through Joule and Mayer, and culminating in Helmholtz and Maxwell. But the linear narrative was one of Thomson's own devising. As Crosbie Smith notes, he neglects to mention his own relatively recent doubts about Davy, the fact that Mayer and Joule were largely ignored and even scorned by their respective German and English scientific peers, and that Thomson himself had taken four years to come around to accepting Joule's work.[46] It is only with the benefit of hindsight—and Thomson is an archetypal early Whig historian of science—that these largely autonomous developments can be seen as being part of a grand unification of science through the notion of energy. In fact, as Elkana has pointed out, if one looks at the developments surrounding the early history of the notion of conservation of energy, in the period 1840–60 what one finds is that quite different problems were exercising different groups of people in different places, and that accordingly they came up with different answers. The answers turned out to be related in certain respects, but the idea that they were all aiming at the same thing is quite mistaken: there was no uniform movement towards the principle of the conservation of energy.[47]

[43] Ibid., 216.   [44] Ibid., 276–7.

[45] William Thomson and Peter Guthrie Tait, *Treatise on Natural Philosophy* (Oxford, 1867). Four volumes were originally envisaged, but only the first ever appeared, and revisions and additions in subsequent editions did not add material from the proposed later volumes.

[46] Smith, *The Science of Energy*, 8–9.

[47] Yehuda Elkana, 'The Conservation of Energy: A Case of Simultaneous Discovery?', *Archives internationales d'histoire des sciences* 90.1 (1970), 31–60.

THE NATURE OF UNIFICATION

It is not just the reconstruction of the various developments into a single story that is questionable. The elaboration—in Helmholtz, Maxwell, and Thomson and Tait—of very general mechanical conceptions that underpin an interconvertibility of forces from virtually the whole of the physical sciences, can be understood in a number of different ways. One common way in which they have been taken is in terms of a unification of the physical sciences. But there are larger questions at issue here, for the unification of the physical sciences is really a prelude to a far more ambitious unification programme. Helmholtz's colleague du Bois-Reymond wrote to Carl Ludwig in January 1848, for example, that it was only because of Helmholtz's work that physics, 'having become a science, has received a goal', where the goal was the methodological basis of progress in science, something that would reveal its fundamental unity, namely the reduction of all phenomena, organic and inorganic, to theoretical mechanics.[48] Thirty-eight years later, Virchow was writing that it was through mechanics that 'we attained to intimately connecting the organism and the processes of life to knowledge of the physical and chemical things and processes, from which they differ only through the make-up and inner variety of the arrangements and effects'.[49] Nor did this fundamental unity stop at the life sciences. As Rabinbach has argued:

The image of the body as the site of energy conservation and conversion also helped propel the ambitious state-sponsored reform of late nineteenth- and early twentieth-century Europe.... The dynamic language of energy was also central to many utopian social and political ideologies of the early twentieth century.... [M]odern *productivism*—the belief that human society and nature are linked by the primacy and identity of all productive activity, whether of laborers, of machines, or of natural forces—first arose from the conceptual revolution ushered in by nineteenth-century scientific discoveries, especially thermodynamics.... Helmholtz, a pioneer of thermodynamics, argued that the forces of nature (mechanical, electrical, chemical, and so forth) are forms of a single, universal energy, or *Kraft*, that cannot be either added to or destroyed. As Helmholtz was aware, the breakthrough of thermodynamics had enormous social implications. In his popular lectures and writings he strikingly portrayed the movements of the planets, the forces of nature, the productive force of machines, and of course, human labor as examples of the principle of conservation of energy. The cosmos was essentially a system of production whose product was the universal *Kraft* necessary to power the engines of nature and society, a vast and protean reservoir of labor power awaiting its conversion to work.[50]

---

[48] See Anderton, 'The Limits of Science', 62–3. For a statement of du Bois-Reymond's position see his 'Über die Lebenskraft' (1848) in du Bois-Reymond, *Reden*, i. 1–26.

[49] Quoted in Anderton, 'The Limits of Science', 67–8.

[50] Anson Rabinbach, *The Human Motor: Energy, Fatigue, and the Origins of Modernity* (Berkeley, CA, 1992), 2–3.

This doesn't mean of course that the larger unification programmes follow on from the idea of the unification of the physical sciences, but it does mean that, given the ramifications, we need to be clear what the unification of the physical sciences would amount to, and whether 'unification' is the correct characterization of what is happening. On the first question, consider the case of energy. If energy did indeed secure a unification of the physical sciences—and we should note that the idea of energy as the basis of physics ('energetics') rapidly went into decline at the end of the nineteenth century, if only because of its inability to throw any light on electromagnetism[51]—what the unification consists in is not perhaps what all advocates of unification have in mind, and it is distinct from, and indeed in opposition to, earlier notions of the unification of science in terms of Newtonian central forces. In his attack on vitalism, du Bois-Reymond targeted the notion of a vital force unifying the whole of science. The focus of this attack was not so much the vital nature of this force as how force itself was conceived. Forces, he argued, do not exist as such, and in particular are not ontologically distinct causes of motion.[52] 'Ontologically distinct causes' were exactly what vitalists needed, and du Bois-Reymond characterized this as a mistaken attempt to emulate the terminology of physics.[53] But force and energy are abstract ideas, somewhat like that of money, which is only realized in concrete forms such as euros, dollars, pounds, etc.[54] There is no 'money' apart from these, and what we need to know is what each is worth in terms of the others. Similarly with energy: we need to know rates of exchange between the different forms, and this is what Joule, Mayer, Helmholtz, and Ostwald established. In other words, unification in terms of energy is a matter of intertranslatability,[55] not the identification of some single hypostatized power or force.

On the second issue, the question that needs to be raised is whether what is being presented as a programme of unification is in fact something different, simply the reconciliation of disparate doctrines. The physicist Paul Dirac, for example, suggested that the appropriate goal in making fundamental connections between disciplines in physics was that of removing inconsistencies, not attempting to unite theories that were previously disjoint. The former, he argued, led to

---

[51] See Russell McCormmach, 'H. A. Lorenz and the Electromagnetic View of Nature', *Isis* 61 (1970), 459–97; Christa Jungnickel and Russell McCormmach, *Intellectual Mastery of Nature: Theoretical Physics from Ohm to Einstein* (2 vols, Chicago, 1986) ii., 211–53.

[52] Du Bois-Reymond, 'Lebenskraft', *Reden*, i. 13.

[53] It was not only vitalists who sought to ontologize force/energy: Mayer, the co-discoverer of the conservation of force/energy was also inclined to do this. See Kenneth L. Caneva, *Robert Mayer and the Conservation of Energy* (Princeton, NJ, 1993), 25–33.

[54] I take the analogy from David Knight, *The Making of Modern Science* (Cambridge, 2009), 89.

[55] Note Joseph Larmor's remark in 1907 that, along with its establishment of inorganic evolution, the great achievement of the doctrine of energy is one of mensuration: it has 'furnished a standard of industrial values which has enabled mechanical power . . . to be measured with scientific precision as a commercial asset'. Joseph Larmore, 'Lord Kelvin', *Proceedings of the Royal Society* 81 (1908), iii–lxxvi: xxix.

brilliant successes, as in Maxwell's investigation of an inconsistency in the electromagnetic equations, Planck's resolution of inconsistencies in the theory of black-body radiation, and Einstein's resolution of inconsistencies between his theory of special relativity and the Newtonian theory of gravitation. By contrast, the 'top down' method of attempting to unify physical theories that were previously disjoint, he noted, had produced nothing of significance.[56] Of course, whatever the original motivation, removal of inconsistencies may in some cases effectively result in a unified body of theory (depending on how we think of unification), but the crucial point is that in others it may establish overlap rather than reduction—limited translatability for specific local purposes rather than anything universal.[57]

[56] Quoted in Ian Hacking, 'The Disunities of the Sciences', in Peter Galison and David J. Stump, eds, *The Disunity of Science* (Stanford, 1996), 37–74: 54.

[57] Such cases are explored, in the context of post-war microphysics, in Peter Galison, *Image and Logic: A Material Culture of Microphysics* (Chicago, 1997).

5

The Autonomy of the Material Sciences

The problems of unification are compounded when we turn from the physical sciences to the natural sciences more generally, as becomes evident if we concentrate on the next science down the hierarchy in the nineteenth-century unified-science models, namely chemistry. It is worth reminding ourselves from the outset that a hierarchy of the sciences based in mathematics or mechanics was not the only option on offer in the eighteenth and early nineteenth centuries. At the turn of the century, proponents of *Naturphilosophie*, for example, adopted a model of polarities that put the life sciences at the centre of the natural realm. Schelling explicitly subordinated chemistry to the life sciences, writing that 'the *purposeful formation* of animal matter...can be explained only by a principle that lies beyond the sphere of chemical processes', which are 'incomplete processes of organization'.[1] This was a standard theme of *Naturphilosophie*, with its leading representative after the death of Schelling, Ignaz Döllinger, announcing in an influential article that: 'One should not delude oneself into thinking that one could find the basis of life in the component parts that chemistry has been able to decipher.'[2] Ørsted for one was influenced by this tradition,[3] but chemists generally were unimpressed. For them, chemistry was autonomous with respect to both the physical and the life sciences. Gmelin's influential chemistry textbook of 1789, for example, explicitly promotes this autonomous standing of chemistry in its opening pages,[4] and twelve years earlier the chemist and naturalist Torbern Bergman had written:

The Science of Nature seems to have three degrees. The first fixes our attention to the outsides, and teaches us to collect external characters, in order to enable us to distinguish various natural bodies; and that this is the proper object of NATURAL HISTORY. If we penetrate still deeper by our contemplation, and examine the *general qualities* of matter (its

---

[1] Friedrich W. J. Schelling, *Von der Weltseele: Eine Hypothese der höheren Physik zur Erklärung des allgemeinen Organismus* (Hamburg, 1798), 215.
[2] Ignaz Döllinger, 'Über den jetzigen Zustand der Physiologie', *Jahrbücher der Medicin als Wissenschaft* 1 (1805), 119–42: 127.
[3] See Olaf Breidbach, 'The Culture of Science and Experiments in Jena Around 1800', in R. M. Brain, R. S. Cohen, and O. Knudsen, eds, *Hans Christian Ørsted and the Romantic Legacy in Science* (Dordrecht, 2007), 177–216.
[4] Johann Friedrich Gmelin, *Grundriß der allgemeinen Chemie zum Gebrauch bei Vorlesungen* (2 vols, Göttingen, 1789), i. 2–3.

Civilization and the Culture of Science: Science and the Shaping of Modernity, 1795–1935. Stephen Gaukroger, Oxford University Press (2020). © Stephen Gaukroger.
DOI: 10.1093/oso/9780198849070.001.0001

extension, impenetrability and *vis inertiæ*) in regard to its peculiar relations; it is that which is commonly called NATURAL PHILOSOPHY (*Physica*). But CHEMISTRY is the *innermost part*, since it examines the material elements, their mixtures, and proportions to one another. The first teaches us the elementary rudiments, the alphabet of the great book of nature; the second instructs us in spelling; and the third, to read distinctly. The two first therefore are no more than subsidiary sciences, which conduct us to the last, as the proper great subject.[5]

Chemists like Bergman and Gmelin were concerned with the material world, by contrast with the physical world. The importance of the distinction goes back to the problematic attempts of seventeenth- and early eighteenth-century mechanists to account for that behaviour that we would now label chemical in terms of the mechanically described behaviour of postulated micro-corpuscles. The problems are if anything more severe when we turn to the analytical mechanics of the late eighteenth and nineteenth centuries. What is at issue is whether material properties and physical properties are intertranslatable, in particular whether the material properties (we are confining our attention to non-living things here) can be given a full physical description.

THE PHYSICAL AND THE MATERIAL

The basic difficulty in thinking of chemistry in terms of the resources of physics was that the kinds of physical entities with which mechanics works are quite distinct from material substances. Mass points do not behave remotely like the material components of things, for example,[6] and provide no basis on which to account for the behaviour of substances of interest to chemists: the persistent identity of certain substances, why their properties are typically quite different from any of their components, why they cannot be broken up by mechanical means into their original components, why substances combine in definite simple proportions, or even how they can combine in the first place. The nature of reactivity was probably the most central concern of chemistry, yet it was something that atomism was unable even to address. For those who thought that mechanics, of one kind or another, was the ultimate or only real science, anything that fell outside its purview was dismissed as mere classification, awaiting its eventual and inevitable transformation into a fully mathematized physical science.

---

[5] Carl Wilhelm Scheele, *Chemical Observations and Experiments on Air and Fire . . . with a Prefatory Introduction by Torbern Bergman, translated from the German by J. R. Forster* (London, 1780), xv–xvi. Originally published as *Chemische Abhandlung von der Luft und dem Feuer* (Uppsala and Leipzig, 1777).

[6] This is the problem that Newton encountered (in a non-chemical context) in Book II of the *Principia*, where he starts with a definition of a fluid in terms of mass points, but halfway through Book II he suddenly shifts to an experimental and purely phenomenological account in terms of the ease with which a substance yields to a force. See Gaukroger, *The Collapse of Mechanism*, 72–6.

But the problem with this approach was that, by the second half of the eighteenth century, mechanics was not getting very far at all in terms of producing new theories of natural phenomena, whereas chemistry, flourishing as an autonomous discipline, was very successful, to such an extent that it was starting to be taken as a model for some of the physical sciences. In 1789, Johann Samuel Gehler was writing that attempts at mechanical explanations of heat were fruitless, for example, and that if one wanted 'to correctly order these phenomena and bring them under specific laws, one had to speak the language of the chemist'.[7] The credentials of mechanics, by comparison with chemistry, did not exactly put it in a superior position, and it was not as if it was offering a plausible alternative account of material properties.[8] Consequently, as we are about to see, the problem of unification as it affected chemistry was less one of unification with the physical sciences, but rather one of the unification of chemistry itself, on a par with the question of the internal unification of the physical sciences, where the problems were equally complex and where just as much was at stake.

Chemistry in eighteenth-century France had begun, briefly, with an atomist agenda.[9] Fontenelle's reorganization of the Académie des Sciences in 1699 had meant that the theme of a unified natural philosophy along Cartesian lines became very much a general desideratum, but in the experimentalist tradition of Homberg and Lémery in the first decades of the eighteenth century such considerations were subservient to practical concerns. The move to remove chemistry altogether from the domain of a supposed unified science, in which it was subservient to atomist explanations, and to establish it as an autonomous area of study, effectively began in 1718, when Étienne Geoffroy published the first version of his table of chemical affinities, that is, a table indicating, in the form of a ranking, the degrees to which different substances are found to combine with and displace one another.[10] As Klein and Lefèvre note in their detailed account of how eighteenth-century chemistry was actually pursued: 'As to eighteenth-century chemists' ways of identifying and classifying substances and their writing histories of substances, atomism and corpuscularianism were totally irrelevant.'[11] Micro-corpuscularianism was a dead end, just as ancient atomism had been.

---

[7] Johann Samuel Gehler, *Physikalisches Wörterbuch* (5 vols, Leipzig, 1787–92), iv. 549. Cited in Peter Hanns Reill, *Vitalizing Nature in the Enlightenment* (Berkeley, 2005), 72.

[8] For discussion of such an implausible attempt, see Stephen Gaukroger, 'Kant and the Nature of Matter: Mechanics, Chemistry, and the Life Sciences', *Studies in History and Philosophy of Science* 58 (2016), 108–14.

[9] See Mi Gyung Kim, *Affinity, That Elusive Dream: A Genealogy of the Chemical Revolution* (Cambridge, MA, 2003), ch. 2.

[10] See Gaukroger, *The Collapse of Mechanism*, 206–17, and *The Natural and the Human*, 78–101. But cf. Victor Boantza, *Matter and Method in the Long Chemical Revolution: Laws of Another Order* (Farnham, 2013), who shows that there was nevertheless an earlier strong resistance to corpuscularianism in writers such as Du Clos.

[11] Ursula Klein and Wolfgang Lefèvre, *Materials in Eighteenth-Century Science: A Historical Ontology* (Cambridge, MA, 2007), 62.

In the absence of the classificatory strategies of atomism, chemists returned to two perennial sets of categories. Not only did French chemistry textbooks organize their material around the division into mineral, vegetable, and animal (a classification central to Linnaeus' division of the 'empire of nature' into three 'kingdoms') for example, but, more importantly, the traditional categories of earth, air, fire, and water still played a role, and there was a fundamental assumption that any comprehensive understanding of the natural world had to account for the four basic types of substance. When, towards the end of the eighteenth century, there was a move away from considering the traditional elements as components of matter responsible for the properties of bodies, to considering them rather as vehicles or instruments of chemical change, there nevertheless remained four basic types, though now they began to be conceived not as different substances but as different states of substances corresponding to the traditional elements: solids (earth), liquids (water), gases (air), and combustion (fire).

An equally distinctive feature of chemistry was its thoroughly experimental nature. This is a characteristic that stood at the forefront of eighteenth-century chemistry, but it was also the feature of chemistry that marked it out from analytical mechanics in the first half of the nineteenth century. As Nye notes: 'At the very beginning of the nineteenth century, *Physik* was taught in the German university philosophical faculty, and *Chemie* in the medical faculty . . . . . At this time the idea of a physics laboratory hardly existed; the word *laboratory* implied research in chemistry.'[12] This character of chemistry as an empirical discipline was set from the 1720s onwards, as attention came to be directed exclusively to the phenomenal level of chemical behaviour. In the long entry on chemistry in the *Encyclopédie*, a sharp contrast was drawn between the uncertain and speculative nature of basic 'physics' (i.e. mechanics) and the careful experimental and observational results of chemistry.[13] But it is not just the claims to evidential certainty that are of interest here, it is the way in which experiment provides the resources for the conceptual framing of the discipline, indeed provides a set of resources in direct competition with those of mechanics. This is nowhere more evident than in a fundamental difference in what is taken to be the primitive state of matter for the purposes of analysis: solids or liquids. In Euler, for example, one starts from mass points, builds up the mathematical resources that take one to rigid bodies, then builds up new mathematical resources to take one from rigid bodies to flexible bodies, repeating the process to take one to elastic bodies, and finally to fluids. Solids provide the starting point for understanding matter. One moves from mass points to rigid bodies, and finally on to fluids, not directly from mass points to elastic bodies or fluids for example. With the advent of field theory this process becomes a little murkier, at least if one assumes that the physically describable

---

[12] Nye, *Before Big Science*, 9.
[13] 'Chimie', in Diderot and d'Alembert, *Encyclopédie*, viii. 12–63.

states such as rigid bodies and fluids correspond to material states such as solids and liquids, and perhaps gases. But such correlations are largely lost in field theory: the visualizing models might work in terms of flexible tubes and incompressible liquids, but those models are not designed to capture the differences between real material states.

In chemistry, by contrast, the paradigm form of material substances was liquids, not solids. This was because the operational path whereby the properties of matter were identified were thought of in terms of the action of fluids. There was no theoretical reason within eighteenth-century chemistry why the focus should be on fluids, as there had been a theoretical reason in eighteenth-century mechanics to take solid bodies as the primary form of matter for example. Eighteenth-century chemists did not start from a conception of what the ultimate constituents of substances must be and try to account for the properties of these substances in terms of the pre-identified constituents. They moved in the opposite direction, taking chemically characterized substances and asking how they could be treated, for example through heating or by being dissolved, so as to yield constituents. One key constraint in chemistry was purely instrumental, being determined by contingencies such as the forms of instrumentation available (furnaces and glassware). There were two historical areas, deriving from the sixteenth and early seventeenth centuries, that were formative here: extraction of precious metals from ores, which was pursued through the use of solvents, and the preparation of pharmaceutical products from plants, which employed distillation. To discover or isolate the constituents of something, one did not physically break it down into its homogeneous atoms, but rather began by liquefying it, dissolving it in an acid or alkali: its properties could be revealed only once it was in a liquid form. Alternatively, one could heat the substance to be analysed, breaking it down in this way, but for this to be instructive from the point of view of analysis, the nature of heating had to be understood. The models that were developed to understand the action of heating in chemistry, by contrast with the primitive mechanist understanding in terms of increased motion of atoms, worked in terms of fluids: heat was conceived along the lines of a fluid; or heating was thought of in terms of matter being energized by an active subtle fluid; or it was thought of as flowing between bodies due to the pull of attractive forces; and so on.

In the course of the nineteenth century, there was a gradual move from liquids to vapours as the analytical paradigm of substances. In 1826 Jean-Baptiste Dumas had shown how to measure vapour densities and use these to determine atomic weights,[14] and the use of this method had become so fundamental that a French chemist in the late nineteenth century was recorded as telling his students that

---

[14] Jean-Baptiste Dumas, 'Memoire sur quelques Points de la Théorie atomistique', *Journal de Chimie Physique* 33 (1826), 337–91; Jean-Baptiste Dumas, 'Dissertation sur la Densité de la Vapeur de quelques corps simples', *Journal de Chimie Physique* 50 (1832), 170–8.

'bodies which are not volatile do not have molecular weight'.[15] Just as in the eighteenth century, getting substances into a liquid form was a prelude to analysis, so in the nineteenth century getting them into a vapour form gradually began to play this role, and in the twentieth century getting them into a crystalline form was the key to analysis. In each case, this was for purely instrumental reasons that had nothing to do with any intrinsic chemical interest that liquids, vapours, or crystals might have had, but was due rather to the purely contingent development of techniques that turned out to be revelatory when applied to the questions of chemical structure that concerned chemists. Such instrumental constraints are not peculiar to chemistry, and in particular do not mark it out from the physical sciences: different, but nevertheless instrumental, constraints are very evident in experimental disciplines such as the study of electricity. In fact, in one sense it is misleading to refer to these as 'constraints', if this is taken to suggest that their contingency indicates that they are added extras which can be abstracted out, in order that one might arrive at the conceptual core. This translation of scientific enquiry into a philosophical—purely conceptual—form is misguided if only because the instrumental 'constraints' provide the means by which results are arrived at: the material steps in the argument as it were. The legitimacy of the result depends on the choice of (contingent) resources, and the ways in which they are used. If these are jettisoned so is the rationale for the result.[16]

CHEMICAL STRUCTURE

As well as the material 'constraints' on chemical investigation, there was also a conceptual one, that of what one wanted the chemical process to yield, and this gave rise to an operational partitioning of the subject matter of chemistry into two: mineral chemistry (inorganic chemistry) and carbon chemistry (organic chemistry). The analysis of organic matter was pursued via destructive distillation, whereby the 'essence' (active ingredients) of the substance could be isolated. Given the pharmaceutical importance of recovering the essence of a substance in the production of medicines, what mattered was the end product, and it was irrelevant that the original material was lost in the process. Analysis via solvents, by contrast, was developed for extracting minerals in metallurgy, and it was not a question of extracting the essence and simply discarding everything else, but rather one of dissolving substances and recovering of substances from solution.

---

[15] Quoted in Nye, *Before Big Science*, 40.

[16] The jettisoning of stages in a procedure on the assumption that this enables one to get to the result unencumbered by how one arrived there, is mirrored in seventeenth- and eighteenth-century (mis)understandings of logic, notoriously in Descartes, who believed that, having reached a conclusion by deductive means, one should jettison the chain of argument in order to grasp the connection between the premises and the conclusion in an unmediated way. See Stephen Gaukroger, *Cartesian Logic: Descartes' Conception of Inference* (Oxford, 1989).

For this, reversible processes were of paramount importance. Substances were treated through heating or by being dissolved so as to yield constituents and then these same processes were used to resynthesize substances from the separated components.[17]

The partitioning into mineral/inorganic and carbon/organic chemistry was the formative feature of nineteenth-century chemistry. What chemists learned from the two about the nature and structure of material substances was quite different. In very abstract terms, it would seem that inorganic chemistry must provide the foundations for a more general chemistry because it identifies the elements from which substances are made—the ultimate chemical constituents of all substances—and investigates how they combine with and displace one another. Organic chemistry, by contrast, is just concerned with substances containing one of these elements, carbon, and these are almost exclusively limited to carbon compounds containing one or more of three other elements: hydrogen, oxygen, and nitrogen. In short, it looks like inorganic chemistry is chemistry per se, and that organic chemistry is really little more than a particularly restricted sub-speciality of chemistry. But there is a crucial feature of organic chemistry that propelled it to the forefront of theoretical chemistry in the nineteenth century. As far as the kinds of reactions that are investigated in inorganic chemistry are concerned, for the most part one can get by perfectly well by providing what are in effect recipes that tell you how to identify and mix ingredients: how much of which elements a compound contains (ideally a simple ratio), and how much of each element one has to combine to produce a new compound. But the ingredients are very limited in organic compounds, and many different compounds have exactly the same ingredients in exactly the same proportions. What organic chemistry requires, therefore, is an account of the internal structure of the compound, since that is where the difference lies. It's not that inorganic compounds don't have an internal structure: it is rather that one didn't have to know what this internal structure was for the questions of concern to inorganic chemists. Organic chemistry, by contrast, cannot get off the ground without an account of structure, and because it was structure that revealed the key to how substances react with one another, it provided the resources for a general account of chemistry, one that ultimately allowed connections to be made with the physical sciences.

From the early decades of the eighteenth century, when the failure of mechanism to provide foundations for chemistry became evident, it had a strong degree of autonomy with respect to the physical sciences, but nineteenth-century developments in both inorganic and organic chemistry subsequently allowed chemists to propose connections between the chemical and physical sciences. As far as inorganic chemistry and its relation to the physical sciences is concerned, there are

---

[17] See Klein and Lefèvre, *Materials in Eighteenth-Century Science.*

two programmes of note in the early decades of the century: Berthollet's attempt to model chemical reactions in mechanical terms, and Dalton's attempt to construe chemical elements in terms of hard spherical atoms.

Berthollet begins his 1803 *Essai de statique chimique* with the claim that the forces that bring about chemical phenomena 'all derive from the mutual attraction between the molecules of bodies. The name affinity has been given to this attraction so as to distinguish it from astronomical attraction. It is probable that both are one and the same property'.[18] He proceeded to study the effect of a number of varying factors—particular reacting masses, heat, solutions, and cohesive forces—on the course of the reaction, in an attempt to establish a parallel with astronomy. This approach was central to what can be identified as a Laplacean programme, which dominated French physical theory from the beginning of the century until the 1820s,[19] and Laplace himself had suggested as early as 1796 that chemical reactions were the result of attractive forces between the 'molecules'.[20] Forces were treated by analogy with the Newtonian account of gravitation—something considered to be in no need of further explanation and which could be treated in mechanical terms—although both positive and negative forces were allowed, and they did not always have to obey a strict inverse square law. The approach was bolstered by an ingenious 1806 paper by Biot and Arago in which they proposed that experimental investigation of optical refraction could bear directly on the forces that governed the course of chemical reactions—the latter being a phenomenon that was experimentally intractable—because the short-range forces that caused the refraction of light were also the forces of chemical affinity.[21]

The Laplacean programme was coming under heavy criticism after 1815, however, notably in the hands of Fourier, Fresnel, and others, and it did not survive this criticism. Berthollet's rigorous analysis only served to illustrate, as

[18] Claude Louis Berthollet, *Essai de statique chimique* (2 vols, Paris, 1803), 1. Note that 'molecules' here are just smallest parts.

[19] See Robert Fox, 'The Rise and Fall of Laplacian Physics', *Historical Studies in the Physical Sciences* 4 (1974), 89–136.

[20] Pierre-Simon Laplace, *Exposition du système du monde* (2 vols, Paris, 1796), ii. 196–8.

[21] Jean-Baptiste Biot and François Arago, 'Mémoires sur les affinités des corps pour la lumière, et particulièrement sur les forces réfringentes des différens gaz', *Mémoires de la Classe des Science Mathématiques et Physiques de l'Institut National de la France* 7 (1806), 301–87: 327–33. A brief explanation may be helpful here as it is still often assumed that refraction is a purely physical phenomenon. Up to the 1750s it had been assumed that refraction was due to the 'optical density' of the medium, a physical property, and the aim was to overcome the chromatic aberration produced in lenses by the divergence of different coloured rays, by using lenses composed of glasses with different refractive indices, which were glued together. To this end, Newton and then Euler had proposed physical formulas describing how homogeneous light rays passing through one medium (air) to an optically denser medium (glass) would diverge. But when Dollond produced the first compound achromatic lens in 1757, it gradually became clear that it was not just a matter of physics, but involved the chemical structure of the different glasses (see the summary account in Gaukroger, *The Collapse of Mechanism*, 325–6). Biot and Arago are exploring the consequences of this chemical understanding.

Thackray puts it, 'the profound divide between the comparative simplicity and order displayed by astronomical phenomena and the ceaseless variety and change with which chemistry must deal'.[22] At the same time, Berthollet's chemistry fell victim to Dalton's much simpler atomic theory. Dalton had been working on an atomistic theory from around 1803,[23] in the context of his theory of gases, but his first comprehensive account was in his *New System of Chemical Philosophy* of 1810.[24] Elements had been understood throughout the eighteenth century to be the isolable components of bodies that had resisted all attempts at further analysis. In particular, they were not treated as the postulated ultimate constituents of bodies. In his determination of the weights of elements, Dalton reintroduced atoms, and he worked with two rules.

The first was an extrapolation from experiments: the law of multiple proportions, which stated that elements combine in definite and constant proportions, which are in a ratio of small whole numbers. The law of simple proportions underlay Dalton's theory, but Berthollet's work indicated that variations were possible depending on the amounts of the initial reactants. Compared to Berthollet's approach, Dalton's extrapolation was far too simple and abstract, ignoring the concerns of a number of earlier writers on chemistry about the effect of initial conditions on chemical reactions. Bergman, for example, had been particularly alert to the role of the environmental conditions in which a chemical reaction takes place, noting that it can alter and direct the specific form of the product, and Lavoisier was well aware that reactions could proceed in steps of saturation levels, each step producing different qualities. Berthollet gave the question of the conditions under which reactions took place a central role in understanding chemical processes. Dalton's and Berthollet's approaches could not have been more different. As Reill notes,

Berthollet's approach virtually transformed chemistry into a form of natural history, whereby the chemist had to deal with each reaction both as a type, illustrating a general law, and as a specific event in which individual conditions play a major role. Not only was the chemical universe made up of many unique individuals whose number was being multiplied enormously through chemical research, but the individuals themselves were influenced directly by the conditions under which they were formed and existed. Thus, as

---

[22] Arnold Thackray, *Atoms and Powers: An Essay on Newtonian Matter-Theory and the Development of Chemistry* (Cambridge, MA, 1970), 232.

[23] In 1805 he writes that he has 'lately been prosecuting an enquiry' into 'the relative weights of the ultimate particles of bodies': John Dalton, 'Absorption of Gases by Water and Other Liquids', *Memoirs of the Literary and Philosophical Society of Manchester*, 2nd series, 1 (1805), 271–87: 286. Dalton's 1803 notebooks are now lost but were seen in 1896: Henry E. Roscoe and Arthur Harden, *A New View of the Origins of Dalton's Atomic Theory* (London, 1896), 26–9.

[24] John Dalton, *A New System of Chemical Philosophy* (3 vols, Manchester, 1808–27). Note that, unlike Epicurean atoms, Dalton's atoms are each surrounded by a cloud of caloric, which is added to as the substance is heated, causing expansion. The caloric theory was, as his contemporaries realized, one of the weakest points in his account: see Robert Fox, 'Dalton's Caloric Theory', in D. S. L. Cardwell, ed., *John Dalton and the Progress of Science* (Manchester, 1968), 187–201.

in natural history, chemical genera, species, and varieties could be grasped only through a procedure combining diachronic and synchronic understanding.[25]

Although the law of simple proportions was generally considered by Dalton's contemporaries to have been established beyond doubt by Berzelius in 1811, there remained an issue about just how simple simple proportions were. As an indication of this, note that chemists now distinguish between 'berthollides' and 'daltonides'. The former comprise nonstoichiometric compounds, above all those common solid inorganic compounds—principally metal oxides and sulphides— in which the number of atoms in the elements present cannot be expressed as a ratio of small whole numbers, because a few atoms are typically missing or packed too tightly into a lattice. Ferrous oxide, for example, is usually designated FeO, but it is actually closer to $Fe_{0.95}O$. Elementary inorganic chemistry could proceed in abstract terms without probing the structural complexity of the arrangement of molecules, but once structural questions began to be raised, a whole host of complicating factors arose, and these mean that the law of simple proportions has to be abandoned as an account of the macroscopic properties of molecules. In short, despite Dalton's confidence, there remained an issue about just how simple simple proportions were: the question depended on a number of factors such as what state the substance was in, what quantity it was in, how the atoms or molecules in the sample were packed, and more generally in what physical and chemical environment—levels of salinity, hydration, temperature, pressure— reactions took place.

The second rule is that of simplicity, stating that when two elements form only one compound, the compound is binary, i.e. it contains a single atom of each. Dalton's justification for this seemingly arbitrary rule was that the most stable arrangements are those which minimize the proximity of atoms.[26] If the atoms form two compounds, however, then one is binary and the other ternary, i.e. two atoms of one and one of the other, or vice versa (see Fig. 5.1). And so on. We are given no idea of what holds the atoms in compounds together,[27] but Dalton did consider that there is a repulsive force between like atoms, but nothing, either repulsive or attractive, between unlike ones. The repulsive force sets a limit on the number of like atoms that can exist in a compound, and would evidently prevent the existence of groups of like atoms forming alone.

The crucial move was Dalton's application of these rules to the discovery of atomic weights. Taking hydrogen to be the lightest element, it was given a

---

[25] Reill, *Vitalizing Nature*, 87. Cf. Kim, *Affinity*, 439–55.

[26] John Dalton, 'Dr. Bostock's Review of the Atomic Principles of Philosophy', *Journal of Natural Philosophy, Chemistry and the Arts* 29 (1811), 143–51.

[27] See Paul Needham, 'Has Daltonian Atomism Provided Chemistry with any Explanations?', *Philosophy of Science* 71 (2004), 1038–48; and Alan Chalmers, *The Scientist's Atom and the Philosopher's Stone: How Science Succeeded and Philosophy Failed to Gain Knowledge of Atoms* (Dordrecht, 2011), 177–82.

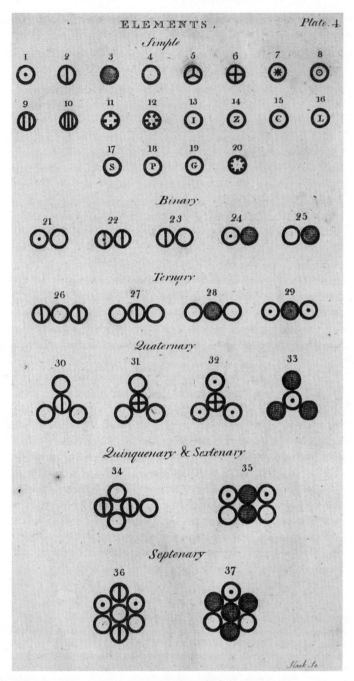

Fig. 5.1. Dalton's atoms, *A New Chemical Philosophy* (1810)

notional weight of 1. The rule of simplicity dictated that water was composed of one atom of hydrogen and one atom of oxygen, and experiment revealed that it contained 85 per cent oxygen and 15 per cent hydrogen by weight, so the atomic weight of oxygen was 5.66 times that of hydrogen. In this way, he was able to set out a table of atomic weights.

The atomist interpretation was part of a corpuscularian programme associated with Newtonianism in Britain, and as such it offered a bridge between basic chemical theory and physical corpuscularianism, albeit one very different from that of the Laplacean programme. Dalton's atoms were indivisible, spherical, and, although his notation represented them as circles of equal size, they differed in bulk and in weight, the latter in proportion to the former.[28] But even though atomism may look like an intuitively appealing way of visualizing the kind of physical structure that made sense of the rule of multiple proportions and that of simplicity, Dalton's contemporaries and successors, such as Woolaston, Davy, and Prout, did not accept it, considering the choice of one atomic weight over another equally probable one arbitrary and unnecessary.[29] For practical purposes, such as the all-important calculation of equivalent weights, they did not need it: the ontologically non-committal language of 'equivalents' accomplished everything that the language of atomism did. Nevertheless, the atomic hypothesis did have some support and it was open to development, although it was to prove problematic in a number of respects. We can confine our attention to two issues: the conflict with the results of experimental work on combining volumes of gases; and the implications of particular chemical notations, a question that goes to the heart of why the resources available in inorganic chemistry are inadequate for, and can act as an obstacle to, understanding chemical structure.

On the first question, in Dalton's model of gases, weights and volumes of gases are in proportion to their atomic weights.[30] His was a chemistry of combining

---

[28] Contrary to traditional atomism, the indivisibility of Dalton's atoms was purely physical, not conceptual, since an oxygen atom, for example, on his reckoning would presumably be about six times the size of a hydrogen atom, so it was conceptually divisible into six. Mere physical indivisibility is problematic, however. As Dumas, anticipating a distinction between physical and chemical atomism, asked in 1836, what possible difference could it make to the facts of chemistry if elementary chemical masses were able to be 'cut up infinitely by forces independent from chemistry'? Jean-Baptiste Dumas, *Leçons sur la philosophie chimique* (Paris, 1837), 233–4. Cited in Mary Jo Nye, *From Chemical Philosophy to Theoretical Chemistry: Dynamics of Matter and Dynamics of Disciplines, 1800–1950* (Berkeley, 1993), 65.

[29] On the nineteenth-century atomism debates, see William H. Brock and David M. Knight, 'The Atomic Debates: "Memorable Evenings in the Life of the Chemical Society"', *Isis* 56 (1965), 5–25.

[30] Dalton, *A New System of Chemical Philosophy*, i. 163. This was quickly questioned, however. In 1809, Gay-Lussac, on the basis of experiments with Humboldt on the combustion of hydrogen and oxygen, published a paper showing that volume ratios in gaseous reactions are whole numbers: Joseph-Louis Gay-Lussac, 'Mémoire sur la combinaison des substances gazeuses, les unes avec les autres', *Mémoires de physique et de chimie de la Société d'Arcueil*, 2 (1809), 207–35. This indicated that, at equal pressure and temperature, equal volumes of gases contain the same number of particles. The result was reinforced by Avogadro in an 1810 paper, who showed that the number of particles in

weights, which took him to a theory of the weights of what he considered to be the combining atoms. Note however that what is at issue here is chemical atomism, by contrast with physical atomism.[31] There is no question of Dalton's atoms being the kind of materialized mass points that some physicists, for example, were proposing. Reduction of chemistry to physics was not on the agenda, and if it had been, Dalton's atoms would not have provided a remotely plausible route. On the second question, Dalton's atomistic explanation of the law of constant proportions acted as a significant obstacle to further progress, because it gave the impression that chemical activity was a function of the quantities of different atoms, much as cooking is a function of the quantities of the different ingredients. His representation of the elements in the form of solid atoms was meant to picture microscopic reality, and hence we would expect his reaction to forms of representation in which letters stood in for elements to be hostile. In 1814 Berzelius proposed such a notation, with letters for elements, such as O for oxygen, and combinations of these for compounds, such as CuO for copper oxide and H^2O for water (Fig. 5.2). Dalton rejected these out of hand, not least in the meetings of Section B of the British Association for the Advancement of Science. A report on the Fifth Meeting of the Association in 1835 notes 'Dr. Dalton depreciating, in the strongest manner, the "ridiculous and absurd" system in use among continental writers, and prophesying the downfall of the Association, if it sanctioned a system so much calculated to interfere with the advancement of knowledge'.[32] Two years later, in rejecting a paper for publication which used, and seemingly because it used, Berzelian notation, he wrote that

Berzelius' symbols are horrifying: a young student in chemistry might as soon learn Hebrew as make himself acquainted with them. They appear like a chaos of atoms. Why not put them together in some sort of order? [They] equally perplex the adepts of science, discourage the learner, as well as to cloud the beauty and simplicity of the Atomic Theory.[33]

Note that Berzelius did not present his account as a substantive alternative to Dalton's atomist theory, and his examination of combining weights and volumes yielded the same results as the atomistic account, and indeed demonstrated the

a given volume at a given temperature and pressure is independent of their individual size or mass: Amedeo Avogadro, 'Essai d'une manière de déterminer les masses relatives des molécules élémentaires des corps, et les proportions selon lesquelles elles entrent dans ces combinaisons', *Journal de Physique* 73 (1810), 58–76.

[31] See Alan Rocke, *Chemical Atomism in the Nineteenth Century: From Dalton to Cannizzaro* (Columbus, OH, 1984), esp. 10–15.

[32] *The Athenaeum* (1835), 619. See the discussion in Morrell and Thackray, *Gentlemen of Science*, 485–90; and Timothy L. Alborn, 'Negotiating Notation: Chemical Symbols and the British Society, 1831-1835', *Annals of Science* 46 (1989), 437–60.

[33] Quoted in Chalmers, *The Scientist's Atom and the Philosopher's Stone*, 193, q.v. The editor of the *Philosophical Transactions* subsequently sought a second opinion, and on the basis of this accepted the paper for publication.

NOMS DES SUBSTANCES.	FORMULES.	POIDS DE L'ATOME.		CONTIENT POUR CENT.		
		O = 100.	H = 1.	Base.	Acide.	$\ddot{\mathrm{H}}$
Acide manganique......	$\ddot{\mathrm{M}}\mathrm{n}^3$	1937,66	155,27			
margarique.......	$\overline{\mathrm{Mr}} = \mathrm{H}^{67} \mathrm{C}^{35} \mathrm{O}^3$	3393,38	271,91	C=78,84	O=8,84	H=12,32
	$\overline{\mathrm{Mr}}^2$	6786,76	543,83			
	$\overline{\mathrm{Mr}}^3$	10180,14	815,74			
molybdique.......	$\dot{\mathrm{Mo}}$	898,52	72,00	66,61	33,39	
	$\ddot{\mathrm{Mo}}^2$	1797,04	144,00			
	$\ddot{\mathrm{Mo}}^3$	2695,56	216,00			
mucique.........	$\overline{\mathrm{M}} = \mathrm{H}^{10} \mathrm{C}^6 \mathrm{O}^8$	1321,02	105,85	C=34,72	O=60,56	H=4,72
	$\overline{\mathrm{M}}^2$	2642,05	211,71			
	$\overline{\mathrm{M}}^3$	3963,07	317,55			
nitreux.........	$\dot{\mathrm{N}}$	477,04	38,22	37,11	62,89	
	$\ddot{\mathrm{N}}^2$	954,07	76,55			
	$\overset{\cdots}{\mathrm{N}}^3$	1431,11	114,68			
nitrique..........	$\dot{\mathrm{N}}$	677,04	54,25	26,15	73,85	
	$\ddot{\mathrm{N}}^2$	1354,07	108,50			
	$\overset{\cdots}{\mathrm{N}}^3$	2031,11	162,75			
oléique..........	$\overline{\mathrm{Ol}} = \mathrm{H}^{120} \mathrm{C}^{70} \mathrm{O}^5$	6599,40	528,81	C=81,08	O=7,58	H=11,34
	$\overline{\mathrm{Ol}}^2$	13198,80	1057,63			
	$\overline{\mathrm{Ol}}^3$	19798,20	1586,45			
osmique.........	$\ddot{\mathrm{Os}}$	1644,49	131,77	75,68	24,32	
	$\ddot{\mathrm{Os}}^2$	3288,97	263,55			
	$\overset{\cdots}{\mathrm{Os}}^3$	4933,46	395,32			
oxalique.........	$\ddot{\mathrm{C}}$	452,87	36,29	33,76	66,24	
	$\ddot{\mathrm{C}}^2$	905,75	72,58			
	$\overset{\cdots}{\mathrm{C}}^3$	1358,62	108,87			
oxichlorique.......	$\ddot{\mathrm{C}}\mathrm{l}$	1142,65	91,56	38,74	61,26	
	$\overset{\cdots}{\mathrm{C}}\mathrm{l}^2$	2285,30	183,13			

Fig. 5.2. Berzelius, *Théorie des Proportions Chimiques et Table Synoptique des Poids Atomiques* (1835)

universality of Dalton's procedure for determining the weights of elements by extending it to complex organic substances such as oxalic acid. For Berzelius, the strength of his own account lay in the fact that, unlike Dalton's atoms, it does not make unprovable assumptions about the underlying material structure of chemical substances, and he even used 'atom' and 'volume' as interchangeable expressions. On the other hand, Dalton's atomism would seem to have the advantages of allowing one to visualize the underlying structure of chemical compounds. But, as

Klein has pointed out, the atomic figures appear only three times throughout his works, and they never act as heuristic aids in the way that Berzelian-type formulas did from the 1840s onwards, above all because they allowed one to understand how elements might replace one another in compounds, something Dalton's representations cannot do.[34] In inorganic chemistry in the early decades of the nineteenth century, a Daltonian atomistic interpretation of chemical reactions was not especially misleading, and the law of constant proportions, for example, held whether it was given an atomistic interpretation or not. Berzelius himself says of his representation and that of Dalton that they are 'absolutely the same thing'.[35] But as attention begins to turn to organic chemistry, the fact that different compounds may have the same number of elements in the same proportion, together with the fact that some reactions in organic chemistry inevitably result in by-products, making a Daltonian-type calculation extremely difficult, meant that the constraints offered by Daltonian atomism begin to become a real obstacle, and its standing collapsed. The new generation of chemists in Section B of the British Association were not persuaded by Dalton's outburst two years earlier, for example, and adopted the Berzelian notation from 1837.

Berzelius' names for compounds reflected their chemical nature, and they are classified in terms of their behaviour in electrolysis. The impact of the discovery, in 1800, of the electrolytic decomposition of compounds was especially significant for the development of nineteenth-century chemistry, and Davy had reacted almost immediately, arguing that elements in compounds were held together electrically, and, correlatively, that chemical changes were essentially electrical in nature. In electrolysis, electropositive substances migrate to the cathode and electronegative ones to the anode, and Berzelius classified all elements into a series depending on their electrochemical properties, the most electronegative being oxygen, and the most electropositive being potassium. In his 1819 *Essai sur la théorie des proportions chimiques*, Berzelius had argued that the laws regulating chemical combinations in organic chemistry were very different from those in inorganic chemistry, and that there were so many combinations in the former that no determinate proportions existed. But in 1837 Dumas, in proposing the doctrine of chemical 'radicals', the binary components of compounds, suggested that the only difference between inorganic and organic chemistry was that the radicals in the former were simple, whereas in the latter they were complex.[36] Here there seemed to be the basis for a fundamental unity in chemistry, in the

[34] Ursula Klein, *Experiments, Models, Paper Tools: Cultures of Organic Chemistry in the Nineteenth Century* (Stanford, 2003), 23–40; Chalmers, *The Scientist's Atom and the Philosopher's Stone*, ch. 9.

[35] J. Jacob Berzelius, 'An Address to Those Chemists Who Wish to Examine the Laws of Chemical Proportions, and the Theory of Chemistry in General', *Annals of Philosophy* 5 (1815), 122–31: 127.

[36] Jean-Baptiste Dumas and Justus Liebig, 'Note sur l'état actuel de la Chimie', *Comptes Rendus des Séances de l'Académie des Sciences* 5 (1837), 567–72. The paper was actually by Dumas alone, who, probably mistakenly, thought he was also representing Liebig's view.

sense of providing a single account of organic and inorganic substances, and one which reinforces its autonomy from the physical sciences. In the 1820s and 1830s, electrochemistry had been routinely applied to organic and inorganic chemistry, but there emerged a number of problems with electrochemical dualism, which Berzelius had proposed as the universal form of chemical bonding. Its universality slowly collapsed as problems arose. Hydrogen began to pose a difficulty as it was discovered that it showed both electropositive and electronegative polarities, and there was the problem of how similar atoms could be combined in an electrochemical account. But it was in the realm of organic chemistry that the most intractable problems arose. However well electrochemical dualism turned out to work with inorganic compounds, it failed significantly with organic ones.[37]

As attention came to be focused on organic chemistry, there was a reversal of the priority of the laws of chemistry, with organic coming to take precedence over inorganic chemistry.[38] In an attempt to deal with the failure of electrochemical theory of radicals to account for organic reactions, Dumas replaced it with a theory of 'types'.[39] In organic chemistry, types are substances that contain the same number of equivalents united in the same way and showing the same properties. For example, when hydrogen in acetic acid is replaced by chlorine (or bromine, or iodine), the chlorinated product—chloroacetic acid—shares its properties with acetic acid: the 'type' remains the same. Indeed, developments of the theory by Gerhardt and by Williamson led to the elaboration of basic types, and Williamson set out a basic 'water type' which he believed was sufficient for all inorganic and most organic compounds.[40] Here the central oxygen atom holds the rest of the molecule together, and one builds up a series of molecules of increasing complexity in this way:

H	C_2H_5	C_2H_5	C_2H_3O
O	O	O	O
H	C_2H_5	H	H
water	ether	alcohol	acetic acid
and so on			

[37] In modern terms, bonding in inorganic compounds is predominantly ionic: the metal loses electrons to become a positively charged cation, while the nonmetal accepts these electrons to become a negatively charged anion. By contrast, in organic compounds bonding is predominantly covalent: electron pairs are shared between atoms, either one electron coming from each atom, or both from one atom.

[38] See Colin Russell, *The History of Valency* (Leicester, 1971), ch. 4; and Klein, *Experiments, Models, Paper Tools*, ch. 7.

[39] Jean-Baptiste Dumas, 'Premier mémoire sur les types chimiques', *Annales de Chimie et de Physique* 73 (1840), 73–103.

[40] Alexander Williamson, 'Theory of Etherification', *Journal of the Chemical Society* 4 (1852), 229–39.

There were a number of further developments of the theory, but the ultimate prize was to account for the complicated behaviour of carbon. In 1858 Kekulé offered a new fifth 'type' to Gerhardt's four—ammonia, water, hydrochloric acid, and hydrogen—namely a 'marsh gas' type,[41] of which the first four in the series are:

C	H	H	H	H	marsh gas
C	H	H	H	Cl	methyl choloride, etc.
C	H	Cl	Cl	Cl	chloroform, etc.
C	(NO$_4$)	Cl	Cl	Cl	chloropictirin
and so on					

However, rather than just indicating similarity of property, which is what Gerhardt had confined himself to, Kekulé treats the compounds as being related by mutual transformation. Moreover, whereas Gerhardt had taken the theory of types to be based on double decomposition, Kekulé stretches the framework to encompass other types of reaction as well,[42] reworking the theory of radicals and reconciling it with his theory of types. This enables Kekulé to build up a fourfold account in which: it is established that elements/atoms have a saturation point (i.e. a limit to the room an atom has for the attachment of a fixed number of other atoms); this saturation point is crucial for the atoms that hold together different parts of the molecule, particularly two or more radicals; the application of this to carbon leads to recognition of a fourfold saturation capacity; and calculation of this saturation capacity shows how its atoms can be self-linking.[43] With these developments we have an elementary theory of valency, that is, a theory of how atoms combine with one another to form molecules or compounds, and with this the basis for the enunciation of the periodic law of elements. This is really the first theory of chemical bonding that applies without distinction to the inorganic and organic realms, but it could only have originated in organic chemistry, because it was only in organic chemistry that one needed to understand structure before one could understand the composition and behaviour of the substances.

A further aspect of Kekulé's innovation hinges on the question of structure. Gerhardt had shown that the properties of a compound depend exclusively on the relationships between all the atoms of the molecule. Structure comes to be the defining feature of chemical substances, replacing a classification based on the origin of samples, in particular whether it comes from something living or non-living (which is why 'carbon chemistry' is a more accurate term than 'organic chemistry'). Gerhardt was concerned with how structure shows similarity of property, and the formulas did not have any other significance for him: they

---

[41] Friedrich August Kekulé, 'Ueber die Constitution und die Metamorphosen der chemische Verbindungen und über die chemische Natur des Kohlenstoffs', *Annalen der Chemie und Pharmacie* 106 (1858), 129–59.

[42] Russell, *The History of Valency*, 66–80.   [43] Ibid., 110–13.

were just a systematic way of representing different types. But they came to take on real structural significance, and it was in this context that Kekulé—treating compounds as being related by mutual transformation, thereby giving the model a dynamic character—was worried about the extent to which a linear representation could be said to capture the actual positions of atoms in the molecule for example. In this way, a programme to picture how atoms are arranged in complicated molecular structures was built up.

Does this new characterization of chemistry in terms of the investigation of the structural basis for how atoms and molecules behave mean that inorganic and organic chemistry could now form part of a unified field? In the first Faraday lecture given to the Chemical Society in 1869, Dumas, after being adamant at the beginning of the decade that there were two separate chemistries, now writes that with regard to organic compounds, 'chemistry need no longer hesitate. Such substances are formed in the same manner as mineral matter, they exhibit all the conditions of its composition, its structure and its properties.'[44] But after 1870 organic and inorganic chemistry separated again, as a result of a number of factors, a key one of which was valency.

Valency had been treated by Kekulé and others as a fundamental property of the atom, on a par with its atomic weight. The concept had been developed in organic chemistry, where fixity of valencies does indeed seem secure (despite some problems with aromatic and unsaturated compounds). But in inorganic chemistry matters were different. Mendeleev's periodic table was designed to show the dependency of two variables, atomic weight and valency, but in inorganic chemistry there were numerous compounds where the notion of fixed valencies broke down at a fundamental level. One concerted attempt to establish the unification of organic and inorganic chemistry came in 1905 with Werner's treatise on inorganic chemistry.[45] The crucial move was denial of fixed valency of elements, and he developed his idea of a 'coordination' number, introduced twelve years earlier, to counter fixed valencies. The coordination number is the number of atoms or radicals attached to a central atom. When complex inorganic compounds react, for example when platinum chloride accepts two molecules of ammonia, the traditional approach was to represent the formula of the resultant compounds in a way that retained platinum's fourfold valency. The price of doing this was the representation of compounds that had very similar properties in terms of very different structures, dictated by the need to save platinum's tetravalency. Werner argued that the constancy of properties could be accounted for if platinum were given a coordination number of 6: that is, if six atoms or radicals were allowed contiguous with a central platinum atom. Coordination numbers had no basis in traditional valency theories, and Werner had to devise a somewhat

[44] Quoted in Colin A. Russell, *The Structure of Chemistry* (Milton Keynes, 1976), Part II, 24.
[45] Alfred Werner, *Neuere Anschauungen auf der Gebiete der anorganischen Chemie* (Braunschweig, 1905).

ad hoc notion of 'auxiliary valencies'—by contrast with the atom's primary valencies—to account for them.[46]

In sum, once we get down to the details of nineteenth-century and early twentieth-century chemistry, it becomes evident that there are no grounds for simply assuming either the unity of chemistry and physics, or even the unity of chemistry. Faraday's discovery that a given quantity of electricity liberates elements from their compounds in amounts that are proportional to their calculated equivalent weights may have initially suggested an identity between electrical force and chemical affinity. But electrochemical dualism soon hit insuperable obstacles, especially in organic chemistry, as we have seen, and was abandoned. Dalton's postulation of atoms—which had very different properties from one another and which, when combined, form compounds that typically had very different properties from the component atoms—owed nothing to conceptions of atoms in the various branches of physical theory, so presented no opening for unification. With the major developments in chemistry later in the century, far from a trend towards unification, what we find is quite the opposite: the growth of specialist inorganic chemistry; the growth of specialist organic chemistry; and the emergence of physical chemistry, which rather than healing the breach between the two actually acted to separate them further.[47] And when Werner provided a new model of structure which accounted for those complex inorganic reactions that the old structural theory had been unable to explain, he precipitated a deep bifurcation of chemistry, in which different structural models had to be used in organic and inorganic chemistry. At a general level, one way to describe what is at issue here is in terms of the modularity of science. It consists of relatively discrete areas, some of which may resist connections with others. By contrast, unity imposed through a philosophical/metaphysical homogenization of science loses explanatory richness and a sense of how results are generated in the first place.

EQUATIONS AND MODELS

One of the problems that arises from thinking in terms of unification is that it distracts us from what the concrete aims of scientific investigation in the period were. In the case of chemistry, for example, Alan Rocke identifies the actual concerns in these terms:

for those who actually lived through the period, the dominating story in chemistry of the 1860s, 1870s, and 1880s was neither the periodic law, nor the search for new elements, nor the early stages of the study of atoms and molecules as physical entities. It was the

---

[46] Consequently, it is not surprising that, despite their subsequent importance, it took another forty years before they played a major role in chemistry. See Russell, *The Structure of Chemistry*, Part III, 18.

[47] Ibid., Part III, 9.

maturation, and demonstration of extraordinary scientific and technological power, of the 'theory of chemical structure'—loosely defined, a set of ideas that enabled one to succeed in tracing and portraying the exact way in which atoms are connected up with each other to form molecules.[48]

In short, in exploring how complex molecular structures were built up, what chemists were interested in was how these structures could be pictured in atomic terms. The point of explanation is to lead to understanding, and picturability played a crucial role in enabling this understanding. Similar strictures held in physics. Thomson, for example, put a high price on the visualizability of mathematical formulae, writing: 'I can never satisfy myself until I can make a mechanical model of a [mathematical] thing. If I can make a mechanical model I understand it. As long as I cannot make a mechanical model all the way through I cannot understand.'[49] Maxwell similarly was particularly committed to the necessity of visualizing physical processes, even if they did not represent (or even set out to represent) actual elements in nature: visualization of these processes was central to understanding them. Moreover, the physicality of such representations, by contrast with abstract mathematical formulations, was crucial, particularly to the extent to which one can represent motions physically. This enables physicists, Maxwell tells us, to learn 'at what a rate the planets rush through space, and they experience a delightful feeling of exhilaration. They calculate the forces with which the heavenly bodies pull at one another, and they feel their own muscles straining with the effort. To such men momentum, energy, mass are not mere abstract expressions of the results of scientific inquiry. They are words of power, which stir their souls like the memories of childhood.'[50]

The crucial point is that the picturability requirements were quite different in physics and chemistry, something obscured by the assimilation of chemists' atoms to physicists' atoms. They relied on wholly different resources, they were accounting for completely different kinds of things, and they were drawing on different kinds of instrumentation.[51] Picturing a physical process to oneself does not enable one to picture a chemical process. Indeed, the autonomy of chemistry in relation to the physical sciences relied to a significant extent on the different demands of picturability requirements of the two. Such considerations, which focused on

---

[48] Alan J. Rocke, *Image and Reality: Kekulé, Kopp, and the Scientific Imagination* (Chicago, 2010), xiv.

[49] William Thomson, *Notes of Lectures on Molecular Dynamics and the Wave Theory of Light* (Baltimore, 1884), 270–1. And as Smith and Wise note, he would reject 'Maxwell's transfer of electrostatics of Fourier's complete geometrical form for heat conduction because it required the invention by analogy of an electrical entity with no empirical referent.' *Energy and Empire*, i. 187.

[50] James Clerk Maxwell, 'Address to the Mathematical and Physical Sections of the British Association for the Advancement of Science' (1870), in *The Scientific Papers of James Clerk Maxwell*, ed. W. D. Niven (New York, 1965), 218.

[51] On picturing specifically chemical structures and processes, see Christopher Ritter, 'An Early History of Alexander Crum Brown's Graphical Formulas', in U. Klein, ed., *Tools and Modes of Representation in Laboratory Sciences* (Dordrecht, 2001), 35–46; and Stephen Weininger, 'Contemplating the Finger: Visuality and the Semiotics of Chemistry', *Hyle* 4 (1998), 3–27.

explanation not on ontology, militated against grand unification phantasies, which were grounded in metaphysics rather than in the working practice of physicists and chemists.[52]

Consider the case of chemical formulas as models. Ursula Klein has argued that in the 1830s, chemical formulas functioned as paper-tools for building models of the constitution of organic compounds and their reactions, showing how the manipulation of these chemical formulas displayed new possibilities for the explanation of chemical reactions that went well beyond the original goals of chemists:

Following the 'suggestions' arising from tinkering with chemical formulas, the French chemist Jean Dumas went a step further than was necessary for his initial purposes of model building. He introduced a new model of the chemical reaction of an organic substance which became a paradigmatic object of the new concept of substitution incorporating chemical atomism. This additional step was not only done *ad hoc*, it also ironically undermined a central aspect of Dumas' goals . . . . [I]t was the chemists' pragmatic application of chemical formulas as paper-tools for traditional modelling which paved the way for the introduction of a new atomistic model of chemical reactions.[53]

Here, the chemical equation, a model of the possible structure of chemical compounds, not only provides a tool for representing structure, but allows one to manipulate this structure in order to investigate the substitution of constituents of the model for one another. In this way, the notions of proportion, compound, and reaction were united in a novel and powerful understanding of chemical structure, one that worked with what Klein calls paper-tools, which not only provided guidance for experiments, but which also functioned autonomously of chemical theory. In short, it was not theory or experiment that was providing the fundamental insights here but manipulation of the model. And if it is manipulation of models that is doing the explanatory work in cases such as these, then we clearly need to rethink how the parts of physical enquiry are connected with one another. Unlike theories, which can be considered to be on a par if they can be translated into an axiomatic form, models are highly heterogeneous, and what

[52] In this respect, the argument of Eugene Ferguson—'The Mind's Eye: Non-Verbal Thought in Technology', *Science* 197 (1977), 827–36—that science and technology are distinguished by the fact that the former is transmitted through written textbooks and journal articles whereas the latter is transmitted in terms of drawings, models, and directing copying skills, is misleading if the difference is taken to be an essential one between science and technology. It concerns a particular mode of presentation in contrast to others in the field, but does not at all reflect the way in which scientists think through problems, which is often closer to the procedures by which those in 'technology' think through problems. We shall return to the science versus technology question in Chapter 10.

[53] Ursula Klein, 'Techniques of Modelling and Paper-Tools in Classical Chemistry', in Mary S. Morgan and Margaret Morrison, eds, *Models as Mediators: Perspectives on Natural and Social Science* (Cambridge, 1999), 146–67: 147–8.

they tell us cannot be reduced to some 'theoretical content' as if they were merely representations of theories.[54]

ATOMS: A UNIFICATION OF CHEMISTRY AND PHYSICS?

If, as I have argued, the physical and material sciences in the nineteenth century were largely modular, particularly in relation to one another but also internally, we can ask to what extent this modularity was overcome in what has typically been promoted as a paradigm case of internally generated unification, namely the early twentieth-century establishment of the atomic structure of matter and the emergence of physical chemistry.[55] After all, it is possible that an advocate of the unity of the sciences might concede that there was a significant degree of modularity in the physical and material sciences in the nineteenth century, but argue that this changed with the establishment of modern atomism, which brought together developments in physics and chemistry, establishing common ground of a fundamental nature in the doctrine of the discrete structure of the microscopic realm. But as we are now about to see, in identifying the specific factors that established modern atomism, we can show that these exhaust the range of contributory factors. There is no remaining role for unification aspirations. We shall see that it is a question of modularized problem solving in disciplines that cannot be assimilated to one another, but which, when the demand arises, can be brought together in an area of overlap in such a way that, in very disparate fashions, they supply resources for an account that goes beyond each of them taken separately.

In the second half of the nineteenth century, the question whether matter had a discontinuous or a continuous micro-structure divided physicists and chemists, but the solution was found in physics, without any input from chemistry. The way in which it was effected is instructive. Elements of wholly independent developments in the theory of gases, in the study of the electromagnetic spectrum, and in experiments on pollen dust by botanists, turned out, when combined in a particular way and applied to a particular problem, to create a whole not only significantly greater than the parts but significantly different from the parts, establishing a discontinuous physical micro-structure.

[54] See Margaret Morrison, 'Models as Autonomous Agents', Mary S. Morgan and Margaret Morrison, eds, *Models as Mediators: Perspectives on Natural and Social Science* (Cambridge, 1999), 38–65; Davis Baird, *Thing Knowledge: A Philosophy of Scientific Instruments* (Berkeley, 2004), ch. 2.

[55] On the problems of basing modern chemistry on physical principles see Kostas Gavroglu and Ana Simões, *Neither Physics Nor Chemistry: A History of Quantum Chemistry* (Cambridge, MA, 2012). As the authors point out, Dirac's 1929 statement that: 'The underlying physical laws necessary for the mathematical theory of a large part of physics and the whole of chemistry are thus completely known' is 'a theoretically correct but practically meaningless dictum' (9).

Let us begin with the question why there were no developments in chemistry that contributed to the eventual establishment of a discontinuous physical structure in matter: why Dalton's atoms were not the 'first steps' in a process of establishing atomism that culminated in the discoveries of Perrin et al. The first thing to note is that chemists after Dalton hardly ever used the language of atoms before the 1890s, preferring instead to talk in terms of 'equivalences'. As Davy had put it, the atom was simply a unit of chemical reaction, not a material entity.[56] It might be thought that the discovery of the periodic table changed this, but as Colin Russell notes, there is in the literature an overstatement of the importance of the periodic table for suggesting lines of research. Most nineteenth-century discoveries, as he points out, owed little to the periodic table and actually offered it an embarrassing challenge.[57] Even stoichiometry, the study of the amounts of chemical agents needed to yield a given amount of product in reactions, which had been the proving ground of Dalton's chemical atomism, and had led directly to the doctrine of equivalents, could no longer be seen as supporting atomism, Ostwald argued, because 'chemical dynamics' had made 'the atomic hypothesis unnecessary for this purpose and has put the theory of stoichiometrical laws on more secure ground than that furnished by a mere hypothesis'.[58] The picture was complicated even further by the fact that 'chemical atomism' had to accommodate molecules and ions as well as atoms by the middle of the nineteenth century. The establishment, in physics, of a discontinuous micro-structure would in itself have been of little help in this respect to chemists, for whom functional differences between atoms, molecules, and ions were paramount. To complicate matters even further, there were moves to construe molecules, ions, and atoms as simply different forms of energy since it was energy, Ostwald argued, that was ultimately what persisted through various changes, not its molecular, ionic, or atomic manifestations.[59] This was somewhat speculative on Ostwald's part, but it does indicate how, once physical considerations are brought to bear, 'chemical atoms', which many believed crucial to the understanding of chemical structure and reactions, could quickly evaporate.

Nevertheless, it remained the case that in chemistry pure substances could not be ranged by insensible gradations into a continuous series, and that the law of definite proportions, as well as the existence of isomerism, only seemed to make sense in terms of a discontinuous material micro-structure. Strictly speaking, such material micro-structure did not have to be at the fundamental level assumed by advocates of atomism such as Dalton. One possibility was that, if chemical atoms had any reality at all, they could perhaps be meso-level phenomena, to be resolved

[56] See *Collected Works of Sir Humphry Davy* (9 vols, London, 1840), vii. 96–7.
[57] Russell, *The Structure of Chemistry*, Part III, 14.
[58] Wilhelm Ostwald, 'Elements and Compounds', in C. S. Gibson and A. J. Greenaway, eds, *Faraday Lectures 1869–1928* (London, 1928), 185–201: 187.
[59] Wilhelm Ostwald, *Die Energie* (Leipzig, 1912), 126–7.

at the more fundamental physical level into differing energy levels, for example, in the micro-structure. In this way, chemistry could retain a commitment to atomism independently of what physics identified as the nature of the micro-structure of matter: all advocates of chemical atomism needed was that, at some level appropriate to chemistry, matter manifested a discontinuous structure. The postulated discontinuous matter of chemistry was not in a position to dislodge physical considerations, and developments in chemistry had little impact on the physical understanding of the nature of matter, or on the developments that led to the establishment of the physical reality of the atom.

In physics, the question whether matter has a discontinuous or a continuous micro-structure had precedents in physical optics, where the corpuscular theory of light dominant from the end of the seventeenth century had been replaced with the establishment of a wave theory of light by Young and Fresnel at the beginning of the nineteenth century. Physical optics was (and remains) a fundamental part of physics, and the establishment of the continuity of matter in this context had general significance for the understanding of the nature of matter. More generally, as Nye notes, not only did the classical interpretations of the first and second laws of thermodynamics suggest a continuous structure for matter, but 'the pursuit of physical problems, such as the observation of the critical point and liquification of gases, along with studies of chemical equilibrium and the application of Gibbs' phase rule, Le Chatelier's principle, and Duhem's thermodynamic potential, only confirmed a tendency to view matter as continuous in nature and to question critically the particulate, mechanical models of the past'.[60] Note also that, as far as particulate theories in physics were concerned, there were a number of competing atomic models. There was no atomism per se: 'atomism' covered a number of approaches. In particular, the kinetic theory of gases, whereby atoms move and collide in empty space, did not have the field to itself, and was in fact a relative latecomer, revived only in the 1850s. Between 1780 and 1860 the dominant atomic theory of gases described in physics textbooks was the atmospheric atom model. Here there were two kinds of atom, matter atoms and ether atoms. Matter atoms attracted each other and attracted ether atoms, whereas ether atoms repelled each other. The theory invoked imponderable particulate fluids to explain electrical, magnetic, thermal, and gravitational properties.[61] This was the kind of atomism that was taught at universities, and Mach's teacher Andreas Ritter von Ettingshausen, for example, was the author of one of the standard textbooks elaborating this theory.[62]

---

[60] Mary Jo Nye, *Molecular Reality: A Perspective on the Scientific Work of Jean Perrin* (London, 1972), 14.
[61] See Stephen G. Brush, 'Mach and Atomism', *Synthese* 18 (1968), 192–215: 194.
[62] Andreas Ritter von Ettingshausen, *Anfangsgründe der Physik* (Vienna, 1844).

In 1894 the chemist Thomas Thorpe wrote that:

Corpuscular theories are now altogether banished from certain domains of physics. Indeed, the most weighty attacks yet made on the atomic theory have been delivered by those who are mainly occupied with the problems and abstract conceptions of energy. It is significant that Faraday, who began his scientific career as a chemist, seems to have gradually loosened his hold on the atomic theory as he became more and more absorbed in the contemplation of purely physical phenomena.[63]

If the idea of continuous matter was to be questioned, the questioning had to take place in physics. We can identify three disparate developments, elements of which were able to be judiciously put together in such a way that they could be used to establish the existence of a discontinuous physical structure of matter in the first decade of the twentieth century. The first derived from the application of statistical techniques to the understanding of the behaviour of gases, the second from work on interpreting the radiation spectrum, and the third from a largely forgotten series of botanical experiments dating from the 1820s which took on a new physical significance in the last decade of the century.

The theory of gases had effectively begun in the seventeenth century with Boyle and Hooke's account of the relation between temperature, pressure, and volume in a gas. But the mechanisms responsible for these macroscopic phenomena remained unknown. In 1738 Daniel Bernoulli advanced a kinetic theory of gases,[64] an account on which pressure in gases is due, not to repulsive forces between particles ('atoms'), but rather to the mechanical effect of elastic particles, which are moving in all directions at great speed, striking against a solid. The theory was able to make sense of some elementary properties of gases, namely compressibility, tendency to expand, rise in temperature in compression and fall in temperature in expansion, and the trend towards uniformity over time. There were a number of problems with developing this kind of account however. One especially intractable difficulty was that the number of particles involved was unknown because there was no estimate of their size, but it was clearly enormous, so an account in terms of the position and velocity of every individual particle was impossible. Rather, what was needed was a statistical account which related macroscopic phenomena such as temperature, velocity, pressure, density, stresses, and heat flow, to averages of quantities characteristic of the microscopic state. But not only was an estimation of the average velocities and average sizes of the particles not forthcoming, there was a deeper conceptual problem concerning the use of statistical formulations. Although from 1805 statistical procedures were used in astronomy to estimate the accuracy of the final result of a large number of observations from an account of the occurrence of single errors in observation,

---

[63] Thomas E. Thorpe, *Essays in Historical Chemistry* (London, 1894), 370.
[64] Daniel Bernoulli, *Hydrodinamica sive de viribus et motibus fluidorum commentarii* (Strassburg, 1738).

using Legendre's 'method of least squares',[65] the procedures here operated as an aid to correlating and correcting observations, not as a replacement for strict deterministic laws. Yet it was the latter that were considered constitutive of the outcome of physical enquiry, and there was judged to be something deeply unsatisfactory about mere statistical accounts of physical phenomena, which were deemed to concern themselves with measures of ignorance rather than some attribute of nature.

Others had explored the kinetic account of gases in the 120 years between Bernoulli and Clausius, but it was the latter's paper on 'the kind of motion we call heat' of 1857[66] that initiated significant interest among physicists in the kinetic theory. Clausius imagined that the particles making up the gas all moved with the same speed, but when Maxwell built a theory of gases on the basis of Clausius' work, in a paper published in 1860,[67] he was able to deal with particles with a range of speeds, developing an account not only of the average speeds of the particles but also of the distribution (deviation from the average) of those speeds in a given volume of gas at a given temperature.[68] Maxwell had the advantage of already having developed skills in this kind of analysis. In 1859 he had offered a solution to the nature of Saturn's rings. These clearly could not be solid since, due to differences in distance of the parts of the rings from the planet, gravity would then cause the parts to rotate at different speeds, giving rise to stresses which no solid material could withstand. Consequently, he developed a statistical procedure to show how the rings would behave if different collections of particles revolved at different speeds: if the size and number of the particles in the ring corresponded to waves in the ring's density, then it could be demonstrated that it would remain stable.[69]

The significant advance here from the point of view of our present concerns is the ability to set up the problem in such a way as to deploy statistical techniques to capture the aggregate behaviour of the particles making up a gas. The calculations in Maxwell's gas paper were formal and there was no attempt to provide any underlying physical story however. Yet clearly for Clausius and Maxwell the speeding particles could not be the infinitesimal points of analytical mechanics,

---

[65] Adrien Marie Legendre, *Nouvelles méthodes pour la détermination des orbites des comètes* (Paris, 1805).

[66] Rudolph Clausius, 'Ueber die Art der Bewegung, welche wir Wärme nennen', *Annalen der Physik* 100 (1857), 353–79.

[67] James Clerk Maxwell, 'Illustrations of the Dynamical Theory of Gases. Part I. On the Motions and Collisions of Perfectly Elastic Spheres', *Philosophical Magazine* 19 (1860), 19–32; 'Part II. On the Process of Diffusion of Two or More Kinds of Moving Particles Among One Another', *Philosophical Magazine* 20 (1860), 21–37. See Elisabeth Wolfe Garber, 'Clausius and Maxwell's Kinetic Theory of Gases', *Historical Studies in the Physical Sciences* 2 (1970), 299–319.

[68] As early as 1823 Fourier had pointed out the statistical inadequacy of simply offering mean values and had argued that it was also necessary to find the limits within which the mean was confined: Joseph Fourier, 'Mémoire sur la population de la ville de Paris dépuis la fin du XVIIe siècle', in Joseph Fourier, *Recherches statistiques sur la ville de Paris et le Département de la Seine* (2 vols, Paris, 1821–3), i. p. xx.

[69] James Clerk Maxwell, *On the Stability of the Motion of Saturn's Rings* (London, 1859).

since infinitesimal points cannot collide. In an 1868 paper Boltzmann started to flesh out the underling physical story, which he explored in terms of atomism.[70] He took a well-understood case, the rise of a volume of gas in the earth's gravitational field, and showed that Maxwell's formula correctly predicted how the number of particles with a particular energy would change in the process. In a critical development, in 1865 Loschmidt had devised a model that enabled him to come up with an estimate for the size of air molecules/atoms by starting from a liquid, whose volume he assumed to be roughly the volume of the individual molecules multiplied by the number of molecules, then, using the experimental establishment of the volume of gas generated by the evaporation of a liquid, calculating what the size of the gas molecules must be.[71] A physical dimension for air molecules/atoms—10^{-9} metres—gave materiality to these. Nevertheless, this materiality was far from decisive. Lindley sums up the situation well:

To enthusiasts for atoms and the kinetic theory of gases, a quantitative estimate of the dimensions of atoms was yet another piece of evidence that they were on the right track. Showing that invisible objects must have a particular size made them, perhaps, a little more real. To critics, on the other hand, Loschmidt's analysis still didn't prove anything. The fact that some mathematical expressions and experimental data could be knitted together in such a way as to yield a number corresponding to the diameter of a purely hypothetical object was in no way evidence that the hypothetical entities were real . . . . The argument against kinetic theory was that in the absence of tangible evidence that atoms existed, it was mere mathematics, empty theorizing.[72]

This was the problem that the most determined defender of the physical reality of atoms, Boltzmann, continued to face. Despite showing in great detail how the formula that he and Maxwell had worked out for the distribution of speeds of particles in a gas was not only consistent with the empirical data but was the only possible solution,[73] the question remained of whether he was simply engaged in a mathematical exercise.

Nevertheless, there was one significant obstacle that he was able to overcome: the apparent conflict between a mechanical atomist account and the irreversibility of physical processes formulated in the second law of thermodynamics. At a conference organized by Boltzmann in Lübeck in 1895, Ostwald argued that 'the irreversibility of natural phenomena proves the existence of processes that

[70] Ludwig Boltzmann, 'Studien über das Gleichgewicht der lebendigen Kraft zwischen bewegten materiellen Punkten', Wiener Berichte 58 (1868), 517–60.

[71] Johann Josef Loschmidt, 'Zur Grösse der Luftmoleküle', Sitzungsberichte der kaiserlichen Akademie der Wissenschaften Wien 52 (1865), 393–413.

[72] David Lindley, Boltzmann's Atom: The Great Debate that Launched a Revolution in Physics (New York, 2001), 27–8.

[73] Ludwig Boltzmann, 'Weitere Studien über das Wärmegleichgewicht unter Gasmolekülen', Wiener Bericht 66 (1872), 275–370.

cannot be described by mechanical equations'.[74] In an 1866 paper Boltzmann had in fact proposed a purely mechanical demonstration of the second law,[75] but it proved unconvincing. By 1872 however, and especially in an 1877 paper, he was able to associate the second law with probabilistic processes,[76] and in 1896 he argued that mechanics alone was insufficient, and that a statistical understanding regulated by probability theory was needed. The combination of the two showed conclusively that the irreversibility of thermal processes could be accounted for fully in probabilistic, statistical terms,[77] enabling him to reduce the description of the behaviour of matter to a set of equations describing the probabilistic motion of atoms. But while this established the compatibility of an atomistic interpretation with basic physical laws, it did not in itself establish the reality of atoms.

The issue of the physical reality of atoms underwent a transformation in the first decade of the next century with two developments. The first hinged on the interpretation of the radiation spectrum of electromagnetic oscillations as set out by Max Planck in 1900.[78] In investigating the spectrum of radiation emitted by a body at a particular temperature, Planck had found it necessary to divide up the energy of the electromagnetic oscillations into smaller units. In doing this, he used a procedure which Boltzmann had devised of dividing up the energy of a gas into smaller units. Neither Boltzmann not Planck considered these units to have any fundamental physical meaning: the idea of discrete amounts of energy was just a mathematical device that enabled calculations. But in a 1905 paper, Einstein gave a physical interpretation of these discrete amounts of energy, arguing that luminous energy can only be absorbed or emitted in discrete amounts of definite size: in the propagation of a light ray, the energy is not distributed continuously over steadily increasing spaces, but rather consists in a finite number of energy quanta.[79] While this development in itself was not decisive in favour of atomism, one of the mainstays of the advocacy of a continuous material micro-structure had lain in the wave theory of light, reinforced in Maxwell's equations. The establishment of light quanta—photons—now effectively put discontinuous and continuous micro-structures more on a par as far as the physics was concerned.

[74] See the discussion in Robert Deltete, 'Helm and Boltzmann: Energetics at the Lübeck Naturforschversammlung', *Synthese* 119 (1999), 45–68.

[75] Ludwig Boltzmann, 'Über der mechanische Bedeutung des Zweiten Hauptsatzes der Wärmetheorie', *Wiener Berichte* 53 (1866), 195–220.

[76] Ludwig Boltzmann, 'Über die beziehung dem zweiten Haubtsatze der mechanischen Wärmetheorie und der Wahrscheinlichkeitsrechnung respektive den Sätzen über das Wärmegleichgewicht', *Wiener Berichte* 76 (1877), 373–435.

[77] Ludwig Boltzmann, 'Entgegnung an die wärmetheoretischen Betrachtungen des Hrn. E. Zermelo', *Annalen der Physik* 57 (1896): 772–84.

[78] See Lindley, *Boltzmann's Atom*, ch. 11; and for full details, Thomas Kuhn, *Black-Body Radiation and the Quantum Discontinuity, 1894–1912* (Chicago, 1978).

[79] Albert Einstein, 'Über einen die Erzeugung und Verwandlung des Lichtes betreffenden heuristischen Gesichtspunkt', *Annalen der Physik* 17 (1905), 132–48.

In the same year as the light quantum article, Einstein also published a paper on Brownian motion,[80] which established atomism far more decisively. Indeed, the catalyst for the decisive shift to atomism in the first decade of the twentieth century was the rethinking of the significance of a series of experiments from the 1820s on reproduction in plants, experiments that had been largely forgotten in the intervening decades, but whose significance became highly contested in the 1890s. In 1827, the botanist Adolphe Brongniart reported on the rapid motion of small pollen granules in water, identifying them as 'animalcules', and noting that larger granules remained stationary and that the rate of motion increased with a rise in temperature.[81] The claim that the granules' motions were vital phenomena was immediately challenged by the chemist François-Vincent Raspail, who identified as possible causes surface capillary, evaporation, and the agitation of the air.[82] Brongniart replied, reporting on experiments in which the motion continued despite the water being enclosed in a glass capsule covered with mica, preventing surface capillary evaporation and agitation of the water.[83] In 1828, Robert Brown extended Brongniart's observations by using a variety of different pollen granules.[84] Brown accepted Brongniart's answer to Raspail, that the phenomenon could not be due to external causes influencing the ambient liquid, but his experiments had two surprising results. First, some of the pollen granules had been dried and preserved for up to twenty years, yet still underwent rapid motion. Second, suspended particles of inorganic substances, such as rocks and earth, underwent the same kind of rapid motion. Experiments having effectively ruled out vital forces and external causes, various internal sources of the motion were sought. Brongniart proposed electrical attraction as the most promising candidate, but Brown was able to show that the motion continued in circumstances where electrical attraction was negligible.[85] Light, magnetism, polarity, and radiating caloric were all suggested, but with no success, as interest gradually petered out.[86]

Jean Perrin had been exploring an atomist interpretation of Brownian motion, whereby it was the interaction between atoms and the pollen granules that caused

---

[80] Albert Einstein, 'Über die molekularkinetischen Theorie der Wärme geforderte Bewegung von in ruhenden Flüssigkeiten suspendierten Teilchen', *Annalen der Physik* 17 (1905), 549–60.

[81] Adolphe Brongniart, *Mémoire sur la génération et le développement de embryon dans les végétaux phanérogames* (Paris, 1827).

[82] Published in English translation a year after its appearance in French as François-Vincent Raspail, 'Observations and Experiments Tending to Demonstrate that the Granules which are discharged in the Explosion of a Grain of Pollen, Instead of being Analogous to Spermatic Animalcules, are not even Organised Bodies', *Edinburgh Journal of Science* 10 (1829), 96–106.

[83] Adolphe Brongniart, 'Nouvelles recherches sur le pollen et les granules spermatiques des végétaux', *Annales des Sciences Naturelles* 15 (1828), 381–90.

[84] Robert Brown, 'A Brief Account of Microscopical Observations Made in the Months of June, July, August, 1827, on the Particles Contained in the Pollen of Plants; and on the General Existence of Active Molecules in Organic and Inorganic Bodies', *Philosophical Magazine* 4 (1828), 161–73.

[85] Robert Brown, 'Additional Remarks on Active Molecules', *Philosophical Magazine* 6 (1829), 161–6.

[86] See the discussion in Nye, *Molecular Reality*, 9–13.

the motion of the latter, from the 1890s, very much as an experimentalist.[87] In 1908 he published three papers on Brownian motion that were informed by his experimental work, but did not take account of Einstein's papers.[88] But once he had considered these latter, he realized his own meticulous experimental work measuring the Brownian motion of translation and rotation confirmed Einstein's theoretical results. He was able to show that, in Brownian motion, each particle is agitated by molecular collisions and the mean energy maintained by particles in these collisions is equal to that of the molecules themselves.[89]

In sum, before the 1890s there were no compelling physical reasons to accept the discontinuity of matter, but by the Solvay conference of 1911, and particularly with the publication of Perrin's *Les Atoms* of 1913, atomism had become firmly established, and as Nye notes, the remaining small minority 'who condemned the molecular viewpoint as pure speculation in 1913 continued to do so until their death'.[90] Note, however, that we cannot consider that the issue here is simply a question of whether atoms exist or not: as simply a methodological question of positivism versus realism. For those who continued to hold out against atomism after 1913, such as Mach, who died in 1916, the issue of the existence of atoms was embedded in the context of a dispute over whether a continuist or a discontinuist account of the nature of matter best met the explanatory demands of physics. The continuist theory of matter had been advocated by its proponents on the grounds that there was an implicit continuity of matter in the equations of mechanics, and it was the most compelling interpretation of the series of second-order differential equations in terms of which mechanics expressed physical processes.[91] The circumstances under which a discontinuist theory of matter prevailed offer a paradigm case of the complexities of reconciling competing, and well-established, conceptions of matter, fought out on theoretical and experimental grounds, and drawing, in this case, on disciplines as diverse as botany and electromagnetism. It is difficult to conceive what rationale there could be for an interpretation of these developments as a successful realization of a unification of science agenda, still less of an ontological agenda, or what such approaches could contribute to our understanding of them.

---

[87] See Charlotte Bigg, 'Evident Atoms: Visuality in Jean Perrin's Brownian Motion Research', *Studies in History and Philosophy of Science* 39 (2008), 312–22; Charlotte Bigg, 'A Visual History of Jean Perrin's Brownian Motion Curves', in Lorraine Daston and Elizabeth Lunbeck, eds, *Histories of Scientific Observation* (Chicago, 2011), 156–79.

[88] Jean Perrin, 'L'agitation moléculaire et le mouvement brownien', *Comtes rendus de l'Académie des Sciences* 146 (1908), 967–70; Jean Perrin, 'La loi de Stokes et le mouvement brownien', *Comtes rendus de l'Académie des Sciences* 147 (1908), 475–6; Jean Perrin, 'L'origine de mouvement brownien', *Comtes rendus de l'Académie des Sciences* 147 (1908), 530.

[89] See the discussion in Nye, *Molecular Reality*, ch. 3.   [90] Ibid., x.

[91] Note, however, Boltzmann's criticism of this view on the grounds that the differential equations were originally obtained by taking the limits of discontinuous points: 'Über die Unentbehrlichkeit der Atomistik in der Naturwissenschaft', in Ludwig Boltzmann, *Populäre Schriften* (Leipzig, 1905), 141–57: 144–5.

Moreover, in case one were to think that such problems were distinctively nineteenth- and early twentieth-century ones, subsequently resolved with the move to fundamental particle physics, consideration of two issues will indicate that this is far from being the case. First, we must beware of a kind of Democrateanism that offers itself as a description of fundamental particles. The analogy between atoms and bricks—the idea that atoms/fundamental particles are like the bricks out of which buildings are made—is a prime culprit. The image of atoms as bricks seems to have been quite common in the late 1920s and early 1930s, its main currency being in popular articles in newspapers such as *The Observer* (1925) and the *Manchester Guardian* (1932 and 1933),[92] although C. P. Snow, who should have known better, also used it in an article on chemistry in 1933.[93] Since then it has been revived a number of times, and was given a new lease of life by the philosopher J. J. Smart.[94] But this picture, where atoms or particles have intrinsic properties that account for the macroscopic properties of the larger bodies which they compose, bears no relation at all to modern elementary particles (which are not even strictly speaking 'particles' but wave functions which can manifest as waves and/or particles), which do not behave in the way they do in producing macroscopic events solely or even largely because of some intrinsic quality.

There are a number of difficulties with the resort to fundamental particles. As things stand at present, for example, the 'Standard Model' of elementary particle physics tells us that there are twelve particles (six quarks and six leptons) and four forces (electromagnetism, strong force, weak force, and the Higgs field) at the most fundamental physical level. But a little probing reveals the explanatory economies that one might expect from 'fundamental' particles are illusory. As regards the particles themselves, the Standard Model count does not include the eight different 'colours' of gluons, or currently undetected partners for all elementary particles, such as sfermions, postulated by supersymmetry theory. Nor does it include the many different extra particles that have been postulated in an effort to overcome difficulties in the Standard Model, such as axions, branons, cornucipons, cuscutons, dyons, erebons, flaxions, giant magnons, macros, magnetic monopoles, maximons, planckons, preons, simps, skyrmions, sterile neutrinos, wimps, wimpzillas, and so on.

---

[92] See Jeff Hughes, 'Unity Through Experiment?', 78 n.32.

[93] C. P. Snow, 'Chemistry', in H. Wright, ed., *University Studies: Cambridge 1933* (London, 1933), 97–121: 98.

[94] J. J. Smart, *Philosophy and Scientific Realism* (London, 1963), ch. 3. In fact, physics does not *necessarily* have to anchor science in the microscopic realm at all. The project of geometrodynamics developed a macro-reductive programme whereby the properties of 'fundamental' particles are to be accounted for ultimately in terms of the large-scale structure of space-time. On this account, it is the macroscopic features of space-time geometry that explain the properties of the whole physical realm, including particles. See John A. Wheeler, *Geometrodynamics* (New York, 1963).

More importantly, confining our attention to the particles of the Standard Model, once we ask how these interact, we need to invoke some twenty free (or 'adjustable') parameters specifying the properties of the particles (masses, phases, and mixing angles) and the strengths of the particles' interactions (the 'coupling constants'). The Standard Model itself furnishes us with no clues as to why these parameters have the values they do: they are simply the values that experiments reveal. Indeed, it is possible to identify a variety of 'constants'—including the number of spatial dimensions, the ratio of fundamental energies, the cosmological constant, and the number of electrons and protons in the observable universe—which we are unable to derive from any mathematics or principles of physics, and which look like quite contingent features of our universe. The crucial point is that there is very radical fine-tuning in these constants, in that slight changes in the assignment of values lead to worlds that could not possibly develop the kind of structures we find in ours: the intrinsic properties of fundamental particles may remain the same, but chemical elements, life, and galaxies would no longer be able to form.[95] At the same time, these features depend on there being more matter than antimatter in the universe, but we do not know why there is this imbalance. To make things worse, the Standard Model only describes 15 per cent of matter in the universe, since the rest is made up of 'dark matter' which the model does not describe: nor can it describe the dark energy that makes up 68.3 per cent of the energy-matter content of the universe.

In the light of these considerations, recourse to the properties of fundamental particles hardly provides a model of complete and comprehensive explanation even for those macroscopic phenomena with which we are familiar (which make up less than 5 per cent of the total energy-matter content of the universe). Indeed, the search for an absolute 'fundamental' level, a carry-over from Democrateanism where atoms were deemed to have a few simple intrinsic properties on which everything else rested, begins to look more and more like a metaphysical dead end.

Second, it is true that appeal to micro-explanation is how we often establish basic connections between phenomena in the physical sciences, and that in many cases this can be taken as an explanatory strategy that is independent of any metaphysical agenda. But we can accept that micro-reduction might be the natural strategy in some areas of the physical sciences without this meaning that micro-reduction is constitutive of physical explanation in any sense. There is no shortage of cases where micro-explanation has proved counterproductive. Note, in this context, a relatively recent episode in the history of physics: the bitter debates over whether physics funding should be devoted to an expensive super-collider in the USA in the mid-1990s. In their opposition to this proposal, condensed matter physicists argued that their discipline (the largest in physics by numbers of practitioners, dwarfing fundamental particle physics) was

---

[95] See Martin Rees, *Just Six Numbers: The Deep Forces that Shape the Universe* (London, 1999); John D. Barrow and Frank J. Tipler, *The Anthropic Cosmological Principle* (Oxford, 1986).

autonomous with respect to the guiding principles of particle physics, and flatly denied that all physics followed in some way from the fundamental laws of particle physics.[96] From whatever perspective one looks at it, explanatory reductionism—the idea that there is some fundamental physical level that underlies the behaviour of material bodies and which must be invoked in any ultimate explanations of that behaviour—is an empty, fruitless, and counterproductive assumption.

---

[96] See Peter Galison, 'Introduction: The Context of Disunity', in Peter Galison and David J. Stump, eds, *The Disunity of Science* (Stanford, 1996), 1–33: 2.

6

The Autonomy of the Life Sciences

As we turn to the implications of the unification of science for the life sciences, we might begin by asking why it was the physical sciences that were taken as the model for science as a whole, and not the life sciences. It would seem that the natural sciences have always been favoured over the life sciences in this respect in the West, particularly since the seventeenth century, but not necessarily elsewhere: not in China for example. And one significant difference between these two is the lack of a conception of 'nature' in the latter. As Geoffrey Lloyd, in a comparison of ancient Greek and ancient Chinese science, notes, there is no straightforward equivalent to the classical Greek term for nature, *phusis*, in classical Chinese:

Of course, a variety of terms *tian* (heaven), *wu* (things), *xing* (character), *li* (pattern), *dao* (the way), *zi ran* (spontaneity . . .), do survive perfectly adequately, in different contexts, to express ideas where the English translation would be in terms of nature, the Greek in terms of *phusis*, or their cognates. Yet that is still not to say that there is a concept of nature in classical Chinese, just the one, nor that it was a major preoccupation. Rather, the evidence from ancient China shows how well they got along *without* any such central preoccupation. So what we must at all costs avoid is the assumption that there is a single concept of nature towards which both Greeks and Chinese were somehow struggling, let alone that it was *our* concept of nature as in 'natural science'.[1]

How we think of 'nature' will shape what we consider the most basic form of understanding of natural processes to be. In the absence of our distinctive Western understanding of 'nature',[2] medicine would seem to be a more promising candidate for example, one that would offer something very different from the physical sciences, not least in conceiving the aim of the exercise in practical terms (health), and which would put the life sciences, as adjuncts to medicine, centre stage. In the West it was with the rise of mechanism, in the form of biomechanics,

---

[1] G. E. R. Lloyd, *Adversaries and Authorities: Investigations into Ancient Greek and Chinese Science* (Cambridge, 1996), 6–7 (I have omitted Chinese characters from the quotation). Nevertheless, the idea that Greek philosophy was 'inquiry into nature' requires a very broad understanding of nature: see André Laks, *The Concept of Presocratic Philosophy: Its Origin, Development, and Significance* (Princeton, NJ, 2018), ch. 1. 'Nature' is a complex notion, and what is at issue in classical Greek thought should not be identified with the 'natural realm', which is something that emerged in the Middle Ages as a contrast with the supernatural realm.
[2] See Philippe Descola, *Beyond Nature and Culture* (Chicago, 2013).

Civilization and the Culture of Science: Science and the Shaping of Modernity, 1795–1935. Stephen Gaukroger, Oxford University Press (2020). © Stephen Gaukroger.
DOI: 10.1093/oso/9780198849070.001.0001

that the idea that resources that had been developed for purely physical processes could be used to explain those in living things. This approach did not completely disappear with the collapse of mechanism, and although straightforwardly reductionist projects did become less prevalent, this did not mean that attempts to find a place for the life sciences in a hierarchical conception of the unity of the sciences ceased. Rather, a new and less transparent form of commitment to the unity of the sciences emerged with the idea that the relation between the physical and the life sciences was one of increasing 'complexity', in which new properties emerged at separate tiers or levels as one moved away from the physical base. The postulation of emergent properties is certainly not confined to the life sciences, but consideration of the development of the life sciences enables us to highlight the way in which they have been caught up in questions about the unity of science in a context in which straightforward reductionism was widely recognized as an unpromising strategy. At the same time, I want to suggest that, if one is seeking a general model for science, then the life sciences offer something better than the physical sciences. Their lack of a grounding in a fundamental ontology and their explanatory pluralism have typically been seen as something to be remedied by assimilating them to a physical sciences model. But, as we shall see, these are actually strengths of the life sciences, and it is from them that we can learn lessons for the physical sciences, and science in general.

The nineteenth-century life sciences emerged in large part out of a fundamental reworking of natural history.[3] From as early as Aristotle's account of the sciences, natural history had always been treated as a merely descriptive form of classification, with no legitimate aspirations to scientific standing.[4] Such classifications tended to vary locally, and there had been attempts at various times to render natural-historical disciplines into a scientific form, one with some claim to universality. One of the earliest such attempts was Cesalpino's *De Plantis* (1583), which set out to provide a new rationale for botanical taxonomy which would transform it, along Aristotelian lines, into a genuine causally orientated form of natural philosophy. But this attempt, and similar ones, came to nothing.[5] In the early nineteenth century, however, a debate broke out between Geoffroy Saint-Hilaire and Cuvier on whether the animal body reflected a basic structure that could be understood through comparative anatomy, or whether it was a

---

[3] Generally, see: Wolf Lepenies, *Das Ende der Naturgeschichte: Wandel kultureller Selbstverständlichkeiten in den Wissenschaft des 18. und 19. Jahrhunderts* (Munich, 1976); Dietrich von Engelhardt, *Historisches Bewußtsein in der Naturwissenschaft von der Aufklärung bis zum Positivismus* (Freiburg im Breisgau, 1979).

[4] Two standard textbooks in the German university system entrenched this distinction from the second half of the eighteenth century: Johann Christian Polycarp Erxleben, *Anfangsgründe der Naturgeschichte* (Göttingen, 1768); Johann Christian Polycarp Erxleben, *Anfangsgründe der Naturlehre* (Göttingen, 1772).

[5] On the complexities of the debates over classification in the eighteenth and nineteenth centuries, see Harriet Ritvo, *The Platypus and the Mermaid and Other Figments of the Classifying Imagination* (Cambridge, MA, 1997).

matter of function rather than form, the structure of the body being a product of the functions that it performed.[6] Here was something that was beginning to look like a debate over the natural principles underlying the differentiation of kinds of animals: and a debate with a political edge, for if Geoffroy were right it was possible that all animal life (including humans) could be part of a continuous developmental series. At the same time, as part of a repackaging of the life sciences, there was a reduction of the scope of natural history, which had traditionally included animals, plants, and rocks, classified along the same general lines into families, genera, and species. The study of rocks was gradually being separated from the other areas of natural history classification, and by the early decades of the century animals had become the focus of enquiry, marginalizing botany from theoretical developments, despite the fact that at a practical level the identification of plants remained crucial since botany was one of the most commercially important scientific enterprises of the century.[7] The natural history of animals thus forms an appropriate starting point, and we can distinguish three main forms: the study of the 'life history' of animals; the study of the external and internal (anatomical) form and structure; and systematics, that is, the relationship between species and their organization, both past and present, into a larger systematic natural order.

The move from the natural history of animals to a 'scientific' zoology was manifested in a number of ways. Natural historians had traditionally not confined their attention to description and ordering, but had also concerned themselves with functional questions, such as reproduction and nutrition. But early in the nineteenth century, physiological investigation became increasingly separated from traditional museum and field studies. As Farber notes, researchers 'practiced natural history in museums, government agencies, and some zoology and botany faculties of universities, whereas work in physiology primarily took place in medical settings and in certain zoological or botanical departments and institutes', and the split led to different approaches to the study of life and to different conceptions of what the unity of the life sciences meant.[8] The split can be

---

[6] See Toby Appel, *The Cuvier–Geoffroy Debate: French Biology in the Decades before Darwin* (Oxford, 1987).

[7] As Janet Browne notes, 'botany during the nineteenth century was the most significant science of its day, creating and destroying colonial cash crops according to government policy and building the economic prosperity of a nation.' *Darwin's Origin of Species: A Biography* (London, 2006), 91–2. For details, see Lucile H. Brockway, *Science and Colonial Expansion: The Role of the British Royal Botanic Gardens* (New Haven, CT, 2002); and more generally Richard Drayton, *Nature's Government: Science, Imperial Britain, and the 'Improvement' of the World* (New Haven, CT, 2000).

[8] Paul Lawrence Farber, *Finding Order in Nature: The Naturalist Tradition from Linnaeus to E. O. Wilson* (Baltimore, MD, 2000), 73.

illustrated in the shift of emphasis in German biology journals in the 1830s and 1840s.[9] In 1835 the new journal *Archiv für Naturgeschichte* covered plants, animals, and fossil remains. Life histories and descriptive zoology were central topics, as were the modes of life and geographical distribution of species. But within a decade, the next generation of researchers sought a more 'scientific' approach to the subject. The *Zeitschrift für wissenschaftliche Zoologie*, founded in 1848, and which took as its field comparative anatomy, histology, embryology and morphology, and physiology, set out to have 'the most scientific character possible', and to this end excluded announcements of new species and genera unless they contributed directly to such areas as life histories and systematics. In the following year the physiologist Carl Vierordt was contrasting the purely 'descriptive' discipline of anatomy with the 'explanatory' discipline of physiology, writing that the former was of no use in solving physiological problems.[10] Carl Ludwig's influential handbook of human physiology published three years later begins with the statement that 'scientific physiology has the task of determining the functions of the animal body, and so of deriving them from its elementary conditions.'[11] As Nyhart notes, by shifting the emphasis from the general properties and appearances of living bodies to the determination of function, 'Ludwig's definition called for shifting physiology's task and also for abandoning its existing relationship to the study of form. Instead of focusing on the organization of the material substrate that made function possible, physiologists would concentrate on the physicochemical basis of function.'[12] Indeed, for Ludwig, morphology had nothing to offer, and in an 1854 article he completely rejected an anatomically based theory of reflex transmission, arguing that it was simply a matter of electrical impulses and was independent of the organism's anatomical organization.[13]

This bifurcation of the life sciences reinforces a division into two areas: on the one hand, there were those that depended on laboratory experimentation and microscopy, which claimed to offer causal accounts of the phenomena, and whose prospects for placement within a unified conception of science looked promising; on the other hand, there were areas traditionally deemed to have a purely descriptive non-causal standing, such as accounting in historical terms for rock

---

[9] See Lynn K. Nyhart, 'Natural History and the "New" Biology', in N. Jardine, J. A. Secord, and E. C. Spary, eds, *Cultures of Natural History* (Cambridge, 1996), 426–43.

[10] Carl Vierordt, 'Über die gegenwärtigen Standpunkt und die Aufgabe der Physiologie', *Archiv für physiologische Heilkunde* 8 (1849), 297–316.

[11] Carl Ludwig, *Lehrbuch der Physiologie des Menschen* (2 vols, Leipzig, 1852–6), i. 1.

[12] Lynn K. Nyhart, *Biology Takes Form: Animal Morphology and the German Universities, 1800–1900* (Chicago, 1995), 69. More generally, see Thomas H. Broman, *The Transformation of German Academic Medicine, 1750–1820* (Cambridge, 2002), ch. 6; and Andrew Cunningham, 'The Pen and the Sword: Recovering the Disciplinary Identity of Physiology and Anatomy Before 1800. I: Old Physiology—the Pen', *Studies in History and Philosophy of Biology and Biomedical Sciences* 33 (2002), 631–55.

[13] Carl Ludwig, 'Zur Ablehnung der Anmuthungen in Herrn R. Wagner in Göttingen', *Zeitschrift für rationelle Medizin* 5 (1854), 269–74. See Nyhart, *Biology Takes Form*, 70–2.

strata, or the study of the distribution and diversity of species.[14] The former studies were generally designated 'physiology', and until well into the nineteenth century biology and physiology were virtually synonymous expressions. As Coleman notes, to a numerous and rapidly expanding group of biologists, 'historical explanation was of little interest and probably irrelevant. Their subject was physiology'. Their aim was 'to probe ever more deeply into functional operations of the living creature. Virtually every phenomenon to greet their eye was caught up in the ceaseless flux of life, and the determination and control of that flux was the physiologist's grand objective'.[15] Morphology had also claimed this standing at the centre of biology, but the 'historical' approaches with which it was associated were, of necessity, pursued outside what were considered the core explanatory disciplines. By the 1850s, however, German morphologists were pitting themselves explicitly against reductionist physiologists such as Ludwig, Helmholtz, Virchow, du Bois-Reymond, and Brücke, defending the scientific standing of study of the organism as a teleologically defined whole.[16] And indeed in the wake of late nineteenth-century attempts to link developmental morphology and evolutionary theory, the impact of morphology ultimately outstripped any other eighteenth- or nineteenth-century scientific developments, and prompted questions about human self-understanding that were unprecedented in a scientific context. In this respect, the different paths taken by the life sciences in the nineteenth century allow us to explore just what benefits are offered by working outside unification strategies.

THE PHYSIOLOGICAL PATH

In 1801, Bichat had noted sarcastically, in the context of physiology, that 'it would certainly be a subject of very interesting research to determine how molecules, from being foreign to the vital properties, and possessing only physical ones, come by degrees to have the rudiments of vital properties'.[17] By mid-century, however, physico-chemical reduction had become well established in physiology. In an 1858 essay, Virchow writes: 'If we cannot do otherwise than think mechanistically about natural processes, we ought not to be reproached when we apply this kind of thinking to the course of *all* natural events.'[18] But

[14] But see Charles Wolfe, '"Cabinet d'histoire naturelle", or: The Interplay of Nature and Artifice in Diderot's Naturalism', *Persepectives on Science* 17 (2009), 58–77.
[15] William Coleman, *Biology in the Nineteenth Century: Problems of Form, Function, and Transformation* (New York, 1971), 11.
[16] See, for example, Oscar Schmidt, *Entwicklung der verleichenden Anatomie: Ein Betrag zur Geschichte der Wissenschaft* (Jena, 1855). See the discussion in Nyhart, *Biology Takes Form*, chs 2 and 3.
[17] Xavier Bichat, *Anatomie Générale, appliquée à la Physiologie et à la Médicine, Première Partie. Tome Premier* (Paris, 1801), lxviii.
[18] 'On the Mechanistic Interpretation of Life', in *Disease, Life, and Man*, 132.

matters are not so simple. In one respect, there is more at stake in the reduction of the life sciences to the physical sciences than there is in the reduction of the material sciences to the physical ones. Failure to reduce chemistry to the physical sciences could be treated by reductionists as a failure of chemistry to rise above the status of classification. Physicists had occasionally dismissed it as simply a matter of recipes and mere 'cooking',[19] and Rutherford is quoted as saying that all science is either physics or stamp collecting.[20] The last jibe works less well in the case of the life sciences. Chemistry could conceivably be dismissed as being incomplete without a mechanist reduction, but if the life sciences were treated non-mechanically they were in direct conflict with any reductionist aspirations that physics may have harboured. This is because, by contrast with chemistry, anti-mechanist advocates in biology had at least one clear alternative, a form of teleological holism. Du Bois-Reymond, in an 1876 talk on 'Darwin versus Galiani', while defending the mechanist alternative, sees this clearly:

For us the only knowledge is mechanical knowledge, no matter how poor a surrogate for true knowledge this may be, and consequently there is only one true scientific form of thought, the physical-mathematical. Because of this, there can be no mistake more serious than to believe that one has explained the purposefulness of organic nature by appealing to an immaterial intelligence, conceived in our image and acting purposefully . . . . As soon as one leaves the realm of mechanical necessity one enters the boundless foggy regions of speculation, gaining nothing.[21]

Compare this with the comments of a no less distinguished biologist, J. D. Bernal, thirty years later, in his presidential address to the Physiological Section of the British Association:

That a meeting point between biology and physical science may at some time be found, there is no reason for doubting. But we may confidently predict that if that meeting-point is found, and one of the two sciences is swallowed up, that one will not be biology.[22]

Clearly both views are compatible with the idea of an eventual unification of physics and the life sciences, but they are diametrically opposed on what such a unification would mean. Conceptions of the unity of science can vary radically, depending on the hierarchy that they deploy: in the simplest case, whether it is bottom up as in atomism, or top down as in holism.

---

[19] See Nye, *From Chemical Philosophy to Theoretical Chemistry*, 57.

[20] Quoted in J. D. Bernal, *The Social Function of Science* (London, 1939), 9. Rutherford's view of what physics included was also narrow, telling C. P. Snow that spectroscopy was just 'putting things into boxes': quoted in C. P. Snow, *Variety of Men* (London, 1939), 9.

[21] *Reden*, i. 560–1. The title of the paper comes from the fact that it begins with a debate between d'Holbach and Galiani, a guest of d'Holbach's French salon and opponent of his materialism.

[22] J. S. Haldane, 'An Address on the Relation of Physiology to Physics and Chemistry, Delivered before the Physiological Section of the British Association for the Advancement of Science, Dublin, 1908', *The British Medical Journal* (12 September 1908), 693–6: 696.

The question of alternative, and opposing, unification strategies goes back to the eighteenth century. As with chemistry, where micro-corpuscularian mechanical theories could not even begin to account for the chemical reactions that chemists were interested in, so in physiology, mechanical accounts could offer no plausible resources for understanding basic physiological processes. A good example is the action of glands. On the biomechanical account pursued by reductionists, glandular activity was simply the result of the compression of the glands by surrounding bones and muscles. But in 1752 Haller reported a number of experiments showing that the excretion of liquids from glands where they had been stored could not be the result of pressure because the space around the gland could not be altered in size. Glands were lodged in such a way that no muscles could exert pressure on them: a water-soaked sponge, for example, inserted into the parotid gland could not expel any water no matter how strenuous the movement of the jaw.[23] But a more general problem was the impossibility of a mechanical account providing a plausible distinction between the living and the non-living. The debates tended to hinge around the question of sensibility, with Whytt looking for a centralized source of sensibility in the organism, whereas Bordeu and others located sensibility, and life, in each individual organ, arguing that it is the collective activity of individual organs that make the organism something living.[24]

In 1797 we find Humboldt calling for a new branch of science, 'vital chemistry', investigating the chemical changes that take place in matter during the exercise of vital functions.[25] This turned out to be far more difficult than could have been imagined. In the early decades of the nineteenth century, chemists such as Berzelius sought, unsuccessfully, to produce relatively simple organic substances such as sucrose in the laboratory,[26] reinforcing a belief—supported by Lamarck for example[27]—that only life can produce organic compounds. Wöhler's 1828 synthesis of an organic compound, urea, from an inorganic one, ammonium cyanate, has traditionally been taken to be a decisive blow against such a view.[28] But in fact neither he nor his contemporaries were much interested in whether

[23] See the discussion in Elizabeth Williams, 'Sciences of Appetite in the Enlightenment, 1750–1800', *Studies in History and Philosophy of Biological and Biomedical Sciences* 43 (2012), 392–404.

[24] See Gaukroger, *The Collapse of Mechanism*, 394–402.

[25] Alexander von Humboldt, *Versuche über die gereizte Muskel- und Nervenfaser* (2 vols, Posen and Berlin, 1797–9), ii. 41–2. Ursula Klein has pointed out to me that, earlier, in his *Aphorismen aus der chemischen Physiologie der Pflanzen* (Leipzig, 1794), Humboldt had argued for a vital force that differs from all known forces, but in the *Versuche* he is clear that *Lebenskraft* ('life force') is the result of all known material forces. And Charles Wolfe has reminded me that, in his long entry on 'chymie' in the *Encyclopédie*, Vernel had called for a vital chemistry.

[26] See Bent S. Jørgensen, 'Berzelius und die Lebenskraft', *Centaurus* 19 (1964), 258–81.

[27] Lamarck, *Recherches sur les causes des principaux faits physiques* (2 vols, Paris, 1794), ii. 274.

[28] For example in Erik Nordenskiöld, *The History of Biology* (New York, 1928), 406–7.

this posed a problem for vitalism:[29] it was the isomorphism of the two substances that caught their attention.[30] In any case, ammonium cyanate was hardly a canonical inorganic compound, since it was (rightly) regarded as a decomposition product of organic material, and Wöhler had in fact obtained it through processing of material derived from hoofs and horns.[31] But there remained the more general question of what marked out the living from the non-living. Since what was at issue here were organic chemical compounds, not organic bodies, it might seem that there was no reason why the elimination of vitalism from chemistry should automatically have ruled it out in physiology. But matters are not so simple. At the end of the eighteenth century Kant, in his *Kritik der Urtheilskraft*, had set out the options for how we conceive of an apparently non-mechanical teleologically guided living realm in terms of four alternatives: life can be accounted for in terms of a Spinoza-like *conatus* in matter; a divine teleology regulating the world; a purely mechanical world in which teleology is a feature of the way in which we must conceive of living things and not something inherent in them; and finally hylozoism, the doctrine that living powers inhere in matter.[32] To these we might add a straightforward mechanical reduction that made no explanatory concessions to teleology at all. Since the question at issue was one in matter theory, namely whether matter contained vital powers or not, and since matter theory was chemistry, it was not possible to separate vitalist chemistry and vitalist physiology. As far as chemistry was concerned, if physiology was vitalist, it was because its chemistry was vitalist.

But linking physiology and the physical sciences did not have to proceed through chemistry: there were other possible routes. There was one traditional area that had a curiously intermediate standing between physics, chemistry, and the life sciences, namely electricity. As we have seen, the discovery of the electrolytic decomposition of chemical compounds in 1800 had an immediate effect in chemistry, leading Davy and others to argue that elements in compounds were held together electrically, and that chemical changes were essentially electrical in nature. Throughout the eighteenth century there had been a number of writers identifying electricity with a life force. In Britain, Rackstrow for example

[29] See John Hedley Brooke, 'Wöhler's Urea and its Vital Force?—A Verdict from the Chemists', *Ambix* 15 (1968), 84–114; Peter J. Ramberg, 'The Death of Vitalism and the Birth of Organic Chemistry: Wöhler's Urea Synthesis and the Disciplinary Identity of Organic Chemistry', *Ambix* 47 (2000), 170–95; and Ursula Klein, *Experiments, Models, Paper Tools*, 212–13.

[30] Isomers agree in chemical composition but differ with respect to their proximate components. In this case, one arrangement of the atoms of N_2H_4CO, namely ammonium cyanate, $[NH_4^+]$ $[NCO^-]$, is transformed into another form, urea, $(NH_2)_2CO$. The interest in isomerism lay in the fact that it was hard to account for unless one assumed real, structural molecules.

[31] See Russell, *The Structure of Chemistry*, Part I, 21.

[32] Immanuel Kant, *Gesammelte Schriften* (29 vols, Berlin, 1900 onwards), v. 388; Immanuel Kant, *Critique of the Power of Judgement* (Cambridge, 2000), 259. See the discussion in Gaukroger, 'Kant and the Nature of Matter'.

explicitly makes this connection,[33] and Berdoe accounts for the development of the embryo and the development of the species in terms of divesting of an electrical fluid.[34] With Galvani's experiments on the electrical stimulation of frogs' legs, electricity had begun to look briefly like a serious contender for a bridge between inert matter and living things.[35] Although Volta was able to show that the electrical stimulation could be described in purely physical terms, physiologists were nevertheless concerned to explore the connections between electrical stimulation and nervous activity,[36] and as late as 1838, the electrical experimenter Andrew Crosse announced the appearance of tiny spiders in the course of an electrocrystallization experiment in which a piece of volcanic rock was connected by two platinum wires to batteries:

At the end of fourteen days, two or three very minute white specks or nipples were visible on the surface of the stone, between the two wires, by means of a lens. On the eighteenth day these nipples elongated and were covered with fine filaments. On the twenty-second day their size and elongation increased, and on the twenty-sixth day each figure assumed the form of a perfect insect, standing on a few bristles which formed its tail. [See Fig. 6.1.] On the twenty-eighth day these insects moved with their legs, and in the course of a few days more, detached themselves from the stone, and moved over its surface at pleasure, although in general they appeared averse to motion, more particularly when first born.[37]

The press announced the 'creation' of insects, labelling them *Acarus crossii*, although Crosse himself considered the possibility that he had activated microscopic insect eggs embedded in his sample.[38]

Around the turn of the century, on the assumption of a distinctively holist commitment to the unity of science, attempts began to be made in the Romantic movement to incorporate the non-living into a general schema focused on the life sciences. In Germany, in the last decade of the eighteenth century, we witness a significant number of concerted attempts by different authors to assimilate, to one

[33] Benjamin Rackstrow, *Miscellaneous Observations, together with a Collection of Experiments on Electricity* (London, 1748).

[34] Marmaduke Berdoe, *An Enquiry into the Influence of the Electric Fluid in the Structure and Formation of Animated Beings* (Bath, 1771).

[35] See, for example, Johann Wilhelm Ritter, *Beweis, daß ein beständiger Galvanismus den Lebenprozeß im Tierreich begleitet* (Weimar, 1798); Johann Wilhelm Ritter, *Das elektronische System der Körper* (Leipzig, 1805); Christoph Heinrich Pfaff, *Über thierische Elektricität und Reizbarkeit* (Leipzig, 1795); Alexander von Humboldt, *Versuche über die gereizte Muskelfaser* (1797).

[36] See Naum Kipnis, 'Luigi Galvani and the Debates on Animal Electricity, 1791–1800', *Annals of Science* 44 (1987), 107–42; and Marie Jean Trumpler, 'Questioning Nature: Experimental Investigations of Animal Electricity in Germany, 1791-1810', unpublished PhD dissertation, Yale University, 1992.

[37] Andrew Crosse, 'On the Production of Insects by Voltaic Electricity', *Annals of Electricity* 1 (1836–7), 242–4: 242–3.

[38] See Iwan Rhys Morus, *Shocking Bodies: Life, Death and Electricity in Victorian England* (London, 2011), 141.

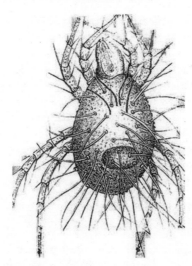

Fig. 6.1. Andrew Crosse, electrical insect

degree or another, physical and vital forces into a single notion of force or energy, nature itself becoming a holistic interplay of polarities.[39] In one of the classic texts of Romantic *Naturphilosophie*, Henrich Steffens begins his account with the formation of rocks via chemical processes, which he argues express two fundamental vital powers: a carbon-based power of vegetation and a nitrogen-based power of animation.[40] Moving through various forms of mineral and chemical polarities, and from the plants and animals formed from these materials, he concludes with an account of how a different type of force predominates at each level: reproductive in lower plants and animals, irritability in insects, sensibility in higher animals. The aim of the exercise, he makes explicit, is to grasp the unity of forces in nature and ourselves.[41]

---

[39] In just the last five years of the eighteenth century, for example, there are: Johann Christian Reil, 'Von der Lebenskraft', *Archiv für die Physiologie* 1 (1795), 8–162; Joachim Dietrich Brandis, *Versuch über die Lebenscraft* (Hannover, 1795); Christoph Wilhelm Hufeland, *Ideen über Pathologie und Einfluß der Lebenskraft auf Entstehung und Form der Krankheit* (Jena, 1795); Jacob Fidelis Ackermann, *Versuch einer physikalischen Darstellung der Lebenskräfte organischer Körper* (2 vols, Frankfurt-on-Main, 1797–1800); Friedrich W. J. Schelling, *Einleitung zu dem Entwurf eines Systems der Naturphilosophie. Oder über den Begriff der Speculativen Physik und die innere Organisation eines Systems dieser Wissenschaft* (Jena and Leipzig, 1799). The idea of polarities is due to Goethe. As Zammito notes: 'The two concepts of *Polarität* (polarity) and *Steigerung* (intensification) have come down to us as the quintessence of Goethe's vision of morphology and, more than that, the fountainhead of much that we associate with *Naturphilosophie*.' John H. Zammito, *The Gestation of German Biology: Philosophy and Physiology from Stahl to Schelling* (Chicago, 2018), 288.

[40] Henrich Steffens, *Beyträge zur inneren Naturgeschichte der Erde* (Freiberg, 1801).

[41] Henrich Steffens, *Grundzüge der philosophischen Naturwissenschaft* (Berlin, 1806).

Steffens' is a particularly elaborate account of the basic force underlying natural processes, full of highly speculative details. More streamlined understandings can be found in a number of other writers. In the first decade of the nineteenth century, Lamarck, one of the earliest defenders of the idea of species transmutation, denied that there were any fixed species, genera, orders, or families of animals or plants.[42] Buffon had denied that there were fixed species boundaries in nature, that they merge into one another, but Lamarck was making a different point: it is not that classifications are conventional, but that what holds at one time cannot be expected hold at another. Nature is dynamic, constantly changing, and matter is inherently active, operating via the medium of subtle fluids such as heat and light, which render the organism responsive to the environment. On this holist conception, ever-varying changes in the environment act on a wholly malleable organism, transforming it continuously. Here a conception of matter as something essentially and intrinsically living, and responsive to its environment, allows Lamarck to make sense of the transformationist idea of species descent with modification. At the same time Oken likewise starts from the assumption of the unity of matter, and asks how this can generate the diversity of the living world.[43] The solution, he argues, is to postulate an original mucous-like substance, which forms primitive spherical vesicles. These vesicles—which he terms 'infusoria'[44]—come together to form progressively more complex organisms, regulating not just the structure and function of organic bodies, but their development as well. The latter will prove to be especially important, for, on these holist notions, it is a single force that shapes all living things. As the physician and botanist Gottfried Treviranus puts it: 'We will therefore, first, view living nature as a whole that has always been, still is, and ever will be involved in making steady transformations; but also, second, assume in these transformations a fixed, lawful path.'[45] It is reasonable to expect that this lawful path is revealed in the full range of developmental processes. Oken writes that: 'Even should the account of these parallels not be everywhere correct, it follows clearly that there is a complete

---

[42] Jean-Baptiste-Pierre-Antoine de Monet de Lamarck, *Recherches sur l'organisation des corps vivans et particulièrement sur leur origine, sur la cause de ses développemens et des progrès de sa composition, et sur celle qui, tendant continuellement à la détruire dans chaque individu, amène nécessairement sa mort; précédé du discours d'ouverture du cours de zoologie, donné dans le Muséum national d'Histoire Naturelle* (Paris, 1802); Jean-Baptiste-Pierre-Antoine de Monet de Lamarck, *Philosophie zoologique, ou exposition des considérations relatives à l'histoire naturelle des animaux; à la diversité de leur organisation et des facultés qu'ils en obtiennent; aux causes physiques qui maintiennent en eux la vie et donnent lieu aux mouvemens qu'ils exécutent; enfin, à celles qui produisent les unes le sentiment et les autres l'intelligence de ceux qui en sont doués* (2 vols, Paris, 1809).

[43] Lorenz Oken, *Die Zeugung* (Bamberg, 1805); Lorenz Oken, *Lehrbuch der Naturphilosophie* (3 vols, Jena, 1809–11).

[44] The term usually referred to freshwater micro-organisms.

[45] Gottfried Reinhold Treviranus, *Biologie, oder Philosophie der lebenden Natur* (6 vols, Göttingen, 1802–22), iii. 4–5.

parallelism between the development of the foetus and that of the animal kingdom.'[46] And Meckel, one of the staunchest early supporters of the 'biogenic principle', the doctrine that ontogeny recapitulates phylogeny, writes in 1821 that 'the same organism, from its first appearance to a certain stage of its existence, runs through the most important levels of which the series of organic forms consists.'[47] The recapitulation doctrine—prominent in the early decades of the century,[48] then suffering something of a decline, but revived with a vengeance after the publication of *The Origin of Species* in 1859[49]—did not need a holist notion of a dynamically guided active nature, but there can be little doubt that this conception provided a foundation for both transformationism and recapitulation theory, shaping how they were thought of.

But to the extent to which the notion of an active nature took the form of a vitalist notion of matter, there were two developments that bore directly on, and transformed, debates over the nature of life, the one questioning vitalism, the other decisively destroying the idea that the issues had anything to do with the nature of matter. The first was the discovery by Helmholtz and Mayer that body heat and muscular action produced by animals could be derived from the physico-chemical process of oxidation of foodstuffs. In other words, the energy expended in muscular actions can be captured in physico-chemical, ultimately mechanical, terms. The work done by animals, as Helmholtz puts it, is an 'operation comparable in every respect with that of the steam-engine'.[50] This was particularly significant because of the central role played by force/energy in many vitalist accounts: it was a distinctive form of force/energy, after all, that marked out the living from the non-living. To have it shown that in muscular action—the paradigm case of the expenditure of force/energy in a living thing—nothing more than a mechanical process was involved, was a severe blow to the idea that the matter of living things exhibited irreducibly vital forces.

The second development was the emergence of cell theory.[51] Throughout the eighteenth century, in disputes between mechanists, hylozoists, and others, the question of life was located on the terrain of matter theory. As I have indicated, many of those working in the life sciences in the eighteenth century took the

---

[46] Oken, *Lehrbuch der Naturphilosophie*, ii. 387.

[47] Johann Meckel, *System der vergleichenden Anatomie* (5 vols, Halle, 1821), i. 9. See the discussion in Sander Gliboff, *H. G. Bron, Ernst Haeckel, and the Origins of German Darwinism: A Study in Translation and Transformation* (Cambridge, MA, 2008), ch. 1.

[48] It was particularly prominent in Britain in the 1820s and 1830s, where it was associated with Lamarckism: see Adrian Desmond, *The Politics of Evolution: Morphology, Medicine, and Reform in Radical London* (Chicago, 1989), 81–92.

[49] The revival lasted from the 1860s to around 1900, then going into decline again in the period from 1900 to the 1930s: see Nicolas Rasmussen, 'The Decline of Recapitulationism in Early Twentieth-Century Biology: Disciplinary Conflict and Consensus on the Battleground of Theory', *Journal of the History of Biology* 24 (1991), 51–89.

[50] Helmholtz, 'The Aims and Progress of Physical Science', *Science and Culture*, 216.

[51] See François Duchesneau, *Genèse de la théorie cellulaire* (Montreal and Paris, 1987).

irreducibility of the phenomenon of life to indicate that it was an intrinsic property of matter, and argued that the study of lifeless matter was simply a sub-species of matter theory, not a model for it. In this respect, there had been considerable empirical success in showing that a biomechanical approach to living things was mistaken, as in Haller's work. But matter theory had been successfully colonized by chemistry in the course of the eighteenth century, and as long as the life sciences were incorporated into matter theory, they could exercise no autonomy, constantly facing the threat of reductive approaches.

This changed in the early decades of the nineteenth century, with the elaboration of cell theory. From the mid-seventeenth century onwards, the earliest microscopists, notably Hooke, Grew, Malpighi, and Leeuwenhoek, were aware of the existence of cells, but for them cells were hollow chambers surrounded by solid walls, and their role was a purely structural one, Hooke arguing that cells were responsible for the buoyancy of cork for example. Grew and Malpighi had shown that plant tissues were largely composed of cells, and in 1808 Brisseau-Mirbel offered a comprehensive account of the cellular structure of plants, in which vessels and tubes are merely elongated cells.[52] With Dutrochet's work in the 1820s,[53] and that of Raspail in the 1830s,[54] cells, bolstered by the growing realization that all cells had nuclei, took on a physiological role, as the basic units of metabolic exchange. Tissues, which had played such a crucial role in physiology up to this point, gradually gave way to cells as the ultimate units of life.

Cell theory meant that the question of where the separation between the living and the non-living lies could be taken out of the realm of matter theory of the type envisaged by writers of Kant's generation, thereby transforming the kinds of resources to which those studying living things could appeal. At the same time, because cell theory arose from a generalization from similarities between the constituents of animals and plants, it produced a new form of unification of the study of living things, establishing a significant degree of autonomy for the life sciences. As Schwann put it in the Preface to his *Mikroskopische Untersuchungen*, the work that announced the discovery of cell function: 'The object of the present treatise is to prove the most intimate connexion of the two kingdoms of organic nature, from the similarity in the laws of development of the elementary parts of animals and plants.'[55]

---

[52] Charles-François Brisseau-Mirbel, *Exposition et défense de ma théorie de l'organisation végétale* (The Hague, 1808).

[53] René Joachim Henri Dutrochet, *Mémoires pour servir à l'histoire anatomique et physiologique des végétaux et des animaux* (Paris, 1824).

[54] François-Vincent Raspail, *Nouveau système de physiologie végétale et de botanique fondé sur les méthodes d'observation, qui ont été dévelopées dans le nouveau système de chimie organique* (Paris, 1833).

[55] Theodor Schwann, *Microscopical Researches into the Accordance in the Structure and Growth of Animals and Plants* (London, 1847), ix. The German original appeared as *Mikroskopische Untersuchungen über die Uebereinstimmung in der Struktur und den Wachstum der Thiere und Pflanzen* (Berlin, 1839).

Schleiden and Schwann's work made it clear that living things were made up from cells and cell products, and that cells possess all the properties of life: they are the basic units of living things. One fundamental question remained however, that of how cells arose in the first place. In the *Mikroskopische Untersuchungen* Schwann stated that: 'The principal result of this investigation is, that one common principle of development forms the basis for every separate particle of all organized bodies, just as all crystals, notwithstanding the diversity of their figures, are formed according to similar laws'.[56] It was Schleiden who had developed this account, whereby new cells derive from an undifferentiated chemical substance in the nucleus, which he called cytoblasm: small slime granules coalesce and form separate nucleoli which grow and become more sharply defined, emerging finally as cytoblasts over which an egg-shaped wall grows, following the analogy with the growth of crystals.

This account received less support than the alternative theory, which held that all cells arise from cell division, and the latter had become established by the 1850s,[57] but whatever the disputes surrounding the origin of cells, one thing was clear: from the 1830s onwards physiology was increasingly cell physiology. Neither mechanist dreams of reducing physics to atomistic point masses, nor those of chemists such as Dalton of reducing chemistry to hard, spherical chemical atoms, had borne fruit, but finally here was a discipline that made a plausible claim to identifying its functionally basic constituents. Moreover, unlike contemporaneous physical and chemical atoms, the understanding of cells did real work: all the various healthy and morbid functions of the body were the outward expressions of the activities of cells. Physiology and pathology were transformed forthwith.

One effect of the emergence of cell theory and the principle of conservation of energy—and one respect in which they can be seen to work in tandem—was a shift in the way in which vitalism was thought of and defended. A common eighteenth-century view was that espoused by Lamarck, for example, at the end of the century, when he wrote that all non-living material bodies were just 'the debris of living bodies, either as originally deposited or as subsequently altered by nature'.[58] But the abandonment of the assumption that life is a property of matter—a property which it loses when the organism dies, and which takes the form of a force or energy that is distinct from anything physical or chemical—did not mean that the vitalist programme was abandoned. It had slowly begun to take a different form, as early as Berzelius at the beginning of the century. The new separation was between the development of the embryo—the production of organic form—and the maintenance and regulation of ongoing physiological processes in the fully formed organism. If the latter could be accounted for fully

---

[56] Schwann, *Microscopical Researches*, ix.
[57] See Henry Harris, *The Birth of the Cell* (New Haven, CT, 1999), ch. 14.
[58] Lamarck, *Recherches sur les causes des principaux faits physiques*, ii. 376.

in physico-chemical terms, then vitalism could be restricted to embryonic development, although whether this can still be called 'vitalism' is open to question, for if the vital powers regulating embryonic development are no longer active in the living adult, it would seem that such 'vital' powers lose at least any straightforward connection with life. As regards the development of the embryo, no one was denying that there were physico-chemical processes at work, but for the anti-reductionist, the causality had to be of a different order from these.

PHYSIOLOGY AND CHEMISTRY: THE FAILURE OF REDUCTION

William Harvey bequeathed to his followers in the mid-seventeenth century three problems—bodily heat, respiration, and the nature of the blood[59]—and from the eighteenth century onwards respiratory physiology effectively became constitutive of physiology.[60] Lavoisier, building on the work of earlier experimenters such as Priestley, over a twenty-five-year period developed a theory of respiratory physiology that established a fundamental connection between respiration and combustion. He summarized the connection in these terms in a 1790 paper:

Respiration is just the slow combustion of carbon and hydrogen, similar in every respect to that which takes places in a lamp or a lighted candle, and from this viewpoint breathing animals are really combustible bodies that burn and are consumed. In respiration, as in combustion, it is the atmospheric air that supplies the oxygen and the caloric; but since in respiration what is involved is the very substance of the animal, it is the blood that supplies the combustible material, and if animals do not habitually replenish through taking food what they lose through respiration, the oil would soon run out in the lamp, and the animal would perish, just as a lamp would be extinguished when it runs out of fuel. The proofs of the identical effects of respiration and combustion can be deduced directly from observation. The air used up in respiration no longer has the same quantity of oxygen when it is expelled from the lungs; it contains not only carbonic acid gas [carbon dioxide] but also much more water than it had before it was inhaled.[61]

In both respiration and ordinary combustion, the combustible agent takes oxygen from the air and returns carbon dioxide. In neither case does nitrogen, despite being found in abundant quantities in the air, play any part in the process. 'Animal heat', long a source of puzzlement, was now not only definitively

---

[59] See Robert G. Frank Jr, *Harvey and the Oxford Physiologists* (Berkeley, 1980).

[60] See Charles A. Culotta, 'Tissue Oxidation and Theoretical Physiology: Bernard, Ludwig, and Pflüger', *Bulletin of the History of Medicine* 44 (1970), 109–40; and Everett Mendelsohn, *Heat and Life: The Development of the Theory of Animal Heat* (Cambridge, MA, 1964). On respiratory physiology up to the first half of the twentieth century see David Keilin, *History of Cell Respiration and Cytochrome* (Cambridge, 1966).

[61] Antoine-Laurent de Lavoisier, *Oeuvres de Lavoisier publiées par les soins du Ministère de l'Instruction Publique* (6 vols, Paris, 1864–1893), ii. 691–2.

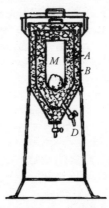

Fig. 6.2. Lavoisier and Laplace, ice calorimeter, engraved by Mme Lavoisier

identified as a product of chemical combustion, but was subjected to measurement. Lavoisier and Laplace devised an instrument, the calorimeter, to measure the quantity of heat produced in a given period of time per unit of carbon dioxide produced (Fig. 6.2). The machine contained a store of ice, and the heat was measured by the amount of ice it melted.

A number of problems with Lavoisier's account came to light in the early decades of the nineteenth century however.[62] First, the location of the supposed combustion was far from clear, with the lungs, the blood, and tissues all being suggested.[63] Second, experiments to demonstrate the presence of oxygen and carbon dioxide in the blood gave conflicting results.[64] Third, experiments reported in 1811 and 1812 indicated that the nervous system makes a significant contribution to animal heat.[65] Finally, experiments by Dulong and Despretz in the early 1820s measuring exhaled carbon dioxide failed to account for all the heat

[62] See Caneva, *Robert Mayer and the Conservation of Energy*, 49–68, to which I am indebted here.

[63] Lavoisier was convinced that the lungs were the site of respiration, but Spallanzani showed that insects and snails, which had no lungs, respired, and that even dead specimens were able to produce carbon dioxide, from which he concluded that respiration must occur in tissues: Lazzaro Spallanzani, *Mémoires sur la respiration* (Geneva, 1803), 65–70.

[64] Fifty years later, Gustav Magnus' reports of his experiments on absorption of gases in the blood finally cleared up outstanding issues: Gustav Magnus, 'Ueber die im Blute enthaltenen Gase, Sauerstoff, Stickstoff und Kohlensäure', *Annalen der Physik und Chemie* 40.4 (1837), 583–605.

[65] Benjamin Collins Brodie, 'The Croonian Lecture, on some Physiological Researches, respecting the Influence of the Brain on the Action of the Heart, and on the Generation of Animal Heat', *Philosophical Transactions of the Royal Society of London* 101.1 (1811), 36–48; Benjamin Collins Brodie, 'Further Experiments and Observations on the influence of the Brain on the generation of Animal Heat', *Philosophical Transactions of the Royal Society of London* 102.2 (1812), 378–93.

supposedly generated in terms of the combustion of oxygen.[66] In his influential textbook of physiology, Johannes Müller concluded that however probable the oxygen theory of animal heat was, 'one must nevertheless not forget that the production of heat does not depend on chemical processes alone, but is also subject to the influence of living parts'.[67] Müller, as a physiologist, was concerned with aspects of respiration that chemists, who regarded the animal simply as the site of a combustion process, ignored. This physiological approach was widely shared in the early decades of the century—even such an enthusiastic a supporter of Lavoisier as Magendie had to concede that respiration was only the principal source of animal heat, not its sole source[68]—and in 1830 the physiologist Friedrich Tiedemann set out the issue in these terms:

> As regards the cause of the generation of heat in animals, few phenomena of life have had so many diverse theories advanced about them as precisely these. . . . The physiologists who embrace the modern chemical doctrines think to find the cause of animal heat in the circumstances of respiration, which they likened to a combustion process taking place between the constituents of the venous blood and the oxygen of the inspired air, and whereby the blood liberated combines with the arterial blood and is distributed throughout the body. Other physiologists seek the source in digestion, nutrition, secretion, even the nervous system. Without here entering into the examination of those theories . . . so much is certain, that all of them afford no satisfactory explanation.[69]

Nevertheless, with the publication of Liebig's treatise on 'animal chemistry' in 1842, a purely chemical understanding of animal heat returned, and one a good deal more nuanced than that of Lavoisier.[70] Two years before the animal chemistry volume, Liebig had written a treatise on agricultural chemistry, devoted principally to the pressing problem of soil degradation, in which he treated decay as a form of slow combustion. He noted that, in the decay of wood, the proportion of carbon increases, whereas if the atmospheric oxygen had reacted directly with the carbon, then it should decrease. He concluded that it is the hydrogen in the wood that is oxidized by the atmospheric oxygen, while the carbon dioxide comes from elements of the wood. Carbon, he notes, never combines with oxygen at normal temperatures to form carbon dioxide.[71] So too

[66] Reports of the experiments were published as: César-Mansuète Despretz, 'Recherches expérimentales sur les Causes de la chaleur animale', *Annales de chimie et de physique* 26 (1824), 337–64; Pierre-Louis Dulong, 'Mémoire sur la chaleur animale', *Annales de chimie et de physique*, 3rd series, 1 (1841), 440–55.

[67] Johannes Müller, *Handbuch der Physiologie des Menschen für Vorlesungen*, vol. 1, pt. 2 (4th edition, Coblenz, 1841), 81.

[68] François Magendie, *Précis élémentaire de physiologie* (2 vols, Paris, 1836), ii. 519–20.

[69] Friedrich Tiedermann, *Physiologie des Menschen. Erster Band* (Darmstadt, 1830), 477. Quoted in Caneva, *Robert Mayer*, 56.

[70] Justus Liebig, *Die organische Chimie in ihrer Anwendung auf Physiologie und Pathologie* (Braunschweig, 1842).

[71] Justus Liebig, *Die organische Chemie in ihrer Anwendung auf Agricultur und Physiologie* (Braunschweig, 1840), 282.

in respiration: it is water, not carbon dioxide, that is formed in the lungs, with some of the uncombined oxygen being absorbed into the blood. As for the problem that the oxygen content of the carbon dioxide that is exhaled is greater than the amount of oxygen which is inhaled, Liebig solves the problem in terms of a theory of fat storage: oxygen must be removed from non-nitrogenous foodstuffs before it can be stored as fat, the purpose of which is to protect the body against atmospheric oxygen. Caneva points out that fundamental to Liebig's understanding of the animal economy was his conception of oxygen 'not as the sustainer of life but as an agent against whose destructive tendencies the vital processes are in constant struggle. Thus the carbonaceous materials oxidized in the body serve not only to generate heat, but also to ward off the potentially harmful effects of oxygen on the body as a whole.'[72]

Liebig put chemistry at the centre of physiology, but made it clear that this was not a question of a reduction of physiology to chemistry for him. Setting out his position in an 1846 article, he writes that the view that the explanation of 'vital phenomena can make do with chemical and physical forces alone' is:

the extreme consequence of a reaction against one which preceded it. In the era of philosophical physiology not yet long past one explained *everything* by means of the vital force. The reaction entirely rejects the vital force and believes in the possibility of being able to reduce all vital processes to physical and chemical causes. In the living animal body—so one said forty years ago—*other* laws hold sway than in inorganic nature, all processes are of *another* nature. Many of today's physiologists, on the contrary, consider them to be of the *same* nature. The unprofitability of both views for us lies in the fact that neither then nor now has one attempted to determine or discover the deviations in the effects of the vital force and in the actions of organic forces, or their similarity or identity.[73]

The issues in dispute among physiologists nevertheless hinged on chemistry. On Liebig's account, it is the oxidation of carbohydrates and fats that produce animal heat, and it is the 'plastic elements of nutrition' (proteins) which are broken down, yielding not only material products such as carbon dioxide, water, and the nitrogenous substances found in the urine, but also a force responsible for vital activity, manifested above all in muscular action. Because fats and carbohydrates only produce heat, the fundamental tissues of the organism must be made up of nitrogenous substances. The excretion of nitrogenous products thereby becomes a measure of muscular work. The problem was that there were clear counterexamples to the association of work with nitrogen excretion: bees, for example, seemed to survive wholly on sugar, which is nitrogen-free. Far from nitrogen excretion

---

[72] Caneva, *Robert Mayer*, 63.
[73] Liebig, 'Das Verhältniß der Physiologie und Pathologie zur Chemie und Physik, und die Methode der Forschung in diesen Wissenschaften', *Deutsche Vierteljahrs Schrift* 3 (1846), 169–243: 238. Quoted in Caneva, *Robert Mayer*, 217.

being a measure of muscular activity, everything indicated that it was instead directly related to oxygen consumption and carbon dioxide production.[74]

Liebig differed from Lavoisier on the metabolic role of oxidization, but not on its centrality to the processes of respiration. An essentially Lavoisian model seemed the only option, and the issue was posed in terms of an experimental physico-chemical approach being the only fruitful way of proceeding.[75] A year before Liebig's description of physico-chemical reductionism as an overreaction to vitalism, Mayer published his account of role of the conservation of energy in the life sciences.[76] As with Helmholtz's more mathematically elaborate statement of the conservation of energy published a year later, the construal of respiration, nutrition, and excretion in terms of energy exchange, the construal of muscular action in terms of energy, and the application of the principle of the conservation of energy to these interactions, established a physical understanding of energy at the core of physiology.[77] The principle of conservation of energy, in particular, established a quantitative equivalence between inputs and outputs. This does not definitively rule out some vital force being the source of the metabolic processes, but such a force would have to obey the physically formulated principle of conservation of energy, and this was beginning to make vital forces look redundant, which is the conclusion that Mayer and Helmholtz came to. Even more damaging to the idea of vital forces here is du Bois-Reymond's point that forces are not ontologically distinct causes of motion,[78] and as we have seen, conservation of force/energy is more properly described in terms of the intertranslatability of forms of energy rather than of some single hypostatized power or force. In this case, talk of a 'vital force' being involved would lose its rationale.

Physiology had become constitutive of biology for many physiologists by mid-century, especially in Germany, where Ludwig, Helmholtz, Virchow, du Bois-Reymond, and others pursued a reductionist programme. This programme was built on in various ways. Du Bois-Reymond, for example, in a series of elaborate experiments,[79] showed that the nerves and muscles produce electric currents that can be measured precisely, arguing that phenomena that had been taken to manifest vital forces most clearly, namely the activity of nerves and muscles, were in fact fully accountable in physical terms.[80] The trouble was that, when it

---

[74] See Coleman, *Biology in the Nineteenth Century*, 133.

[75] See, for example, Gabriel Gustav Valentin, *Lehrbuch der Physiologie des Menschen* (2 vols, Braunschweig, 1844) i. 151; Karl Vierordt, *Physiologie des Athmens, mit besonderer Rücksicht auf die Außcheidung der Kohlensäure* (Karlsruhe, 1845), 226.

[76] Julius Robert Mayer, *Die organische Bewegung in ihrem Zusammenhange mit dem Stoffwechsel. Ein Beitrag zur Naturkunde* (Heilbronn, 1845).

[77] The definitive experimental demonstration of the conservation of energy in living things came in a classic 1894 experiment by Max Rubner: see his *Kraft und Stoff im Haushalte der Natur* (Leipzig, 1909), 27.

[78] Du Bois-Reymond, 'Lebenskraft', *Reden*, i. 13.

[79] See the account of the experiments in Finkelstein, *Emil du Bois-Reymond*, 65–75.

[80] Du Bois-Reymond, *Untersuchungen über thierische Elektricitricität. Erster Band* (Berlin, 1848).

came to setting out these physical terms, he relied on the idea that the physical constituents exhibited an electrochemical dualism of the kind advocated by Berzelius, and as we have seen this turned out to be untenable in the case of organic compounds. Nevertheless, the reductionist programme, which treated the animal body as simply a site of physical and chemical processes, seemed to offer the best chances of successful explanation. Following what it identified as the distinctive experimental nature of the physical and chemical sciences, it was motivated to establish elaborate experimental programmes, and its proponents believed that, even if it left some things unexplained, it was nevertheless, by contrast with vitalist accounts, the only genuine explanatory enterprise. As du Bois-Reymond put it in a passage we have already cited, but which is sufficiently revealing that it is worth quoting again: 'For us the only knowledge is mechanical knowledge, no matter how poor a surrogate for true knowledge this may be, and consequently there is only one true scientific form of thought, the physical-mathematical.'[81]

But the strengths of an experimental physico-chemical approach in areas such as respiratory physiology were not matched in developmental physiology, which had always presented a dilemma for physico-chemical explanation. Descartes, for example, in offering a strictly mechanist account of the development of the foetus, had been faced with the problem of how, if the foetus is formed from inert undifferentiated matter which is the same in every animal, it can develop into one kind of animal in one case and a completely different kind in another. The answer, on his account, lay in the physically characterizable environment of the womb, which differs from one species to another, and causes variations in development. This understandably struck many as implausible, and his succes- sors, such as Malebranche, who was a staunch Cartesian in other respects, did not accept that epigenesis could explain how a complexly differentiated organism of a particular species could result from such a process, and concluded that preform- ation was the only solution.[82] Preformation theories triumphed over epigenetic accounts from the 1670s until the 1740s,[83] when Maupertuis revived epigenesis on a chemical, as opposed to a physical, basis. But Maupertuis does not simply substitute chemistry for physics, he adopts a new attitude to developmental questions. Jacques Roger has remarked of Maupertuis' work that 'it would be impossible to overemphasize the crucial influence of the reintroduction of dur- ation, an essential dimension, into the science of life. In the area of nature in its totality, Maupertuis found himself led to an examination of the data of heredity, to the idea of the history of life, and to the affirmation of a generally operant transformationism.'[84] It is with Maupertuis that the life sciences develop a

[81] Du Bois-Reymond, *Reden*, i. 560–1.
[82] See Gaukroger, *The Emergence of a Scientific Culture*, 337–46.
[83] See Gaukroger, *The Collapse of Mechanism*, 356–65.
[84] Jacques Roger, *The Life Sciences in Eighteenth-Century French Thought* (Stanford, 1997), 393.

distinctive developmental model, one which becomes firmly established with the publication of Buffon's immensely popular *Histoire naturelle*, the first volume of which appeared in 1749.[85]

Such questions deal with form and structure, and so fall under morphology, by contrast with physiology which, broadly defined, deals with function. That is, they fall on the 'wrong' side of the move to make biology 'scientific': they are, in their critics' view, descriptive rather than being concerned with seeking out causes. Yet they overlap with physiological questions in a number of areas, especially in embryology. And it is in those areas of biology that stand outside physiology that vitalist challenges to the standing of physiology as the ultimate form of biology arose. In the eighteenth and early nineteenth centuries, muscular action had been considered the paradigm case of vital forces. But the elaboration of the principle of the conservation of energy in a physiological context had seriously undermined vitalist aspirations to account for the regular functioning of animal bodies. The development of the embryo was a different matter, and in the course of the nineteenth century it replaced muscular action as the paradigm area in which the presence of non-physical natural forces, if they existed, could be detected. The idea was that this presence was manifested in the teleological guidance that regulated embryonic development.

The question of teleology in the formation of animals was not a new one, and it had been given a novel solution in Kant towards the end of the eighteenth century. Kant rejected any notion of goal-directedness in matter, arguing that the behaviour of everything in the natural realm is mechanical, but that living things also need to be accounted for in terms of final causes, which means offering teleological explanations. We cannot escape teleology in fully accounting for the behaviour of living things but, he argues, the need for teleological explanations tells us something about us, something about how we make sense of the world, not something about the world: we need to remove any element of teleology from matter itself, and to account for it instead in terms of how we perceive and make sense of particular kinds of activity.[86] The problem was that this strategy of removing teleology from the living realm appealed to no one: certainly not to his idealist successors, but neither to those thinkers who saw themselves as representing the continuation of the Kantian tradition, such as Trendelenburg and Lotze, who insisted that although the world is governed by mechanical laws, as Kant had argued, it is also governed by teleology or purpose, something they considered necessary on a number of grounds, not least the need to sustain ethical values in the face of science.[87]

The more realist version of teleology took a number of different forms. In its main vitalist version, an agent that actively selects and arranges matter in the

[85] George Louis Buffon, *Histoire naturelle, générale et particulière* (15 vols, Paris, 1749–67).
[86] Immanuel Kant, *Gesammelte Schriften* v. 359–83; *Critique of the Power of Judgement*, 233–55.
[87] See Frederick C. Beiser, *Late German Idealism: Trendelenburg and Lotze* (Oxford, 2013).

organism was postulated, and this was the view of Stahl, influential in the eighteenth century. Alternatively, purposeful forces could be disallowed, but phenomena such as irritability and sensitivity were treated as vital forces, not reducible to physical or chemical agents, as in Haller and Bichat. There was also a non-vitalist but still teleological reading whereby what was at stake was an 'emergent property' dependent on the specific order and arrangement of the components, a position taken up by Blumenbach, Reill, and Kielmeyer.[88] Kielmeyer provides a good statement of the position in an 1806 lecture:

All mixtures in the inorganic realm are pure chemical works, capable of being explained merely by the laws of chemical affinity as products of the affinity of matter. The mixtures in the organic realm on the other hand are either contrary to the laws of affinity which are observed to hold outside of organized bodies, or at least they are not formed according to them. The only exception to this general observation occurs in cases where the material of the organic body is expelled as a dead substance as in urine and even in lifeless bones. Here in the excreted parts of the organic body the normal affinities begin to reappear.[89]

But this leaves open the crucial question, namely how and by what means are the usual laws of affinity circumvented? It is at best mysterious how one can have both restriction to physico-chemical laws and a circumvention of these laws.

There is a clear dilemma here. On the one hand, simply insisting that living things can only be explained on the basis of the procedures built up for the explanation of non-living ones—the physical and chemical sciences—was not going to work. Even those reductionist physiologists such as du Bois-Reymond who insisted that physico-chemical explanation was the only acceptable kind of explanation admitted that this did not account for everything: it left some regions of the understanding of living things unexplained, perhaps on a permanent basis. On the other hand, vitalists, particularly in the wake of Helmholtz's elaboration of the principle of the conservation of energy, had no shared or coherent explanatory strategy, simply pointing to areas, such as foetal development, that could not be explained other than in goal-directed terms. But this in itself could hardly foster vitalism, since foetal development could just as easily be explained in terms of external guidance, that is, in traditional supernatural terms.

---

[88] See Timothy Lenoir, *The Strategy of Life: Teleology and Mechanics in Nineteenth-Century German Biology* (Chicago, 1982), ch. 1; and Eve-Marie Engels, 'Die Lebenskraft—metaphysiche Konstrukt oder methodologisches Instrument? Überlegungen zum Status von Lebenskräften in Biologie und Medizin im Deutschland des 18. Jahrhunderts', in K. T. Kanz, ed., *Philosophie des Organischen in der Goethezeit* (Stuttgart, 1994), 127–52.

[89] Carl Friedrich Kielmeyer, 'Allgemeine Zoologie, oder Physik der organische Körper', unpublished manuscript, quoted in Lenoir, *The Strategy of Life*, 51. Kielmeyer had set out his programme in an influential 1793 lecture, published as *Über die Verhältniße der organischen Kräfte unter einander in der Reihe der verschiedenen Organisationen, die Gesetze und Folgen dieser Verhältniße* (Stuttgart, 1793). On Kielmeyer see John H. Zammito, *The Gestation of German Biology*, ch. 9.

ESTABLISHING THE AUTONOMY OF PHYSIOLOGY

Finding a way out of this dilemma required that two basic questions be rethought. The first was whether the scientific standing of physiology is dependent on the extent to which it can be reduced to the physical and material sciences. The second was whether the distinctiveness of the life sciences, what marks them out as autonomous with respect to the physical and material sciences, depends on their ability to secure the existence of some non-physical force which distinctively shapes and guides biological processes. From the mid-1850s, Claude Bernard offered an account which effectively answered both these questions in the negative. What Bernard showed was that the scientific standing of physiology can be established on its own terms, without recourse to reductionism, and that its autonomy does not require the introduction of life-specific irreducible forces.

In important respects, Bernard's approach was not unprecedented, and the title of his main work—'an introduction to the study of *experimental* medicine'—gives an indication of its genre. It should be seen as continuous with the experimental natural philosophy tradition of the seventeenth and eighteenth centuries. Experimental natural philosophy dealt with connections between the phenomena by exploring the relations between these at the phenomenal level, rather than seeking to explain them in terms of something more fundamental. For its exponents, such as Boyle, Newton, Gray, and Geoffroy, the goal was to show that such forms of explanation were complete in themselves, and not merely a preliminary to some ultimate micro-reductive account. Geoffroy, construing chemistry as an intrinsically experimental discipline, used such an approach to bring unity to chemistry: it became the study of affinities between chemical substances. Bernard also construed his discipline, physiology, as intrinsically experimental and showed that this not only brought an organizational coherence to physiology, but that in the process it secured its standing as an autonomous area of scientific enquiry. Physiology had taken a decidedly experimental turn by the middle of the nineteenth century. As Kremer points out, in the case of physiologists such as du Bois-Reymond, Brücke, and Helmholtz, by 'cultivating relationships with instrument makers, military engineers, physicists, and mathematicians, these "organic physicists" in the second half of the nineteenth century made the kymograph, galvanometer, nonpolarizing electrodes, thermocouple, mercury manometer, and the frog's isolated gastronemius muscle into veritable icons of experimental physiology'.[90]

---

[90] Richard L. Kremer, 'Physiology', in Peter Bowler and John Pickstone, eds, *The Cambridge History of Science, Volume 6: The Modern Biological and Earth Sciences* (Cambridge, 2009), 342–66: 346.

In the first instance, it was autonomy from medicine that was at issue for physiologists, and Bichat played a crucial role here.[91] In the late eighteenth and early nineteenth centuries, physiologists learned their skills in hospitals, and these were surgical skills rather than theoretical ones. By contrast, between 1798 and his untimely death in 1802, Bichat pursued an elaborate experimental programme at the Hôtel-Dieu, Paris' largest and oldest hospital, using his highly refined surgical skills to explore the structural and functional commonalities between organs. He argued that it was not in the organs themselves that analogous normal and pathological phenomena could be found however, but at a more basic level, that of the tissues making them up, and he identified twenty-one different kinds of tissues, each characterized in terms of their degrees of sensibility and irritability. These tissues were not just anatomical features, but had physiological significance—such significance that they effectively became the defining subject matter of physiology.

Bichat draws a sharp distinction between the physical and chemical sciences and the study of life, although he sees an analogy between physical properties such as gravitation and chemical properties such as affinities, on the one hand, and vital properties on the other.[92] These vital properties, irritability and sensitivity, are what allow the living body to counter the physical and chemical forces that would destroy it. In the course of his studies, he moves the location of irritability and sensitivity from the organs to the tissues making up the organs. The tissues, he tells us, are the real organizing elements of our bodies, whose nature is constantly the same wherever they may be found:[93] it is in the examination of specific tissues, in a pathological state and subjected to therapeutic agents, that the future of pathological anatomy and therapeutics lie.[94]

Bichat had defined life as the ensemble of functions which resist death, and the task for physiologists was to identify what resources one needed to draw upon to establish what these functions were and how they operated. Bernard had significant advantages over Bichat in this respect. By the early 1820s, the comparative physiologist Henri Dutrochet, one of Bichat's younger contemporaries, began to pursue microanatomy, and he included both animals and plants in his studies.[95] Tissue theory proved problematic in the long term, and the problems were compounded by Bichat's rejection of the use of microscopes in anatomy and physiology.[96] Developments in microscopy in the 1830s made microscopes an

[91] See Erwin Ackernecht, *Medicine at the Paris Hospital, 1794–1848* (Baltimore, 1967).
[92] See Bichat, *Anatomie Générale*, xxxvii: 'In creating the universe God endowed matter with gravity, elasticity, affinity . . . as well, one part received as its share sensibility and irritability.' See Charles Wolfe, 'On the Role of Newtonian Analogies in Eighteenth-Century Life Science: Vitalism and Provisionally Inexplicative Devices', in Z. Biener and E. Schliesser, eds, *Newton and Empiricism* (Oxford, 2014), 223–61.
[93] Bichat, *Anatomie Générale*, lxxix–lxxx.   [94] Ibid., xxxv–xxxviii
[95] John V. Pickstone, 'Vital Actions and Organic Physics: Henri Dutrochet and French Physiology During the 1820s', *Bulletin of the History of Medicine* 50 (1976), 191–212.
[96] Bichat, *Anatomie Générale*, 51.

indispensable tool of anatomy, physiology, and pathology. As we have seen, in 1837 Schleiden and Schwann announced a fundamental anatomical commonality when they showed that all animals and plants shared a cellular structure, and with the investigations of Ducrochet and others into cell physiology, the cell replaced tissues as the physiological building blocks of living things.

In the 1850s, Bernard reinforced the emphasis on animal physiology and the mix of experimental and pathological approaches that he had inherited from his teacher, Magendie.[97] In his 1865 *Introduction à l'étude de la médecine expérimentale*,[98] he rejects traditional nosological approaches to disease as passive and descriptive. The assumption behind these traditional approaches was that diseases are distinct entities, to be identified through pathological anatomy. Rather, Bernard sets out to show that they are in fact distorted physiological functions. In this respect, he owed a debt to Virchow, who pursued most vigorously the identification of disease as pathologies of particular sites in the organism, urging that these sites not be identified with tissues but rather with cells.[99] These developments were crucial for Bernard's conception of pathology as a branch of experimental physiology.[100] At the same time, the *Introduction* demonstrates a move away from his early experimental researches on higher mammalian functions towards a concern with providing a unitary explanation for the phenomena of life common to animals and plants: in other words, a general theory of the organism.[101]

Bernard's account of the organism has sometimes been assimilated to positivism, in that he rejects a metaphysical search for underlying causes and insists that physiology must confine itself to observable relations between the phenomena: indeed he rejects the search for a definition of life as metaphysical, and insists instead that we learn from observations and experiment what life is. Alternatively, his account has been placed ambiguously between vitalism and physico-chemical reductionism: vitalism because he insists that the laws governing the behaviour of the organism are not those of physics or chemistry, physico-chemical reductionism because the activity consists of nothing but physico-chemical processes. But these attempts to assimilate his understanding of the organism to what are taken

[97] See the account of Bernard's relation to Magendie in John E. Lesch, *Science and Medicine in France: The Emergence of Experimental Physiology, 1790–1855* (Cambridge, MA, 1984); see also the exhaustive account of Bernard's career in physiology in Frederic Lawrence Holmes, *Claude Bernard and Animal Chemistry: The Emergence of a Scientist* (Cambridge, MA, 1974).

[98] Claude Bernard, *Introduction à l'étude de la médecine expérimentale* (Paris, 1865).

[99] See, for example, Rudolph Virchow, *Die Cellularpathologie in ihrer Begründung auf physiologische und pathologische Gewebelehre* (Berlin, 1858).

[100] See Georges Canguilhem, *Etudes d'histoire et de philosophie des sciences* (Paris, 1975), 127–71; and Mirko D. Grmek, *Raisonnement expérimental et recherches toxicologiques chez Claude Bernard* (Geneva, 1972).

[101] As well as the *Introduction*, the other crucial works of Bernard on these questions are: *Leçons sur les phénomènes de la vie communs aux animaux et aux vegetaux* (2 vols, Paris, 1878–9); *Cahiers de notes, 1850–1860*, ed. M. D. Grmek (Paris, 1965); and *Leçons de physiologie opératoire* (Paris, 1879).

to be defining positions in the philosophy of the life sciences mask what is in reality both the tradition in which his work stands, and the novelty of the way in which he reworks this tradition. The insistence on sticking to observable causes is crucially connected with the idea that explanation consists in discovering the conditions under which the phenomena being investigated are manifested. Experiment is observation elicited under specified constraints. Discovery of the conditions under which the phenomenon is produced means that one should ideally be able to work backwards: the phenomena should be reproducible by reproducing these conditions. Understanding the phenomena ultimately takes the form of the ability to manipulate them, certainly a goal of experimental physiology if it is to serve medical ends, but also something of more general import, for this goal is not something specific to physiology. It is a central feature of the experimental natural philosophy tradition, and Bernard's concern with manipulation in physiology exactly mirrors Gray's concern with manipulation in his experiments on electricity over a hundred years earlier, which were similarly motivated by a concern to break away from fruitless concerns with some 'ultimate' level of explanation.[102]

The explicit assumption in Bernard's treatment of vital processes in the organism is that what is at issue is a straightforwardly causal process, so to the extent to which a commitment to vitalism is believed to require non-determinist forces, there is a clear disavowal of vitalism. The processes are deterministic, and they are physico-chemical, because there is nothing else they could be. But an understanding of physics and chemistry will not in itself yield an understanding of the behaviour of the cellular constituents of organisms. Such an understanding rests on a crucial distinction that Bernard makes between the external environment of the organism and its internal environment (*milieu intérieur*).[103] The external environment in which the organism lives affects it in various ways, and for the organism to live an independent life it needs regulatory mechanisms that secure the stability of the internal conditions of the organism. Some kind of equilibrium feedback mechanism has to be involved. In higher animals, plasma mediates between the cells and the external environment and secures the integration of the chemical and physical processes that occur in the organism, an integration which no physical or chemical laws could regulate. Bernard has no doubt that manifestations of vital phenomena are determined by physical and

---

[102] See Gaukroger, *The Collapse of Mechanism*, 203–6. There is a significant philosophical literature construing causation in terms of manipulation: for example R. G. Collingwood, *An Essay on Metaphysics* (Oxford, 1940); Georg Hendrik von Wright, *Explanation and Understanding* (Ithaca, NY, 1971); Peter Menzies and Huw Price, 'Causation as a Secondary Quality', *British Journal for the Philosophy of Science* 42 (1991), 157–76; and James Woodward, *Making Things Happen: A Theory of Causal Explanation* (Oxford, 2003).

[103] See the discussion in Joseph Schiller, *Claude Bernard et les problèmes scientifiques de son temps* (Paris, 1967), ch. 11.

chemical conditions: what he is denying is that their behaviour can be captured in terms of physical and chemical laws.

In the *Introduction*, Bernard offers a comprehensive statement of his general approach. The sciences serve as instruments for one another: mathematics is an instrument for physics, chemistry, and biology, while physics and chemistry are instruments for physiology and medicine. But 'physicists and chemists are not mathematicians because they make calculations; physiologists are not chemists or physicists because they make use of chemical reagents or physical instruments, any more than chemists and physicists are physiologists because they study the composition or properties of certain animal or vegetable fluids or tissues'. He continues:

> Each science has its problem and its point of view which we may not confuse without risk of leading scientific investigation astray. Yet this confusion has often occurred in biological science which, because of its complexity, needs the help of all the other sciences. We have seen, as we still often see chemists and physicists who, instead of confining themselves to the demand that living bodies furnish them suitable means and arguments to establish certain principles of their own sciences, try to absorb physiology and reduce it to simple physico-chemical explanation. They offer explanations or systems of life which tempt us at times by their false simplicity, but which harm biological science in every case, by bringing in false guidance and inaccuracy which it then takes long to dispel. In a word, biology has its own problems and its definite point of view; it borrows from other sciences only their help and their methods, not their theories. This help from the other sciences is so powerful that, without it, the development of the science of vital phenomena would be impossible. Prior knowledge of the physico-chemical sciences is therefore decidedly not, as is often said, an accessory to biology, but, on the contrary, is essential to it and fundamental. That is why I think it is proper to call the physico-chemical sciences allied sciences, and not sciences accessory to physiology.[104]

Bernard's approach differs from those of other eighteenth- and nineteenth-century writers on the life sciences in a number of respects, but there is one that stands out: he does not see ontological considerations as having any role in our understanding of vital phenomena. In this respect he differs not only from reductionist physiologists and philosophers, but also from what Timothy Lenoir has identified as a Kantian programme that lies at the origins of biology in the nineteenth century.[105] Lenoir argued that the members of the 'Göttingen School'—Treviranus, Blumenbach, Reill, and Kielmeyer—developed a research programme in the life sciences that used Kant's conception of the life sciences to steer a path between mechanism and vitalism, which had generally been treated as standing in strict opposition to one another. The solution was to treat teleological forces as a necessary heuristic device, while being committed to mechanism alone at the ontological level. As a general programme for saving a commitment to

---

[104] Claude Bernard, *An Introduction to the Study of Experimental Medicine* (New York, 1957), 95.
[105] Lenoir, *The Strategy of Life*.

mechanism, Kant's approach was unsuccessful, if only because mechanism failed not only in the life sciences but also in chemistry and even in physics, where teleology was not at issue.[106] But the fundamental problem is not so much the assumption that mechanism is ontologically primitive, but that ontology has anything to do with how we explain the properties of living things.

BEYOND EMERGENT PROPERTIES

The assumption behind general unification projects that take the physical sciences as a model is that it is because the physically characterizable constituents of things are as they are that the things of which they are the constituents have the properties they do. In other words, physics is the common denominator of all natural phenomena, and accordingly it should be possible to construct a ranking or tiering of phenomena on the basis of how close they are to physics. But we have seen that this approach is, at best, a gross oversimplification of how even the parts of the physical sciences themselves are related. Leaving to one side the problems that arise for reductionism once we take seriously the fact that it is models, not theories, that do much of the explanatory work in physics from the nineteenth century onwards, there are significant difficulties involved in the reconciliation of conflicting theories within the discipline of physics, and these are compounded and effectively made irresoluble once projects for the unity of science are mapped on to them. The problem is that such metalevel unification projects envisage something that operates at a level of abstraction where any empirical difficulties are treated as mere temporary obstacles on the route to unification. There are two (unstated) assumptions of the metalevel conception that are crucial to its resilience in the face of the failures of reductionism. The first is the assumption that there *must* be a unity of science: that it is not just an empirical matter of fact that science is a unified enterprise. This effectively gives the unity of science an a priori status, meaning that failures to unify have no evidential bearing on its general standing. The second assumption provides the rationale for the first. It is that the natural

---

[106] See Gaukroger, 'Kant and the Nature of Matter'. A number of critics have pointed out that the various members of the Göttingen School were in fact far more closely indebted to vitalism than Lenoir allows: see Robert J. Richards, 'Kant and Blumenbach on *Bildungstrieb*: A Historical Misunderstanding', *Studies in the History and Philosophy of Biological and Biomedical Sciences* 31 (2000), 11–32; Reill, *Vitalizing Nature*; John H. Zammito, 'The Lenoir Thesis Revisited: Blumenbach and Kant', *Studies in History and Philosophy of Biological and Biomedical Sciences* 43 (2012), 120–32; Andrea Gambarotto, 'Vital Forces and Organization: Philosophy of Nature and Biology in Karl Friedrich Kielmeyer', *Studies in History and Philosophy of Biological and Biomedical Sciences* 48 (2014), 12–20. Moreover, it seems clear that Kielmeyer, although a student of Blumenbach's, developed his ideas independently of Blumenbach: see Frank Dougherty, 'Über den Einfluß Johann Friedrich Blumenbachs auf Kielmeyers feierliche Rede von 1793: Mit einer Anhang über Kielmeyers Göttinger Lektüre', in K. T. Kanz, ed., *Philosophie des Organischen in der Goethezeit* (Stuttgart, 1944), 50–80.

realm itself is unitary: as if there were some master plan behind nature that makes unification a universal requirement of a complete explanation.

Without reduction, can we still have the unity of science? The doctrine of emergent properties has been supposed to fill in the gaps where reduction ceases to work. The claim is that as one moves from one tier or level to the next, one moves to a new level of 'complexity', and new properties emerge which are not reducible to the supposedly more fundamental properties but which in some way still result from them. In his discussion of the prospects for reduction in the life sciences, Ernst Mayr writes that his discussion of reductionism 'can be summarized by saying that the analysis of systems is a valuable method, but that attempts at a "reduction" of purely biological phenomena or concepts to laws of the physical sciences has rarely, if ever, led to any advance in our understanding. Reduction is at best a vacuous, but more often a thoroughly misleading and futile, approach. This futility is particularly well illustrated by the phenomenon of emergence.'[107] Emergence, he continues, 'is a thoroughly materialistic notion. Those who deny it . . . are forced to adopt pan-psychic or hylozoic theories of matter.'[108] There are two issues at stake in the doctrine of emergent properties here. The first is an ontological point about the respective merits of vitalism and materialism in the life sciences: Mayr's view is that unless we adopt the latter we are de facto committed to the former. The second is a point about the lack of explanatory power of reductionism, a lack supposedly made good by the postulation of emergent properties.

The doctrine of emergent properties has become the preferred option in explaining why we need to keep in place the hierarchy of sciences—something that is constitutive of the unity of science—while acknowledging the failure of strict reductionism. Its function is to account for the inability of physical explanation to provide a complete account of the phenomena at what we might term the higher levels, or levels of greater 'complexity', while retaining the primacy of the physical level of description. An overriding motivating factor in the advocacy of the unity of science relies on an appeal to ontology, specifically on a mapping of ontological onto explanatory considerations. Emergent properties are offered as an alternative to reductionism, an alternative that promises the benefits of reductionism without any of the costs. The shared assumption is that it is because physics is as it is that everything else is as it is: the ultimate constituents from which everything is composed are just the kinds of things described in physics, and it is the peculiarities of the arrangements and states of these constituents that give rise to the contents of the different disciplines. But whereas the reductionist believes that this means that all we need to know about, ultimately, is physics, the advocate of emergent properties by contrast argues that, at the explanatory level,

---

[107] Ernst Mayr, *The : Diversity, Evolution, and Inheritance* (Cambridge, MA, 1992),63.
[108] Ibid., 64.

we need to invoke different laws and different kinds of descriptions, for example for chemistry, biology, and consciousness.

Mayr's characterization of the benefit of emergent properties involves a wholly unjustified and unnecessary ontologization of the issue of how the life sciences could be autonomous with regard to the physical and material sciences and yet eschew vitalism. What does his claim that we need a commitment to materialism amount to? Materialism was defended in the *philosophes* of eighteenth-century Paris and the German 'scientific materialists' as part of a programme of reduction of the mental to material processes, a denial of any spiritual realm, i.e. atheism, and was often accompanied by a commitment to radical politics. This is not what is being advocated by Mayr. Rather it is a view of materialism as something whose legitimacy derives from the physical sciences. But, in a situation bearing an analogy to the lack of fit between the physical and the material sciences, materialism in a literal sense has been left behind in the course of the twentieth century as developments from geometrodynamics to string theory have postulated very unmaterial-looking constituents of the physical universe. It is the physical sciences, not the material sciences, that bear the ontological load, and as a consequence materialism has become physicalism. In other words, the commitment to materialism translates into an open-ended commitment to the physical sciences. But what does this open-ended commitment amount to? It can only be a commitment to the priority of the physical sciences, and this in turn can only be an explanatory priority.

But if we need different kinds of explanations for these different kinds of phenomena, we can ask what the rationale is for labelling them 'emergent', as opposed to autonomous. The term suggests that, even though everything is ultimately just physics, the natural realm is layered in a way not describable in physics.[109] Consider the idea of 'supervenience', which in this case amounts roughly to the claim that there can be no difference in the emergent properties without a corresponding difference in the basic physical ones. There are certainly *prima facie* cases where relations such as supervenience hold, but the cases that we have been concerned with in our discussion of the natural and the life sciences are not like this. The core problem for emergent properties is the suggestion that the new properties come into existence in a way not describable either by physics or by the 'emergent' disciplines, which simply take the new properties as given. The way in which these new phenomena 'emerge' out of purely physical ones should surely appear problematic to advocates of materialism or physicalism. But the

---

[109] Jaegwon Kim argues that reductionism and emergent properties are incompatible: 'if emergent properties exist, they are causally, and hence explanatorily, inert and therefore largely useless for the purpose of causal/explanatory theories. If these considerations are correct, higher-level properties can serve as causes in downward causal relations only if they are reducible to lower-level properties. The paradox is that if they are so reducible, they are not really "higher-level" any longer.' 'Making Sense of Emergence', *Philosophical Studies* 95 (1999), 3–36: 33.

crucial point is that what generates these emergent properties is not an explanatory requirement but an ontological one.

The ontological requirement is physicalism: the doctrine of emergent properties assumes the truth of physicalism, as a statement of ontological faith, and then sets out cases where it deems the resources of physicalism to be insufficient. The term 'emergent properties' may be relatively recent, but the phenomenon of hierarchical ordering and the appreciation of its problems isn't. The doctrine resembles a certain kind of property classification that was abandoned in the early modern period, namely the hierarchical classification of things into plants, animals, and humans in terms of the emergence of various kinds of 'soul': vegetative souls, responsible for lower functions such as nutrition; sensitive souls, conferring the power of sensation; and rational souls. Descartes was one of the most effective critics of this kind of approach, pointing out that it was no more than an exercise in labelling.[110] The vegetative and sensitive 'souls' are merely names invented for what are considered to be significant qualitative differences. These labels have no explanatory value, but they were presented as if they furthered our understanding in some way. This is a charge to which emergent properties accounts are similarly susceptible: labelling its problems as if this were a way of coming to terms with them.

Theories of emergent properties and supervenience nevertheless have a general resilience, one that confers on them something like an a priori standing. We need to distinguish between two types of developments here. The first are those within disciplines that are designed to resolve inconsistencies or discrepancies between different theories, in the course of which deeper connections may (or may not) be revealed. The second operate not at the level of science as such, but at that of metascience, whose functions include providing an overriding rationale for science, providing a legitimation of the generality of its claims, and offering an account of the relation between scientific and non-scientific questions. At this metascientific level, ontology—a general account of what kinds of things there are, in the modern era typically taking the form of a speculative matter theory

---

[110] Even explicit exercises in hierarchical classification which have no metaphysical pretensions encounter profound difficulties. Linnaeus, for example, had offered a natural-historical distinction: 'Minerals are unorganized; vegetables are organized and live; animals are organized, live, feed and move spontaneously'. Richard Owen, in his *Lectures on the Comparative Anatomy and Physiology of the Invertebrate Animals* (2nd edition, 1855), starts with this definition of Linnaeus, and by contrast with an increasing number of anatomists from Buffon onwards who insisted that the distinction between plants and animals was merely conventional, he was clear that this division, at least, was fixed in nature. But he was unable to secure the distinction. As Jan Sapp points out: 'Movement, as the mark of an animal, was belied by some of Ehrenberg's Polygastria, which possess locomotion like an animal but released oxygen like a plant. Friedrich Wöhler had reported in 1843 that some of the free and locomotion Polygastria eliminate pure oxygen; and on the other hand, mushrooms and sponges exhale carbon dioxide. "Chemical antagonism", Owen said, "fails as a boundary line where we must require it—viz., as we approach the confines of the two kingdoms."' Jan Sapp, *The New Foundations of Evolution: On the Tree of Life* (Oxford, 2009), 22. It is clear that translation of the Linnean distinction into emergent properties would have been wholly counterproductive and misleading.

which offers itself as a philosophical reflection on selected developments in the physical sciences—plays a crucial role. Advocates of the unity of science tend to run these two together, effectively taking the second as simply a rationalization of the first. But the two make very different kinds of claims, and their motivations and goals differ radically. Unless we distinguish them we will not be able to understand just what is actually doing the work in attempts to unify the sciences.

If it were the case, as modern proponents of ontology evidently believe, that science sets out ultimately to tell us about the nature of reality, identifying the basic kinds of things that exist in the world, then ontology might provide some guidance in understanding the relations between the sciences. But how seriously can we take this view of what science does? Note that the strategy it relies upon is not new, but, conceptually speaking, a reworking of scholastic metaphysics, where it was developed most fully in the context of exploring the relation between natural-philosophical and theological theories, which had different sources and employed different procedures, and which occasionally seemed to conflict. It was considered that the conflict could not be real, however, for surely there must be a unity of understanding, a unity that reflected the fact that the natural world had been created by a single creator with consistency of purpose. The approach worked reasonably well in that context, at least before the sixteenth century, but its revival in a secularized form as a theory about the relation between the sciences does its greatest damage—whatever other misconceptions it engenders, and whatever other dead ends it leads us to—by locking us into a conception of the unity of the science as an unquestioned assumption behind the whole project, because the unity of science, which is contentious, is treated as if it were a reflection of the unity of 'reality', which is supposedly uncontentious. But the latter is neither contentious nor uncontentious: it is simply meaningless in a secular context, for what on earth could the 'unity of reality' refer to? What would a 'disunity of reality' look like? This misconceived way of proceeding simply begs the question of the unity of science, at the same time indicating why it could have been treated as something with a virtually a priori standing. The whole point of the historical approach that we are taking is to bring to light just how contingent the emergence of the idea of the unity of science has been in the modern era, and how the motivations behind it, and the forms of legitimation that have been used to defend and encourage it, turn out, with a little probing, to have been anything but the kinds of thing that we would expect of something a priori or conceptually fundamental.

Once we confine our attention to explanation, it is impossible to conceive how emergent properties could be anything more than an exercise in labelling. Stripped of their ontological rationale, what could their role be other than indicating where the kinds of explanation that are considered fundamental cease to explain, and in that case what exactly is 'fundamental' about these? The history of science is littered with examples of labelling masquerading as explaining, and one can easily imagine the doctrine of vegetative and sensitive souls, for example, being subsumed under, and perhaps even being a model for, the theory of emergent properties for plants and animals. In accounting for the fact that in different scientific areas very different kinds of resources may be called upon—if only because the point of the explanatory exercise varies radically—recourse to emergent properties tells us absolutely nothing, not even providing elementary guidance. If the life sciences don't use resources developed in the domain of physical-mechanical problems, it is not because they are 'more complex' (particle physics is at least as complex as anything in the life sciences) or because they involve emergent properties, but because these are inappropriate to their subject matter. Similarly in chemistry, as we saw in Chapter 5, and similarly in geology, as we shall see in Chapter 7: it is not a matter for physics whether movements of the earth's crust, volcanic eruptions, and tsunamis are unchanging geological processes, for example, but something requiring a completely different set of explanatory resources. Once we have abandoned reductionism, there is no mystery that emergent properties must be invoked to resolve.

7

The Unity of the Life Sciences

One thing that emerges from Bernard's account of 'experimental medicine' is the idea of physiology as a point of intersection of various scientific projects, where he identifies the means of intersection having at least two variants: alliance and accessory. The development of the life sciences in the nineteenth century provides a more general illustration of this phenomenon that is doubly striking. It not only involved the meeting of trajectories along two distinct and seemingly autonomous paths, the physiological and the historical, but the latter, to which we now turn, involved at least three distinct forms of enquiry which came to align themselves with one another, in a process of qualified mutual reinforcement rather than assimilation.

Two traditional areas of natural history, the study of mineral strata and the investigation of the distribution and diversity of species, despite being excluded by advocates of physiology as falling outside the life sciences, continued to flourish, in different arenas, in the early and middle decades of the nineteenth century. Joined with another discipline of contested scientific standing, developmental morphology, from the 1860s onwards they provided the resources for a new account of the transmutation of species.[1] Geology established a vastly expanded timeline which provided evolution through natural selection with enough time for the transmutation processes to take place. The detailed study of the distribution and diversity of species, and the study of the variation within species, allowed for a conception of species transmutation on a secure empirical basis, by contrast with speculative evolutionary accounts that had predominated up to that point. And close study of the development of the embryo suggested to researchers a growth from a primitive to an advanced state which mirrored that being proposed in the account of the evolution of species, where the experimental observation of the foetal stages was considered to allow a reconstruction of the evolutionary stages, which were otherwise accessible only in a very piecemeal way through the fossil record.

---

[1] These are the disciplines that Darwin identifies in the Introduction to the *Origin of Species*: 'In considering the Origin of Species, it is quite conceivable that a naturalist, reflecting on the mutual affinities of organic beings, on their embryological relations, their geographical distribution, geological succession, and such other facts, might come to the conclusion that each species had not been independently created, but had descended, like varieties, from other species.' *On the Origin of Species by means of Natural Selection, or the Preservation of Favoured Races in the Struggle for Life* (London, 1859), 3. This first edition of the *Origin* appears as volume 15 of *The Works of Charles Darwin*, ed. Paul H. Barrett and R. B. Freeman (29 vols, London, 1986).

Civilization and the Culture of Science: Science and the Shaping of Modernity, 1795–1935. Stephen Gaukroger, Oxford University Press (2020). © Stephen Gaukroger.
DOI: 10.1093/oso/9780198849070.001.0001

NATURAL HISTORY AND THE AGE
OF THE EARTH

In Linnaeus' classification, the natural realm had been divided into three 'kingdoms'—mineral, vegetable, and animal—and a single classificatory scheme had been applied to all three: classes, orders, families, genera, species, and varieties. In each case, classification was designed to reveal affinities between things, and it proceeded on the basis of diagnostic characters. In mineralogy, these included such features as crystalline form, hardness, colour, and cleavage (splitting along structural planes). The classification of rock strata was on the face of it the most mundane exercise imaginable, but in the course of the eighteenth century it quickly gave rise to basic questions. The first was that of the age of the earth. In the second half of the century, there were a number of geological observations suggesting that the biblical timescale was simply not of the right order of magnitude.[2] Volcanoes were a striking case of the inadequacy of a geological history of four or five thousand years, for if, as seemed reasonable, they had been built up by a succession of eruptions at the same rate and on the same scale as those recorded from Pliny onwards, the total age of the volcano stretched well beyond recorded history. Another indication of the longevity of the earth was river valleys, at least some of which could be attributed to erosion, and the time required for such erosion was again immense. Finally, there was the existence of secondary rock strata, which earlier investigators had assumed could have been laid down all at the same time, but by the late eighteenth century it had become evident that they must have been deposited layer by layer, in which case the timescale involved was again orders of magnitude greater than the biblical one.

The second set of questions concerned the fossils found embedded in rock strata. There were two main issues: their distribution, and what they were fossils of. On the question of distribution, fossils of what were clearly sea creatures were found in rock strata on mountain tops, far from the sea. One possible explanation for their presence was that they had been deposited there in a large-scale elevation and subsidence of great blocks of land relative to the sea, in a time frame that was orders of magnitude greater than that of the biblical Flood.[3] James Hutton had advocated such a view, and argued that observation of present-day processes of erosion and sedimentation provided indications of the mechanisms by which

---

[2] See Martin J. S. Rudwick, *Bursting the Limits of Time: The Reconstruction of Geohistory in the Age of Revolution* (Chicago, 2005), 119–24.

[3] On critics of the idea that the Flood had been responsible—as William Buckland had argued in his *Reliquiae diluvianae: or, observations on the organic remains contained in caves, fissures, and diluvial gravel, and on other geological phenomena, attesting the action of an universal deluge* (London, 1823)—see Martin Rudwick, *Worlds before Adam: The Reconstruction of Geohistory in the Age of Reform* (Chicago, 2008), 82–6, and on the persistence of the view in the 1820s, 177–80.

this happened.[4] On the question of what exactly the fossils were fossils of, many of them did not resemble modern animals and plants, as, for example, the ubiquitous trilobites (Fig. 7.1). This raised the question of the nature and origins of these animals and plants.

In 1795 the geologist Horace-Benédict de Saussure argued that investigators needed to find out whether there were fossil shells found in older mountains but not in newer ones, and to estimate the relative ages and dates of the species; they needed to compare fossil shells, plants, etc. with living analogues, to determine whether or not there was an exact resemblance, and if not whether it was a case of different varieties or different species.[5] Cuvier took up Saussure's challenge to verify the claim that the fossil bones of quadrupeds such as elephants, rhinoceroses, oxen, and deer were different from those of their extant counterparts. Comparative anatomy, he argued, showed beyond any doubt that these were different species from the extant ones.[6] Building on Cuvier's work, Eugène-Louis-Melchior Patrin, in a comprehensive treatise on mineralogy published in 1801,[7] argued that the oldest organized bodies were generated spontaneously in the sea, and with the gradual fall in sea level more complex bodies arose. He then went on to develop Cuvier's thesis that the older the stratum, the greater the difference between the fossils and bones found there and modern species.

The study of rocks might have been excluded from the 'scientific' part of natural history by those who considered physiology constitutive of biology, and its separation from the study of living things might have seemed a logical move for those concerned with living things, if only because of the non-cellular nature of rocks, but it came to have radical consequences for the life sciences. In 1823 Humboldt, in the process of showing how dating of rocks depended on the relative position in the strata,[8] had been the first to systematically relate geology, surface conditions, and the distribution of plants and animals. But the radical consequences of geology became most evident with the publication of Charles

---

[4] James Hutton, 'Theory of the Earth: or an investigation of the laws observable in the composition, dissolution, and restoration of land upon the Globe', *Transactions of the Royal Society of Edinburgh* vol. 1, part 2 (1788), 209–304.

[5] Horace-Benédict de Saussure, 'Agenda, ou tableau général des observations et des recherches dont les résultats doivent servir de base à la théorie de la terre', *Journal des mines* 4.20 (1795), 1–70: 32–3.

[6] Initially in Georges Cuvier, 'Mémoire sur les espèces d'éléphants vivantes et fossiles. Lu le premier pluviose an IV', *Mémoires de l'Institut national des sciences et des arts. Sciences mathématiques et physiques* 2 (1799), 1–22; then in detail in Georges Cuvier, *Recherches sur les ossemens fossiles de quadrupèdes, où l'on rétablit les caractères de plusieurs espèces d'animaux que les révolutions du globe paroissent avoir détruites* (4 vols, Paris, 1812) On Cuvier's palaeontology, see William Coleman, *Georges Cuvier Zoologist: A Study in the History of Evolution Theory* (Cambridge, MA, 1962), ch. 5; and Rudwick, *Worlds before Adam*, ch. 1.

[7] Eugène-Louis-Melchior Patrin, *Histoire naturelle des minéraux, contenant leur description, celle de leur gîte, la théorie de leur formation, leurs rapports avec la géologie ou l'histoire de la terre, le détail de leurs propriétés et de leurs usages, leur analyse chimique* (5 vols, Paris, 1801).

[8] Alexander von Humboldt, *Essai géognostique sur le gisement des roches dans les deux hémisphères* (Paris, 1823).

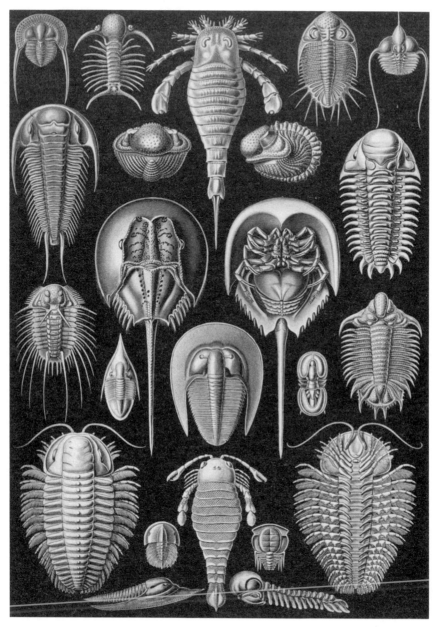

Fig. 7.1. Trilobites: from Ernst Haeckel, *Kunstformen der Natur* (1904)

Lyell's *Principles of Geology*, the first volume of which appeared in 1830.[9] The subtitle—'an inquiry how far the former changes of the Earth's surface are referable to causes now in operation'—indicates the method of investigation. One main claim is that there is no evidence at all that the earth was formed in stages, as biblical accounts maintained. Rather, it constantly undergoes countless tiny accumulative changes which are the result of natural forces operating uniformly over very long periods, and these changes gradually build up empirically observable large-scale effects. Darwin took Lyell's volumes on his voyage on the Beagle, and when writing up his observations on his return to England, he published three volumes of geological observations that built on Lyell's account, and offered revisions to details of it.[10]

Lyell took the earth to be effectively ageless, and subject to very long-term geological cycles. What was at issue was not just changes in the arrangement of rocks but the plant and animal distribution that was intimately connected to surface conditions. The biogeography of the earth was slowly changing in concert with its geological changes. Here Lyell was drawing a direct connection between the life sciences and the earth sciences, and it was one to which Darwin was alert: particularly the mechanism that Lyell was proposing, cumulative small changes having very substantial effects. But his account harboured a deep problem. His treatment of geological changes was gradualist. Rocks are very slowly but continuously changed. One obvious inference that might be drawn from this was that the animals and plants on the surface must likewise undergo a continuous transmutation. But Lyell was completely opposed to transmutation: animals and plants were created in a piecemeal fashion, without any continuities. He did not explain why, given that he had established such a close connection between what happened in the mineral realm and the distribution of plants and animals in the living realm, the modality of change should be so different. Nor did he defend the idea of the piecemeal creation of species in its own right. Rather, he mounted a full-scale attack on the standard account of transmutation, that of Lamarck.

Although Lamarck's transmutationism was attractive to many, those who studied fossils adopted Cuvier's concept of discrete species, each well adapted to a particular mode of life. But Cuvier introduced a new ingredient into the issue

⁹ Charles Lyell, *Principles of Geology: being an Inquiry how far the former changes of the Earth's surface are referable to causes now in operation* (3 vols, London, 1830–3).

¹⁰ Charles Darwin, *The Structure and Distribution of Coral Reefs. Being the first part of the geology of the voyage of the Beagle, under the command of Capt. Fitzroy, R.N. during the years 1832 to 1836* (London, 1842); Charles Darwin, *Geological Observations on the Volcanic Islands visited during the voyage of H.M.S. Beagle, together with some brief notices of the geology of Australia and the Cape of Good Hope. Being the second part of the geology of the voyage of the Beagle, under the command of Capt. Fitzroy, R.N. during the years 1832 to 1836* (London, 1844); Charles Darwin, *Geological Observations on South America. Being the third part of the geology of the voyage of the Beagle, under the command of Capt. Fitzroy, R.N. during the years 1832 to 1836* (London, 1846). The books are available in a modern edition as volumes 7 to 9 of *The Works of Charles Darwin.*

Fig. 7.2. Henri De La Beche, *Duria Antiquior* (1830)

Fig. 7.3. George Nibbs, *The Ancient Weald of Sussex* (1838)

when he insisted that the fossil record showed the reality of the extinction of species, and indeed by the 1830s there were numerous imaginative reconstructions of the fauna and flora of the prehistoric world, building on recent discoveries of gigantic bones and other fossilized material (Figs 7.2 and 7.3). Just as the times when extinctions occurred could be determined roughly from the fossil record, and perhaps even their causes outlined, so too the times of the origins of species could be discerned, but there was absolutely no understanding of how and why these occurred. Geoffroy St. Hilaire had suggested that some fossilized species could have been transmuted into currently living ones under the pressure of severe environmental change, whereas other fossilized species may not have undergone any transmutation and had become extinct. Unlike Lamarck's own account, on which there was transmutation but no extinction, for Geoffroy St. Hilaire one could have both transmutation and extinction. Nevertheless, as Rudwick notes, in practice it made little difference to geologists how species had originated:

New species might have appeared by transmutation, either continuously and insensibly gradually (as Lamarck supposed), or rapidly at times of environmental upheaval (as Geoffroy suggested); or they might have been formed by some other natural process as yet unknown; or even conceivably by immediate divine action. Old species might have disappeared as a result of environmental change, either sudden or catastrophic (as Cuvier and Buckland maintained), or slow and gradual (as Fleming and the Auvergnat naturalists suggested); or by some internal process analogous to old age and mortality in individuals (as Brocchi had proposed); or in some other way as yet undiscovered. These were unsolved problems, intriguing and important in their own right; but the burgeoning practice of geohistory did not have to wait for their successful resolution. What mattered more immediately to geologists was to determine *when* species had appeared or disappeared, and in what manner: whether suddenly and many at once, or in a piecemeal fashion, one by one.[11]

Lyell's *Principles*, with its incremental continuist account of geological change, brought the question of transmutationism to the fore. The *Principles* begin with a long attack on scriptural geology, followed by criticisms of those who had unintentionally given succour to it by maintaining that the history of the earth had been marked by occasional catastrophic events (such as the Flood) which were absent in the modern world. Such a view, he argues, has been fostered in the context of a wholly inadequate understanding of the geological timescale, and only by ignoring the submarine and subterranean processes that were at least as important in understanding geological change as were terrestrial processes. His methodological commitment to the idea that present geological events must be used to model ones in the past encouraged him in the view that there had been in the past no events of a greater magnitude than those observed in the present, and this reinforced his continuism. The most significant problem facing his account,

---

[11] Rudwick, *Worlds before Adam*, 249.

however, came from the fact that the fossil evidence apparently gave strong evidence of directional change, for example for a dramatic cooling of climate, whereas he was arguing for a state of dynamic equilibrium. Building on Humboldt's introduction of isothermal lines, which showed how climate was dependent not simply on latitude but also on the distribution of land and sea, he offered a defence of dynamic equilibrium: the apparent cooling of the global climate, he argued, was in fact due not a cooling of the deep interior, as Buffon and others had assumed, but to changes in the distribution of land and sea, causing global climates to swing from warmer to cooler and then back to warmer over a very long timescale. As with climate, so with life. He rejected a progressive history of life, of the kind offered by Lamarck for example, on the grounds that it was not properly supported by the evidence of what was a very imperfect fossil record. But his application of known processes to the past, in itself applauded, to yield a steady-state geohistory was rejected by a number of contemporary critics, including Sedgwick and Whewell. As Sedgwick notes:

To assume . . . that volcanic forces have not only been called into action at all times in the natural history of the earth, but also, that in each period they have acted with equal intensity, seems to me merely a gratuitous hypothesis, unfounded on any of the great analogies of nature, and I believe also unsupported by the direct evidence of fact. This theory confounds the immutable and primary laws of matter with the mutable results arising from their irregular combination . . . . Volcanic action is essentially paroxysmal; yet Mr. Lyell will admit of no greater paroxysms than we ourselves have witnessed—no periods of feverish spasmodic energy, during which the very framework of nature has been convulsed and torn asunder. The utmost movements that he allows are a slight quivering of her muscular integuments.[12]

The criticisms are of particular interest in that some of the questions facing geologists were similar to those confronting physiologists. Just as many physiologists accepted that physiological processes are physico-chemical in nature, while denying that they were regulated exclusively by physico-chemical laws, so geologists had no doubt that the processes that they described obeyed physico-chemical laws, but they did not think that it was a matter of chemistry and physics whether movements of the earth's crust, volcanic eruptions, and tsunamis were unchanging geological processes, or whether they had formerly operated with a greater intensity. Sedgwick and Whewell criticized Lyell on just this point. As Rudwick points out, they claimed that Lyell, in arguing for the uniformity of nature, 'had confused the highly complex process of geological agency with the basic physico-chemical "laws of nature" on which they were founded'.[13] The latter, they argued, were indeed stable throughout eternity, but the former varied significantly in power and intensity.

---

[12] Adam Sedgwick, 'Address of the President', *Proceedings of the Geological Society of London* 20 (1831), 281–316: 301–7.
[13] Rudwick, *Worlds before Adam*, 345.

The importance for Lyell of dissociating his continuist account of geology from any suggestion of a continuist account of species is indicated by the fact that his explicit criticisms of Lamarck take up a significant portion of the second volume of the *Principles*. Application of the principle that 'former changes of the Earth's surface' be 'referable to causes now in operation' to animal species, is now taken to indicate that since no transmutation of species has been detected over the period of human records, there are no grounds to infer that it would have occurred over longer periods. This is despite his having earlier criticized scriptural geologists on the grounds that they have no understanding of the immensity of the geological timescale. Without transmutation, he faced a significant challenge in accounting for the processes responsible for the introduction and elimination of species of flora and fauna. As one would expect from his earlier discussion, dynamic equilibrium is offered as the solution: in a 'steady state', under the pressure of environmental change, the emergence of new species was constantly balanced by the extinction of species. This left one final problem, that of the apparent directionality of the fossil record, to which Lyell gives a decidedly ad hoc answer, namely, that the fossil record is a very imperfect source of evidence, suggesting that the preponderance of aquatic fossils is due to terrestrial organisms being far less likely to be preserved than aquatic ones. This is indeed the weakest link in Lyell's general system, and it is not particularly surprising that Spencer, for example, was converted to Lamarckism by reading Lyell's exposition and unconvincing criticisms of it.

THE DISTRIBUTION AND DIVERSITY OF SPECIES

These issues take us to the second of the traditional areas of natural history, the study of the distribution and diversity of species. On the Cuvier model, represented in England in the work of the anatomist Richard Owen, individual organisms varied, but every organism was an approximation of an ideal type. Accordingly, what naturalists and systematists were interested in were the individual typical forms, not the exceptions. In particular, because they were not interested in the exceptions, they had no motivation to consider them as indications of species variation. The Cuvier model was committed to the idea of limits to species variation, but these limits were simply set by nature, and investigating them would not have contributed anything to the identification of the ideal types. So, for example, it was the gaps between fossil organisms, not their progressive development in the strata, that commanded Cuvier's attention.[14] However, as variation gradually became an object of study in its own right, the idea of natural

---

[14] Rudwick, *Bursting the Limits of Time*, 392–9.

limits to variation came into question, and species were transformed from discrete, immutable, essential types into the nominal conveniences of systematists.

On his return from his voyage on the Beagle, Darwin collected his observations of fossil mammalia, extant mammalia, birds, fish, and reptiles in five volumes of zoological observations,[15] and three of these turned out to be particularly significant. Among his observations on fossils, he noted the remains of large extinct mammals unearthed in Patagonia. These were sent to museum naturalists in London, who identified them as members of previously unknown species. Darwin noted however that there seemed to be a continuity of 'type' with extant animals, suggesting a lineage. Among his observations on birds, two turned out to be significant. The first concerned the rhea, related to the ostrich and the emu. Darwin noted the geographical distribution of the two kinds of rhea: the southern animals were smaller than the northern ones, suggesting family links across different geophysical areas. The second observation was not accorded immediate significance. It concerned the birds of the Galápagos Islands, fourteen islands not too distant from one another but separated by impassable channels. Darwin collected a number of bird specimens from the islands, noting differences but, the islands being so close to one another, he did not suspect that the individual locations mattered: the birds seemed as unrelated as blackbirds and wrens. A year after Darwin's return to London, however, the ornithologist John Gould identified several different species of ground finches in the collection. The huge variations in beak size and shape turned out to correlate with the different food sources on the different islands, and were the result of niche specialization, and it seemed that the birds had diversified from a common ancestor finch on the mainland.

In speculating on whether species were fixed, or whether they were subject to transmutation, Darwin was already familiar with the writings of Lamarck and his own grandfather Erasmus Darwin, who had argued that plants and animals are spontaneously generated from inorganic materials, and are constantly being transmuted, adapting themselves to different environmental conditions. Their views had attracted widespread criticism, however, not least because they had included human beings in their evolutionary schemas. Nevertheless, one crucial thing that Darwin learned from them was to question the absoluteness of the distinction between species and varieties. This was important because the production of very different varieties from one another was relatively common-place,[16] so if the distinction between the two could be shown to be a fluid one, or a purely conventional one, then one was on the path to the transmutation of

---

[15] Charles Darwin, *The Zoology of the Voyage of H.M.S. Beagle* (5 vols, London, 1839–43). The books are collected in volumes 4 to 6 of *The Works of Charles Darwin*.

[16] Nevertheless, Darwin's focus on pigeon breeding downplayed two other forms of production of varieties—crossing of varieties and inbreeding—which many breeders believed to be essential to the production of varieties: see Bert Theunissen, 'Darwin and his Pigeons: The Analogy between Artificial and Natural Selection Revisited', *Journal of the History of Biology* 45 (2012), 1–34.

species. The production of varieties was an artificial form of transmutation, however, under the guidance of breeders looking for a particular outcome. The *Origin of Species* begins with a chapter on 'variation under domestication', stressing the power of breeders' selective activities, with examples drawn from horticulture and farming but particularly from the breeding of domestic pigeons. It sets out to show how artificial selection works and points out the difficulties of distinguishing species from varieties, and the unwarranted assumptions on which the distinction has been made. In the second chapter, the transition is made to 'natural' selection, that is, selection in the wild, drawing out the parallels between domestic breeds and wild species, and showing that it is even more difficult to separate species and varieties in this natural case.

The first two chapters of the *Origin of Species* defend a theory of species descent with modification, something earlier transmutationists had argued for. The next two chapters offer what Lamarck and Erasmus Darwin lacked: a plausible account of the mechanism in nature by which the transformations were effected. The mechanism by which variation under domestication worked was clear enough: breeders selected for the characteristics they wanted to introduce. But natural 'selection' was manifestly not like this. There was no one doing the 'selecting' and consequently it was difficult to understand just what the 'selection' amounted to. What was needed was an account of the mechanism by which the relevant changes occurred in animals and plants. In 1838 Darwin read Malthus' essay on population, published forty years earlier.[17] As with Wallace twenty years after him, reading Malthus (or in Wallace's case suddenly seeing the significance of what he had read in Malthus years earlier) proved revelatory. The central argument of *On the Principle of Population* was that while food production tends to increase arithmetically (increments are generated by adding), population tends to increase naturally at a geometric rate (increments are generated by multiplying), meaning that, *ceteris paribus*, food production will not keep up with population growth. Consequently, there is a struggle for existence, which means that there must be ways of checking the population: late marriage, contraception, emigration, or reduction of numbers through events over which the general population had no control, such as wars and epidemics. In this way, population numbers and available resources are kept balanced. Darwin translated this into biological terms: there are too many individuals born, there is a struggle for existence, and the better adapted survive. At this point, Lyell's idea of tiny incremental changes ultimately resulting in large-scale transformations locked in. The phenomenon of better-adapted organisms surviving is a cumulative process, with the cumulative result that organisms become increasingly better adapted to their conditions. This

[17] Thomas Malthus, *On the Principle of Population, as it affects the Future Improvement of Society. With Remarks on the Speculations of Mr Godwin, M. Condorcet, and other writers* (London, 1798). On Darwin's reaction to Malthus, see Hale, *Political Descent*, ch. 1; and Janet Browne, *Charles Darwin: Voyaging* (London, 1995), 384–90.

is the process that Darwin calls natural selection: unlike the evolutionary schemes of Lamarck and Erasmus Darwin, at least as they were interpreted, it is not guided by some fixed end but solely by the environmental conditions in which the organism finds itself, conditions that may change radically (for example with geological changes) and which would then mean that natural selection would result in a change of direction, as it were, in the development of the species. Darwin always considered a number of mechanisms to act together in evolution—use and disuse of parts (the most important mechanism of adaptive evolution after natural selection); sexual selection (introduced later in *The Descent of Man*); directed variation; correlated variation; spontaneous variations; and family selection[18]—but it was natural selection that held the key to his account of evolution.

By 1844, Darwin had a 200-page essay on natural selection[19] which was effectively a draft of *The Origin of Species*, and which, at the theoretical as opposed to the evidential level, had everything significant that *The Origin of Species* would contain except for the theory of beneficial effects of the diversification of species. But Darwin did not publish it. In the same year, Chambers' *Vestiges of the Natural History of Creation* appeared, prompting extensive popular debate on evolution. It was quickly taken up by the reading public, and it prompted an immediate and predominantly hostile response from all quarters, from the Anglican clergy to scientists. Sedgwick, who was both, in an 1845 letter to Lyell wrote that the *Vestiges* starts 'from principles which are at variance with all sober inductive truth. The sober facts of geology shuffled, so as to play a rogue's game; phrenology (that sinkhole of human folly and prating coxcombry); spontaneous generation; transmutation of species; and I know not what; all to be swallowed, without tasting or trying, like so much horse-physic!! Gross credulity and rank infidelity joined in unlawful marriage, and breeding a deformed progeny of unnatural conclusions!'[20] The most striking feature of *Vestiges* was that it was not just an account of the transmutation of plant and animal species, but a wholly materialist history of the cosmos described in transmutationist terms. In the more modest parts of the book, there was, as Darwin realized, some overlap with his own version of transmutation. Chambers set out to defend an account of common ancestors with a number of examples drawn from anatomy—the giraffe has in its neck the same number of bones as the pig, the limbs of all vertebrates follow the same plan, and so on—and he postulated the existence of intermediate animals to bridge the

[18] See William B. Provine, 'Adaptation and Mechanisms of Evolution after Darwin: A Study in Persistent Controversies', in David Kohn, ed., *The Darwinian Heritage* (Princeton, NJ, 1985), 825–66: 827–33.
[19] Published posthumously as *The Foundations of the Origin of Species: Two Essays written in 1842 and 1844 by Charles Darwin*, ed. Francis Darwin (Cambridge, 1909). Reprinted in volume 10 of *The Works of Charles Darwin*.
[20] John Clarke and Thomas Hughes, *The Life and Letters of the Reverend Adam Sedgwick* (2 vols, Cambridge, 1890), ii. 83–5.

different classes. In the other parts, from the account of the sun and the planets coalescing from an initial fire, and the origins of life beginning with a spark acting on a 'globule', to the continuing evolution of human beings, transmutation was presented as part of a general metaphysical view of the cosmos, a picture of continual transformation with particularly radical consequences insofar as it bore on human evolution.

Darwin, while criticizing the geology and zoology of the book, and ridiculing the author's idea of fish being transmuted into reptiles, was not completely hostile to the _Vestiges_, writing to Huxley that: 'I must think such a book, if it does no other good, spreads the taste for Natural Science. But I am perhaps no fair judge, for I am almost as unorthodox about species as the _Vestiges_ itself, though I hope not quite so unphilosophical.'[21] Nevertheless, _Vestiges_ clearly threatened the standing of transmutationist theories, and Darwin's extensive work on plant and pigeon breeding experiments, seeking out inherited traits and reversion to ancestral type, and his eight-year observations of every species of barnacle, in which he discovered an unexpectedly rapid rate of variation,[22] must be seen as a concerted attempt, not to develop or revise natural selection theory, but to build up a comprehensive body of evidence.

The reaction to the _Vestiges_ by Wallace, who was later to be the co-discoverer of evolution by natural selection, is instructive. When the book appeared, Wallace was simply an experienced naturalist, and he was not only impressed but inspired by the book. In an 1845 letter to his naturalist collaborator Henry Bates, he rejects Bates' view that the theory of progressive development of animals and plants was a 'hasty generalization'. Rather, he urges, it should be taken 'as an ingenious hypothesis strongly supported by some striking facts and analogies, but which remains to be proved by more facts and the additional light which more research may throw upon the problem. It furnishes a subject for every observer of nature to attend to; every fact he observes will make either for or against it, and it thus serves both as an incitement to the collection of facts, and an object to which they can be applied when collected.'[23] Wallace's one initial criticism of the book was in Chambers' lack of distinction between species and varieties, but in reflecting on this he soon began speculating on the relationship between species and varieties in the context of both geography and history. Shortly after reading _Vestiges_, he became an evolutionist[24] and in an 1855 paper—which Lyell had pressed upon

[21] _The Correspondence of Charles Darwin_, ed. F. Burkhardt and S. Smith (9 vols, Cambridge, 1985–94), v. 212–13.

[22] On the importance of the barnacle research for boosting Darwin's confidence in his theory of natural selection, see Roderick Buchanan and James Bradley, '"Darwin's Delay": A Reassessment of Evidence', _Isis_ 108 (2017), 529–52.

[23] Alfred Russel Wallace, _My Life: A Records of Events and Opinions_ (2 vols, New York, 1905), i. 254.

[24] See Michael Schermer, _In Darwin's Shadow: The Life and Science of Alfred Russel Wallace_ (New York, 2002), 54–5.

Darwin in the following year—he was firmly of the view that the distinction between species and varieties was a tenuous one at best.[25] Wallace initially took on board the full *Vestiges* package and, mirroring Chambers' general account of transmutation, he was happy to speculate on an evolutionary move from apes to human beings.[26] This was the sticking point for critics of evolution, and Wallace himself was subsequently to become one of the most prominent defenders of the discontinuity between human beings and other animals, advocating a form of spiritualism to account for the distinctiveness of human beings. Darwin's *The Descent of Man*[27] responded to Wallace on this question in some detail, containing more references to Wallace than to any other source.[28]

EMBRYOS AS ANCESTORS

The theory of natural selection, as we have followed it up this point, has its sources in disparate areas: the study of rock strata; the study of the distribution of plants and animals on the earth's surface; the study of fossils; theories of the transmutation of species in Lamarck, Erasmus Darwin, and (at least for Wallace) Chambers' *Vestiges*; and a theory of the relation between population growth and food resources that was initially developed as a form of political arithmetic. There was one more general area that was drawn upon: morphology, and its sub-discipline embryology.

Morphology studies homologies, that is, the sameness of organs across different species. Much of the initial motivation for it arose from *Naturphilosophie* of various kinds, as in Goethe's search for Ur-forms (archetypes) in botany, generalized to science as a whole,[29] and in Burdach, for whom the aim was the exploration of forms to connect structures and functions with their common bases or 'grounds', which exist at an ideal level.[30] Burdach defines morphology as setting out to investigate the meaning and significance of organic forms, discerning the laws of their formation, understanding the process by which nature organizes them, the origins and progress of their structure, and the effects of this structure.[31] This should enable the morphologist, on Burdach's account, to

[25] Alfred Russel Wallace, 'On the Law Which has Regulated the Introduction of New Species', *Annals and Magazine of Natural History*, 2nd series, 16.93 (1855), 184–96.

[26] See, for example, Alfred Russel Wallace, 'On the Habits of the Orang-Utan in Borneo', *Annals and Magazine of Natural History*, 2nd series, 17.103 (1856), 26–32. See Ross A. Slotten, *The Heretic in Darwin's Court: The Life of Alfred Russel Wallace* (New York, 2004), ch. 7.

[27] Charles Darwin, *The Descent of Man, and Selection in Relation to Sex* (London, 1871).

[28] Slotten, *The Heretic in Darwin's Court*, 289.

[29] Johann Wolfgang Goethe, *Zur Naturwissenschaft überhaupt, besonders zur Morphologie* (Stuttgart, 1817).

[30] See Nyhart, *Biology Takes Form*, 39–50.

[31] Karl Friedrich Burdach, *Über die Aufgabe der Morphologie* (Leipzig, 1817), 31.

understand the extent to which present forms were remnants of extinct ones, and to what extent they seeded future forms.

It was the debate between Cuvier and Geoffroy Saint-Hilaire, lasting throughout the 1820s and coming to a head in 1830 at meetings of the Académie des Sciences, that did most to reshape morphology, raising fundamental morphological issues in zoology. Cuvier argued that the structure of an animal was determined by function, and he distinguished four basic body plans—vertebrate, articulate, mollusc, radiate—on which all animal structure is based.[32] These basic plans reflect the functional integrity of the different animal types. It is physiology, not anatomy as such, that holds the key to morphology: despite anatomical similarities between crustacea and insects for example, the presence of a heart and brachia in crustacea indicate a mode of life different from that of insects. Geoffroy Saint-Hilaire, by contrast, argued that there was a unity of design in all animals, that all animals had a single plan that preceded all the modifications in animals that were determined by functional requirements.[33] The modifications to this plan could be discerned in the development of the vertebrate embryo, which in its early stages manifested the structure of invertebrates. This approach had two consequences. First, it suggested an evolutionary process from simple to complex organisms along lines suggested by Lamarck, that is, one altered by the effects of the environment and the use and disuse of particular structures.[34] Second, species transmutation and succession in the fossil record was construed as analogous to the stages of embryonic development.

In the *Origin of Species*, Darwin had written that morphology is 'the most interesting department of natural history, and may be said to be its very soul'.[35] The interest lies in the fact that 'community in embryonic structure reveals community of descent',[36] and embryology 'rises greatly in interest, when we thus look at the embryo as a picture, more or less obscured, of the common parent-form of each great class of animals'.[37] Darwin had worked assiduously in

[32] Georges Cuvier, *Le régne animal distribué d'après son organisation pour servir de base à l'histoire naturelle des animaux et l'introduction à l'anatomie comparée* (5 vols, Paris, 1829–30). The classification was open to development. In particular, the category of radiates had the appearance of mere leftovers from the other categories. In 1848 the young German morphologist Rudolph Leuckart fundamentally reworked Cuvier's account and replaced radiates with two classes: coelenterates and the echinoderms. Rudolph Leuckart, *Über die Morphologie und Verwandtschaftsverhältnisse der wirbellosen Thiere: Ein Beitrag zur Charakteristik und Classifikation der thierischen Formen* (Braunschweig, 1848).

[33] Étienne Geoffroy Saint-Hilaire, 'Histoire des Makis, ou singes de Madagascar', *Magasin encyclopédique* 1 (1796), 1–48; Étienne Geoffroy Saint-Hilaire, *Philosophie anatomique* (2 vols, 1818–22).

[34] On the indebtedness of Geoffroy Saint-Hilaire to Lamarck see Appel, *The Cuvier–Geoffroy Debate*.

[35] Darwin, *The Origin of Species*, 434. Compare his comment to Joseph Hooker in a letter of December 1859: 'Embryology is my pet bit in my book, & confound my friends no one has noticed this to me.' *The Correspondence of Charles Darwin*, vii. 431–2

[36] Darwin, *The Origin of Species*, 449.   [37] Ibid., 450.

embryology from the mid-1830s onwards,[38] and even though only half of one chapter is devoted to morphology and embryology in the *Origin*, as Nyhart points out, 'in fact the position of the chapter—the very last one before the book's recapitulatory conclusion—offers a clue to its unifying and generalizing quality. This is the chapter in which Darwin connects his theory to the natural system as a whole, in which he argues that indeed the natural system is none other than the genealogy of nature, explained and structured by natural selection. It is the book's punch line.'[39]

If Darwin's account in the *Origin of Species* were correct, then the ultimate relatedness of all animals, as revealed in their adult homologies, should, he reasoned, be equally evident in their embryological development. 'The embryo', he writes, 'is the animal in its less modified state; and in so far it reveals the structure of its progenitor. In two groups of animal, however much they may at present differ from each other in structure and habits, if they pass through the same or similar embryonic stages, we feel assured that they have both descended from the same or nearly similar parents, and are therefore in that degree closely related. This community in embryonic structure reveals community of descent.'[40] Darwin sought to connect the traditional disciplines of embryology, classification, and morphology to his study of variation, inheritance, and selection. But if evolution proceeds by natural selection acting upon variations among organisms, a host of new questions are raised for morphology. He puts the questions in these terms:

> How, then, can we explain these several facts in embryology—namely between the very general, but not universal difference in structure between the embryo and the adult; of the parts in the same individual embryo, which ultimately become very unlike and serve for diverse purposes, being at this early stage of growth alike; of embryos of different species within the same class, generally, but not universally, resembling each other; of the structure of the embryo not being closely related to its conditions of existence, except when the embryo becomes at any period of life active and has to provide for itself; of the embryo apparently having sometimes a higher organization than the mature animal, into which it is developed. I believe that all these facts can be explained . . . on the view of descent with modification.[41]

Many nineteenth-century naturalists assumed that there was a law of development of both individuals and life in general embodied in nature.[42] On this idea, there was an increasing progress and complexity, and this was revealed in the fossil record which paralleled the development of the individual embryo. But there was

---

[38] See Robert J. Richards, *The Meaning of Evolution: The Morphological Construction and Ideological Reconstruction of Darwin's Theory* (Chicago, 1992).

[39] See Lynn K. Nyhart, 'Embryology and Morphology', in Michael Ruse and Robert J. Richards, eds, *The Cambridge Companion to the 'Origin of Species'* (Cambridge, 2009), 194–215: 197.

[40] Darwin, *The Origin of Species*, 449.   [41] Ibid., 442–3.

[42] See Gliboff, *H. G. Bron*; Nyhart, *Biology Takes Form*; Richards, *The Meaning of Evolution*; and the very useful summary in Nyhart, 'Embryology and Morphology', to which I am indebted here.

disagreement on the specifics of how this development took place. There were in effect two competing accounts: linearity and differentialism. The linear view, associated with recapitulationism, was that the brains of mammalian foetuses, for example, passed through the adult stages of the lower vertebrate classes as they developed. Lyell, although critical of the view, summarizes it well in the second volume of the *Principles of Geology*:

Tiedemann found, and his discoveries have been most fully confirmed and elucidated by M. Serres, that the brain of the fœtus, in the highest class of vertebrated animals, assumes, in succession, forms analogous to those which belong to fishes, reptiles, and birds, before it acquires the additions and modifications which are peculiar to the mammiferous tribe. So that, in the passage from the embryo to the perfect mammifer, there is a typical represen-tation, as it were, of all those transformations which the primitive species are supposed to have undergone, during the long series of generations, between the present period and the remotest geological era.[43]

Against this linear account of development, Baer rejected the idea of a single scale of development from the most primitive life form to the most advanced.[44] Instead, he followed Cuvier's division of animals into four basic types, arguing that the embryo in its first stage of development exhibited the type, then successively the more specific characteristics of the class, order, genus, and species. In the early stages, Baer insisted, the embryos were indistinguishable.[45] The zoologist Henri Milne-Edwards developed Baer's account to establish that the more characteristics two organisms had in common during their development, the more closely they were related.[46] This was a particularly significant move because it meant that one could potentially use embryology to determine how closely two organisms should be classified.

The issues are complex. If we ask what it was that the stages being passed through by present-day embryos of higher forms were being compared with, there are various different answers. As Nyhart points out, they might be being com-pared to present-day adults (especially of lower forms), to present-day embryos, to adults of past forms, or to embryos of past forms.[47] Moreover, Darwin himself was equivocal on the question of linear versus differential development.[48] It was nevertheless embryology that carried the standard of evolutionary theory, above all in the work of Haeckel. Haeckel combined the linear evolutionism of Goethe

[43] Lyell, *Principles of Geology*, ii. 400. Étienne Serres was the leading disciple of Geoffroy. In his *Anatomie comparée du cerveau* (2 vols, Paris, 1824–6), he argued that, while it was difficult to identify homologies between the brains of some adult vertebrates, the brains of all vertebrates at an early age were very alike.

[44] Karl Ernst von Baer, *Über Entwickelungsgeschichte der Thiere: Beobachtung und Reflexion* (2 vols, Königsberg, 1828–37).

[45] Ibid., i. 221.

[46] Henri Milne-Edwards, 'Considerations sur quelques principes relatifs à la classification naturelle des animaux', *Annales des sciences*, 3rd series, 1 (1844), 65–99.

[47] See Nyhart, 'Embryology and Morphology', 203–5.   [48] See ibid., 209–15.

and Lamarck with the theory of the *Origin of Species* in his *Generelle Morphologie* of 1866,[49] and then more explicitly in his far more popular *Natürliche Schöpfungsgeschichte* (1868),[50] whose frontispiece depicted the 'species of man' and their animal forebears in a scale of descent (Fig. 7.4).

Both *Naturphilosophie* and the Darwinian theory of evolution are altered in Haeckel's reworking. Haeckel's aim was to reconstruct a general evolutionary history of life, filling in the gaps in the fossil record with stages in embryonic development. The guiding thesis of his evolutionary morphology, one that has come to be associated above all with his name, is the biogenic principle, that ontogeny recapitulates phylogeny: that is, the embryonic development of an

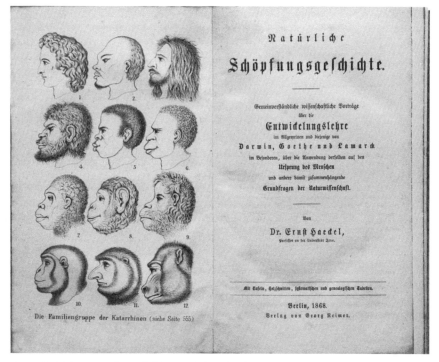

Fig. 7.4. Title page of Haeckel, *Natürliche Schöpfungsgeschichte* (1868)

[49] Ernst Haeckel, *Generelle Morphologie der Organismen: Allgemeine Grundzüge der organischen Former-Wissenschaft, mechanisch begründet durch die von Charles Darwin reformirte Descendez-Theorie* (2 vols, Berlin, 1866). Nyhart has suggested that Haeckel's reconstruction may be less linear that it has appeared: *Biology Takes Form*, 134–5.
[50] Haeckel, *Natürliche Schöpfungsgeschichte* (Berlin, 1868), translated into English as *The History of Creation* (London, 1876). See the discussion in Robert J. Richards, *The Tragic Sense of Life: Ernst Haeckel and the Struggle over Evolutionary Thought* (Chicago, 2008), ch. 7.

organism expresses all the intermediate forms of its ancestors throughout evolution. Nevertheless, as Gliboff notes, ontogeny could never be read as a continuous record of phylogeny for Haeckel. When new characteristics are acquired towards the end of individual development, the earlier portion provided a useable recapitulation of the earlier portion of the phylogeny, but when inserted in the middle of the development, the record is 'falsified' in some way from the insertion point onwards, and there are other eventualities that could obscure the phylogenetic information.[51] Moreover, evolution or development (*Entwicklung*) does not mean a chain of fully predetermined forms as it did for his pre-1860 predecessors, but, in line with Darwin's account, an open-ended and unpredictable process, where not every environmentally induced change would be positive. Heredity was internal and could only repeat in an offspring what had already been present in the parent, whereas variation was triggered outside the organism by independent and unpredictable physical forces. Like Darwin, Haeckel allowed inheritance of acquired characteristics, but, as Gliboff notes, this Lamarckian mechanism produced heritable, mostly favourable variation, upon which Darwinian natural selection still had to act.[52] Haeckel was also forthcoming about the applicability of the theory of evolution to human beings, in a way that Darwin had not been in the *Origin*, writing that 'we find the same law of progress in effect everywhere in the historical development of the human race. Quite naturally! For here, too, in civil and social conditions, the same principle of struggle for existence and natural selection are what drives peoples irresistibly forward and raises them stepwise to higher culture.'[53]

Haeckel's reading of evolution became, along with that of Thomas Huxley in Britain, the public face of Darwinism in the last three decades of the nineteenth century, and the historian of biology Erik Nordenskiöld noted that it was 'the chief source of the world's knowledge of Darwinism'.[54] Many were in accord with Huxley's assessment that, whether one agreed with Haeckel or not, 'one feels that he has forced the mind into lines of thought in which it is more profitable to go wrong than to stand still'.[55] The popularity of his works was partly due to the 'ontogeny recapitulates phylogeny' slogan, but to an even greater extent it was due to his visually striking images of comparative embryonic development, many of which were originally designed as wall charts for use in his lectures, where pictorial immediacy was more important than refinement of detail (Fig. 7.5).[56]

---

[51] Gliboff, *H. G. Bron*, 178. Cf. Richards, *The Tragic Sense of Life*, 232–4.
[52] Gliboff, *H. G. Bron*, 156.
[53] Ernst Haeckel, 'Ueber die Entwicklunstheorie Darwin's', in Ernst Haeckel, *Gesammelte populäre Vorträge aus dem Gebiete der Entwicklungslehre Heft 1* (Bonn, 1878), 1–28: 28.
[54] Nordenskiöld, *The History of Biology*, 515.
[55] Thomas Henry Huxley, 'The Natural History of Creation—by Dr. Ernst Haeckel', *Academy* 1 (1896), 566–80: 41.
[56] See Richards, *The Tragic Sense of Life*, 235–6. There is a comprehensive account of the images in Nick Hopwood, *Haeckel's Embryos: Images, Evolution, Fraud* (Chicago, 2015).

One area in which Haeckel's generalized notion of human evolution was quickly being taken up was in psychology. Haeckel himself had maintained, in his *Generelle Morphologie* of 1866, that of all the branches of anthropology, 'not one is so affected and altered by the theory of descent as psychology . . . . In order to have a proper understanding of the highly differentiated, delicate mental life of civilized man, we must therefore observe not only its gradual awakening in the child, but also its step by step development in lower, primitive peoples and in vertebrates.'[57] In particular, commitment to the biogenic principle became widespread in the 1870s, and, in his influential 1874 textbook on physiological psychology, Wundt takes it as a premise that in embryology, as in comparative anatomy, 'the same law of development is found, in that the earlier stages of the higher vertebrates are similar to the continuing levels of organization of the lower ones'.[58] Wundt combines mechanistic physiology with evolutionary biology, and as Woodward points out, Wundt's account

simply extended, in psychological terms, the biogenetic law . . . In addition, apperception was the mental parallel of interference theory. In keeping with the embryological evidence for a hierarchy of functions in the nervous system, Wundt described the inhibitory and excitatory control of sensation and perception, cognition and volition, by successively higher centers for voluntary 'apperceptive' activity. Just as the interference theory guided physiological experimentation, so the theory of apperception would guide psychological experimentation on the levels of function from reflex to voluntary. Underlying both was the biogenetic law that the development of the individual recapitulates the development of the race.[59]

Freud makes even more ambitious use of the principle. As late as 1917, in his *Introductory Lectures on Psycho-Analysis*, he was advocating the biogenic principle along the lines adumbrated by Haeckel and his popularizer Wilhelm Bölsche, who had argued that sexuality evolved from a primeval saclike organism, the gastrae.[60] On this account, in its first stage, sexuality was oral, but as the gastrae evolved into the gastrointestinal tract the sexual organs became associated with the cloaca, and finally true genitalia developed. Freud writes that, in considering the development of the ego and of the libido,

we must lay emphasis on a consideration which has not often hitherto been taken into account. For both of them are at bottom heritages, abbreviated recapitulations, of the

[57] Haeckel, *Generelle Morphologie*, ii. 434.

[58] Wilhelm Wundt, *Grundzüge der physiologischen Psychologie* (Leipzig, 1874), 43.

[59] William A. Woodward, 'Wundt's Program for the New Psychology: Vicissitudes of Experiment, Theory, and System', in William R. Woodward and Mitchell G. Ash, eds, *The Problematic Science: Psychology in Nineteenth-Century Thought* (New York, 1982), 167–97: 175.

[60] Ernst Haeckel, 'Die Gastrula und die Eifurchung der Thiere', *Jenische Zeitschrift für Naturwissenschaft* 9 (1875), 402–508; Wilhelm Bölsche, *Das Liebesleben in der Natur: Eine Entwickelungsgeschichte der Liebe* (3 vols, Leipzig, 1898–1903). See the discussion in Frank J. Sulloway, 'Freud and Biology: The Hidden Legacy', in William R. Woodward and Mitchell G. Ash, eds, *The Problematic Science: Psychology in Nineteenth-Century Thought* (New York, 1982), 198–227.

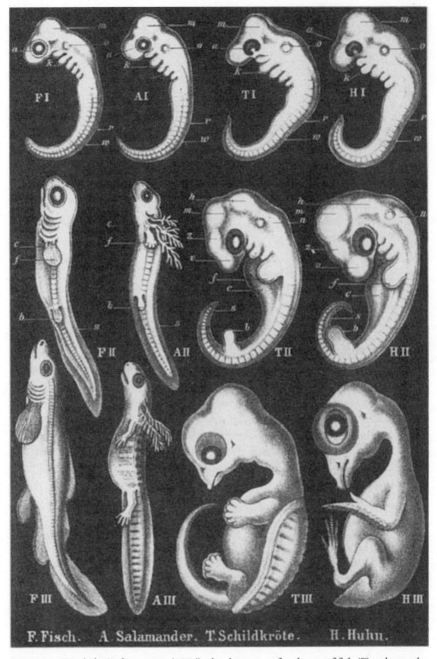

Fig. 7.5a. Haeckel, *Anthropogenie* (1874), development of embryos of fish (F), salamander (A), turtle (T), chick (H)

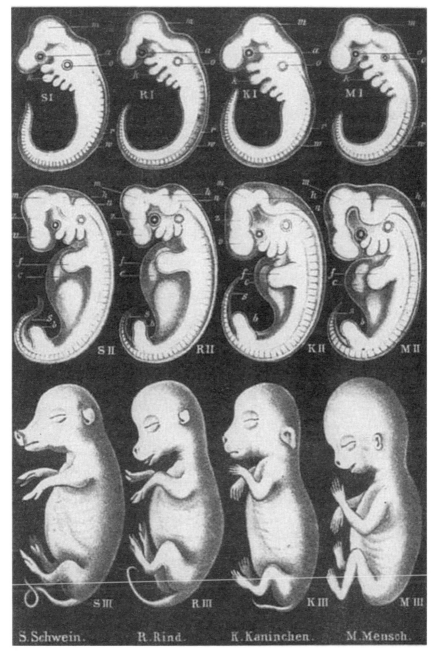

Fig. 7.5b. Haeckel, *Anthropogenie* (1874), development of embryos of pig (S), cow (R), rabbit (K), and human (M)

development which all mankind has passed through from its primeval days over long periods of time. In the case of the development of the libido, this *phylogenetic* origin is, I venture to think, immediately obvious. Consider how in one class of animals the genital apparatus is brought into the closest relation to the mouth, while in another it cannot be distinguished from the excretory apparatus, and yet in others it is linked to the motor organs—all of which you will find attractively set out in W. Bölsche's valuable book. Among animals one can find, so to speak in a petrified form, every species of perversion of the [human] sexual organization.[61]

The extension of the biogenic principle to psychology failed to keep track of developments in embryology itself however. As early as the 1870s, the embryologist Wilhelm His had raised problems with the idea that phylogeny could be the cause of ontogeny.[62] His maintained that any morphological resemblance between embryos or adults of different species could not be taken as evidence of a hereditary relationship, but showed only that they were products of the same or similar mechanical processes and initial conditions of development.[63] His was not merely breaking the relationship between ontogeny and phylogeny, he was also questioning whether a historical account could have explanatory power. Haeckel saw such a historical account as the essence of evolutionary theory, and he resisted any criticism of the Lamarckian idea of inheritance of acquired characteristics. Darwin too had thought that such inheritance had to play a role in evolution, but he died before the most important debate, that between Haeckel and Weismann, broke out. Weismann argued that the development of germ cells—the 'germ-plasm'—could always be traced back to the fertilized egg, with the rest of the body—the 'somatoplasm'—not being involved in the process. On this account, there was no way in which characteristics acquired during development or adulthood by the somatoplasm could be communicated to the germplasm. That is, the environmental effects which Lamarck, and following him Haeckel, had appealed to could not be heritable.[64]

By the beginning of the twentieth century embryology was becoming significantly more remote from evolutionary theory. Work in evolutionary embryology was increasingly showing that there was no simple linear progression from invertebrates to vertebrates: embryos do indeed pass through a series of stages that correspond to those of their ancestors, but not all embryological features

---

[61] Sigmund Freud, *Introductory Lectures on Psycho-Analysis*, in *The Standard Edition of the Complete Psychological Works of Sigmund Freud* (24 vols, London, 1953–74), xvi. 354.

[62] On the criticisms of Haeckel by his contemporaries, see Richards, *The Tragic Sense of Life*, ch. 8. See also Andreas Daum, *Wissenschaftspopularisierung im 19. Jahrhundert: Bürgerlich Kultur, naturwissenschaftliche Bildung und die deutsche Öffentlichkeit 1848–1914* (Munich, 2002), 65–84.

[63] See, for example, Wilhelm His, *Unsere Körperform und das physiologische Problem ihrer Entstehung* (Leipzig, 1874); and Wilhelm His, 'On the Principles of Animal Morphology', *Proceedings of the Royal Society of Edinburgh* 15 (1888), 287–98.

[64] See August Weismann, 'Zur Frage nach der Verebung erworbener Eigenschaften', *Biologisches Centralblatt* 6 (1886), 33–48. On Weismann generally see Frederick B. Churchill, *August Weismann: Development, Heredity, and Evolution* (Cambridge, MA, 2015).

reflect ancestral patterns.[65] With the questioning of the biogenic principle, the fortunes of evolutionary theory no longer hung critically on those of embryology. At the same time, the doctrine of inheritance of acquired characteristics, which seemed so obvious and so crucial to theories of descent by modification, was discarded once the importance of Weismann's work became recognized.

The other area that we identified as being intimately tied to evolutionary theory, geology, also suffered a setback. The issue was the age of the earth. In the first edition of *The Origin of Species*, Darwin had estimated that natural selection and the evolution of human beings would have required a time span of around three hundred million years. This was consonant with his calculation of the age of the Weald, a geological formation in England, on the basis of estimates of the rate of erosion. William Thomson, however, calculated, on the basis of the principles of thermodynamics and the principle of energy conservation, that the earth could not possibly be that old.[66] Thomson—bringing together, in a general theory, geological elevation, earthquakes, volcanoes, the spheroidal shape of the earth, the earth's precession and nutation, measurements of terrestrial properties of matter, and energy dissipation in accord with the second law of thermodynamics—investigated the thermal properties of rocks and made calculations of the temperature at the centre of the earth, which led him to conclude that the earth must have been molten some time between twenty million and a hundred million years ago. In the second edition of *The Origin of Species*, Darwin reluctantly conceded the point. It turned out that Thomson relied on a false assumption about the source of the heat of the earth and the sun. The only significant source that he (or anyone else at the time) could envisage was gravitational contraction, but it subsequently emerged that nuclear fusion was the primary source of the heat, and this yielded a much greater age for the earth and the sun than even that which Darwin was assuming.[67]

[65] See Francis M. Balfour, *A Treatise on Comparative Embryology* (2 vols, London, 1880); Walter Garstang, 'The Theory of Recapitulation: A Critical Restatement of the Biogenic Law', *Zoological Journal of the Linnean Society* 35 (1922), 81–101; Brian K. Hall, 'Balfour, Garstand and de Beer: The First Century of Evolutionary Biology', *American Zoologist* 40 (2000), 718–28. In recent decades evolution and embryonic development have been reunited in the new doctrine of evolutionary development: see Sean B. Carroll, *Endless Forms Most Beautiful: The New Science of Evo Devo and the Making of the Animal Kingdom* (London, 2005); and the contributions to M. D. Laubichler and J. Maienschein, eds, *From Embryology to Evo-Devo: A History of Developmental Evolution* (Cambridge, MA, 2007), especially Manfred D. Laubichler, 'Does History Recapitulate Itself? Epistemological Reflections on the Origins of Evolutionary Developmental Biology', 13–33; William C. Wimsatt, 'Echoes of Haeckel? Reentrenching Development in Evolution', 309–55; and Günter P. Wagner, 'The Current State and the Future of Developmental Evolution', 525–45.

[66] William Thomson, 'On the Secular Cooling of the Earth', *Philosophical Magazine* 25 (1863), 1–14. For an account of the controversy, see Smith and Wise, *Energy and Empire*, ii. ch. 16 and 17.

[67] It has to be said that Thomson was poor at backing winners. Six years before the Wright brothers started to achieve sustained and powered flight, Thomson, in a letter of 8 December 1896, declining an invitation to join the Aeronautical Society, wrote that he had 'not the smallest molecule of faith in aerial navigation'. Quoted in Charles Gibbs-Smith, *The Aeroplane: An Historical Survey of Its Origins and Development* (London, 1960), 35.

There is a difficult issue here. However little connection one might assume between basic physics and the details of evolution, they cannot conflict with one another, and thermodynamics was a rigorously formulated and well-established discipline, whereas evolutionary theory was having teething problems attempting to reconcile disparate theories: in the distribution of species, embryology, geology, and in identifying the mechanisms by which natural selection occurs, not to mention the fact that it emerged in a context that had been shaped by seventy years of radical and highly speculative accounts of human evolution in the work of Erasmus Darwin, Lamarck, Chalmers, and Spencer, among others. It would be a mistake to see physics as more 'fundamental' than evolutionary theory, however. That is just not the kind of relation in which they stand to one another. After all, the thermodynamic approach had had its own teething problems. In the 1840s dynamical geology faced a significant challenge from Agassiz's theory of glaciation: theories of internal heat seemed unable to account for the changes in temperature evidenced in recent geological periods.[68] By contrast, estimates of rates of erosion, which is what Darwin's original account depended on, were well established.

It is a feature of evolutionary theory that it is formed through a coalescing of theories having different provenances, which can be stripped away or revised independently, while the theory itself remains coherent, if not necessarily quite as robust as it was previously. The reaction of physiologists to Darwin's theory of natural selection is interesting in this respect. They welcomed the Darwinian account enthusiastically. This was unsurprising, for as Haeckel was to put it emphatically: 'We see in Darwin's discovery of natural selection in the struggle for existence the most striking evidence for the exclusive validity of mechanically operating causes in the entire field of biology. We see in it the definitive death of all teleological and vitalistic interpretations of organisms.'[69] At the same time, natural selection was explicitly proposed as a basis for the unification of the sciences, with the first scientific Darwinist journal, the first issue of which appeared in 1877, bearing the title *Kosmos, Zeitschrift für eine einheitliche Welt-anschauung auf Grund der Entwicklingslehre* (or *Kosmos, Journal for a Unified Worldview on the Basis of a Theory of Development*). For Haeckel, advocate of unification par excellence, the route to a mechanical understanding tied together recapitulationism, descent with modification, an end to teleology, and a form of monistic materialism. In his criticisms of His, he writes:

With this statement an unbridgeable chasm is defined, which separates the older, teleo-logical and dualistic morphology from the newer, mechanistic and monistic one. If the physiological function of heredity and adaptation are shown to be the sole causes of organic form, then therewith, at the same time, every source of teleology, of dualistic and

[68] See Martin J. S. Rudwick, 'The Glacial Theory', *History of Science* 8 (1969), 136–57; and Crosbie Smith, 'William Hopkins and the Shaping of Dynamical Geology: 1830-1860', *British Journal for the History of Science* 22 (1989), 27–52.
[69] Haeckel, *Generelle Morphologie der Organismen*, i. 100. Original in italics.

metaphysical points of view, will be removed from the field of biogeny. The sharp opposition between the two guiding principles is thus clearly defined. *Either a direct, causal connection between ontogeny and phylogeny exists or it does not exist.* Either ontogeny is a condensed excerpt of phylogeny or it is not. Between these two assumptions there is no third one![70]

Consider, given this, what it was that the German reductionist physiologists saw as the value of evolutionary theory. Helmholtz indicates how the issues were perceived by this group:

Before the time of Darwin only two theories respecting organic adaptability were in vogue, both of which pointed to the interference of free intelligence in the course of natural processes. On the one hand it was held, in accordance with the vitalistic theory, that the vital processes were continuously directed by a living soul; or, on the other, recourse was had to an act of supernatural intelligence to account for the origin of every living species. The latter view supposes that the causal connection of natural phenomena had been broken less often, and allows of a strict scientific examination of the processes observable in the species of human beings now existing; but even it is not able to entirely explain away those exceptions to the law of causality, and consequently it enjoyed no considerable favour as opposed to the vitalistic view, which was powerfully supported, by apparent evidence, that is, by the natural desire to find similar causes behind similar phenomena. Darwin's theory contains an essentially new creative thought. It showed how adaptability of structure in organisms can result from a blind rule of a law of nature without any intervention of intelligence.[71]

Helmholtz is clearly not envisaging a potential unification of the life sciences here. He and his colleagues such as du Bois-Reymond and Virchow were very selective. They embraced natural selection because it gave them a way to avoid teleology, and they weren't too concerned with other aspects, which, in the main, they left to morphologists and others to sort out. But there was one exception, one question which did divide evolutionary theory and a discipline such as physiology. Haeckel, in all his talk of causal connections, wasn't using the term 'cause' in quite the same way as the physiologists, for as well as mechanical causality, he also, as a good Darwinist, advocated historical causality. In his response to His and the zoologist Alexander von Götte, for example, he ridiculed their belief in a purely mechanical explanation of growth in place of one based on the animal's phylogenetic past: their interpretation of embryology undermined the theory of evolution by dispensing with the relevant embryological evidence.[72] Biology must, he argued, be above all a historical science. One of the strongest reactions to this came from du Bois-Reymond in an 1872 lecture on the limits of

---

[70] Ernst Haeckel, 'Die Gastrea-Theorie, die phylogenetische Classification des Thierreichs und die Homologie der Keimblätter', *Jenaische Zeitschrift für Naturwissenschaft* 8 (1874), 1–55: 6.

[71] Hermann von Helmholtz, in David Cahan, ed., *Hermann von Helmholtz, Science and Culture* (Chicago, 1995), 218–19.

[72] Ernst Haeckel, *Ziele und Wege der heutigen Entwickelungsgeschichte* (Jena, 1875).

knowledge.[73] Du Bois-Reymond had read and digested Darwin before even Haeckel, focusing on the blind nature of Darwinian evolution, which he believed made it compatible with mechanism. He placed the limits of knowledge at the limits of mechanical explanation, so could not count historical explanations as genuine forms of knowledge, even though he recognized that the theory of natural selection as it stood relied on them. Haeckel, on his view, had transgressed the limits of knowledge by advocating pluralism in biological explanation.

THE 'MODERN SYNTHESIS'

In 1909, the fiftieth anniversary of the publication of *The Origin of Species*, a commemorative collection of essays entitled *Darwin and Modern Science* was published under the editorship of the botanist A. C. Seward who, in his Preface, notes 'the divergence of views among biologists in regard to the origin of species'.[74] As Ernst Mayr has pointed out, twenty years later advocates of genetics and evolution by natural selection were still wholly at odds with one another, talking different languages, asking different questions, and adhering to different basic assumptions.[75] By 1942, however, Julian Huxley could write in his *Evolution, The Modern Synthesis*:

Biology in the last twenty years, after a period in which new disciplines were taken up in turn and worked out in comparative isolation, has become a more unified science. It has embarked upon a period of synthesis, until today it no longer presents the spectacle of a number of semi-independent and largely contradictory sub-sciences, but is coming to rival the unity of older sciences like physics, in which advance in any one branch leads almost at once to advance in all other fields, and theory and experiment march hand-in-hand. As one chief result, there has been a rebirth of Darwinism.[76]

Huxley did not hesitate to draw out the political consequences of his 'modern synthesis'—a fusion of Mendelian genetics, population studies, natural history,

[73] Emil Du Bois-Reymond, *Über die Grenzen des Naturerkennens: Die Sieben Welträtseln* (Leipzig, 1916).

[74] A. C. Seward, ed., *Darwin and Modern Science* (Cambridge, 1909), vii.

[75] Mayr, *The Growth of Biological Thought*, 566.

[76] Julian Huxley, *Evolution, the Modern Synthesis* (London, 1942), 503–4. As well as Huxley's own *Evolution*, the formative works were: Theodosius Dobzhansky, *Genetics and the Origin of Species* (New York, 1937); Ernst Mayr, *Systematics and the Origin of Species* (New York, 1942), which showed how systematics could be refounded on the basis of the new synthesis; George Gaylord Simpson, *Tempo and Mode in Evolution* (New York, 1944), which showed how the same evolutionary processes occurred on both the macro and the micro scale, specifically in both the horse and the fruit fly; Bernhard Rensch, *Neuere Probleme der Abstammungslehre* (Stuttgart, 1947), which established the consistency of macro and micro evolution; and G. Ledyard Stebbins, *Variation and Evolution in Plants* (New York, 1950), which showed in detail how natural selection accounted for evolution in plants. See the summary discussion in Agar, *Science in the Twentieth Century and Beyond*, ch. 9.

and Darwin's theory of natural selection. In contrast to the nineteenth-century national unification programmes of du Bois-Reymond, Virchow, and others, his was staunchly internationalist and part of a concerted push for world government. It was a crucial part of Huxley's overall programme that evolution was progressive—he considered it to be nothing less than a remedy for the ills of the modern world—but he conceived of selection as strictly mechanical, so evolution itself could not have a goal or end point. This distinguished him from earlier proponents of evolution such as Haeckel, who believed that evolution was not just restricted to the species, but that there was evolution of the psyche as well. Mach, for example, had also argued along these lines,[77] and had construed science as an evolutionary adaptation. Its task, as he puts it in his 1886 *Beträge zur Analyse der Empfindung*, is to 'provide the fully developed human with as perfect a means of orientating himself as possible'.[78] Ten years later he writes that the aim of 'scientific economy is to provide us with a picture of the world as complete as possible—connected, unitary, calm and not materially disturbed by new occurrences: in short a world picture of the greatest possible stability. The nearer science approaches this aim, the more capable will it be of controlling the disturbances of practical life, and thus serving the purpose out of which its first germs were developed.'[79]

There is much here that Huxley would have ascribed to, but his route was of necessity more demanding. As one commentator has noted, articulating 'as non-teleological a version of natural selection that could still somehow give direction and make possible progressive evolution, and at the same time adhering to selection as a mechanistic—and therefore legitimate—scientific principle was the challenge that Huxley faced'.[80] Huxley's solution, not unlike that of Mach, was to turn to humans' ability to modify their own environment through the conscious wilful use of intelligence. As he writes in the closing chapter of *Evolution*: 'The future of progressive evolution is the future of man. The future of man, if it is to be progress and not merely a standstill or a degeneration, must be guided by a deliberate purpose. And this human purpose can only be formulated in terms of the new attributes achieved by life in becoming human: his purpose must take account of his unique features as well as those he shares with other life.'[81] Even more explicitly, in the pamphlet he published when he took over as the first full-time director of UNESCO in 1946, he writes that UNESCO 'must work in the context of what I call *Scientific Humanism*, based on established facts

---

[77] Ernst Mach, *Knowledge and Error: Sketches on the Psychology of Enquiry* (Dordrecht, 1976), 171.

[78] Ernst Mach, *The Analysis of Sensations: And the Relation of the Physical to the Psychical* (New York, 1959), 37.

[79] Ernst Mach, *Principles of the Theory of Heat: Historically and Critically Elucidated* (Dordrecht, 1986), 337. First published as *Die Principien der Wärmelehre* (Leipzig, 1896).

[80] Vassiliki Betty Smocovitis, *Unifying Biology: The Evolutionary Synthesis and Evolutionary Biology* (Princeton, NJ, 1996), 144.

[81] Huxley, *Evolution, the Modern Synthesis*, 577.

of biological adaptation and advance, brought about by means of Darwinian selection, continued into the human sphere by psycho-social pressures, and leading to some kind of advance, even progress, with increased human control and conservation of the environment and natural forces'.[82]

Huxley was not the only evolutionist to uphold the doctrine of evolutionary progress in the 1940s: Dobzhansky, Mayr, Simpson, and Stebbins all addressed the question of 'the future of mankind' in evolutionary terms. What had happened since the beginning of the century to change the fortunes of Darwinism, from a marginalized account of evolution to something so secure that it could act as a programme for social and political reform? The change can be summarized in two developments. The first stage began with the discovery, around 1900, that in hybridization experiments particular parental characteristics, which may be lost in the first generation, will reappear in the second generation if the offspring are crossed in the specific ratio of 1:3. This was contrary to the prevailing view that, in hybridization, such characteristics were simply blended in offspring. In fact, Mendel had already discovered this forty years earlier, but his work remained largely unknown (the pages of Darwin's copy of Mendel's 1865 paper on hybridization remained uncut), and for those who were familiar with it, its implications remained dormant because it was presented as a work on hybridization rather than inheritance. Among those who simultaneously rediscovered Mendelian ratios around the turn of the century was the Dutch botanist Hugo de Vries, who offered a 'mutation theory'.[83] Although he had been a staunch supporter of Darwinism, his 'gene' theory, as it came to be known, indicated that evolution may occur by large discontinuous jumps, and this 'saltationist' account was in direct conflict with Darwin's commitment to the slow accumulation of small steps, which his reading of Lyell had convinced him at an early state of his thinking must be the mechanism of evolution. The problem with Darwin's account was that it was hard to see how the natural selection of individual characteristics could have the power to create new species. There was widespread agreement that saltation was needed for this, and De Vries' account was not only backed up experimentally but fitted the palaeontological record.

By 1918, as Provine notes, 'most prominent geneticists had accepted Mendelism and Darwinism as complementary . . . , and believed natural selection of small Mendelian differences was the mechanism of evolution'.[84] Within a few years, a number of biologists—notably Fisher, Haldane, and Wright[85]—realized that unit

---

[82] Julian Huxley, *Memories* (2 vols, New York, 1970–3), 15. See Gregory Blue, 'Scientific Humanism and the Founding of UNESCO', *Comparative Criticism* 23 (2001), 173–200.

[83] Hugo de Vries, *Die Mutationstheorie. Versuche und Beobachtungen über die Entstehung von Artem im Pflanzenreich* (2 vols, Leipzig, 1901–3).

[84] Provine, 'Adaptation and Mechanisms of Evolution after Darwin', 841.

[85] R. A. Fisher, *The Genetical Theory of Natural Selection* (Oxford, 1930); J. S. Haldane, *The Causes of Evolution* (London, 1932); Sewall Wright, 'Evolution in Mendelian Populations', *Genetics* 16 (1931), 97–159.

genes could explain continuous variation if several of them influenced the same characteristic. This was the first phase of the 'synthesis' identified by Huxley, and he explains that it consisted in the recognition that Mendelian principles are operative in all organisms, that small-scale continuous Darwinian variability has a Mendelian basis, and that it can be demonstrated mathematically that small selection pressures acting on minor genetic differences can cause evolutionary change.[86] The second phase came in 1937 with Dobzhansky's translation of the abstract mathematics of gene selection theory into practical propositions of use to fieldworkers, in the process connecting a number of biological subdisciplines to the 'synthesis' of the first phase.[87]

Two things are worth noting about this 'synthesis'. The first is that various areas of study were absent. Embryology was one. As Mayr, looking back on the period in 1993, noted: 'The representatives of some biological disciplines, for instance, developmental biology, bitterly resisted the synthesis. They were not left out of the synthesis, as some of them now claim, but they simply did not want to join.'[88] The study of microbes was another area that was missing. The issues here were more fundamental, and hinged on a deep division between traditional evolutionary biology and molecular phylogenetics. Proponents of the former—notably Dobzhansky, Simpson, and Mayr—resented the (well-funded) rise of the latter from the 1960s onwards, and asserted the unique importance of Darwinian principles in understanding evolution.[89] Here the threat came from the apparently Lamarckian behaviour of bacteria. As Jan Sapp explains:

Bacterial evolution seemed to lie beyond neo-Darwinian conceptions. Bacterial adaptations to their physical environment, their resistance to lethal viruses, and the transformations resulting from the acquisition of what some considered to be infectious genes from the environment that then exist in symbiosis within their hosts, were viewed to belong to a neo-Lamarckian realm. As Stanford bacteriologist W. H. Manning commented in 1934: 'About the only conventional law of genetics and organic evolution that is not definitely challenged by current bacteriologists is the nineteenth century denial of the possibility of spontaneous generation of a bacterial cell. Even this is questioned by certain recent theorists . . . Whether or not future refinements in immuno-chemical technique can or will bridge the gap between the apparent Lamarckian world of bacteriology and the presumptive Darwinian world of higher biological science is beyond current prophecy.'[90]

The peculiar behaviour of bacteria from the perspective of the 'modern synthesis' was recognized in passing by Huxley, who writes that we must 'expect that the

[86] Huxley, *Evolution, the Modern Synthesis*, 25.

[87] Dobzhansky, *Genetics and the Origin of Species*.

[88] Ernst Mayr, 'What Was the Evolutionary Synthesis?', *Trends in Ecology and Evolution* 8 (1993), 31–4: 32.

[89] Michael Dietrich, 'Paradox and Persuasion: Negotiating the Place of Molecular Evolution Within Evolutionary Biology', *Journal of the History of Biology* 31 (1998), 87–111.

[90] Sapp, *The New Foundations of Evolution*, 78–9. Cf. Jan Sapp, *Evolution by Association: A History of Symbiosis* (Oxford, 1994).

processes of variation and evolution in bacteria are quite different from the corresponding processes in multicellular organisms'.[91] But what was at issue was an anomaly, behaviour so bizarre that it simply could not be incorporated into thinking about the nature of evolution.

Second, Gould has pointed out that whereas the first phase included a vigorously pluralistic range of permissible mechanisms, which remained in place early in the second phase, by the 1950s things had changed, as natural selection had become the orthodoxy, 'often accompanied by a strong and largely rhetorical dismissal of dissenting views'. This hardening of views, he argues, 'extended beyond overconfidence in adaptation to a more general, and sometimes rather smug, feeling that truth had now been discovered, and that a full account of evolution only required some mopping up and adumbration of details'.[92] These developments were highly overdetermined, with politics playing a significant role. After the Second World War, there emerged new evidence of non-Mendelian inheritance based on cytoplasmic entities, which, normally transmitted sexually, could be transmitted artificially by infection. Yet as Sapp notes, the evidence for non-Mendelian heredity 'became caught in cold war rhetoric between communists led by T. D. Lysenko in the Soviet Union, who denied the existence of genes and who advocated the inheritance of acquired characteristics, on the one hand, and Western geneticists who insisted that chromosomal genes in the nucleus were the sole source of hereditary change, and who denied the inheritance of acquired characteristics, on the other'.[93]

This either/or attitude subsequently changed, and it is now coming to be generally accepted that pluralism is inevitable in accounting for evolutionary mechanisms. As Alexander Rosenberg, for example, has noted:

That there are no laws of biology to be reduced to laws of molecular biology, and indeed that there are no laws of molecular biology, can be shown by the same considerations that explain why genes and DNA cannot satisfy reduction's criterion of connection. The individuation of types in biology is almost always via function: to call something a wing, or a fin, or a gene is to identify it in terms of its function. But biological functions are naturally selected effects. And natural selection for adaptations—that is, environmentally

---

[91] Huxley, *Evolution, the Modern Synthesis*, 132.

[92] Stephen Jay Gould, *The Structure of Evolutionary Theory* (Cambridge, MA, 2002), 505. On the plurality of views, see Provine, 'Adaptation and Mechanisms of Evolution after Darwin', 842–53.

[93] Sapp, *The New Foundations of Evolution*, 120. There were pre-war sensitivities about non-Mendelian heredity in the politics of the biological community. In 1934 a Harvard geneticist wrote: 'There are several types of phenomena where there is a direct transfer, from cell to cell, of alien matter capable of producing morphological changes. It is not to be supposed that modern biologists will cite such instances when recognized, as examples of heredity. But since an earlier generation of students used them, before their cause was discovered, to support arguments on the inheritance of acquired characteristics, it is well to be cautious in citing similar, though less obvious, cases as being illustrations of non-Mendelian heredity.' Edward M. East, 'The Nucleus-Plasma Problem', *American Naturalist* 68 (1934), 289–303, 402–39: 431.

appropriate effects—is blind to differences in physical structure that have the same or roughly similar effects. Natural selection 'chooses' variants by *some of their effects*, those that fortuitously enhance survival and reproduction. When natural selection encourages variants to become packaged together into larger units, the adaptations become functions. Accordingly, the structural diversity of the tokens of a given Mendelian or classical or population system or structure is inevitable. And no biological kind will be identical to any single molecular structure or manageably finite number of sets of structures.[94]

But the problems are not confined to matching structure and function. Explaining the difficulties encountered by philosophers and others in trying to account for how the Mendelian gene is reduced to molecular biology, Griffiths and Stotz remark on how the experimental practices that are used to identify genes are not some added extra, but integral to our understanding of genes:

Although the new, molecular identity of the gene was now its dominant identity, the other, instrumental identity did not simply go away. The original role of the Mendelian gene continues to define the gene in certain areas of biological research . . . . Reductionists are correct that the gene turned out to be grounded in DNA, but they fail to recognize that the development of genetics has left us with more than one scientifically productive way of thinking about DNA and the genes it contains. This is in large part because they have failed to recognize how the different identities of the gene are anchored in different experimental practices.[95]

There has in fact always been a good claim for pluralism in the life sciences. Whether or not they thought it a good thing, as the nineteenth century progressed most physiologists recognized, particularly as evolutionary theory and embryology developed, that there was a significant degree of explanatory pluralism in biology. In 1929 the biologist Joseph H. Woodger complained that:

If we make a general survey of biological science we find that it suffers from cleavages of a kind and to a degree which is unknown in such a well-unified science as, for example, chemistry. Long ago it has undergone that inevitable process of subdivision into special branches which we find in other sciences, but in biology this has been accompanied by a characteristic divergence of method and outlook between the exponents of the several branches which has tended to exaggerate their differences.[96]

J. J. Smart makes a similar claim, suggesting that what marks out the physical from the biological sciences is that the physical sciences form a unified self-contained enterprise, whereas the biological sciences are a heterogeneous mishmash of laws and assumptions borrowed from other areas, and brought together

---

[94] Alexander Rosenberg, 'Reductionism (and Antireductionism) in Biology', in David Hull and Michael Ruse, eds, *The Cambridge Companion to the Philosophy of Biology* (Cambridge, 2007), 120–38: 122–3.

[95] Paul Griffiths and Karola Stotz, *Genetics and Philosophy: An Introduction* (Cambridge, 2014), 4.

[96] Joseph H. Woodger, *Biological Principles: A Critical Study* (London, 1929), 84.

more in the way in which we might conceive of a technology rather than a science.[97] We saw in earlier chapters that the first part of this claim, on the physical sciences, is false. We have no reason to quarrel with Smart's character-ization of the biological sciences however, only with the presumed contrast with the physical sciences, and with the implicit assumption that the biological sciences are thereby deficient.

Reversing Smart's assessment, I suggest that the biological sciences, conceived not as part of a unified enterprise but as harbouring an inevitable element of unruliness by comparison with the physical 'core', might act as a better model for science generally, and the physical and material sciences more particularly.[98] The choice is not so much between unity and pluralism, as between homogenization and pluralism. Pluralism allows us to make links where appropriate, but there is no a priori requirement (explanatory or otherwise) that everything must ultim-ately be connected, and a fortiori there is no requirement that the links established must ultimately take a form of reduction. At the same time, pluralism provides a means of targeting explanations. The targeting of explanations is not something peculiar to the life sciences, but pervades the whole of scientific enquiry, and the absence of a sense of the importance of targeting is perhaps the greatest casuality of unity of science conceptions. Thinking in terms of targeting is a way to capture the exploratory, active nature of explanation, rather than envisaging explanation as an umbrella under which we passively bring phenomena.

Putting to one side the question whether or not it is misconceived as a conception of how science actually proceeds, it is worth asking at this point whether the unity of science might bring a fruitful—if possibly idealized—sense of purpose to the pursuit of science, for example, or provide a vantage point which enables one to grasp connections that would otherwise escape one's attention. There is no doubt that the idea of a unified understanding of nature has acted as a heuristic spur to enquiry. As Kirchhoff, supporter of the Helmholtz/de Bois-Reymond mechanist reduction programme, noted in the context of the reduction of the life sciences in 1865: 'The aim of the natural sciences will never be fully achieved, but already the very fact that it has been identified as being the aim offers a certain satisfaction, and in this approach lies the greatest enjoyment that the study of the phenomena of nature can impart.'[99] But what exactly does this approach consist in in practical terms? I have emphasized the crucial difference

---

[97] Smart, *Philosophy and Scientific Realism*, ch. 3. Cf. Daniel Dennett's more nuanced comment that some 'would go so far (I am one of them) as to state that what biology is, is the reverse engineering of natural systems.' Daniel C. Dennett, *Brainchildren: Essays on Designing Minds* (Cambridge, MA, 1998), 256.

[98] A similar approach is taken in Sandra D. Mitchell, *Unsimple Truths: Science, Complexity, and Policy* (Chicago, 2009), but Mitchell pursues an epistemological route which I shall be arguing in Chapter 10 fails to capture the source of the need for pluralism. In Chapter 9 I shall also question the fruitfulness of epistemological programmes.

[99] Gustav Kirchhoff, *Über das Ziel der Naturwissenschaften* (Heidelberg, 1865), 24–5.

between attempts to reconcile disparate theories, particularly in cases where these appear to be at odds with one another, and the project of unification, which encompasses a different kind of aspiration, one that extends far beyond overcoming inconsistencies between theories. But at a practical level, is there anything more to the drive to unification than the drive to reconcile disparate theories? And if there is, is there a cost to following such a fictional route, namely that the direction in which it leads one may be fruitless and counterproductive, fostering aims and goals that are wholly imaginary, defensible only on a dogmatic basis? The cost is clear in the grander unification projects. Comte, for example, was openly hostile to experimental research because it could lead to awkward discoveries that could only result in further uncertainties.[100] His terminology is revealing because what such unity of science projects share is a concern with certainty, as opposed to understanding for example, and certainty is something that can militate against revision, as the history of religious belief amply testifies.

Note that even on Smart's characterization, any general project for the unity of the sciences is out of the question. The life sciences cannot be included in the unification project for the reasons that he gives. This leaves what he refers to as the physical sciences, within which I have distinguished the physical sciences proper (physics) and the material sciences (chemistry). These were distinguished precisely because, as I have argued, they cannot be incorporated into a single enterprise, by contrast with Smart, who clearly assumes that they are part of the same thing. Nevertheless, on both Smart's reading and on ours, there is no unified enterprise that includes the physical sciences (in either sense) and the life sciences. The question now arises: once we rid ourselves of the hierarchical assumptions that the unity of the sciences carries with it, is there some more fruitful way of characterizing scientific practice, of the kind that we have encountered in this and earlier chapters?

The alternative, pluralism, has a number of varieties in the literature on physical and biological explanation. Some argue, for example, that a preference for micro-explanations cannot be justified, and does not capture how explanations in physics work,[101] whereas others hold that macro- and micro-explanations when integrated give a better explanatory account of the phenomenon of interest than they would if taken separately.[102] In many cases, it may be possible to work with a hybrid. The historian of hydrodynamics Oliver Darrigol, for instance, in his examination of nineteenth-century fluid mechanics, has noted that there was no way to conceive of the physical nature of fluids in such areas as elasticity, heat,

---

[100] See Kent, *Brains and Numbers*, 61.

[101] See, for example, Nancy Cartwright, *How the Laws of Physics Lie* (Oxford, 1983); Nancy Cartwright, *The Dappled World* (Cambridge, 1999).

[102] See, for example, Sandra Mitchell, *Biological Complexity and Integrative Pluralism* (Cambridge, 2003). See also John Dupré, *The Disorder of Things* (Harvard, MA, 1995), and Caterina Marchionni, 'Explanatory Pluralism and Complementarity: From Autonomy to Integration', *Philosophy of the Social Sciences* 38 (2008), 314–33.

electricity, magnetism, and optics, and that 'molecular intuitions . . . were combined with more phenomenological reasoning. Viscous-fluid and elastic-body
theorists similarly hybridized molecular and continuum physics.'[103] In other
cases, such as that of mid-twentieth-century microphysics, it became possible to
make connections between particular disparate areas by means of an artificial
language, as it were, designed specifically for making localized connections.
Galison, for example, has invoked the idea of a pidgin language as a way of
connecting the relevant parts of two different physics-engineering projects, which,
by means of wholly different instrumental procedures, offer pictorial and statistical representations respectively in microphysics. These cannot be combined into
anything remotely resembling a single theory, but artificial areas of overlap can be
constructed which meet the demands of a general theory of microphysics.[104]
Another way of dealing with the issues was the idea, widespread in the later
decades of the nineteenth century and drawing its inspiration from Maxwell, that
knowledge in science consists in the establishment of analogies.[105]

On such approaches, pluralism is a permanent feature of scientific explanation,
not a temporary measure: explanations are not judged with respect to some
ultimate form of explanation, such as those proposed by physics. On the face of
it, this puts pluralism at a disadvantage when faced with the necessary task of
distinguishing between genuine explanations and merely purported explanations
which, free of any anchoring in established standards, are justified in cognitively
irrelevant or inappropriate ways.[106] But the alternative route is in effect to resolve
the question by fiat, taking physical explanations to be the kind to which all others
must ultimately be referred, and this cannot work. There are many cases where
preferring explanations outside physics over basic physical ones has proved
productive. The most famous case was that of the conflict between Copernicus'
heliocentric model of the solar system and the underlying Aristotelian physics that
he strictly adhered to, a conflict that Copernicus himself was unable to resolve,[107]
but in response to which he preferred a geometrical construction over the physics
of a universally accepted natural philosophy. And we have just seen an example
of the converse case, where deferring to physics turned out to be the wrong
strategy: Darwin's reluctantly adapting the timescale required for evolutionary
processes on earth to a physical understanding of terrestrial processes turned out
to be unwarranted.

[103] Olivier Darrigol, *Worlds of Flow: A History of Hydrodynamics from the Bernoullis to Prandtl*
(Oxford, 2005), 101–2.
[104] Galison, *Image and Logic*. I compare this with the eighteenth-century natural philosophy
tradition, to which it bears instructive similarities, in *The Collapse of Mechanism*, 294–304.
[105] See René Dugas, *La Théorie Physique au Sense de Boltzmann et ses Prolongements Modernes*
(Paris, 1959), 78–80. For a strong defence of the role of analogies, see Mary Hesse, *Models and
Analogies in Science* (Notre Dame, IN, 1966). The central role of analogy in comparative anatomy
was established in the Discours préliminaire to the first volume of Geoffroy Saint-Hilaire, *Philosophie
anatomique*.
[106] See Gaukroger, *The Emergence of a Scientific Culture*, ch. 7.   [107] Ibid., 169–95.

Once we have discarded the notion that ontology provides an overriding regulative constraint on how ultimate explanations are to be layered, then we are free to abandon the restriction that such connections must be fitted into a prior universal structure or grid if they are to exhibit systematic relations. Rather, there are just different kinds of explanatory resources that we need to draw on in explaining different kinds of things, and from the point of view of explanation there is no a priori ranking of generality in these resources. In some cases, we will find that this allows us to establish a fundamental connection of interconvertibility between two fields, such as in the case of electricity and magnetism. In others, as in the case of 'historical' versus 'mechanical' causality in embryology, we may accept these as complementary, without believing that there could be some underlying commonality that will secure unification. In those cases in between these, which I suspect are the majority of cases that pose problems for the sciences, we may find that there is insufficient commonality for straightforward intertranslatability.

Finally, note that when various disciplines or subdisciplines are brought together to offer a single explanatory resource, the motivation behind this may have little to do with the intrinsic characteristics of the disciplines. Molecular biology, for example, is a grand fusion of the methods, techniques, and concepts of organic chemistry, polymer chemistry, biochemistry, physical chemistry, X-ray crystallography, genetics, and bacteriology.[108] As the historian of molecular biology Lily Kay notes, it did 'not just evolve by the natural selection of randomly distributed disciplinary variants, nor did it ascend solely through the compelling power of its ideas and its leaders. Rather the rise of the new biology was an expression of the systematic cooperative efforts of America's scientific establishment—scientists and their patrons—to direct the study of animate phenomena along selected paths toward a shared vision of science and society.'[109]

RECOVERING A LOST UNITY

I have stressed the difference between establishing consistency between theories, in cases of overlap, where this is achieved through establishing intertranslatability of theories (or relevant parts of theories), on the one hand; and, on the other, what we might term a jigsaw or puzzle model, where there is already a complete picture out there which one aims to reconstruct, so that it is a question of finding the right pieces and organizing them into the right places. This model of putting the pieces

[108] See Yasu Furkawa, 'Macromolecules: Their Structures and Functions', in Mary Jo Nye, ed., *The Cambridge History of Science, volume 5: The Modern Physical and Mathematical Sciences* (Cambridge, 2003), 429–45: 439.

[109] Lily E. Kay, *The Molecular Vision of Life: Caltech, the Rockefeller Foundation, and the Rise of the New Biology* (Oxford, 1993), 3.

together to form a comprehensive picture of the world worked quite well when there was considered to be a single divine plan behind creation, in terms of which one made sense of the world. But such a view cannot survive the transformation of science into a modern secular enterprise. We have no reason to suppose that there is a secular version of such a divine plan, a self-organizing natural realm which successful explanation simply mirrors. The history of the natural-philosophical or scientific disciplines, as I have tried to show in this and earlier volumes, exhibits no convergence on anything like a single unified picture of the world.

In the nineteenth-century German push for national unity, as reflected in the unity of science, and in the push to revive a cosmopolitan culture through the unity of science in the work of writers like Julian Huxley and Neurath in the interwar years, the deep connections between socio-political programmes and conceptions of the role of science become explicit. These connections may at first seem somewhat artificial. But once we consider the way in which science began to occupy the void opened up by the demise of Christianity in Europe in the second half of the nineteenth century, it will start to become a little clearer regarding how it might have begun to act as a model for our general understanding of the world and our place in it. The deep traditional connections between socio-political programmes and religious ones are undeniable. Although the Christian West, unlike the Islamic world, was never a theocracy, Christian values nevertheless took a leading role in the formulation of political and social norms. At the same time, the collapse of cosmopolitan values in the wake of the Great War combined with the manifest universality of science—by contrast with religious, political, and social divisions—opened up a clear path for science to take over a leading role in providing a new set of values. The values of science were defended as free from any prior ideologies and prejudices, and they matched the criterion of universalisability of moral values that both Kant and the utilitarians, in their different ways, promoted: they were available to everyone in exactly the same way. As a member of the Mental Health Institute at the University of Michigan announced in 1957: 'the ethical system derived from scientific behaviour is qualitatively different from other ethical systems—is, indeed, a "superior" ethical system'.[110] In some ways this is simply part of the advance of a scientific culture for, as David Hollinger notes, as the autonomy of science from external influences and demands was increasingly urged and defended, 'it became all the more important that moral qualities for which science was ostensibly a vehicle be seen as intrinsic to science', for then 'society could rest more comfortably with the expansion of science'.[111]

What I want to suggest is that there arose an expectation that science would take over a task for which Christianity had been found wanting, namely that of providing a cohesive unity of understanding, without any thought being given as

[110] Anatol Rapoport, 'Scientific Approach to Ethics', *Science* 150 (1957), 796–9: 797.
[111] David A. Hollinger, *Science, Jews, and Secular Culture: Studies in Mid-Twentieth Century American Intellectual History* (Princeton, NJ, 1996), 84.

to whether such an aim could stand up to scrutiny. Nevertheless, considerable fine-tuning is needed if we are to understand how science took on this new role. Everything hinges on the naturalization of human behaviour, that is, on science taking over from religious conceptions in accounting for human behaviour, with the focus in the initial stages on economic and moral behaviour. But it also depends crucially on the transformation of philosophy into a metatheory of science, something that provides a means of placing science at the centre of any form of understanding. It is to these questions that we now turn in Part III.

PART III

THE EXPANSION OF SCIENTIFIC UNDERSTANDING

8

The Problem of the Human Sciences

In Part II, we investigated the idea of the unity of science, exploring its origins and the various forms that its claims have taken. The aim has been to provide an understanding of how the relations between the sciences were transformed in the course of the nineteenth century, how attempts were made to establish hierarchies, and how the idea of science as a unified enterprise was conceived and promoted. We have seen that extra-scientific considerations—philosophical, social, political, ideological—played a crucial role, and this meant that science was increasingly called upon to offer a novel and exclusive way of making sense of the world and our place in it. Had these developments been confined to the natural and life sciences, debates over the nature of science and the extent of scientific understanding would still have been a feature of nineteenth-century thought—it is hard to imagine how the arguments over evolution could have proceeded without raising such questions for example—but they would have been much more limited in scope than those which in fact took place, and which have shaped thinking about scientific questions since. The catalyst was the emergence of the 'human sciences'. These challenged traditional religious and philosophical thinking about a range of political, moral, and economic issues, and in doing so went to the core of concerns about the limits of science. The problem was that, as a scientific culture developed, the issues turned not just on understanding the world, but increasingly on understanding our place in the world. The human sciences bore directly on the latter, and it was here that the conflicts with religious and philosophical forms of understanding were greatest. What resulted from these conflicts—which included, among other things, intractable perennial questions about the nature of morality and the nature of consciousness—was the emergence of comprehensive metascientific theories. These set out to establish the legitimacy of the newly expanded range of scientific explanation, and in the process they reshaped just what was expected of science.

Among the new issues that arise once we turn to the human sciences, paramount is the question of what to include in the sciences. The unification of science requires the exclusion of some disciplines from the rubric of science as much as it requires the inclusion of others, and what is and what is not included is inevitably a fraught and contested question, not the kind of thing that can typically be settled on internal grounds. As a result, the imposition of a coherence based on a postulated unity of the sciences could only be artificial. I am less

Civilization and the Culture of Science: Science and the Shaping of Modernity, 1795–1935. Stephen Gaukroger, Oxford University Press (2020). © Stephen Gaukroger.
DOI: 10.1093/oso/9780198849070.001.0001

concerned to pursue this artificiality, however, than to explore how it can be used to reveal the ambitions of unification, and the motivations behind it. These become particularly evident as we move away from the natural and life sciences into the more contentious area of the scientific treatment of human behaviour. And although in Part II we have touched in a general way on questions of the hierarchy of the sciences, we should remember that the natural and the life sciences have traditionally given the impression of being natural partners, so there was a presumption in favour of unity, whereas the human sciences are a different matter. Indeed, it is their resistance to placement in the hierarchy that, more than anything else, makes their participation in a unified science questionable.

It was above all the human sciences that prompted new questions about the nature and standing of science, and investigation of these will provide us with a means of opening up questions about how science became associated with particular values and how it promoted them. This in turn will enable us to start examining the struggles between science and other disciplines over ownership of these values. Traditionally this struggle would have been with religious teaching, but this had been eclipsed by the middle decades of the nineteenth century, and the principal protagonist was now philosophy, as it in effect stood in for other disciplines that made claims about the nature of the world and our place in it. I am going to be arguing, in this and in the next chapter, that philosophy, to retain any standing, had to become incorporated in one way or another into science: it had to be reformed so as to display scientific credentials.

By the middle of the nineteenth century, a series of unprecedented conflicts had arisen over the questions of under what circumstances, and to what extent, an exclusively scientific conception was appropriate in understanding the world and our place in it, and what role remained for philosophy. The two most significant sets of disputes centred in Britain and in Germany. In the British case—our concern in this chapter—the issues arose due to advocates of the newly emerging human sciences setting out to resolve issues that had traditionally been the preserve of humanistic or religious inquiry, raising questions about whether these latter had now been superseded. In the case of Germany—our concern in Chapter 9—claims that science exhausted everything that could be known led to the view that philosophy could no longer continue to have any role in our understanding of the world. What is at issue in the two cases is different, but when we examine the two most popular and influential solutions offered—Mill's utilitarian rationalization of the human sciences in the one case and Cohen's Neo-Kantian attempts to counter reductionism by establishing a new role for philosophy in the other—we encounter a very significant overlap. In particular, the question of the unity of science is central to both, and in both cases the solution took the form of attempting to rework philosophy into a scientific discipline, so that it now aspired to become part of the science that it sought to unify.

THE QUANTIFICATION OF HUMAN BEHAVIOUR

In mid-nineteenth-century England, there arose a deep conflict between science and the traditional understanding of human values and behaviour. The source of the conflict was not the physical sciences, or the natural sciences more generally, but the 'human sciences', that is, those disciplines that set out to account for distinctively human behaviour in empirical and/or quantitative terms. If the values claimed for science were not to be abandoned, the question of what to include and what to exclude from its rubric had to be resolved, and the integrity of the unity of science depended on where its borders were located and how effectively they were policed. This was to become one of the major preoccupations of reflection on science from the 1830s onwards, and indeed for many it was virtually constitutive of metascientific enquiry in its philosophical form. It was regarded as central to the standing of science, and it provided continuity over a century of changes to the understanding of science.

One important initial trigger for the disputes in England was the publication of Ricardo's *Principles of Political Economy and Taxation*, which appeared in 1817. The *Principles* provided an axiomatic, deductive account of economic theory, so that even though it was not formulated mathematically, it was quantitative, and came close to satisfying the requirements of a mathematically formulated subject matter that Somerville and others had seen as the mark of a scientific discipline.[1] By contrast with earlier forms of political economy, such as that of Adam Smith, the presentation was spare and formal, which seemed to some critics inappropriate to the extent to which, they believed, it touched on questions better dealt with in moral and religious terms, a consideration confirmed by what these critics took to be the potentially radical nature of some of its conclusions (a radicalism that was to become explicit in Marx's Ricardo-inspired economics).

The problems were compounded by the emergence of utilitarianism at the end of the eighteenth century, gathering strength throughout the nineteenth century. Utilitarianism embraced consequentialism, a quantitative form of ethical decision making designed to replace traditional moral philosophies, which were to a greater or lesser degree Christian, and which had construed ethics in terms of the intentions and the character of the agent, considerations discarded in consequentialism. Political economy and utilitarianism were in the vanguard of the

---

[1] There is debate over just what Ricardo's contribution was, and I shall be following a Sraffa-inspired 'corn-model' reading which stresses the deductive character of his work, at the expense of his empirical studies of bullion for example, because my concern is primarily with what was at issue in Whewell's criticism of him. For an alternative reading see Terry Peach, *Interpreting Ricardo* (Cambridge, 1993). See also Ryan Walter, 'The Enthusiasm of David Ricardo', *Modern Intellectual History* 15 (2018), 353–80, who reads Ricardo as being motivated primarily by 'rational religion'.

naturalization of economic, political, and moral discourse, and it was the greatest of the utilitarians, John Stuart Mill, who came to the defence of political economy as a legitimate, autonomous form of scientific enquiry, going beyond Ricardo in unifying ethics and political economy into a single scientifically inspired utilitarian project. Having translated ethics and economics into a 'scientific form', the utilitarian unification formed what was perhaps the most ambitious of any of the unity of science projects of the nineteenth century: on Mill's conception, the human sciences formed a fully integrated whole of the kind that would have been the envy of physical reductionists. Moreover in carrying out this project, Mill offered an integrated account not only of ethics and political economy, but also of the empirical sciences, logic, and mathematics, and thereby opened up new and unprecedented questions about the nature and limits of scientific understanding, questions that were to preoccupy philosophy to such an extent that enquiry into the standing of science came to be one of its defining features from the second half of the nineteenth century onwards.

The first statement of the issues facing members of the British Society for the Advancement of Science on the standing of the human sciences in its early years was given by Adam Sedgwick in his 'Concluding Presidential Address' at the 1833 Cambridge meeting, in his defence of the inclusion of a new 'statistical' section (Section F) in the Association:

Some remarks may be expected from me in reference to the objects of this Section, as several members may perhaps think them ill fitted to a Society formed only for the promotion of natural science. To set, as far as I am able, these doubts at rest, I will explain what I understand by science, and what I think the proper objects of the Association. By science, then, I understand the consideration of all subjects, whether of a pure or mixed nature, capable of being reduced to measurement and calculation. All things comprehended under the categories of space, time, and number properly belong to our investigations; and all phænomena capable of being brought under the semblance of a law are legitimate objects of our inquiries. But there are many important subjects of human contemplation which come under none of these heads, being separated from them by new elements; for they bear upon the passions, affections and feelings of our moral nature. Most important parts of our nature such elements indeed are; and God forbid that I should call upon any man to extinguish them; but they enter not among the objects of the Association. The sciences of morals and politics are elevated far above the speculations of our philosophy. Can, then, statistical inquiries be made compatible with our objects, and taken into the bosom of our Society? I think they unquestionably may, so far as they have to do with matters of fact, with mere abstractions, and with numerical results. Considered in that light they give what may be called the raw material to political economy and political philosophy; and by their help the lasting foundations of those sciences may be perhaps ultimately laid. These inquiries are, however, it is important to observe, most intimately connected with the moral phænomena and economical speculations,—they touch the mainsprings of passion and feeling,—they blend themselves with the generalizations of political science; but when we enter on these higher generalizations, that moment they are dissevered from the objects of the Association, and must be abandoned by it, if it means not to desert the secure ground which it has now taken.[2]

Sedgwick had in fact initially rejected the proposal for the new section on the grounds that it was politically controversial, and the moving spirits behind the proposal were instead Richard Jones, Malthus, Babbage, Quetelet, and Whewell.[3] Jones was at the centre of the group, and four months before the establishment of the new Section F, he was calling for the formation of a statistical society. His concerns, which go back as early as the mid-1820s, are summed up in a lecture that he gave on political economy at King's College, London, in February 1833:

If we wish to make ourselves acquainted with the economy and arrangements by which different nations of the earth produce and distribute their revenues, I really know of but one way to attain to our object, and that is, to look and see. We must get comprehensive views of the facts, that we may arrive at principles that are truly comprehensive. If we take a different method, if we snatch at general principles, and content ourselves with confined observations, two things will happen to us. First, what we call general principles will often be found to have no generality; we shall set out with declaring propositions to be universally true, which, at every step of our further progress, we shall be obliged to confess are frequently false; and, secondly, we shall miss a great mass of useful knowledge, which those who advance to principles by a comprehensive examination of facts, necessarily meet with on their road.[4]

The facts that Jones refers to come, he tells us, from 'history and statistics, the story of the past, and a detail of the present condition of the nations of the earth'.[5] Ricardo and his followers, by contrast, urged that precisely formulated assumptions and definitions, and rigorous deduction from these, were the keys to economic analysis. The Ricardian Richard Whatley, for example, argued that political economy required very little empirical information for the formulation of its basic principles, and that premature empirical work was 'a mere waste of time and toil'.[6] This does not mean that opponents of this view, namely Whewell and those in his circle, rejected the use of formal argument in economics. As John Lubbock, in a pamphlet on currency, quoting Whewell, remarks:

'When our object is to deduce results from general principles, mathematical processes always afford the safest method, since by then we can most easily overcome difficulties and

[2] *Report of the Third Meeting of the British Association for the Advancement of Science, held in Cambridge in 1833* (London, 1834), xxiii.

[3] See Lawrence Goldman, 'The Origins of British "Social Science": Political Economy, Natural Science and Statistics, 1830–1835', *The Historical Journal* 26 (1983), 587–616; and Victor L. Hilts, 'Aliis Exterendum, or, the Origins of the Statistical Society of London', *Isis* 69 (1978), 21–43: 32–5.

[4] Richard Jones, *Literary Remains, consisting of Lectures and Tracts on Political Economy* (London, 1859), 568–9.

[5] Ibid, 570.

[6] Richard Whately, *Introductory Lectures on Political Economy* (London, 1832), 230.

perplexities which may occur in consequence of any complexity in the line of deduction, and are secure from any risk of vitiating the course of our reasoning by tacit assumptions or unsteady applications of our original principles.' I think with Mr Whewell, that many will find the doctrines of Political Economy, when put in a mathematical shape, more clear, compendious, and manageable. By the use of letters, particular cases may be included in one equation, which cases may be treated separately if the reasoning is carried on with numbers.[7]

Lubbock is quoting here from a paper in which Whewell translates Ricardo into mathematical terms so as to refute him.[8] Whewell and Jones had set out to promote an inductivist school of economics in contrast to Ricardo's mathematical approach. Why, then, would Whewell undertake this mathematical exercise? Henderson has neatly summarized what is behind this turn to mathematics in these terms, noting that Whewell 'developed a strategy for promoting an inductive political economy that involved the use of mathematics as a device for overthrowing the Ricardian deductive political economy that had achieved dominance. Whewell's strategy was to use the mathematical approach to point out faults in the deductive reasoning of the Ricardians, while relying on Jones to supply the inductive alternative'.[9] In other words, the strategy is essentially negative, designed to point out logical errors and faulty conclusions in Ricardo. Political economy is essentially inductive for Whewell, and the point of the mathematical criticisms of Ricardo is to show that a non-inductive theory of political economy cannot be made to work.[10] But Whewell also has a view on the general standing of political economy: as with Adam Smith, for Whewell the concern with political economy had its roots in moral philosophy. In his 1860 book, *On the Philosophy of Discovery*, Whewell raises the question of whether the method of discovery employed in the natural sciences can be applied in 'ethical, political, or social knowledge', noting that there 'the instrument of observation is consciousness', and the problem is that 'observations and ideas are mingled together, and act and react in a peculiar manner'.[11] For Whewell, it must remain an open question whether, given their complexity, the substantive moral and political issues raised by political economy even belong in its ambit in the first place. As Yeo notes, the persistent moral dimension in Whewell's writings can be interpreted as a sophisticated response to the need for a philosophy of science that guaranteed the principles of natural theology and the values of Christianity.[12]

---

[7] John W. Lubbock, *On Currency* (London, 1840), iii–iv.

[8] William Whewell, 'Mathematical Exposition of Some Leading Doctrines in Mr. Ricardo's *Principles of Political Economy and Taxation*', *Transactions of the Cambridge Philosophical Society* 3 (1831), 192–230.

[9] James Henderson, *Early Mathematical Economics* (Boulder, Colorado, 1996), 19.

[10] See ibid., ch. 5.

[11] William Whewell, *On the Philosophy of Discovery* (London, 1860), 292–3.

[12] Richard Yeo, 'William Whewell, Natural Theology and the Philosophy of Science in Mid-Nineteenth Century Britain', *Annals of Science* 36 (1979), 493–512: 498.

This provides an important context for understanding not just Whewell's motivation for taking up some of the positions that he does, but also for locating the debates more generally.

The core question that prompted much philosophy in the early to midnineteenth century in England was the naturalization and/or quantification of human motivation and behaviour. By 'naturalization' here I mean subjecting these phenomena to empirical investigation, on the model of the sciences: though without reduction to the physical or life sciences, by contrast with the various forms of materialism for example. By 'quantification' I mean the application of formal quantitative techniques to human behaviour and motivation.[13] Such approaches called into question traditional metaphysical and religious treatment of these issues, particularly in the case of moral and political questions.

Two kinds of project focused primarily on quantification: political economy and utilitarian ethics. They originated in different contexts, and although Ricardo had started under the influence of Bentham and James Mill, he broke away at an early stage from their influence in developing his own economic theory.[14] This theory laid claim to a scientific standing and insisted on being judged as such.[15] Utilitarian ethics, by contrast, was explicitly associated with a radical politics. But they both used quantification as the key to understanding human behaviour, and they were mutually reinforcing in many respects. In particular, both of them bypassed traditional understandings of human behaviour, whether political, economic, or moral.

In Ricardo's hands, political economy became transformed into a formal study of economic relations. His initial thinking focused on the role of profits in agriculture,[16] and he devised a model in which a single commodity, corn, forms both the capital—in this case the subsistence necessary for workers—and the product.[17] To determine the profit, all that is needed is to determine the difference between the two, the crucial point being that no valuation of the corn is needed.[18] But corn is in a unique position: no other trade is in the position of not employing the products of other trades, with all other trades needing to employ its product. Consequently, if there is to be a uniform rate of profit in all trades, corn becomes the standard, because no changes in value can alter the ratio

[13] On the naturalization and quantification of human behaviour in the eighteenth and early nineteenth centuries, see Gaukroger, *The Natural and the Human*.

[14] See Murray Milgate and Shannon C. Stimson, *Ricardian Politics* (Princeton, NJ, 1991).

[15] There is a good account in Margaret Schabas, *The Natural Origins of Economics* (Chicago, 2005).

[16] Agriculture was not only by far the largest sector of the economy at the time, but in Ricardo's view it was less susceptible to changes in fashion, to wars, and to new taxes: David Ricardo, *The Works and Correspondence of David Ricardo*, ed. P. Sraffa and M. Dobb (11 vols, Cambridge, 1951), i. 263.

[17] Here I follow Sraffa's helpful introduction to *The Works and Correspondence*, i. pp. xii–lxii.

[18] Hence the title of that classic of neo-Ricardian economics, Piero Sraffa, *Production of Commodities by Means of Commodities* (Cambridge, 1960).

of product to capital, both of which consist of the same commodity. It is the exchange values of the products of other trades relative to their own capital (corn) that need to be adjusted to yield the same rate of profit as the growing of corn. The advantage of this ingenious procedure, as Sraffa points out, is that it makes possible an understanding of how the rate of profit is determined without the need for a method of reducing a heterogeneous collection of commodities to a common standard.[19] The next step, which forms the keystone of the *Principles*, is to abandon the simplification that wages consist only of corn, and to treat them as a variety of products, a generalization made possible by substituting labour for corn, so that the rate of profit is now determined by the ratio of the total labour of a country to the labour required to produce the necessities for that labour.

With this account in hand, Ricardo then began to build a general theory of value.[20] Stressing the difference between causes which affect the value of money and those that affect the value of commodities, at the same time showing that the common assumption of the invariability of precious metals as a standard of value is without foundation, he was able to demonstrate that a change in wages could not in itself alter the price of commodities, as Smith and others had assumed, because a rise in wages would affect the owner of the gold mine, for example, as much as other industries. Consequently it was the relative conditions of production of gold and other commodities that determined prices, not the remuneration of labour. The value of a thing was determined not by the remuneration of the labour required to produce it, but by the quantity of labour required for its production.

Ricardo in effect established the norms of economic theory—the setting out of basic assumptions and the deduction of conclusions from these—but the conclusions that followed from that theory conflicted with traditional conservative political views, particularly on the relationship between landlords and tenants. The problem for those holding such traditional views was that the scientific status of the arguments looked strong. As a result, there arose, among those who wished to defend the conservative position, notably Whewell and his colleagues, a questioning of just what science was, and what it could and could not achieve. Whewell's interest was initially methodological, but his reading of Kant encouraged him also to think in more epistemological terms, and what resulted was a form of philosophical enquiry into the nature of science: the combination of methodological and epistemological concerns yielded what was in effect the first philosophy of science.

---

[19] Sraffa, 'Introduction' to Ricardo, *The Works and Correspondence*, i. p. xxxii.

[20] There is some controversy over exactly what Ricardo was offering. The generally accepted view is that it was a theory of value, but see Peach, *Interpreting Ricardo*, who argues that what Ricardo is advocating is a theory of distribution.

The second quantification project emerged with Bentham's consequentialist moral theory.[21] The problems to which this was a response can be traced back to Mandeville's *Fable of the Bees*. Mandeville had argued that the qualities, such as avarice, that encouraged the kind of successful commercial society which was to everyone's benefit, were the opposite of those that were universally lauded in the case of individual morality.[22] The basic dilemma was that the demands placed on behaviour, by those collective goals that bring with them economic and political well-being, were in conflict with the moral demands that shape the individual behaviour that we value most. One solution to this conflict was the revision and broadening of the scope of individual morality through the advocacy of forms of moral pluralism, and Hume and Smith followed this path. This was in line with the traditional view which, if only implicit, had been that only individual behaviour has a moral dimension. Individuals are able to form intentions and act upon these, and this would seem a prerequisite of moral behaviour. By contrast, the role of intentionality in collective behaviour is difficult to fathom. The natural home of morality, so to speak, lies in the realm of the individual. On such a view, while it is true that empires, nations, and states can act in various ways that attract moral sanction or approval, and can be subject to moral maxims such as that of the doctrine of just war, moral responsibility is properly ascribed not to a state as such but to the ruler. For the kind of approach taken by Mandeville, by contrast, collective activity has a moral dimension, one that conflicts with individual morality, and it is far from clear that the language of intentionality is applicable to collective or aggregate behaviour.

Unlike distributive properties—those properties of the whole that are the properties of each of the parts—collective properties are not mirrored at the individual level, and they cannot be arrived at by abstraction from individual behaviour. An increasing number of writers had recognized this in the eighteenth century, particularly those composing 'philosophical' histories, such as Voltaire, Hume, and Montesquieu. Hume, for example, in developing the idea that the relation between cause and effect is a matter of inference rather than direct perception, argues that such connections are more easily established when changes in human conditions produce changes in large-scale human behaviour, as in the case of the rise and progress of the arts and sciences, or that of the rise and progress of commerce, rather than on an individual level, and he distinguishes two forms of history corresponding respectively to cultural change and to individual actions.[23] Montesquieu, likewise, in a 1734 'Essay sur les causes qui peuvent affecter les esprits et les caractères', writes that causes become less arbitrary to the extent that they have a more general effect: we know better what shapes the achievements of

[21] I draw here on the account in ch. 6 of my *The Natural and the Human*.
[22] Bernard Mandeville, *The Fable of the Bees* (London, 1714). See Edward Hundert, *The Enlightenment's Fable: Bernard Mandeville and the Discovery of Society* (Cambridge, 1994).
[23] 'Of the Rise and Progress of the Arts and Sciences', in Hume, *Essays and Treatises*, i. 111–38.

societies that have adopted a given way of life than we do what shapes the lives of individuals.[24] Note the stress on causal explanation here. If one wants to account for social, political, or economic questions in causal terms, one has in effect to abandon individual behaviour and turn instead to the behaviour of collective or aggregate entities. And to the extent to which rendering phenomena amenable to causal explanation is part of a process of naturalization, then, in the shift from understanding behaviour at the personal level to the aggregate level, one is moving to a different kind of understanding of human behaviour.

Bentham was concerned to offer a theory having universal application, in that it covered collective and individual morality without distinction. His solution to the problem of how an account of morality could cover both individual and collective behaviour was to abandon individual morality as a norm. Traditional accounts had attempted to extrapolate from the case of individual morality, but Bentham reverses the core case: an account perfectly fitted to resolving problems in the collective case is applied to the individual case. This marked a radical departure from traditional conceptions of morality in that questions of intentions and character, not to mention divine and natural law accounts, were jettisoned in favour of something far more straightforwardly observable, namely the consequences of acts. The account had two distinctive features. First, it opened up morality to a quantitative assessment. The assessment of whether an action is right or not is simply a function of the consequences of that action: and consequences could be assessed in comparative terms, in terms of a calculation. Second, with its abandonment of a role for intention and character, moral decision making began to look decidedly more scientific than it had ever done. It was an empirical and transparent matter what the consequences of actions were, and, at least in principle, much more of an empirical matter (when divorced from considerations of intention and character) which of these consequences deserved praise or condemnation. Moreover, unlike traditional understandings of morality centred on intentionality, moral decision making was no longer something to which the language of cause and effect was alien.

It was Mill who brought the new radical economics and the new radical ethical theory into line with one another. He reworked Ricardian political economy to meet the requirements of a utilitarian ethics, setting out a unified utilitarian programme of very general scope.[25] This unification programme explicitly

---

[24] Charles de Secondat, Baron de Montesquieu, *Oeuvres complètes*, ed. R. Caillois (2 vols, Paris, 1949), ii, 39.

[25] It includes not just ethics and political economy, but to some degree the arts as well. This is clear in his example of the violin: the violin maker transforms wood into a permanent object, and the violin teacher imprints the requisite skills on the pupils' minds. Both of these are productive on Mill's account. But the concert violinist is unproductive because the whole analysis is in terms of utility, and the utility produced by the performer evaporates at the point of its production. *The Collected Works of John Stuart Mill* (33 vols, Toronto, 1963–91), iv. 285. See the discussion in Schabas, *The Natural Origins of Economics*, 128.

engages questions of methodology. In his 1836 essay 'On the Definition of Political Economy and the Method of Investigation Proper to It', Mill raises the question of the difference between political economy and the physical sciences. To understand economic production, he argues, one must study physiology, chemistry, mechanics, and geology, but political economy is the study of the 'laws of mind' rather than the 'laws of matter', and so takes as its domain of enquiry 'the phenomena of mind which are concerned in the production and distribution of those same objects'.[26] Nevertheless, as Schabas points out,[27] in his 1848 *Principles of Political Economy*, Mill writes that whatever is produced must be produced under the conditions 'imposed by the constitution of external things',[28] and that the 'conditions and laws of Production would be the same as they are, if the arrangements of society did not depend on Exchange, or did not admit of it'.[29] It is central to Mill's account that it is the powers of nature, not labour as Ricardo had argued, that do all the work. All labour does is put objects in a certain configuration: the spinner, for example, does not produce the linen cloth, rather it is the natural forces of cohesion that hold the cloth together. What labour produces is not material goods, but utilities, and it is in the production of utilities that the key to exchange lies.

Mill was ambiguous on the question of whether the economy was part of a natural order, and so part of a material science, or a mental phenomenon. Ideally, what he seems to have wanted was the latter,[30] but his conception of a separate economic realm ultimately rested on a theory of the workings of the mind, which he had been unable to develop beyond a very elementary stage. As a mental phenomenon, the economy was the result of rational agency, no longer directly governed by natural forces. Its scientific standing was thus not dependent on its relation with the physical sciences, because its subject matter was of a different kind from theirs. Nevertheless, if it was to retain a standing as a scientific discipline, it had to be reconciled methodologically with the other sciences.

THE MORAL STANDING OF INDUCTION

This is the context within which the dispute between Whewell and Mill took place. The issues at stake were complex but they can be summed up as being between 'intuitionism'—Mill's term for what he took to be Whewell's position—and a form of positivism which Mill himself, initially under the influence of Comte, developed.[31] Simplifying somewhat, although both Whewell and Mill

---

[26] Mill, *Collected Works*, iv. 316–18.
[27] Schabas, *The Natural Origins of Economics*, 126–7.   [28] Mill, *Collected Works*, ii. 199.
[29] Ibid., iii. 455.   [30] See Schabas, *The Natural Origins of Economics*, 128–41.
[31] In his letter to Comte of 8 November 1841, he tells Comte that it was his reading of him that, ten years earlier, freed him from the influence of Bentham. Mill, *Collected Works*, xiii. 334.

were advocates of inductivism, for Whewell morality and science-cum-mathematics ultimately depend on deep and unchallengeable intuitions, whereas for Mill the legitimacy of both science-cum-mathematics and morality must depend solely on their appeal to external criteria. Intuitionism and positivism, at least as understood here, share the belief that what holds for science must hold for morality. This is a defining assumption of the debates. It is an obvious premise of Mill's approach, but also of Whewell's. It might seem that, in contrast to Mill, Whewell is trying to keep questions of morality and politics apart from questions of science. But in fact he is concerned to explore how they are related and to defuse any unwelcome consequences of this relationship, a concern made more urgent by the attempts of political economists and ethical consequentialists to assimilate the two. The overriding concern is the largely negative one that what emerges from science does not invalidate or compromise Christian morality or the political systems considered to be based on it.[32]

The stakes in the dispute between Whewell and Mill could not be higher. Whewell is offering one of the last attempts to provide a comprehensive Christian understanding of the natural and moral realms. The contemporary *Bridgewater Treatises* had embraced this aim,[33] and Whewell was a contributor to these, but like his co-authors, he is at a disadvantage in his project. The era in which it was possible to provide the kind of overarching physico-theological structure for knowledge at which earlier physico-theological works had aimed had passed. The best that could now be hoped for was a construal of science that did not threaten religious orthodoxy.[34] Mill by contrast is offering one of the first comprehensive wholly secular understandings of the natural and moral realms, preceded only on this scale by that of Comte, which it easily surpasses in terms of plausibility. Although it is on very different bases and with very different consequences, both Whewell and Mill are committed to the unity of science.

Because of the moral and political dimensions of their conceptions, struggles over competing models of the unity of science become struggles over the nature of civilization in nineteenth-century England. The context in which this occurred was one in which it was believed that scientific inquiry produced moral effects in the minds of its cultivators. Brougham, in his *Objects, Pleasures and Advantages of Science*, and Herschel in his *Preliminary Discourse*, had both stressed the moral benefits of a scientific education, attracting the ire of High Church theologians who, fearful of the growth of a secular morality based on science, attempted to

---

[32] On these questions more generally, see Boyd Hilton, *The Age of Atonement: The Influence of Evangelicalism on Social and Economic Thought, 1795–1865* (Oxford, 1988), ch. 2.

[33] *The Bridgewater Treatises, on the Power, Wisdom, and Goodness of God as Manifested in the Creation* (8 vols, London, 1834–7).

[34] See Jonathan Topham, 'Science and Popular Education in the 1830s: The Role of the Bridgewater Treatises', *British Journal for the History of Science* 25 (1992), 397–430; Jonathan Topham, 'Beyond the Common Context: The Production and Reading of the *Bridgewater Treatises*', *Isis* 89 (1998), 233–62.

associate science with self-conceit and religious indifference.[35] Mirroring Herschel's defence of the morality of scientists against such criticisms, Whewell made questions of character central to debates about the nature of science,[36] and it was the moral standing of induction around which the issues came to revolve.

The questions first arose in the context of a series of breakfast meetings in Cambridge in 1812–13, with Whewell, Herschel, Babbage, and Jones discussing how to revive a programme of scientific discovery in the physical sciences in England and raise it to a level on a par with the late seventeenth-century work of Newton. They pursued these questions over several decades, and sought the answer in a commitment to what they considered to be the Newtonian methodology of induction. But it was not just a question of fostering the physical sciences. There was also Ricardo's newly developed scientific approach to economics, which as we have seen worked on the basis of deduction from postulates, which was not inductive, but at the same time not that different from the procedure adopted in much of Newton's *Principia*.

Herschel's *Preliminary Discourse*, which we have already looked at in the context of its role in shaping the programme of the British Association for the Advancement of Science, was one of the first steps in the metatheory of science. Its ranking of the sciences set the stage for an account of what the success of the various sciences consisted in, and Herschel's schema was built on by Whewell in his long review that appeared soon afterwards.[37] Botany, mineralogy, and chemistry were identified as the lowest level of science, at 'the outset of their inductive career'. Next came the branches of 'physics', notably electricity and magnetism, then optics, and at the apex mechanics and astronomy, fully mathematized disciplines which had successfully combined induction and deduction. Induction stood at the basis of the sciences and, as they advanced and became increasingly subject to mathematical treatment, deductive relations could be established.

Induction clearly plays a key role, but what Whewell means by induction needs clarification. Bacon had treated induction as the construction of theories on the basis of extensive observation of nature, and this remained a core meaning, providing a contrast with deduction. In his *Philosophy of the Inductive Sciences* (1840), however, Whewell does not allow that bare induction of this kind could be enough for scientific discovery. Rather, 'discoverer's induction', as he termed it, could not be a passive response to nature but must involve both objective experience and ideas, which the enquiry brings to the investigation of nature, and which provide it with direction. His recognition of the need for the latter derives from his reading of Kant, and the basic question that Whewell takes over from Kant is that of how universal and necessary laws, of the kind that one finds in

[35] See Yeo, *Defining Science*, ch. 5.

[36] William Whewell, 'Modern Science—Inductive Philosophy', *Quarterly Review* 45 (1831), 374–407.

[37] Ibid.

mechanics and astronomy, can be generated simply on the basis of induction from observation and experiment. Kant, he argued, had stressed 'ideas' over 'perceptions' thereby ridding science of its objective connections with nature, whereas Locke and the 'sensationists' had implausibly reduced science to passive observation.

The fundamental ideas that he set out, while influenced by Kant, served a different purpose from concepts such as Kant's 'forms of the intuition'—space, time, and causation. Whewell was not interested just in something that was a permanent and fixed feature of the mind, but also wanted to include ideas fundamental to particular scientific disciplines. So while space and time figure in his fundamental notions, so too do chemical affinities, because he believed that chemistry could not proceed without these. For Whewell, these discipline-specific fundamental ideas might well form some unified scientific understanding at some stage in the future, but the crucial thing is that they are discovered, not innate.

Since fundamental ideas are what guide our empirical investigation of nature, and since, at least in some cases, they are formed in response to prior investigation of nature, it might seem that they must initially come from some kind of induction from basic observations. But Whewell does not allow this, and it is in this context that he makes room for hypotheses. He recognizes neither the inductive vs. hypothetical nor the ideal vs. sensationalist as absolute dichotomies. Moreover, the hypotheses that become fundamental ideas are objective, not just in the sense that they capture something in nature, but because they do this in virtue of corresponding to archetypical ideas existing independently of any particular beliefs that humans have. Here Whewell departs radically from Kant.

In the final edition of his *The Philosophy of the Inductive Sciences*, Whewell added a section arguing that the divine mind contains many archetypical ideas and that God created the universe in accordance with these ideas. He was not alone in this kind of approach. There were parallel developments in anatomy, where the Fellows of the Royal College of Surgeons, such as Joseph Henry Green, a follower of Coleridge, were promoting a science of Platonic archetypes to underpin a divine model for anatomical structure. Interestingly, here, as in Whewell's dispute with Mill, the use of archetypes underlies the struggle between conservatives and radicals, archetypes being proposed to counter radical physicians who, taking up Geoffroy Saint-Hilaire's ideas, postulated a possible evolutionary continuity between animals and humans.[38]

Space is an archetypical idea for Whewell, in that God created physical objects with spatial characteristics and existing in definite spatial relations to one another. God gave us access to these ideas by which he created the universe. He created 'germs' in our minds which, when developed, match or represent divine archetypical ideas, and we come to know the natural world by developing these ideas in

---

[38] See Desmond, *The Politics of Evolution*, ch. 6.

tandem with observation and experiment. It is not unlike the (supposed) discovery of natural kinds in an area such as natural history, where we gradually refine our classifications of families, genera, and species finally to reveal the divisions there in nature. Whewell makes it clear, however, that the process of discerning archetypes is not one that the individual can carry out. Rather, it is the process that scientists undertake in their routine work: it requires discussion and debate based on the evidence, for only this can lead to clarification and progress in identifying and grasping the archetypical ideas.

The moral consequences of inductive and deductive habits of thought are crucial for Whewell. Chapters 5 and 6 of his Bridgewater treatise, *Astronomy and General Physics Considered with Reference to Theology*,[39] are devoted to the moral and religious implications of inductive and deductive habits of mind. Induction was the procedure of 'discoverers', something he associated with a natural-theological process of uncovering God's purpose in the world. Proponents of the idea that deduction was constitutive of scientific enquiry, by contrast, dealt with remote abstractions, treating the truth of basic laws as necessary, never considering that they could have been different and so overlooking the possibility that they had been ordained by God.

The moral standing of induction for Whewell was nowhere more evident than in his discussion of Ricardian economics. In his *On the Principles of Political Economy and Taxation*, Ricardo offered a purely deductive account of rent, wages, and profit that started from a number of seemingly uncontentious basic assumptions, and worked out deductively from these. The deductive procedure could be defended in Newtonian terms, but more specifically it resembled that of eighteenth-century mechanics, where the deduction proceeds from basic definitions of physical concepts. It is in fact close to what Whewell had identified as the highest stages of scientific enquiry, in mechanics and astronomy. But on Whewell's account these areas had gradually been built up from inductive foundations, whereas what Ricardo was proposing was something that ignored induction altogether.

For Whewell, the more successful a science became the more it tended to take a deductive form. Could it be, then, that economics had simply bypassed the inductive stage? Whewell and his circle set out to show the failings that resulted from attempting to do this. The issue was crucial to their understanding of how a scientific discipline was established. Whewell and Jones opposed what they considered to be the way in which Ricardo applied the label 'science' to his economics, which they considered was merely a way of presenting this theory as dogma. In reviewing Jones' *Essay on the Distribution of Wealth and on the Sources of Taxation* (1831), Whewell uses Jones' discussion of different kinds of rents to criticize Ricardo's extrapolation from one kind of rent, farmers' rent, to its

---

[39] William Whewell, *Astronomy and General Physics Considered with Reference to Theology* (London, 1833).

universal form. This, he writes, provides 'the most glaring example of the false method of erecting a science which had occurred since the world has had any examples of the true method'.[40] True principles, he argued, cannot be obtained by 'some transient and cursory reference to a few facts of observation or of consciousness'.[41]

Whewell was particularly concerned with the way in which Ricardians compared their strict definitions of rents, wages, and capital with the procedures of geometry. Relations were explored in geometry on the basis of 'first principles' which made no reference to experience, but were purely the product of thought. By contrast, in the physical sciences, definitions were helpful only if they captured relations that had been discovered to be there in nature. For example, in mechanics and optics, one does not simply start with precise definitions: rather, the definitions emerge as disputes are resolved and new theories established. Exact definitions are not the causes, but the consequences of scientific progress. Understanding of the different stages of development of the sciences, and how they achieved a particular, standing was crucial: political economy could not claim the status of a mature deductive science. Indeed, on Whewell's account, economics was at the bottom of the scientific hierarchy because the complexity of the social realm required extensive empirical research, whereas this exercise had only just begun.

But there was another element in Whewell's criticisms. If Ricardo's account of value and rent is taken in moral terms, then since labour produces the entire product, rent is an unjust institution and landlords are simply parasitic. Ricardo eschewed such a moral reading, at least publicly, considering his work something that should be judged on its scientific merits. But for Whewell questions of rent, taxes, wages, and capital involved passions and interests distancing economics from the canonical forms of scientific enquiry—whether the mathematical or the physical sciences—that must form the standard for enquiry. As a consequence, he believed, Ricardians were prone to unchristian and immoral views of human nature. They committed what he considered to be the same error as the utilitarians, assuming that 'principles of action are known by consciousness and do not require detailed observation'.[42]

In sum, Whewell rejected the implicit moral criticism inherent in Ricardo's theory of rent. He denied that Ricardo employed a scientific method, and he wondered whether the complexity of the mass of empirical work that would need to be done for economics to start on the road to science could ever be achieved. Moreover he doubted that, even if this could be achieved, it could ever have the

[40] William Whewell, 'Jones—on the Distribution of Wealth and the Sources of Taxation', *British Critics* 10 (1831), 41–61: 51–2.

[41] Ibid., 53.

[42] Whewell to Richard Jones, 12 July 1831: in Isaac Todhunter, *William Whewell, DD. Master of Trinity College Cambridge: An Account of his Writings with Selections from his Literary and Scientific Correspondence* (2 vols, London, 1976). ii. 123.

scientific standing to which Ricardo aspired given that, unlike the other sciences, it was not a purely factual or conceptual enterprise, but was caught up in moral and political questions that lay outside its purely empirical basis.

Nevertheless, as his views developed, Whewell saw less of a divide between what he regarded as strict science and other types of enquiry, and he came to consider that the physical sciences and morality did have some structural similarities.[43] As early as 1835 he wrote to Jones that in James Mackintosh's *Dissertation on the Progress of Ethical Philosophy* (1830), he had got a glimpse, 'which I have long been wishing and struggling for, of the inductive history of ethics'.[44] In his 1852 *Lectures on the History of Moral Philosophy in England*, Whewell saw the issue in terms of 'the arbitrary or necessary nature of moral truth', and although he now believed that moral knowledge could not be arrived at by induction, it was like scientific knowledge in that it was regulated by general rules which in his *Elements of Morality* (1845) he tells us must be necessary truths. The rules regulating duty and affections are necessary truths that are as fundamental to human thought as are our understanding of space and time. This does not mean that we grasp them immediately however. Rather, we come to understand them, if only imperfectly at first, through a cultivation of the mind, but once grasped they are self-evident. In mechanics, he explains in a letter to Frederic Meyers of 6 September 1845: 'although we now have axioms, defined conceptions and vigorous reasonings, we can point to persons, and to whole ages and nations, who did not assent to those axioms because the mechanical ideas of their minds were not sufficiently unfolded. In this we have, it seems to me, an answer to the objection that what I assert as moral axioms are not evident to all men. They are as much evident to all men as the axiom that a body will not alter its motions without a cause, or the like. They become evident to men in proportion as all men have their conceptions of the terms of science rendered clear and distinct.'[45] The comparison with mechanics here is not accidental. Whewell is offering a general theory of knowledge that accounts for both scientific and moral knowledge, which establishes the objectivity of both. There are two ingredients in his account, induction and 'fundamental ideas', that play the crucial role in the empirical sciences, but it is fundamental ideas alone that establish objectivity in non-empirical areas such as pure mathematics and morality. Self-evident fundamental ideas cannot be identified by seeking 'the casual opinion of individual men any more than we can determine the axioms of geometry by polling schoolchildren'.[46]

But while it was relatively uncontentious to have induction playing the role that he allocates to it, fundamental ideas were an entirely different matter. Here

[43] Generally, see Laura Snyder, *Reforming Philosophy: A Victorian Debate on Science and Society* (Chicago, 2006).
[44] Whewell to Jones, 9 May 1835: in Todhunter, *William Whewell*, ii. 308.
[45] Janet Mary Douglas, *The Life and Selections from the Correspondence of William Whewell* (London, 1881), 327–8.
[46] William Whewell, *Lectures on Systematic Morality* (London, 1846), 34–5.

was the major point of conflict between Whewell and Mill. The fact that Whewell's 'necessary truths' of morality matched the prevalent Anglican conservatism suggested to Mill that it was not so much a question of Ricardo dressing up his economics as dogma, as it was of Whewell dressing up his conservatism as dogma.

THE HUMAN SCIENCES AND THE BIRTH OF METASCIENCE

Mill's original concern was not with questions of science but with political questions, in particular the elaboration and defence of a radical politics. Whewell was just an apologist for the status quo in Mill's view.[47] But this raised the question of how one should discuss morality. Whewell had located his discussion of morality in the context of knowledge generally, and had contrasted it with scientific knowledge in crucial respects. In taking up questions of scientific knowledge, Mill engages the issues on the terrain that Whewell sets out. At stake were the standing of fundamental ideas and induction.

Mill seeks objectivity in our moral and scientific judgements as much as does Whewell. One basic reason behind this agreement is that both see morality in terms of the exercise of reason, not sensibility as Hume and Smith advocated, for example. Consequently questions of objectivity and truth are paramount. The contrast between the two lies in the fact that induction appeals to an external standard, whereas Whewell's 'intuitionism' seeks a priori support for moral ideals. Mill advocates abandoning the idea of a priori support and making the whole of knowledge subject to external standards:

The contest between the morality which appeals to an external standard, and that which grounds itself on an internal conviction, is the contest of progressive morality against stationary—of reason and argument against the deification of mere opinion and habit. The doctrine that the existing order of things is the natural order, and that, being natural all innovation on it is criminal, is as vicious in moral, as it is now at last admitted to be in physics, and in society and government.[48]

Induction is the external standard in the case of the sciences, consequentialism in the case of morality. Like Whewell, Mill believed in objective standards generally. Consequently the answer to the problem of moral diversity and conflict— between different societies or between individual morality and commercial culture—cannot lie in the kind of moral pluralism advocated by Hume and Smith for example. There must be a single moral code. But to specify this moral code in terms of a particular content, as Whewell does, seemed to Mill to

[47] Actually a somewhat unjust view: see Snyder, *Reforming Philosophy*.
[48] John Stuart Mill, 'Whewell on Moral Philosophy' (1852), *Collected Works*, x. 168.

involve an unjustified, and indeed unjustifiable, choice of content. How then could objectivity be secured? Certainly not by induction, or taking a poll, which Whewell presents as the alternative. Rather it consists in eschewing questions of content and elaborating a rule. Kant had proposed such a rule—universalizability, the criterion by which the morality of an action is that it could become one that everyone could act upon in similar circumstances—but this was an internal rule, whereas Mill, following the utilitarian tradition of Bentham and James Mill, wants something external, namely the consequences of the act, to serve as the criterion by which to decide morality. It was not the character of the person acting, the person's intention, or the universalizability of the act that mattered but the observable consequences. If the act had harmful consequences, then it was immoral to act in that way. A great deal of qualification and fine-tuning was necessary to establish the plausibility of consequentialism between its initial formulation in Bentham and Mill, and that of their successors such as Sidgwick, but the crucial point was that consequences were external and observable. This, combined with the fact that they were calculable, in that one could compare actions in terms of the amount of benefit or harm they caused, gave consequentialist morality a quasi-scientific standing: an independence from particular habits, prejudices, or ill-formed beliefs, and a decision procedure that was not unlike a form of economic calculation. Moreover, since both Whewell and Mill saw an intimate connection between morals and politics, what was at issue was not just differing theories of how moral judgements should be made, but also the standing of the traditional conservative land-owning politics, and radical political reforms.

A great deal was at stake, and what was new was the way in which the standing and application of science lay at the foundation of the debates. Bentham and James Mill had not connected their programmes for moral and political reform with science, nor had defenders of conservative morality and politics. Whewell changed this, propelling the epistemological and methodological standing of science into the centre of the question of how we are to live our lives.

Mill's strategy was to demonstrate that fundamental ideas were not needed either in the physical or the mathematical sciences, so Whewell's proposed justification for his moral theory in the role of fundamental ideas in the sciences fell flat. But Whewell's account had two apparent strengths. First, bare induction, from particular facts to general conclusions, is unlikely to yield fruitful scientific results: the idea that induction might be guided by 'fundamental ideas', indistinctly grasped at first but gradually coming to light in the course of the enquiry, provides a means by which induction might be led along fruitful paths. Second, for Mill, while it is external standards that secure objectivity in the cases of the natural sciences, in the form of induction, and morals and politics, in the form of consequentialism, it is not clear what provides these standards in a deductive discipline such as pure mathematics. Whewell can rely on a variety of fundamental ideas to secure mathematical truths, maintaining that they correspond to ideal archetypes. But these are internal criteria, as is Kant's construal of mathematical truths as synthetic a priori. If Mill wants to make a general case for external criteria, then he has to provide an account of what these would be in the case of an apparently purely deductive endeavour such as mathematics or logic.

This Mill sets out to do, in one of his most ambitious philosophical exercises, offering a novel theory of logic. On his account, inference is the process by which we move 'from known truths, to arrive at others really distinct from them'.[49] Logic 'is the entire theory of the ascertainment of reasoned or inferred truth',[50] and it divides into two forms, induction and deduction (the latter of which he calls ratiocination): when 'the conclusion is more general than the largest of the premises, the argument is Induction; when less general, or equally general, it is Ratiocination'.[51] Both induction and deduction share fundamental characteristics on this conception, and both fall under Mill's general account of truth, whereby 'truths are known to us in two ways: some are known directly, and of themselves; some through the medium of other truths. The former are the subjects of Intuition, or Consciousness; the latter, of Inference'.[52] Directly known truths are sensory experiences, unlike Whewell's archetypical ideas, which are purely intelligible, i.e. non-sensory. They allow scientific inference to meet an external criterion of objectivity. But while they provide the raw material on which induction works, this material is a little too 'raw' to explain how induction might lead to the generation of general theories. It should be noted however that many of the problems here are artificial, and disappear once we stop thinking in terms of a sharp contrast between inductive and hypothetical methods. In the polemical disputes between Newtonians and Cartesians beginning in the last decade of the seventeenth century, the sharpness of this distinction was stressed by Newtonians, who viewed the use of hypotheses as violating objectivity,[53] and to some extent both Whewell and Mill are heirs to this. But induction, as it had been conceived from Bacon onwards, had never been simply a matter of directly inferring general theories from particular facts, which Bacon dismissed as childish. Rather, the crucial role was played by eliminative induction. Here, the idea is that reflection on the facts suggests several possible explanations, and one goes through these invoking evidence to eliminate them one by one until one arrives at an explanation that cannot be eliminated in this way. Such a procedure serves the same function as an explicitly hypothetical method, except that one can give an account of how the hypotheses are generated, namely from facts established on the basis of observation.

But this does not settle the question whether Mill's conception of induction reflects scientific discovery accurately. Whewell, who had a far greater command than did Mill of the history of science and actual practice in the physical sciences,

---

[49] Mill, *A System of Logic* (1872) in *Collected Works*, vii. 163.   [50] Ibid., vii. 7.
[51] Ibid., vii. 162–3.   [52] Ibid., vii. 6.
[53] See Larry Laudan, *Science and Hypothesis: Historical Essays on Scientific Methodology* (Dordrecht, 1981).

had no difficulty identifying cases that fitted his model of discovery. Taking the example of Kepler's discovery that planetary orbits were elliptical (as opposed to circular), Whewell argues that the discovery of new facts was not what was at issue. Rather, it was a question of bringing the data under the appropriate idea. The data in this case had already been collected by Tycho Brahe, who hadn't made sense of it. It was only when Kepler successfully fitted the data into the shape of an ellipse that the true shape of orbital motion was discovered. Mill would have to argue here that the data itself suggested an ellipse, but this is contrary to the details of how the discovery was made: it didn't occur to anyone that orbits could be anything but circular. Whewell treats Kepler's idea of the ellipse as a fundamental idea,[54] something purely intelligible rather than sensory, and discovery arises, on his view, in the application of the intelligible ideas to sensory data.

Whatever the problems with Mill's account of induction, however, it is his account of deduction that has been the most controversial part of his account of knowledge. We can in fact identify two different views of deduction in his writings. One is that it is a particularly abstract form of induction, the other is that logic is a branch of psychology. While there are significant difficulties in reconciling these, to some extent they play different roles, a practical one and a theoretical one.[55] Logic as an art of reasoning provides rules by which to guide our reasoning, rules that cover both truth (and so evidence) and validity. Logic as a science of reasoning analyses the mental processes involved in reasoning, and falls under the rubric of an empirical psychology.

The idea that deduction is a particularly abstract form of reasoning, by contrast with inductive reasoning, which is more concrete, ties deduction and induction together closely.[56] But if induction and deduction are so closely related, and we know that induction rests upon empirical foundations, the question arises what deduction rests upon. From what do we derive our knowledge of a general truth, Mill asks, and answers: 'Of course, from observation. Now, all which man can observe are individual cases. From these all general truths must be drawn, and into these they may again be resolved; for a general truth is but an aggregate of particular truths; a comprehensive expression, by which an indefinite number of individual facts are affirmed or denied at once.'[57] He gives the example of geometry, asking 'what is the ground for belief in axioms—what is the evidence on which they rest? I answer, they are experimental truths; generalizations from observation.'[58]

[54] William Whewell, *Of Induction* (London, 1849). Nevertheless, Whewell seems to be conflating Kepler's account of elliptical orbits with that of Newton. Unlike Newton, Kepler considered that the mathematics delivered circular orbits, and that the ellipses were the result of various physical factors acting on these.

[55] Mill, *Collected Works*, vii. 12.   [56] Ibid., vii. 193.   [57] Ibid., vii. 186.

[58] Ibid., vii. 231.

To appreciate what is at issue here, we need to remember that there was a long tradition of thinking about inference, stretching back at least to Descartes, in which deductive inference was considered trivial.[59] In particular, it was considered that the premises are always contained in the conclusion in a formally valid inference, so the deduction can never go beyond the premises, and can never tell us anything new. This account mirrored contemporary views of the syllogistic, but it made deductions in geometry, which seemed to be an informative discipline, beg the question. One solution (Descartes) was to argue that in geometry it was the analytical problem-solving inferences that did the work, not the synthetic deductive presentation of the kind one gets in Euclid's *Elements*, which was merely a presentational device. Mill moves in the opposite direction, arguing that geometrical proofs are informative, but he cannot invoke either Kant's account of geometry as synthetic a priori, or Whewell's archetypes. Informativeness for Mill means meeting an external standard, and observation is the only such standard. But the informativeness of geometrical theorems is limited because all they can secure is compression of masses of particular truths at varying levels of abstraction. This is revealing because it indicates that Mill's way of thinking about deduction here is really about geometrical truths—what makes them true—not about deductive inference as such. Their truth derives from their compression of many truths in an economical and fruitful way. The traditional question of logicians, namely which inferences are truth-preserving and why, is not really addressed here. Kant and Whewell had both attempted to account for how the truths of mathematics could be at the same time both informative and necessary. Kant thought of our grasp of geometrical theorems in terms of features of the human mind that were not merely universal but necessary, in that they were a condition of possibility of how we conceive of the world. Whewell by contrast thought, in Platonist terms, of their necessity deriving from their correspondence to divine ideal archetypes. Mill effectively abandons necessity and accounts for informativeness.

The second strand in Mill's account is his 'science of reasoning'. But there is a connection between the two strands, for, despite the great gap between them, he moves from the core thesis of his art of reasoning, that 'the sole object of Logic is the guidance of one's own thoughts',[60] to the core thesis of his science of reasoning, that logic is a branch of psychology:

I conceive it to be true that Logic is not the theory of Thought as Thought, but of valid Thought; not of thinking, but of correct thinking. It is not a Science distinct from, and coordinate with, Psychology. So far as it is a science at all, it is a part, or branch, of Psychology; differing from it, on the one hand as a part differs from the whole, and on the other, as an Art differs from a Science. Its theoretic grounds are wholly borrowed from Psychology, and include as much of that science as is required to justify the rules of the art.[61]

[59] See Gaukroger, *Cartesian Logic*.   [60] Mill, *Collected Works*, vii. 6.
[61] Ibid., vii. 359.

In other words, the laws of logic are dependent on contingent facts about human psychology. That these are *contingent* facts is not a worry for Mill. By contrast with those who held that logical truths were analytic and necessarily true, Mill held that they could have been different, but not in the sense that there is variation from person to person, or society to society: contingency in logic doesn't entail relativism, any more than the facts of physics being contingent entails relativism in physics. It is these universal but contingent facts that explain the objectivity of reasoning processes, that is, the objectivity in moving from premises to conclusion by means of a particular chain of reasoning. Objectivity lies here in something about the mind, but the mind is to be accounted for in terms of psychology, an empirical discipline.

How does this empirical psychology proceed? Psychology studies thought, and attempts to discover the laws that regulate thought, and if it is to have a scientific standing, it must be subject to external criteria of evidence. But how could what is involved in mathematical operations—calculations or proofs—be captured in empirical-psychological laws of thought? To claim that what we treat as valid proofs or correct calculations is a fact about how our minds work perhaps makes sense in Kantian terms, where the fact is a conceptual one about what it is to think, but it is difficult to see what rationale there could be for it as an empirical feature of thought. Moreover it is difficult to understand what justification there could be for Mill's claim that we discover logical, geometrical, and arithmetical truths experimentally. But the most pressing questions for the application of psychology to reasoning came not in such areas, but in the psychological assumptions underlying utilitarian economics, for here it starts to become evident just how much a transformation of the human psyche is involved in the utilitarian project.

'A PSYCHOPHYSICAL MACHINE'

Mill's was one of the most comprehensive nineteenth-century schemes for the unification of scientific understanding, encompassing ethics, politics, economics, and logic and scientific reasoning. A common, if problematic, strand in these endeavours was psychology. Millian ethics was consequentialist, so questions of intentionality and character were absent, which might lead one to ask what role could be left for psychology. But the aim of consequentialist ethics is not a description of how people act when they make moral decisions. It is strongly prescriptive, rejecting those eighteenth-century accounts, such as those of Smith and Hume, which had grounded morality in sensibility. In opposition to these, it makes behaving morally a matter of subjecting oneself to the dictates of reason. This is effectively to advocate a reform of behaviour, through a reform of the principles by which we make decisions on moral questions. This in turn requires an understanding of what, in broad terms, we might describe as the psychology of the moral agent. Moreover, because of Mill's psychologistic construal of reasoning, morality ultimately becomes subsumed under rational psychology, that is, a psychology devoted to understanding how reasoning works.

Strictly speaking, for Mill there are two kinds of enterprise that fall under the rubric of psychology: psychology proper and what he terms 'ethology'.[62] Psychology proper, as we have seen, is a science devoted to discovering the universal laws of the mind. Ethology is a 'science of the formation of character', which is derivative, its laws being deduced from the universal laws of psychology, which it extends to the particular circumstances in which people, individually or collectively, find themselves. Although ethology's laws are universal, and depend on psychology's core assumption of associationism, the factors that shape the individual circumstances, such as the individual's life history, cannot be circumscribed in full. The aim of ethology is to make predictions regarding the tendencies which different characters exhibit in different circumstances. It is only on the basis of such information that the moral and social sciences can be developed to a degree of theoretical and practical utility. In a note to Alexander Bain in the autumn of 1843, Mill writes that 'There is no chance, for Social Statics at least, until the laws of human character are better treated.'[63] Yet despite its importance in Mill's schema, he was unable to develop his ethology.[64] The *Logic* went through several editions, but the ethology section was never essentially modified, and in 1843 he wrote to Bain that his 'scheme had not assumed any definite shape',[65] with Bain subsequently noting that Mill had 'dropped thinking of it'.[66]

The lack of progress in ethology reflected problems in Mill's development of his psychology. Unable to advance very far in this respect himself, he greeted Bain's *The Senses and the Intellect* (1855)[67] with enthusiasm, writing an effusive review of Bain's work in 1859.[68] What Bain offered is summed up nicely by Rylance:

Bain's psychological programme was what Mill needed in the late 1850s. It offered a politically compatible psychological theory securely within the associationist (experience/a posteriori) tradition, which took some account of the new physiological psychology, but did not hand the discipline entirely over to it. Bain's work took the weight of the standard critique of associationism (best put, for Mill, by Coleridge), that associationism was too passive in its theorization of the human personality, and attempted to remodel it without losing touch with founding principles. In this way, the post-Benthamite ensemble of liberal Utilitarianism would not be unduly disturbed by the new physiology, for . . . Mill

---

[62] The case is set out in ch. 5 of Book VI of the *Logic*.   [63] Mill, *Collected Works*, xiii. 613.

[64] See David E. Leary, 'The Fate and Influence of John Stuart Mill's Proposed Science of Ethology', *Journal of the History of Ideas* 43 (1982), 153–62.

[65] Mill, *Collected Works*, xiii. 617.

[66] Alexander Bain, *John Stuart Mill: A Criticism with Personal Recollections* (London, 1882), 78.

[67] Alexander Bain, *The Senses and the Intellect* (London, 1855).

[68] Mill, 'Bain's Pychology', *Edinburgh Review* (1859), in *Collected Works*, xi. 339–74.

was personally committed to the old introspective and observational methods in psychology.[69]

Perhaps the most intractable problems in this respect lay in Mill's thinking through the nature of economics. As we have seen, he was ambiguous on the question of whether the economy was part of a natural order or a mental phenomenon, and he vacillates between the two. Margaret Schabas has noted that such vacillation was not unusual in Mill's time, yet the idea that economics needed psychological foundations was common.[70] Richard Jennings for example, in his *Natural Elements of Political Economy* (1855), is keen to stress that, while physical and mental questions are involved, the phenomena with which political economy deals are ultimately mental. John Elliott Cairnes offers a similar assessment:

The expressions 'physical' and 'mental', as applied to science, have generally been employed to designate those branches of knowledge, of which physical and mental phenomena respectively form the subject matter. Thus Chemistry is considered as a physical science because the subject matter on which chemical enquiry is exercised, viz., material elements and combinations, is physical. Psychology, on the other hand, is a mental science; the subject matter of it being mental states and feelings . . . . If this be a correct statement of the principles on which the designations 'mental' and 'physical' are applied to the sciences, it seems to follow that Political Economy does not find a place under either category. Neither mental nor physical nature forms the *subject-matter* of the investigations of the political economist. He considers, it is true, physical phenomena, but in neither case as phenomena which it belongs to his science to explain. The subject-matter of that science is wealth; and though wealth consists in material objects, it is not wealth in virtue of those objects being material, but in virtue of their possessing value—that is to say, in virtue of their possessing a quality attributed to them by the mind.[71]

The ambiguity on the question of whether political economy was a study of the physical or the mental pervaded mid-nineteenth-century thought, and was only resolved with the early marginalists: Jevons, Edgeworth, Wicksteed, and Marshall. For Jevons for example, the whole of economics is based on the investigation of 'the condition of the mind', and all prices are reduced to feelings of pleasure and pain at the margin, in terms of the 'final degree of utility'.[72] Edgeworth, by contrast, combines the psychophysical work of Helmholtz and Fechner with a demand for a thoroughgoing quantification of economics, treating human beings as essentially pleasure-seeking, where utility is a form of energy and, along Helmholtzian lines, can be measured. In an appendix to his *Mathematical Psychics* entitled 'On Hedonimetry', he tells us that 'hedonism may still be in the state of

---

[69] Rick Rylance, *Victorian Psychology and British Culture 1850–1880* (Oxford, 2000), 160–1.
[70] Schabas, *The Natural Origins of Economics,* 134–41.
[71] John Elliott Cairnes, *Character and Logical Method of Political Economy* (New York, 1875), 47–8.
[72] W. Stanley Jevons, *The Theory of Political Economy* (London, 1871), 14–15, 52.

heat or electricity before they became exact sciences'.[73] Although he warns that it 'must be confessed that we are here leaving the *terra firma* of physical analogy', nevertheless he proceeds to the mathematics of pleasures and utilities. He identifies 'pleasure-units' and, considering Jevons' treatment of utility as a quantity in two dimensions, intensity and time, he expounds the problems of measurement:

Suppose one state presents about three pleasure-increments, another about two, above zero, that the rate of the former is double that of the latter, their objective duration being the same, is it better to give two marks to each state, say three and two to the former, two and one to the latter, and then to multiply the marks of each; or by a sort of unconscious multiplication to mark at once six and two—*about*; for the comparison of pleasures as to quantity is here admitted to be vague; not vaguer perhaps than the comparisons made by an examiner as to excellence, where numerical marks are usefully employed. To precise the ideas, let there be granted to the science of pleasure what is granted to the science of energy; to imagine an ideally perfect instrument, a psychophysical machine, continually registering the height of pleasure experienced by an individual, exactly according to the verdict of consciousness, or rather diverging therefrom according to a *law of errors*.[74]

There are two issues here in the development of the economics of utility relevant to our broader purposes. The first concerns the subjects of economic behaviour. For Smith, Ricardo, and the classical economists generally, it was various economically defined classes that were the economic agents. The early neoclassical economists, by contrast, as Schabas notes, 'dissolved economic classes as a unit of analysis and built their model of the world from individuals. Whereas with Ricardo the central question was how to divide the pie between the landlords, the capitalists, and the laborers, with Jevons and Marshall the key question became one of maximizing individual utility and taking the aggregate to arrive at meaningful claims about social welfare.'[75]

Second, I want to draw attention to Edgeworth's 'psychophysical machine, continually registering the height of pleasure experienced by an individual, exactly according to the verdict of consciousness, or rather diverging therefrom according to a law of errors'. Here we can begin to glimpse a novel form of modelling of human faculties which, despite its somewhat fanciful presentation in Edgeworth, coheres closely with those late nineteenth- and early twentieth-century theories that devised statistically driven psychological tests. As Mill's Anglican critics realized, utilitarianism, in economics as much as in morality and politics, was a direct competitor to the prevalent Christian model of civilization. It is worth noting that this was often explicit, with Edgeworth for example drawing on Buffon's argument that pleasure was 'an essential attribute of civilization'.[76] As those pursuing a utilitarian-style project develop more comprehensive views on

[73] Francis Ysidro Edgeworth, *Mathematical Psychics: An Essay on the Application of Mathematics to the Moral Sciences* (London, 1881), 98.

[74] Ibid., 100–1.   [75] Schabas, *The Natural Origins of Economics*, 139–40.

[76] Quoted in ibid, 137.

psychology, we can begin to glimpse the origins of programmes to shape the population into the kinds of people who can occupy a scientifically modelled form of civilization. Unlike the landlords and labourers in terms of which Smith and Ricardo worked, the raw utility-seeking individuals of neoclassical economics have little or no identity in their own right, and are ripe for a modelling along the lines of a rational psychology. But in fact, as far as the development of economics is concerned, this rational psychology turns out to be a *rationalized* psychology, as a brief excursus into the subsequent history of economic theory will make clear.

Economic agents, from neoclassical economics onwards, are agents that act on the basis of rational expectations, termed 'optimization'. Such optimization is considered to be unconstrained by any form of bias: it is purely rational.[77] A question immediately arises, however, as to whether, in offering an economic theory along these lines, it is at all realistic to assume completely rational calculations on the part of the agent. Do agents in fact act to optimize their choices on a purely rational basis?[78] There is, for example, a compelling case to be made that economic decision making is an instance of what the economist Herbert Simon called 'bounded rationality'.[79] That is to say, when agents make choices, their rationality is constrained by such factors as the difficulty of the choice, by their limited cognitive powers, and by the time available to make the decision. Even a simple choice about a purchase may potentially involve an indeterminately large number of possible comparisons. Clearly economic agents not only do not, but could not, make a decision on this basis.

One response to such considerations is to argue that all one needs to assume is that economic agents acted as if they were purely rational. Milton Friedman, for example, has argued that what matters is not whether the assumptions are realistic, but the accuracy of the theory's predictions:

excellent predictions would be yielded by the hypothesis that the billiard player made his shots *as if* he knew the complicated mathematical formulas that would give the optimum direction of travel, could estimate by eye the angles etc., describing the location of the balls, could make lightning calculations from the formulas, and could then make the balls travel in the direction indicated by the formulas. Our confidence in this hypothesis is not based on the belief that billiard players, even expert ones, can or do go through the process described; it derives rather from the belief that, unless in some way or other they were

[77] Since 1944 we have had a formal theory—'expected utility theory'—of how such economic behaviour can be derived from basic premises of rational choice theory, in John von Neumann and Oskar Morgenstern, *Theory of Games and Economic Behaviour* (Princeton, NJ, 1944). This builds on an earlier paper of von Neumann: 'Zur Theorie der Gesellschaftsspiele', *Mathematische Annalen* 100 (1928), 295–320.

[78] Such agents include firms as well as individuals, but that is not relevant to our present considerations.

[79] Herbert A. Simon, *Models of Man, Social and Rational: Mathematical Essays on Rational Human Behaviour in a Social Setting* (Oxford, 1957).

capable of reaching essentially the same result, they would not in fact be expert billiard players.[80]

The claim is that the expert billiard player plays in the same way as would someone who knows all the relevant geometry and physics; and that analogously the purchaser of everyday goods reasons in the same way as would someone who was able to perform (instantaneously) a wide range of optimization calculations. We can think of the billiard player taking on an innate sense of Newton's laws, for example, because we know, on independent grounds, that—from a range of possible laws of physics—these are the laws that regulate the behaviour of bodies in motion on billiard tables. But we have no independent reason to assume that optimization is, from among several possible accounts of motivation in economic decision making, the right explanation of purchasing behaviour. The best we could show would be that an optimization strategy would yield this economic behaviour (without any claims as to the uniqueness of the strategy in this respect). But can we even assume that: are optimization and purchasing behaviour in fact compatible? In particular, does optimization theory yield the 'accurate predictions' that Friedman relies upon? Everything here depends on how we characterize purchasing behaviour, and those working in behavioural economics have argued, on empirical and psychological grounds, that human beings purchasing goods do *not* behave as would optimizers.

The pioneer of behavioural economics, Richard Thaler, reports that in the question and answer period after a talk in which he showed what a poor description of actual behaviour optimization economics offers, a distinguished economist asked: 'If I take what you are doing seriously, what am I supposed to do? My skill is knowing how to solve optimization problems.'[81] What, we might ask, is the payoff here: what is the commitment to optimization supposed to secure? The answer is: a scientific approach. And here concern with 'accurate predictions' becomes subordinated to the mathematical character of optimization theory, on the grounds that it is the latter that allows the rigorous generation of accurate predictions.

In general terms, what is at issue here is the mathematical form of neoclassical economics, and the 'scientific' standing that it derives from this. Jevons is often credited with the transformation of economics into a mathematical discipline through his construal of the degree of utility of a commodity as some continuous mathematical function of the quantity of the commodity available. But, in developing their accounts, did the neoclassical economists simply take over mathematical procedures from physicists, designating their enterprise 'scientific' as a result; or was it just pure coincidence that they took over a mathematical

---

[80] Milton Friedman, 'The Methodology of Positive Economics', in Milton Friedman, *Essays in Positive Economics* (Chicago, 1953), 3–43: 21.
[81] Richard H. Thaler, *Misbehaving: The Making of Behavioural Economics* (London, 2016), 43.

formalism identical to that which physicists had devised, a simultaneous discovery as it were? Mirowski, for one, finds Schumpeter's idea that it was pure accident that exactly the same devices were used by economists in the wake of their successful development in physics wildly implausible:

Neoclassical economics made savvy use of the resonances between body, motion, and value in engaging in a brazen daylight robbery: The Marginalists appropriated the mathematical formalisms of mid-nineteenth century energy physics..., made them their own by changing the names of variables, and then trumpeted the triumph of a truly 'scientific economics'. Utility became the analogue of potential energy; the budget constraint became the slightly altered analogue of kinetic energy; and the Marginalist Revolutionaries marched off to do battle with classical, Historicist, and Marxist econo-mists. Unfortunately, there had been one little oversight: the Neoclassicals had neglected to appropriate the most important part of the formalism, not to mention the very heart of the metaphor, namely the conservation of energy. This little blunder rendered the neoclassical heuristic legacy essentially incoherent; but heedless of that fact, the Margin-alists triumphed under their banner of science.[82]

If this criticism is even remotely along the right lines, then we must think seriously about the role of scientific credentialling in establishing how economics proceeds. Optimization, because it has leant itself to sophisticated mathematical treatment, has carried much of the scientific weight of economic theory. But as Kenneth Arrow, in an essay in which he examined the assumption of economic rationality, has argued, optimization is neither a necessary nor a sufficient condition of good economic theory.[83] As he points out, there are any number of possible—formal, rigorous—theories based on behaviour that economists would not call rational. For example, on the standard theory of consumer prices, when prices change the consumer solves a new optimization problem and chooses a new set of goods that satisfies a budget constraint. But one could easily construct a theory on the basis of habits rather than optimization, where the consumer chooses a set of products closest to what he or she bought previously. Arrow also questions the idea that optimization is sufficient to generate the correct predictions, pointing out that this is impossible without auxiliary assumptions. Some of these, such as the assump-tion that everyone has the same tastes ('utility functions'), are highly questionable, but dropping them would undermine the optimization exercise. Advocates of optimization have traditionally been forced to rule out many features of the behaviour of economic actors as irrelevant, but this has become increasingly implausible as the importance of many of these—especially those which exhibit

---

[82] Philip Mirowski, *More Heat than Light: Economics as Social Physics, Physics as Nature's Economics* (Cambridge, 1989), 9.

[83] Kenneth J. Arrow, 'Rationality of Self and Others in an Economic System', *Journal of Business* 59.4 (1986), S385–99.

bias and consequently undermine the idea that the study of economic decision making can be confined to rational behaviour—has been established.[84]

In this last case, the work has been done by psychologists, notably Kahneman and Tversky, devising what they have termed 'prospect theory'. As Thaler explains it: 'Kahneman and Tversky set out to offer an alternative to expected utility theory that had no pretence of being a useful guide to rational choice; instead, it would be a good predictor of the actual choices real people make. It is a theory about the behaviour of Humans.'[85] More generally, Thaler has explored the way in which economic agents ('Econs') and human agents ('Humans') have become completely separated in economics since the late nineteenth century. To the extent that the former have any psychological content, it is as abstract calculators, perhaps even as miniature scientists, for as Arrow points out, we have the curious situation that, in order for the scientific analysis to be carried out, we need to impute scientific behaviour to its subjects.[86] 'Curious' perhaps, but not at all unexpected.

---

[84] See in particular Daniel Kahneman and Amos Tversky, 'On the Psychology of Prediction', *Psychological Review* 80.4 (1973), 237–51; Daniel Kahneman and Amos Tversky, 'Prospect Theory: An Analysis of Decision under Risk', *Econometrica* 47.2 (1979), 263–91.

[85] Thaler, *Misbehaving*, 29.

[86] Arrow, 'Rationality of Self and Others in an Economic System', S391.

9

Understanding the World

Science versus Philosophy

Metascientific interests were not confined to the England of Whewell and Mill. In the third quarter of the nineteenth century, there were parallel developments in the German-speaking world. As in the British case, a metascientific programme came to shape the understanding of scientific practice in a decisive fashion. Whereas in Britain the catalyst had been a scientific political economy, however, in Germany it was 'scientific materialism'. The claim of scientific materialism was that, with the emergence and consolidation of science, there are no longer any genuine philosophical problems: the speculative approach of philosophy can be replaced entirely by the empirical methods of science. The principal response to scientific materialism in Germany was Neo-Kantianism, which radically reshaped the aims of philosophy, transforming it into a metatheory of science. Not only the response, but also how one conceived of scientific materialism in the first place, were shaped by the context in which it was pursued, and the grounds on which it was advocated. In France, for example, a form of scientific materialism was being advocated at the same time as in Germany, yet it did not provoke a metascientific response: it was seen as a challenge, certainly, but not one that required a philosophical rationalization and response. A brief comparison of the differently contextualized forms of scientific materialism will help us to understand what was distinctive about the German case.

In France, modernity was strongly associated with anti-clericalism, and the questions turned on physical anthropology—the evolution of human beings.[1] Atheism and reductionism were the primary issues at stake. Mortillet's influential *La préhistorique*[2] set out the case for prehistorical human cultures for example, arguing that in the Palaeolithic era people lived a peaceful religion-free existence, while Lefèvre's *La renaissance de matérialisme* offered a polemical attack on the ideas of God and religion.[3] The predominant theme however was that set out in

---

[1] See Jennifer Michael Hecht, *The End of the Soul: Scientific Modernity, Atheism, and Anthropology in France* (New York, 2003), ch. 3, to which I am indebted here.
[2] Gabriel de Mortillet, *La préhistorique* (Paris, 1883).
[3] André Lefèvre, *La renaissance de matérialisme* (Paris, 1881).

Civilization and the Culture of Science: Science and the Shaping of Modernity, 1795–1935. Stephen Gaukroger, Oxford University Press (2020). © Stephen Gaukroger.
DOI: 10.1093/oso/9780198849070.001.0001

Charles Letourneau's books on biology and ethnography from the 1870s and 1880s,[4] arguing for the intimate relation between animals and human beings. The continuity between animals and humans as established in evolutionary theory meant not only that the strict separation between the two that had been a premise of Christian thinking had to be abandoned, but also that the systems of morals based on them had to be discarded. The rethinking of morality that evolution required meant that the development of morality and society needed to be accounted for without any reference to a uniquely human soul. This does not mean that it was considered that there was no difference at all between humans and other animals. Abel Hovelacque, for example, argued that it was language that distinguished the two, and not religion or the soul, which he treated in reductionist terms.[5] Naturalism was ubiquitous, and one of the most radical works to appear was Véron's *L'esthétique* which, answering the complaint that 'no science has suffered more from metaphysical dreaming than aesthetics', set out to rescue art from 'phantasies and transcendental mysteries'. Art was a matter of taking pleasure in lines, forms, colours, movements, sounds, rhythms, and images, and was basically a matter of optics and acoustics, although Véron acknowledged that the explanation of cerebral phenomena was insufficiently advanced to account for these to date.[6]

This late nineteenth-century French movement was to have a profound influence on the development of physical anthropology, and it certainly provoked a response from religious conservatives in France, but it posed a threat to philosophy only insofar as philosophy was associated with the Church. In Germany, by contrast, although there were developments in physical anthropology, these were not part of any unified movement and were not the focus of attention before the 1920s.[7] Rather, the issues were embedded in an active philosophical culture shaped by the failure of Classical German Idealism, and scientific materialism was countered primarily by the rise of Neo-Kantianism, a programme devoted to providing a new basis for understanding the nature and limits of science.

German philosophy effectively had the field to itself in continental Europe in the period from the last decades of the eighteenth century up to Hegel's death in 1831. For most German philosophers, Kantian thought had been surpassed first by that of Fichte and Schelling, and then by that of Hegel. Orthodox Hegelians had considered that Hegel's work was the culmination of philosophy, and so had brought philosophizing to an end. With the rapid decline in the standing of

---

[4] Charles Letourneau, *La biologie* (Paris, 1877) and Charles Letourneau, *La sociologie d'après l'ethnographie* (Paris, 1884).

[5] Abel Hovelacque, *La linguistique* (Paris, 1892).

[6] Eugène Véron, *L'esthétique* (Paris, 1878).

[7] See Joachim Fischer, *Philosophische Anthropologie: Eine Denkrichtung des 20. Jahrhunderts* (Freiburg, 2008). Heidegger came to be influenced by Max Scheler's phenomenological version of physical anthropology in the 1920s: see Peter E. Gordon, *Continental Divide: Heidegger, Cassirer, Davos* (Cambridge, MA, 2010), 69–77. Cassirer lectured on Scheler: see ibid., 115–19.

Hegelian philosophy in the wake of his death however, a vacuum opened up, and it did indeed begin to look like philosophy had come to an end, if not for the reasons that Hegelians had supposed. A certain style of philosophizing, which had begun with Kant, but which had been transformed in quite a radical way in Classical German Idealism, had come to an end—collapsing under its own weight as it were—and there was no alternative with anything like its breadth of vision and standing. In Germany, there were a number of reactions to this, some involving a politically radical materialist reworking of Hegel in the writings of the 'Young Hegelians',[8] others involving a revival of Schopenhauer's largely forgotten idiosyncratic reworking of Kant.[9] The crucial development from the point of view of our concerns, is that the collapse of speculative idealism occurred at the same time as a significant rise in the fortunes of science, and the interplay between these is what distinguishes the developments with which we shall be concerned. This is not a peculiarly German development for, as we have seen, Mill in England and Comte in France offered different kinds of responses to the changing fortunes of philosophy and science. The debates in Germany were particularly influential however, and more than any other developments shaped how science and philosophy were conceived well into the twentieth century.

On the face of it, there is a straightforward division in mid-nineteenth-century German thought between those who considered that philosophy no longer had any role to play, and those who thought that philosophy was needed to rationalize and understand just what science is and what its achievements are. The first was the view that, with the demise of philosophy, science can take over all the questions with which philosophy had concerned itself, and can begin to explore them in an empirical way, with the aim of establishing materialism. The second was a fundamental reworking of Kant in the movement known as Neo-Kantianism, an adaptation of elements of Kantian thought to a situation that went beyond that which Kant himself would have envisaged, and whose most pressing concern was with the relationship between philosophy and science.[10] But matters are more complicated, because what is at issue in this division is not 'philosophy' as such. Both sides reject what they consider as 'metaphysics'. Traditionally, metaphysics, in scholastic and early modern philosophy for example, covered such questions as the nature of substance, the relation between the

---

[8] See, for example, William Brazill, *The Young Hegelians* (New Haven, CT, 1970), and John E. Toews, *Hegelianism: The Path Towards Dialectical Humanism, 1805–1841* (Cambridge, 1980).

[9] See Karl Löwith, *Von Hegel zu Nietzsche: Der revolutionäre Bruch im Denken des 19. Jahrhunderts* (Zurich, 1941); and Frederick C. Beiser, *Weltschmerz: Pessimism in German Philosophy 1860–1900* (Oxford, 2016). Schopenhauer was suddenly to become very popular in France in the 1870s, as translations of his works into French appeared, as well as the first substantial introduction to his thought: Théodule Ribot, *La philosophie de Schopenhauer* (Paris, 1874). Durkheim was to become fascinated with Schopenhauer at this time: see Marcel Fournier, *Émile Durkheim: A Biography* (Cambridge, 2013), 48–50.

[10] On the emergence of Neo-Kantianism as a movement, see Frederick C. Beiser, *The Genesis of Neo-Kantianism, 1796–1880* (Oxford, 2014).

spiritual and the material realms, and the nature and existence of God.[11] But by the nineteenth century, ontology had become separated from questions about the nature of substance, and the term 'metaphysics' had largely shed its traditional connotations and became primarily ontology. Albeit for different reasons, both the scientific materialists and the Neo-Kantians rejected metaphysics. The scientific materialists replaced philosophy, both metaphysics and epistemology, with science—as du Bois-Reymond put it, science can safely ignore 'myth, dogma, and time-honoured philosophy'[12]—while the Neo-Kantians replaced metaphysics with epistemology. In fact, to the tripartite relation between science, epistemology, and metaphysics, we must also add methodology, sometimes assimilated into epistemology, sometimes considered as emerging from science itself.

Consider the most straightforward case, the materialist version of the replacement of philosophy by science. We must distinguish two broad kinds of materialism. First there is the group of self-styled 'scientific materialists', such as Vogt, Moleschott, and Büchner, who argued that there was nothing that could not be explained in materialist—either physico-chemical or physiological—terms. The second group, in which we can include Helmholtz, du Bois-Reymond, Virchow, and others, argued that everything that we can explain could be explained in materialist terms, but that there were limits to knowledge. Consciousness provided a point of division between the two groups. Du Bois-Reymond asks:

what play of carbon, hydrogen, nitrogen, oxygen, and phosphorus corresponds to the bliss of hearing music, what whirl of such atoms answers to the climax of sensual enjoyment, what molecular storm coincides with the raging pain of trigeminal neuralgia.... What conceivable connection exists between definite movements of definite atoms in my brain on the one hand, and on the other hand such primordial, indefinable, undeniable facts as these: *I feel pain or pleasure; I taste something sweet, or smell a rose, or hear an organ, or see something red*, and the certainty that immediately follows: *Therefore I am?* [13]

The lesson is clear: in the case of consciousness, we simply have to concede that some things might be inexplicable.[14] In allowing the idea that there might be areas that fall outside purely scientific understanding, du Bois-Reymond and others were not suggesting that there could be other forms of study that would allow

---

[11] On scholastic conceptions of metaphysics see Gaukroger, *The Emergence of a Scientific Culture*, 80–6; on early modern ones, see Gaukroger, *The Collapse of Mechanism*, 104–15.

[12] Du Bois-Reymond, *Reden*, 461.

[13] *Reden*, 475–8. Quoted in Gabriel Finkelstein, *Emil du Bois-Reymond: Neuroscience, Self, and Society in Nineteenth-Century Germany* (Cambridge, MA, 2013), 266.

[14] Consciousness was a persistent problem. Thomas Huxley, for example, was clear that understanding was scientific understanding, and that a scientific understanding of consciousness was impossible: 'We class sensation along with *emotions*, and *volitions*, and *thoughts*, under the common head of *states of consciousness*. But what consciousness is, we know not; and how it is that anything so remarkable as a state of consciousness comes about as a result of irritating nervous tissue, is just as unaccountable as any other ultimate fact of nature': Thomas Henry Huxley, *Lessons in Elementary Physiology* (3rd edition, London, 1872), 188. Cf. John Tyndall, 'Physics and Metaphysics', *Saturday Review* 10 (4 August 1860), 141.

these areas to be opened up. There was no understanding that was not scientific understanding for them: it was just that such understanding could not be exhaustive. For the scientific materialists, the effective postulation of an unknowable realm came suspiciously close to advocacy of a Kantian noumenal realm. At the same time, the division between the two groups had a political and religious dimension, the first tending towards conservative politics and an unquestioning attitude (at least in public) to religious orthodoxy, the second towards political radicalism and atheism. Moreover the scientific materialists, who were marginalized in the scientific community, thrived in the genre of popular science, by contrast with Helmholtz et al., who had an authoritative voice within the scientific establishment, albeit one that they occasionally had to struggle to maintain as Prussian politics shifted ground.

The materialists were divided in this way among themselves on the question of whether science could answer every question. The degree of opposition is evident in a letter of du Bois-Reymond to the physician Georg Liebig, son of Justus Liebig, in 1857:

You ask for my views on Büchner's *Force and Matter*. Heavens, what babble! What I mean is that he and Moleschott rant on about things that were said by others just as clearly and certainly more concisely; that they deal with philosophical problems, the difficulties of which their views make light of, in a way that isn't suitable for the general public; that most of the time little is gained by popularizing such views; that in any case Vogt is causing more harm than good with his cynical frivolity (and his jokes pale in comparison to Voltaire's); that Moleschott is nothing more than a self-parodying copier of your father's writings.[15]

But the two groups were united on the issue of what explanatory resources were legitimate, namely those limited to science. In particular, they were united in opposing the philosophical speculation that had come to a head in German Idealism, and for them this meant an end to philosophy per se. Philosophy, in the form of German Idealism, had had totalizing aspirations, subsuming all forms of knowledge, including science. For the materialists, these totalizing aspirations were not abandoned with the collapse of German Idealism, but were transferred to science.[16]

The 1850s was the decade in which scientific materialism flourished in Germany. The most prominent member of the group was Karl Vogt, who began his university studies in Giessen as a chemistry student under Liebig, but on moving to Bern took up the study of anatomy and physiology, and in 1839 he became an assistant to Agassiz, working on the later volumes of his account of fossil fish.[17] He moved to Paris in 1844, earning a living as a science correspondent and lecturing,

---

[15] Quoted in Finkelstein, *Emil du Bois-Reymond*, 260.
[16] See Gregory, *Scientific Materialism*; and Kurt Bayertz, 'Dépasser la philosophie par la science: Le matérialisme naturaliste en Allemagne au XIXe siècle', in Charlotte Morel, ed., *L'Allemagne et la querelle du matérialisme (1848–1866)* (Paris, 2017), 67–82.
[17] Louis Agassiz, *Recherches sur les poissons fossiles* (5 vols, Neuchâtel, 1833–45).

while publishing popularizations of physiology and geology.[18] Paris radicalized Vogt, both politically and towards scientific materialism. In 1847 he returned to Giessen as professor of zoology, and stood for the Frankfurt parliament as a radical delegate, but was forced to leave Germany a year later, returning to the family home in Bern a virulent critic of the German university system.[19] Having translated Chambers' *Vestiges of the Natural History of Creation* into German, in 1852 he published his 'pictures of animal life',[20] a scathing attack on the idea of divine creation, in which his main target was the Göttingen physiologist Rudolph Wagner, for whom materialism 'must bring science under suspicion of totally destroying the moral foundations of social order'.[21] Wagner responded in newspaper articles, and at the beginning of 1855 Vogt's response, *Köhlerglaube und Wissenschaft*, appeared, and was so popular that it went through three more editions within six months.[22] Its blunt comparisons of the production of thoughts in the brain with the production of bile in the liver, and urine in the kidneys, produced a sensation.

The impact of Vogt's book was enhanced by the appearance two years earlier of Jacob Moleschott's *Der Kreislauf des Lebens*, a detailed defence of a reductionist physiology.[23] Although he was a medical student, Moleschott gravitated towards philosophy at Heidelberg, and soon became interested in the 'Young Hegelians', assiduously studying Spinoza and Feuerbach.[24] From 1845 he studied physiology at Utrecht, and in 1850 he published a formal account of his research on nutrition and the physiology of digestion.[25] *Der Kreislauf des Lebens* was an altogether different kind of book: a sustained attack on Liebig's defence of vitalism and the existence of the soul, it argued that force was a property of matter that enabled it to move, and that there was nothing that eluded sense perception. Finally, one year after the publication of Vogt's volume, the physician Ludwig Büchner published his *Kraft und Stoff*,[26] a synthesis of the work of Vogt and Moleschott, setting the issues out plainly and drawing out their philosophical consequences.[27] This work became the 'bible of scientific materialism', and by the mid-1850s scientific materialism was established in Germany as the scourge of metaphysical thinking.

---

[18] Karl Vogt, *Physiologische Briefe für Gebildete alle Stände* (3 vols, Stuttgart, 1847); *Ocean und Mittelmeet: Reisebriefe* (2 vols, Frankfurt am Main, 1848).

[19] On Vogt's career, see Gregory, *Scientific Materialism*, ch. 3.

[20] Vogt, *Bilder aus dem Thierleben* (Frankfurt am Main, 1852).

[21] Rudolph Wagner, *Menschenschöpfung und Seelensubstanz* (Göttingen, 1854), 29.

[22] Vogt, *Köhlerglaube und Wissenschaft: Ein Streitschrift gegen Hofrath R. Wagner in Göttingen* (Giessen, 1855).

[23] Jacob Moleschott, *Der Kreislauf des Lebens: Physiologische Antworten auf Liebig's 'Chemische Briefe'* (Mainz, 1852).

[24] See Gregory, *Scientific Materialism*, ch. 4

[25] Jacob Moleschott, *Die Physiologie der Nahrungsmittel* (Giessen 1850).

[26] Ludwig Büchner, *Kraft und Stoff. Empirische-naturphilosophische Studien. In allgemein verständlicher Darstellung* (Frankfurt am Main, 1855).

[27] On Büchner, see Gregory, *Scientific Materialism*, ch. 5.

In the writings of the scientific materialists, reductionism was presented as the way to dispense with the Kantian distinction between the noumenal and phenomenal realms,[28] and Vogt, Moleschott, and Büchner are very much in the ontological tradition of French eighteenth-century materialists such as La Mettrie and d'Holbach, where what was at issue was the ontological question of the primacy of matter. This forms the basis for the association of their programme with atheism and radical anti-clericalism, through the denial of the existence of any spiritual substance, or spiritual realm more generally.[29] But the programme of scientists such as Du Bois-Reymond and Helmholtz had no such radical consequences. On the contrary, it studiously avoided them. Its proponents were not interested in denying the spiritual, but with proceeding in scientific enquiry in a way that demonstrated no need for any appeal to the spiritual.

The idea of an unknowable noumenal realm arises, on the materialist view, because of a refusal to acknowledge that sensation is the sole source of knowledge. But if the way to counter an unknowable noumenal realm was to acknowledge sensation as the sole source of knowledge, that was consistent with a number of approaches. In his 1865 lecture on 'the purpose of the natural sciences'[30] for example, Kirchhoff offered a defence of reductionism that distinguishes it from the kind of approach taken by the scientific materialists. His target is not the scientific materialists as such but Johann Hoffman's *Somatologie*,[31] a work that explored the way in which the internal constitution of material bodies determines how they appear. Kirchhoff rejected Hoffman's premise that physics should focus on the nature of material bodies, arguing that it should concern itself instead with forces. But in arguing this he also rejects what he sees as the claims of both Hoffman and the scientific materialists, and makes it clear that his own commitment to reductionism is a commitment to a methodological programme, not an ontological one.[32] Similarly with Mach. For Mach, science rests on purely sensory foundations because only sense impressions can be scientifically validated, and the greatest obstacle lies in metaphysics: 'all metaphysical elements', he tells us, 'are to be eliminated as superfluous and as destructive of the economy of science'.[33] This 'economy of science' is a methodologically secured unity of science: 'anyone who has in mind the gathering up of the sciences into a single whole, has to look for a

---

[28] See ibid., ch. 7.

[29] This was the traditional understanding of materialism. See the definition of materialism in Johann Georg Walch, *Philosophisches Lexicon* (Leipzig, 1726): 'materialism is a position which consists in denying any spiritual substance and admitting nothing other than corporeal substance' (1735). See the discussion in Charlotte Morel, 'La querelle du matérialisme: présentation historique', in Charlotte Morel, ed., *L'Allemagne et la querelle du matérialisme (1848–1866)* (Paris, 2017), 9–43.

[30] Kirchhoff, *Über das Ziel der Naturwissenschaften*.

[31] Johann Valentin Hofmann, *Somatologie oder Lehre von den inneren Beschaffenheit der Korper auf Grund einer vergleichenden Betrachtung der chemischen, morphologischen und physikalischen Eigenschaften derselben* (Gottingen, 1863).

[32] See the discussion in Oldham, 'The Doctrine of Description', ch. 5.

[33] Mach, *The Analysis of Sensations*, xxxviii.

conception to which he can hold in every department of science . . . . When it is a question of bringing into connection two adjacent departments, each of which has been developed in its special way, the connection cannot be effected by means of the limited conceptions of a narrow special department. By means of more general considerations, conceptions have to be created which shall be adequate for the wider domain.'[34]

Part of the problem for scientific materialism was its speculative nature, especially the use of analogies, like those between the digestive system and cerebral processes, but another part was its often explicit association with radical political programmes. It was as crucial for writers such as Marx to ridicule 'crude materialism', and the crude politics it supported, as it was for conservative critics of atheism and materialism to ridicule it. Vogt for one was aware of the difficulties his promotion of atheism caused for his advocacy of materialism. In an 1880 letter to the French materialist Mortillet, who was compiling his ideologically charged *Dictionnaire des sciences anthropologiques*, Vogt warned Mortillet against indulging in his usual ebullient atheism if he wanted the *Dictionnaire* to have any success.[35]

NEO-KANTIANISM AND THE ROLE OF EPISTEMOLOGY

Attempts to revive an autonomous philosophical culture in Germany relied extensively on Kantianism, that is, developmentally speaking, the point just before the rise of Classical Idealism. Two of the most influential mid-century German philosophers, Trendelenburg and Lotze, had much in common with the German Idealists, not least a commitment to a teleological view of nature, but they considered themselves Kantians nonetheless.[36] There were also a number of thinkers, eclectic to a greater or lesser degree, who set out to revive elements of Kantianism, notably Jakob Fries, Johann Herbart, Friedrich Beneke, Kuno Fischer, Eduard Zeller, Otto Liebmann, Jürgen Meyer, and Friedrich Albert Lange.[37] Different, less eclectic but novel forms of Kantianism emerged in the 1870s and 1880s and these can be grouped under the rubric of Neo-Kantianism, forming two 'schools', the Marburg and the Southwestern. Neither of these simply advocated a return to Kant, but learned crucial lessons from the collapse of German Idealism: particularly on the issues of science and reason. In a culture in which there were various proposals to replace philosophy with science, it is not surprising that serious defenders of philosophy were equally serious in their investigation of the nature of science. At the same time, they learned from

[34] Ibid., 312–13.   [35] Hecht, *The End of the Soul*, 97–8.
[36] See Beiser, *Late German Idealism*; and Charlotte Morel, 'Métaphysique et science chez Lotze: Enjeu d'une position médiane dans la querelle du materialisme', in Charlotte Morel, ed., *L'Allemagne et la querelle du matérialisme (1848–1866)* (Paris, 2017), 129–54.
[37] See Beiser, *The Genesis of Neo-Kantianism*, chs 5–8.

Romanticism and from the Idealist reaction to Kant that 'reason' was a more problematic notion than Kant himself had realized. Above all, the concept of reason needed to be reworked if it was to serve the kind of overarching role that Kant envisaged for it. What unified the two Neo-Kantian movements was the attempt to establish an autonomous role for philosophy while recognizing the centrality of science. What separated them was not so much how they conceived of philosophy but how they conceived of science.

Lange, who held the philosophy chair at the University of Marburg, attracted Herman Cohen there in 1873, and it is with Cohen that the distinctive Marburg form of Neo-Kantianism begins.[38] In 1871, Cohen published an account of the debate between Trendelenburg, an eclectic Neo-Aristotelian, and Fischer, an eclectic Neo-Hegelian, over Kant's transcendental idealism,[39] following it up with a fuller account of the consequences for Kantian doctrine.[40] The issues were driven by the rise of *Erkenntnistheorie*—the theory of knowledge or epistemology—a Kantian-inspired search for the underlying principles of knowledge which had as one of its central concerns the question of scientific knowledge,[41] and which shaped philosophical debates after the fall of Classical Idealism, not just in Germany but elsewhere among those who, like Whewell, reflected on the nature of science. Trendelenburg had taken up developments in empirical psychology pioneered by Herbart, Helmholtz, and Lange, arguing that Kant's account of our grasp of space and time needed to be revised and given a physiological or psychological source. At the same time, he insisted that Kant's construal of space and time as a priori intuitions did not mean that they were purely subjective and had nothing to do with objects outside of possible experience. By contrast, Fischer had defended Kant's doctrine of space and time as forms of a (universal) subjectivity, but had felt free to develop Kant's thought along the lines of his own 'freely constructed method'.[42] Cohen drew the line at a psychologization of our faculties, but also demanded a stricter interpretation of what he took to be the core of the Kantian project. In particular, he did not see the dispute between Trendelenburg and Fischer as restricted to the question of the nature of space and time, but rather treated this question as an expression of the

---

[38] On the origins and development of Marburg Neo-Kantianism, see Ulrich Sieg, *Aufstieg und Niedergang des Marburger Neukantianismus: Die Geschichte einer philosophischen Schulgemeinschaft* (Würzburg, 1994); Helmut Holzhey, *Cohen und Natorp, Band 1: Ursprung und Einheit* (Basle, 1986); and Helmut Holzhey, *Cohen und Natorp, Band 2: Der Marburger Neukantianismus in Quellen* (Basle, 1986). Specifically on Cohen, see Dieter Adelmann, *Einheit des Bewusstseins als Grundproblem der Philosophie Hermann Cohens* (Heidelberg, 1968).

[39] Hermann Cohen, 'Zur Kontroverse zwischen Trendelenburg und Kuno Fischer', *Zeitschrift für Völkerpsychologie und Sprachwissenschaft* 7 (1871), 249–96.

[40] Hermann Cohen, *Kants Theorie der Erfahrung* (Hildesheim, 1871).

[41] On the emergence of *Erkenntnistheorie* and its essentially Kantian nature, see Klaus Köhnke, *The Rise of Neo-Kantianism: German Academic Philosophy between Idealism and Positivism* (Cambridge, 1991).

[42] Adolf Trendelenburg, *Kuno Fischer und sein Kant: Ein Entgegnung* (Leipzig, 1870); Kuno Fischer, *Anti-Trendelenburg* (Jena, 1870).

larger issues of the principles of knowledge. For Cohen, Kant provided a systematic method for philosophizing, and it is the systematic nature of this method that he is eager to preserve. As his thinking progresses, he begins to conceive of the theory of knowledge as something that establishes the underpinnings of science, but by contrast with those who sought logical underpinnings of the sciences, Cohen combines an exploration of the history of the sciences, probing the source of their success, everything resting on a fixed notion of their hierarchy. Starting from the natural sciences, the aim is to discover what gives scientific theories their unity and coherence, reconstructing and analysing them so as to identify the conceptual basis on which we are able to group phenomena together and unify them under general causal laws. These general causal laws are mathematical and logical, and they display the 'presuppositions and foundations' inherent in the facts of science. Cohen sees the uncovering of the basic philosophical principles underlying science not merely as a reflection on science but as contributing to its progress, by clarifying the relation between science and mathematics and identifying their common goals. Such an approach involves a historical element, however. As Lydia Patton notes:

Cohen's approach to the history of science is to identify those mathematical relations embedded in a theory that determine a domain of objects or facts, and to analyze how the theory fits into the evolving structure of scientific explanation. A philosophical history of science will evaluate a single theory by identifying the mathematical relations embedded within it. To evaluate more than one theory, especially in the case of a conflict, Cohen argues that we should adopt an analysis by means of philosophical principles that can assess how the theory contributes to our progressive promotion of the methodological and explanatory goals of the sciences.[43]

In one of his last pieces, his Introduction to Lange's *History of Materialism*,[44] Cohen asks 'what is science?' and 'what is cognition?'. Answering the first question, he supplements the examination of the facts of science as revealed by its history with the provision of a philosophical grounding for science. He then links this to the second question, arguing that providing a philosophical grounding for science is the same as providing a philosophical grounding for cognition. Philosophy, in the form of the new Kantian-inspired discipline of epistemology, and science are now part of the same endeavour. In sum, for Cohen, the natural sciences are no less constitutive of knowledge than they were for the scientific materialists, but there is nevertheless a role for philosophy, in the form of epistemology (*Erkenntniskritik*—'the synthetic principles or foundations of knowledge, from which science builds itself and from which its validity derives'—rather than *Erkenntnistheorie* in Cohen's preferred terminology), namely

---

[43] Lydia Patton, 'Hermann Cohen's History and Philosophy of Science', unpublished PhD dissertation, McGill University, 2004, 65–6.

[44] Hermann Cohen, *Einleitung mit kritischem Nachtrag, zur neunten Auflage von Langes Geschichte des Materialismus* (Leipzig, 1914).

that of providing a systematic understanding of the success of science, and thereby making sense of it, that is, making sense of why a scientific understanding of the world is understanding per se. 'In the final analysis', he asks, 'what does philosophy want to accomplish? It wants to represent the progressive connection of philosophical problems to the whole of human culture.'[45] This progressive connection must be one grounded in science.

Cohen arrives at this conception by means of a radical reworking of Kant. In his 1871 *Kants Theorie der Erfahrung*, he had taken issue with Kant's treatment of things-in-themselves, seeking to move them out of the realm of metaphysics into a programme of an epistemological-methodological search for scientific knowledge. By 1902, in his *Logik der reinen Erkenntnis*, the first volume of his 'System of Philosophy' (the three volumes of which mirror Kant's three *Critiques*), he developed this critique more fully. Things-in-themselves are stripped of any existence outside the mind and construed merely as objects of thought: they become regulative concepts of scientific discovery. In a 1912 paper setting out what the basic tenets of the Marburg School had been since its formation, Cohen's successor, Paul Natorp, stressed the creative function of thinking, the identification of thought with spontaneity, seeing thought as something that 'grows from an infinite basis', free from dogmatic isolation because it should 'be regarded it as an infinite task'.[46] In particular, the thing-in-itself is to be understood 'with reference to an unending progression of experience', and as a concept 'which limits experience just by its own creative law'. It is in this context that we should read Cohen's elaborate account of infinitesimals in mathematics, first developed in 1883,[47] and which in 1902 leads him to argue that infinitesimals mirror things-in-themselves, throwing light on the nature of objects of thought. Infinitesimals originate in and are generated purely by mathematical operations. They are not independent of mathematics, and provide an analogy for how things-in-themselves are not independent of thought, but internally generated limits. The consequences of this dependence on thought are drawn for Kant's distinction between sensible intuitions and concepts, both of which Kant himself insisted were needed for knowledge. But Cohen denies that sensibility inhabits a separate faculty, and it is collapsed into cognition via concepts. This denial follows from a key claim of the Marburg School, that there is nothing that is alien to thought. The consequence of losing a separate faculty of sensibility is that knowledge is no longer a relation between the mind and something independent of it: all knowledge now becomes something internal, a doctrine that Cohen calls 'critical idealism'. Truth is no longer a matter of correspondence, but becomes a

---

[45] Hermann Cohen, *Hermann Cohens Schriften zur Philosophie und Zeitgeschichte* (2 vols, Berlin, 1928), ii. 271.
[46] Paul Natorp, 'Kant und die Marburger Schule', *Kant-Studien* 17 (1912), 192–221: 199.
[47] Hermann Cohen, *Das Prinzip der Infinitesimal-Methode und seine Geschichte: Ein Kapitel zur Grundlegung der Erkenntniskritik* (Berlin, 1883).

question of internal relations. Accordingly, it is the coherence of internal relations—the unity of a philosophically construed science—in which the understanding of science now consists, and neither psychology nor ontology have any place in this understanding.

As with Mill, philosophy is completely reconfigured on Cohen's conception. And again like Mill's reformulation of philosophy, it has aspirations to offering a totalizing form of understanding. This becomes clear in Natorp, who, in a work that first appeared in 1911, following up the theme in Cohen, writes:

The task of philosophy has a necessarily most intimate relation to science, and only through it and in no other way to the totality of culture. 'Culture': Under this we subsume the entire communal work of humanity in which it produces the peculiarity of humanness itself and elevates it even further. But all such creative work is precisely a shaping, forming, creating, that is, of bringing to *unity*. No matter how these unities, at which the work of culture is directed—theory or technology, social organization or artistic creation—may be characterized, . . . their value for the totality of culture is determined precisely through their speciality: Yet all of these special unities strive concurrently and necessarily to even higher unities, that is, to ever more concentrated unities, and to an all-encompassing unity, at least according to the idea [of such a unity] . . . . Precisely because the direction towards this ultimate unity lies in the nature of all cultural activity, this creative work of culture necessarily takes the shape of science. This tendency towards science in cultural productivity as such has firstly come to a determinate consciousness of itself in the peculiar Greek people.[48]

As Lembeck, in his introduction to the modern edition of this text, remarks, it is the aim of philosophy for Natorp to inquire 'into that which constitutes the unity of the sciences as such. In this respect it is the science of science or the doctrine of science.'[49] What the philosophy of science sets out to establish, on the account of Cohen and Natorp, is the basis for the unity of science. Science must be a single form of understanding if it is to occupy the cultural role that they see for it, and if it is a single form of understanding then this is something that must be established beyond dispute. The legitimacy of science as the pre-eminent form of understanding the world, and the understanding of our place in the world—and consequently the role of science as the basis for culture, for civilization—depends on its unity, but it cannot establish this simply by relying on its own resources: for that, philosophy, qua theory of the foundations of science, is needed. Philosophy *completes* science, by establishing its unity on an a priori basis. Without

[48] Paul Natorp, *Philosophie—ihr Problem und ihre Probleme. Einführung in den kritischen Idealismus* (Göttingen, 2008), 42. Translated in Sebastian Luft, *The Space of Culture: Towards a Neo-Kantian Philosophy of Culture (Cohen, Natorp, and Cassirer)* (Oxford, 2015), 87–8.

[49] Karl-Heinz Lembeck, 'Einleitung' to Natorp, *Philosophie*, 7–21: 10. Cf. Neurath: 'If one takes the thesis seriously' (as Neurath certainly does) 'that in the field of knowledge one only has to deal with scientific statements, the most comprehensive field of statements must be that of unified science.' 'Unified Science as Encyclopedic Integration', 18.

philosophy, science would risk decomposing into an unruly mixture of disciplines (which, I have argued, is what in fact it is).

What is at stake here is clearly not just a technical question about the connections between the various scientific disciplines, but something on which rests the standing of science as the legitimate core of modern civilization. As with the earlier aspirations to unity among German scientists that we have looked at, there is a social and political edge to the Neo-Kantian aspiration to unify science. The historian of Germany, Thomas Nipperdey, has pointed out that the Marburg school set out a model for 'a political-social theory of reform, idealistic-socialistic democracy, political reconciliation and general human development'.[50] Gregory Moynahan, noting the impact of the Marburg school on the political and cultural life of the germanophone world in the first third of the twentieth century, elaborates on the point:

The key to understanding this influence was that the Marburg circle was . . . largely aimed at transforming the academic disciplines and their vast administrative application in Germany. Often mistakenly understood as focused entirely on the natural sciences, the central goal of the Marburg school under Cohen was the transformation and critique of all the sciences—including particularly the social and human sciences, or humanities— starting with the institution matrix of the university. The historical model was the influence of Leibniz and Kant on German society through the university system in its pervasive role in state and confessional administration, social organization and cameralism in the eighteenth and nineteenth centuries. In this regard, the Marburg school epitomized what Franz Ringer described as a 'Mandarin' attempt at political change from above and others have defined as 'antipolitics' that sought change through administrative activity outside of party politics.[51]

Despite their broader cultural aspirations, however, it is mathematics and the natural sciences that are doing all the work in Cohen and Natorp: the other disciplines are expected simply to follow on. Cassirer, writing in an overview of Cohen's work in a 1912 issue of *Kant-Studien*, rejects the view that Cohen's critique of knowledge is based on a 'one-sided view of mathematical theories of nature' and argues that the biological sciences, ethics, and aesthetics are equally important for him.[52] But while Cassirer is following up a theme in Cohen's late writings here, a theme that indicates a sea change in thinking about science in the first decade of the new century, it is nevertheless one that Cassirer, who has a more pluralist conception of the sciences than Cohen, was to develop, and in doing so

---

[50] Thomas Nipperdey, *Deutsche Geschichte, 1866–1918: Erste Band: Arbeitswelt und Bürgergeist* (Munich, 1998), 681.

[51] Gregory B. Moynahan, *Ernst Cassirer and the Critical Science of Germany, 1899–1919* (London, 2013), xxi. It is perhaps worth noting the popularity of the Kant Society, founded in 1904, and which, by 1912, had 400 members. The breadth of its membership is reflected in the fact that the industrialist Werner von Siemens was a life member. See Willey, *Back to Kant*, 155–6.

[52] Ernst Cassirer, 'Hermann Cohen und die Erneuerung der Kantischen Philosophie', *Kant-Studien* 17 (1912), 252–73: 256.

he would move away from what certainly had been Cohen's 'one-sided view of mathematical theories of nature', even if Cohen was moving to qualify that view.

In sum, while Mill, in responding in the way that he did to Whewell's attempt to offer philosophical considerations that constrained what science can legitimately seek to explain, might be considered to have been saving science from philosophy, Cohen by contrast was saving philosophy from science, from those who thought that there was no more to knowledge than what the sciences had to offer. But both ended up with a very similar conception of what legitimate understanding consisted in. Science, if not wholly constitutive of understanding, was at the very least the sole model for all understanding.[53] The idea of the unity of science was crucial here. Although the unity of science had no internal rationale in the natural sciences, life sciences, or human sciences, and offered no secure guidance for any of these various enterprises, the way in which all forms of understanding became effectively subsumed under scientific understanding in writers like Mill and Cohen is a development premised on an unquestioning commitment to the unity of science. They took it as axiomatic that science must exhibit a coherence that goes well beyond any connections between the various problem-solving accomplishments in which its explanatory success consisted. The unity of science here represents nothing in scientific practice itself, but is secured through the provision of philosophical foundations. But the provision of philosophical foundations is not an added extra, and it has ramifications outside questions of the unity of the sciences. Above all, it has had a lasting effect on the historical interpretation of science.

One of the problems with a philosophical model for science is the temptation to make science a form of applied epistemology: rather than incorporating philosophy into science, science is here effectively incorporated into philosophy. We can consider two Neo-Kantian readings of the history of science in this respect. The first is Alexandre Koyré's histories of the early modern physical sciences, the second Cassirer's rather more ambitious attempt to subsume the history of science under the general history of 'symbolic forms'.

Koyré's readings of the history of science have been immensely influential. More than anyone else, he was instrumental in transforming the history of science from a register of discoveries into a historiographically and philosophically informed enterprise. Although the philosopher with whom he was closest was Husserl, the phenomenological project in which Husserl engaged was one he considered scientific, which for Husserl meant that it had to be timeless,

---

[53] This development, dating from the middle decades of the nineteenth century, marks the culmination of an abrupt shift in the standing of science which, as I indicated in the Introduction, emerged in the late eighteenth and nineteenth centuries. It persisted well into the twentieth century, with Quine, the dominant voice in analytic philosophy mid-century, summing up the approach with his statement that 'philosophy of science is philosophy enough'. Williard van Orman Quine, 'Mr Stawson on Logical Theory', *Mind* 62 (1953), 433–51: 446.

i.e. ahistorical. By contrast, Koyré's project is on the face of it thoroughly historical, and as such it overlaps significantly with Cohen's approach. But on closer examination it turns out that the historical development of science is simply a manifestation of something that, on Koyré's reading, is essentially unchanging: an a priori epistemology of the most timeless Platonist kind. Koyré's 'histories' are in fact essentially Neo-Kantian in the way that science comes to be treated as if it were a manifestation of philosophy. The route from the one to the other is mathematics, and there is nothing complex or subtle about the transition. It is presented simply as if it were just a matter of fact: 'Motion obeys a mathematical law. Time and space are bound together by the law of number. Galileo's discovery transforms the failure of Platonism into a victory. His science is the vindication of Plato.'[54]

In Koyré's account of Galileo's treatment of falling bodies, early modern science becomes the triumph of a Platonist epistemology of mathematical ideal-izations over an Aristotelian empiricism.[55] Koyré goes to great lengths to deny that Galileo relied on empirical and experimental research in developing his account of falling bodies, arguing that he did not carry out, and given the technology available to him could not have carried out, the experiments he described. Yet in fact, he could and did carry out these experiments.[56] Koyré's attempt to strip back mechanics to something purely conceptual is part of the project of reducing it to philosophy. But Galileo actually puts the objection that he just dealt with mathematical idealizations into the mouth of one of his interlocutors, and in a remarkably sophisticated response to this he provides a detailed demonstration of just what is needed to show how a law that describes a situation that never occurs, the free fall of bodies in a void, can be related to experimentally discovered results about fall in a resisting medium, on condition that one distinguishes various factors—specific weight from absolute weight, the buoyancy effect of the medium from the friction effect—and that one is able to compare results concerning the speeds of several bodies of different specific weights across several media, rather than just the speeds of several bodies in one medium, or one body in several media. Galileo succeeds in providing an experi-mental path into the problem of free fall. Indeed, his account of free fall is the first example of thinking through a complex physical problem in experimental terms, and his kinematics provided the model for natural philosophers such as Huygens and Newton.

In the case of Koyré's no less influential reading of Descartes, a different strategy is at play, one that makes Descartes' denial of the existence of a void an

[54] Alexandre Koyré, *Études galiléennes* (Paris, 1939), 290.

[55] See ibid.; also Alexandre Koyré, 'A Documentary History of the Problem of Fall from Kepler to Newton', *Transactions of the American Philosophical Society* 45 (1955), 329–95.

[56] See Thomas B. Settle, 'An Experiment in the History of Science', Science 133 (1961), 19–32; and Stillman Drake, *Galileo Studies* (Ann Arbour, 1970).

exclusively philosophical issue at the centre of the Cartesian programme.[57] Descartes did provide philosophical arguments against the existence of a void, but well after he had devised an account of dynamics on a hydrostatic basis (to which he adheres throughout his career), where the relevant issue is that understanding the dynamics of bodies is a matter of understanding their interaction with the fluids surrounding them: there is no dynamics of isolated bodies for him, no motion or rest except in relation to contiguous matter. Thinking of dynamics as an extension of hydrostatics allows him access to sophisticated physical and mathematical resources, unlike any other approach of the time. The reasons why there cannot be a void have a purely physical basis, whatever the philosophical rationalization that Descartes subsequently provides years later in the context of a philosophical reformulation of his physical system.[58]

The framing of the term 'the Scientific Revolution', in its modern sense, is due to Koyré, and in discussing his *Études galiléennes*, one commentator has noted that the term is confined here to the activity of no one else but Galileo and Descartes.[59] Koyré's approach, reminiscent of Cohen's epistemological rationalization of science, is, for all its sophistication, ultimately reductive:[60] different approaches to natural science are just expressions of different epistemologies, and once one has discovered the true epistemology—an a priorist one based on mathematics for both Cohen and Koyré—one can rewrite the history of successful natural science in terms of it. The approach is sweeping in its implications. As Floris Cohen notes, Koyré 'simply excluded the life sciences from the Scientific Revolution *as conceived by him* because, being inherently qualitative, their concepts were doomed to remain the nonmathematical ones of Aristotelian logic'.[61]

Cassirer's 'philosophy of symbolic forms', by contrast, set out to account for the position and standing of science within a range of cognitive activities, including art, myth, and religion.[62] Philosophy, Cassirer came to think, was too limited as a

[57] As early as his 1922 *thèse complémentaire*, Koyré insists that Descartes should be taken as a philosopher not a 'physicist': see, for example, his criticisms of Gilson's reading of Descartes' physics: *Essai sur l'idée de Dieu et les preuves de son existence chez Descartes* (Paris, 1922), 5–6.

[58] For the detail of the Galileo and Descartes cases see Gaukroger, *The Emergence of a Scientific Culture*, 403–20.

[59] H. Floris Cohen, *The Scientific Revolution: A Historiographical Inquiry* (Chicago, 1994), 78.

[60] See Gaukroger, 'Alexandre Koyré and the History of Science as a Species of the History of Philosophy: The Cases of Galileo and Descartes', in Raffaele Pisano, Joseph Agassi, and Daria Drozdova, eds, *Hypotheses and Perspectives in the History and Philosophy of Science: Hommage to Alexandre Koyré 1964–2014* (New York, 2017), 179–87.

[61] Cohen, *The Scientific Revolution*, 83. Cf. Catherine Wilson, *The Invisible World: Early Modern Philosophy and the Invention of the Microscope* (Princeton, NJ, 1995), who offers detailed argument against the view that seventeenth-century natural philosophy comprised simply physical science and the laws of motion.

[62] Ernst Cassirer, *Philosophie der symbolischen Formen. Erster Teil: Die Spracht* (Berlin, 1923); Ernst Cassirer, *Philosophie der symbolischen Formen. Zweiter Teil: Das mythische Denken* (Berlin, 1925), Ernst Cassirer, *Philosophie der symbolischen Formen. Dritter Teil: Phänomenologie der Erkenntnis* (Berlin, 1929). Translated as *The Philosophy of Symbolic Forms* (3 vols, New Haven, CT, 1955).

Plate 1. Adriaen van Ostade, 'The Alchemist', 1661
Heritage Image Partnership Ltd/Alamy Stock Photo

Plate 2. Joseph Wright of Derby, 'The Alchemist', 1771
ART Collection/Alamy Stock Photo

Plate 3. First World War US Army Recruitment poster

Plate 4. Albert Robida: Leaving the Opera in 2000 (1902)

Plate 5. The Triumph of a Scientific Culture: The Chicago World Fair, 1933

Plate 6. Frontispiece to J. G. Wood, Common Objects of the Country (1858)

RHYNCOTA. ORTHOPTERA.

Plate 7. Frontispiece to Rev. W. Houghton, Sketches of British Insects (1875)

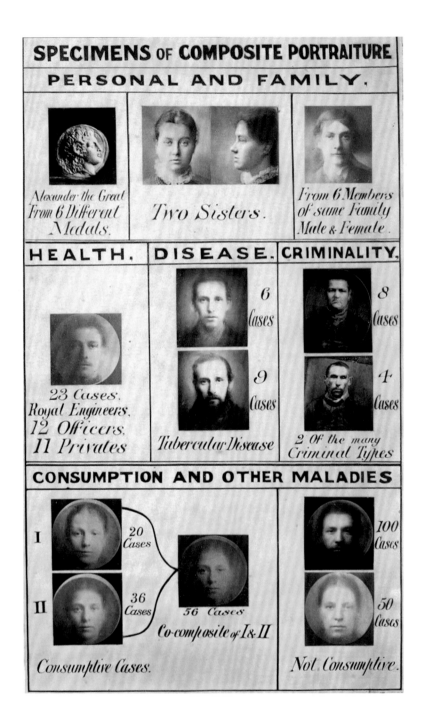

Plate 8. Galton, Frontispiece to Inquiries into the Human Faculty (1883)

form of meta-reflection on science, and had to encompass the full spectrum of human experience, which as well as science included the language, myths, and artistic activities through which human beings came to terms with the world: it had to move from a critique of reason to a critique of culture. These activities are all social constructions on Cassirer's reading and are accounted for in terms of the Kantian doctrine that the world is intelligible to us only because we first subject it to a form of pre-conceptual structuring. Consciousness, he writes, does not just receive impressions from outside but rather each of them is embedded in freely expressed activity. Consequently, 'in what we call the objective reality of things we are confronted with a world of self-created signs and images'.[63] The symbolic expressions are manifestations of the human capacity for working on this pre-conceptual structuring to yield world-formation. Pre-conceptual as the initial structuring may be, however, this world-formation is ultimately regulated by a rational form of conceptual structuring. Religion and art develop from myth, while science develops from language. The form of symbolic meaning in myth is 'expressive' whereas that in language is 'representative'. The last filters out of expressive symbols a world of stable and distinguishable substances. Finally, a third symbolic function, the 'significative', is introduced, and this lies at the basis of the highest form of understanding, science. Science for Cassirer consists in the manipulation of purely relational concepts such as the mathematical abstract notions of space and time: it is a pure system of formal relations. Accordingly, the development of the diverse forms of symbolic expressions turns out to be linear and stadial, where mythical understanding ultimately culminates in scientific understanding, which is the paradigm case of the imposition of a rational conceptual structuring on the world. As Cassirer puts it in the Preface to the second volume of *Philosophie der symbolischen Formen*:

What is commonly called the sensory consciousness, the content of the 'world of perception' . . . is itself the product of abstraction, a theoretical elaboration of the 'given'. Before self-consciousness rises to this abstraction, it lives in the world of mythical consciousness, a world not of 'things' and their 'attributes' but of mythical potencies and powers, of demons and gods. If then, in accordance with Hegel's demand, science is to provide the natural consciousness with a ladder leading to itself, it must first set this ladder a step lower. Our insight into the development of science—taken in the ideal, not temporal sense—is complete only if it shows how science arose in and worked itself out of the sphere of mythical immediacy and explains the law and direction of this movement.[64]

But if science is the upshot of the development of symbolic forms, then on this account it brings a world devoid of colour and meaning, and cannot wholly displace myth in a well-balanced society. In this respect, Cassirer seeks to avoid

[63] Ernst Cassirer, 'Der Begriff der Symbolischen Form im Aufbau der Geisteswissenschaften', *Vorträge der Bibliothek Warburg 1921/22* (!923), 11–39: 15.
[64] Cassirer, *The Philosophy of Symbolic Forms*, ii. p.xvi.

both a myth-inspired *Lebensphilosophie* and a science-inspired positivism.[65] To appreciate the importance of this, we need to turn to a question that had always been a sticking point in disputes among the Neo-Kantians: the human sciences.

THE HUMAN SCIENCES

From Whewell onwards, the understanding of human capacities and behaviour posed a persistent problem for conceptions of the unity of science. Both Whewell and the materialist physiologist group that included Helmholtz, du Bois-Reymond, and Virchow, conceived of the unity of science in such a way that a separate realm was postulated for social, historical, and human phenomena. Whewell sought to exclude the 'moral sciences' from a unified hierarchy of the natural sciences in the confidence that they could be dealt with in terms of a Christianized philosophy, and he explored them in terms of a divinely ordained and Platonically inspired realm of moral archetypes instituted by God. This was a somewhat rearguard defence, a natural theology much diminished—like that of the contributions of his co-authors of the *Bridgewater Treatises*—from its form in earlier centuries, and now reduced to attempting to reconcile a doctrinally fixed Christianity with a continuously developing science. Du Bois-Reymond et al., by contrast, ended up simply sidelining these questions, maintaining that, if we were to remain within the realm of reason, we had to confine ourselves to the natural sciences. Here there was a separate realm, but whether it was conceived in Christian or secular terms was really beside the point as it was unknowable in principle.

Other approaches to the question took an unmistakably secular form. In proceeding in this way, however, the problem they faced was that, in secular terms, science had effectively occupied the explanatory gap left by religion, so it was unclear to what extent there could be any form of understanding that was not scientific. For the 'scientific materialists' this meant that the whole exercise could be confidently collapsed into the natural sciences, but this was an article of faith rather than a research programme. On the other hand, as far as the Marburg school is concerned, it is not clear that Cohen's attempt to reconfigure philosophy as metascience could contribute anything to understanding how the human and the natural sciences were connected.

By contrast with these, we can identify a number of approaches that, working to a greater or lesser degree within the idea of the unity of the sciences, did not shirk from confronting the questions head on. Two of these we have already considered. Comte's conception of the unity of science placed the human sciences firmly within a hierarchy, so that they were construed as a continuation, via

---

[65] See Skidelsky, *Ernst Cassirer*, 122–7.

biology, of the natural sciences, and indeed, the human sciences stood at the head of the unified system, and were treated by Comte as its culmination. Mill, by contrast, rejected both the single hierarchy of Comte and the kind of bifurcation that Whewell had demanded. Although initially influenced by Comte, in his explicit engagement of questions in ethics he abandoned the idea of a single hierarchy of understanding because, as we have seen, he believed that, while objectivity requires that all forms of enquiry must appeal to external quantitative standards, natural science and ethics appeal to different external standards: inductive evidence and observable consequences respectively. The resources they draw upon are different, so there is no question of reduction (or hierarchy), but there is nevertheless a unity of science, because they are subject to the same methodological constraints. Reasoning in the moral and economic realms is not different, at a methodological level, from reasoning in the natural sciences, and indeed the latter is very much a model for the former.

If Mill's approach can be considered to require a little flexibility in what was understood by the unity of science, that of the third approach, the Southwest German form of Neo-Kantianism, of which we can take Windelband as representative, stretched the notion considerably further. The key was the defence of philosophy. Cohen, as we have seen, had defended the autonomy of philosophy against the scientific materialists. Windelband defended it against Zeller's claim that philosophy should adopt the method of the physical sciences.[66] Here Cohen and Windelband are at one. Both are concerned to establish autonomy for philosophy in a distinctively Neo-Kantian genre. But once we go beyond this, significant differences emerge.

Windelband distinguishes between the natural sciences, which establish laws about what is the case, and philosophy, which he argues is a normative realm, determining standards of judgement for what ought to be the case.[67] In an 1882 essay on norms and laws of nature, Windelband argues that these are two activities which treat the object from different perspectives, and so are complementary.[68] There is clearly a significant difference from Cohen here. Cohen is as concerned about reductionism as Windelband, and both wanted philosophy to 'stand over' the sciences as it were, but, methodologically, what Cohen was proposing brought philosophy and the natural sciences very close together, whereas Windelband sees the only road to autonomy to lie in a methodological separation. The empirical sciences that fell outside the traditional natural sciences also play a more central role in Windelband than in Cohen, which is not surprising since it was crucial for

[66] Eduard Zeller, *Über Bedeutung und Aufgabe der Erkenntniss-Theorie. Ein akademischer Vortrag* (Heidelberg, 1862); Wilhelm Windelband, *Geschichte und Naturwissenschaft: Rede zum Antritt des Rectorats der Kaiser-Wilhelms-Universität, Strassburg, geh. am 1. Mai 1894* (Strassburg, 1894).

[67] See his 1882 essay, 'Was ist Philosophie?', in Wilhelm Windelband, *Präludien: Aufsätze und Reden zur Philosophie und ihrer Geschichte Band I/II* (Tübingen, 1924), i. 1–54.

[68] 'Normen und Naturgesetze', *Präludien*, ii. 59–98: 67. See the discussion in Beiser, *The Genesis of Neo-Kantianism*, 502–6, 522–5.

Windelband's project that he distinguished the empirical discipline of psychology from philosophy (in the Kantian understanding of the term). He sets out a detailed characterization of these sciences in terms of a distinction between nomothetic sciences (those which obey general laws) and idiographic ones (which deal with phenomena which are individual and may not be lawlike): psychology, for example, he characterizes as nomothetic, history as idiographic.

The division that Windelband introduces here is problematic however. The idiographic disciplines, in virtue of their non-lawlike character, would seem to be distinct from the sciences proper, yet Windelband's binary distinction requires him to place the human sciences on the science side of the science/philosophy divide. History was a particularly contentious area. In a Comtean vein, Buckle had set out to write a history on 'scientific' principles, offering a comprehensive account of the history of the world on a scientific footing, in which it could be seen to follow laws of historical development in the same way that the motions of the planets are subject to Newton's laws of motion, a project that was much admired but also much criticized. Among the staunchest critics was the German historian Johann Droysen. Droysen's historiographical criticism of Buckle's application of the method of the empirical sciences to history distinguished between three forms of understanding: philosophical understanding (*erkennen*), which was knowledge based on first principles; historical understanding (*verstehen*), which consists in translating something into terms can be understood by someone not necessarily participating in the events; and scientific understanding (*erklären*), which involves subsumption under a mathematical law.[69] Here Windelband's binary distinction is expanded into a tripartite one. Dilthey, drawing on Droysen, reduced his tripartite distinction back into a binary one, but it separated out, not the natural sciences and philosophy, but the natural sciences, *Naturwissenschaften*, which seek to explain (*erklärten*), and the human sciences, *Geisteswissenschaften*, which seek to understand (*verstehen*).[70] The standing of the human sciences thus comes to provide the key to understanding not only the relationship between the sciences and philosophy, but also more generally it generates questions that any metascientific project must now answer. In the context of Neo-Kantianism, this is true not just of Windelband and the members of the Southwestern school, but also of later developments in the Marburg school. Cassirer, for example,

---

[69] Johann Gustav Droysen, *Grundriss der Historik* (Leipzig, 1868). See Frederick Beiser, *The German Historicist Tradition* (Oxford, 2010), 298.

[70] Wilhelm Dilthey, *Einleitung in die Geisteswissenschaften: Versuch einer Grundlegung für das Studium der Gesellschaft und der Geschichte* (Leipzig, 1883). The term *Geisteswissenschaften* derives from Schiel's 1849 German version of Mill's *System of Logic*, where it is offered as the translation of 'moral sciences'. These questions came to play a significant role in Gadamer's work: see the discussion in Hans-Georg Gadamer, *Truth and Method* (New York, 1982), 5–39, and Hans-Georg Gadamer, 'Philosophy or Theory of Science?', in his *Reason in the Age of Science* (Cambridge, MA, 1983), 151–70.

attempted to offer an account of the relation between the natural and the human sciences in his 'philosophy of symbolic forms'.[71]

Here we have the first of two challenges to the standing of scientific explanation. The basic claim raised by proponents of the distinction between *Naturwissenschaft* and *Geisteswissenschaft* is that the kinds of understanding that we seek in accounting for human behaviour are different from those we seek in the natural sciences, because the kind of thing that we want to understand is different. One way in which this contrast can be expressed is in terms of the distinction between reasons and causes: giving the reasons someone has for doing something—interpreting their behaviour—and giving the causes of their behaviour are two different things. The difference is that between interpretation of the behaviour and explanation of it. The former has to capture how the actors themselves conceive of what they are doing, whereas the latter does not. While it might be appropriate to 'stand back' from phenomena in the natural sciences to achieve objectivity, this is inappropriate in the case of the human sciences, where we are dealing not with an objectified realm, but with human beings who have intentional states, emotions, the ability to exercise judgement, etc. These are attributes that they share with the investigator, and which the investigator is therefore in a position to interpret and make sense of. This is clearly something quite different from what we do in the natural sciences. The argument is that the natural sciences have been wrongly taken to provide a model of objectivity per se, something that can simply be exported to any other area of study. This is not to deny that objectivity in the sense of impartiality is appropriate in the human sciences. Rather, what is at issue is whether criteria of objectivity, or guides as to how objectivity might be achieved, can simply be imported from the natural sciences into cases where it is not a question of discovery of causes but the interpretation of behaviour.[72] Mill's proposal for the unity of the sciences, for example, is clearly in jeopardy if such arguments go through. But so too is not only the Marburg account, but Cassirer's extension of it to cover 'symbolic forms' other than science, because these ultimately converge on science, conceived on the model of the natural sciences.

What the arguments reveal is that there are different ways of accounting for phenomena, depending on the nature of the phenomena in question. What they do not tell us is why there should only be two types of account, one for natural and one for human phenomena. This simple bifurcation is unwarranted. In particular, the arguments assume that there is a single kind of approach to explanation in the natural sciences: we have questioned whether this is the case (and will return to it as we explore the source of the 'unruliness' of explanation in the physical sciences further in Chapter 10). Moreover, it seems at least equally

[71] See the discussion in Moynahan, *Ernst Cassirer*.
[72] See the discussion in Stephen Gaukroger, *Objectivity: A Very Short Introduction* (Oxford, 2012), ch. 8.

unlikely that psychoanalysis, history, and social anthropology, for example, will share some basic (and non-trivial) feature common to the *Geisteswissenschaften*. The contrast between the natural and the human sciences simply brings this variety to light in a particularly striking way. It is not constitutive of the varieties of explanation, nor does it exhaust them. In fact it has every appearance of regimenting modes of explanation and interpretation along the lines of a rigid distinction between the mental and physical.

THE 'CRISIS OF THE EUROPEAN SCIENCES'

This brings us to the second challenge to the standing of scientific explanation, where what is in question is not the scope of scientific explanation, but whether science has the legitimacy claimed by its advocates. This reflects a still deeper division than that between the Marburg and the Southwestern approaches which we have just looked at, and it lies not in the distinction between *Naturwissenschaft* and *Geisteswissenschaft*, but that between *Naturwissenschaft* and *Wissenschaft*, a far more general and comprehensive notion of understanding, in which *Naturwissenschaft* had no automatic priority over historical and other forms of knowledge.[73]

The concern with *Wissenschaft* as the core of philosophy can be traced back to the late eighteenth and early nineteenth centuries, to Herder and above all to Wilhelm von Humboldt's conception of the relation between the natural sciences and human culture, as manifested in his work in history and linguistics.[74] Particularly with the establishment of the University of Berlin in 1810 on Humboldt's model, *Wissenschaft* was conceived as critical reflection on any kind of cognitive endeavour, and the social, cultural, and natural sciences were explicitly seen as the means of transforming society and developing its confessional and political structure.[75] As Merz writes, *Wissenschaft* 'applied alike to all the studies which are cultivated under the roof of the alma mater; it is an idea specially

---

[73] The terminology is not fixed, however. In an early criticism of Southwestern Neo-Kantianism, the Marburg Neo-Kantian Alois Riehl distinguishes between *Wissenschaft* (science) and *Weltanschauuung* (worldview) in a way that mirrors the distinction between *Naturwissenschaft* and *Wissenschaft*. Alois Riehl, *Über wissenschaftliche und nichtwissenschaftliche Philosophie: eine akademische Antritte* (Freiberg, 1883).

[74] Wilhelm von Humboldt, 'Über die Aufgabe des Geschichtsschreibers' [1821] in Wilhelm von Humboldt, *Gesammelte Schriften* (17 vols, Berlin, 1903–1936), iv. 35–56. Wilhelm von Humboldt, *Über die Kawi-Sprache auf der Insel Java, nebst einer Einleitung über die Verschiedenheit des menschlichen Sprachbaues und ihren Einfluss auf die geistige Entwickelung des Menschengeschlechts* (3 vols, Berlin, 1836–9). See Peter Hans Reill, 'Science and the Construction of the Cultural Sciences in Late Enlightenment Germany: The Case of Wilhelm von Humboldt', *History and Theory* 33 (1994), 345–66.

[75] Note, however, that the University of Berlin also taught technical subjects: see Ursula Klein, *Humboldt's Preußen: Wissenschaft und Technik in Aufbruch* (Darmstadt, 2015), esp. ch. 29. See also Ursula Klein, *Nützliches Wissen: Der Erfindung der Technikwissenschaften* (Göttingen, 2017).

evolved out of the German university system, where theology, jurisprudence, medicine, and the special philosophical studies are all said to be treated "scientif-ically", and to form together the universal, all-embracing edifice of human knowledge'.[76] Patton points out that Dilthey's version of the Humboldtian ideal of *Wissenschaft* is one of 'a collective endeavour to investigate not only knowledge, but also human experience and the meaning of human action in history'.[77] What is distinctive about Dilthey's hermeneutic conception is his rejection of the search for an a priori ground for our knowledge of human culture. Knowledge is always bound up with the culture that it attempts to know: there is no absolute starting point for knowledge. In a Kantian context, the focus is on metaphysics rather than hermeneutics, however, and the distinction between *Naturwissenschaft* and *Wissenschaft* now effectively becomes transformed into that between epistemology—in Neo-Kantian terms a metatheory of science— and metaphysics respectively. The contrast with the Marburg School is nowhere more evident than in Rickert's statement that the importance of Kant lies not in his treatment of the empirical sciences but in his treatment of the problems of metaphysics, which 'provide the foundation for an all-encompassing theoretical worldview that culminates in the handling of issues in the philosophy of religion. The theory of mathematics and physics is merely preparatory to the treatment of these questions.'[78]

It is instructive to note here that the Southwestern School found a place for metaphysics in Kant's insistence on the importance of the distinction between practical and theoretical reason, on the grounds that morality, which for Kant is a form of practical reasoning, could not be subordinated to epistemology and required a metaphysical understanding.[79] At this point the two forms of Neo-Kantianism come apart at the seams, for the acknowledgement of a need for metaphysics, something anathema not just to the Marburg School but to the whole of pre-war Neo-Kantianism, where epistemology is constitutive of philoso-phy, shifts Kantianism in a completely different direction. The parameters of the debate remain Kantian, and there had indeed been a metaphysical strain within Kantianism from the mid-nineteenth century, with the publication of a new second volume in the second (1844) edition of Schopenhauer's *Die Welt als Wille und Vorstellung*. The first edition (1819) had had no impact and Schopen-hauer reworked and expanded it twenty-five years later to great effect, as it became influential in its rejection of the language of science in favour of more literary genres, but particularly for its existential pessimism, something directly contrary

[76] Merz, *A History of European Thought*, i. 170.

[77] Lydia Patton, 'Methodology of the Sciences', in Michael Forster and Kristin Gjesdal, eds, *The Oxford Handbook of German Philosophy in the Nineteenth Century* (Oxford, 2015), 595–606: 603.

[78] Heinrich Rickert, *Kants al Philosoph der moderner Kultur. Ein geschichtesphilosophischer Versuch* (Heidelberg, 1924), 153.

[79] See Gordon, *Continental Divide*, 58–60.

to ethos of the Marburg school. The reading of Kant as primarily a metaphysician was gradually built up from the later decades of the nineteenth century, in writers such as Alois Riehl, Friedrich Paulsen, and Nicolai Hartmann.[80]

The epistemology/metaphysics division, in the form it takes in this reworking of Kantian questions, is a deep one, and it comes to a head decades later, in 1929, in the famous 'working session' (*Arbeitsgemeinschaft*) at Davos with Cassirer and Rickert's one-time pupil, Heidegger. As Peter Gordon has noted, both Heidegger and Cassirer, in their different ways, saw their work as emerging from and being responsive to 'the crisis of the modern condition', and 'they found themselves locked in a deep and ongoing philosophical discussion that returned again and again to a single question: *What is it to be a human being?*'[81] Intimately tied up with this question was the nature and extent of scientific understanding: whether scientific understanding was the route to our more general engagement with the world, or whether it blocked any such engagement. The way the issues were dealt with at Davos was one that hinged on the interpretation of Kant, on Kant as epistemologist versus Kant as metaphysician,[82] on a cosmopolitan versus a distinctively German Kant. But before we look at the Kantian formulation, it will be helpful to pose the question in broader terms that show its connection with issues that we have dealt with in earlier volumes.

What is fundamentally at stake in the dispute is the relation between propositional understanding—knowledge that something is the case, typically and paradigmatically taking the form of science—and non-propositional understanding, which takes the form of an engagement with the world in terms of raw beliefs, fears, hopes, aspirations, plans, anxieties, concerns, expectations, commitments, and so on. The issues are whether propositional understanding is the only genuine form of understanding, and if not whether propositional and non-propositional forms of understanding are independent of one another. From the Enlightenment onwards, there was much dispute about these questions.[83] The translation of the questions into Neo-Kantian terms raises deeper and more complex problems. Simplifying considerably, these are whether (as Cassirer

[80] Alois Riehl, *Der philosophische Kritizismus, Geschichte und System* (Leipzig, 1878); Friedrich Paulsen, *Immanuel Kant, sein Leben und seine Lehre* (Stuttgart, 1898); Nicolai Hartmann, 'Diesseits von Idealismus und Realismus. Ein betrag zur Scheidung des Geschichtlichen und Übergeschichtlichen in der Kantischen Philosophie', *Kant-Studien* 29 (1924), 160–206. See Rudolph Malter, 'Main Currents of the German Interpretation of the *Critique of Pure Reason* since the Beginnings of Neo-Kantianism', *Journal of the History of Ideas* 42 (1981), 531–51: 544–6.

[81] Gordon, *Continental Divide*, 4. Cf. Michael Friedman, *A Parting of Ways: Carnap, Cassirer, and Heidegger* (Chicago, 2000). More generally, see Kurt Töpner, *Gelehrte Politiker und politisierende Gelehrte: die Revolution von 1918 im Urteil deutscher Hochhschullehrer* (Göttingen, 1970).

[82] Among the questions on which this issue turns was that of which edition of the *Kritik der reinen Vernunft* one used: Schopenhauer and Heidegger both rejected the second edition as not doing justice to the aims of the first edition. See Malter, 'Main Currents of the German Interpretation of the *Critique of Pure Reason*', 547–8.

[83] See, for example, Gaukroger, *The Natural and the Human*, ch. 1.

believed) propositional understanding can be radically revised so as to include what had traditionally been taken to be non-propositional forms, without losing the centrality of scientific understanding; or whether (as Heidegger believed) non-propositional understanding is not just our only possible starting point in engaging with the world, but our only means of engaging with the world per se. This dichotomy determines our mode of investigating the world and our place in it: through suitably reformulated epistemology in the former, and radically reformulated metaphysics in the latter. And these different modes of investigation offer different ways of coming to terms with the world, the one through rational enquiry, the other through what can be described as a form of philosophical contemplation.

The argument, as I've indicated, is formulated in Kantian terms. Kant had argued that sensibility must precede the application of the categories to experience. There must be 'an orientation in thought', on a par with our orientation in space and time, which provides us with a starting point as it were, from which our understanding of the world begins. This starting point is prior to the application of the categories which we use to come to an understanding of the world. The categories are the work of reason. But this does not mean that our orientation in thought, subsequently intimately connected with sensibility, is pre-rational, or not rational in some other way. Kant himself does not provide a definitive answer, though he proceeds as if it were guided by reason. Cohen offers a clear-cut solution, ridding sensibility of any separate role, and subjecting all aspects of experience to the dictates of reason. Such a move makes epistemology the sole arbiter of experience, and, as we have seen, epistemology is essentially a metatheory of science. But as Cassirer realized, and as Cohen in his last work was beginning to realize, to restrict the defence of the rule of reason and cosmopolitanism to the defence of science is too narrow. Cassirer, in his 'philosophy of symbolic forms', therefore attempted to include a wide range of attempts to come to terms with the world, from myth and religion through aesthetic forms, which he brought under a general theory. The general theory ended up as a genealogy of reason however, with the various forms of engagement with the world following a historical sequence and gradually being superseded, so that what we end up with is science at the pinnacle. Although the exercise is supposed to show how non-propositional forms of understanding can be accommodated to a general model, and despite not only Cassirer's insistence that no one region of human experience could claim priority over any other but also his emphasis on the aesthetic dimension of Kant's work, it becomes an exercise in the progress of reason, and the progress of reason could have no other outcome but science. Both critics and supporters of Cassirer had problems reconciling the idea that myth, for example, was autonomous on Cassirer's conception of science. His student Joachim Ritter, reviewing *Philosophie der symbolischen Formen* in 1930, writes that the standing of 'the theory of knowledge and the logic of science as the fundamental philosophical discipline has now been put into question, or at the very least has become problematic'.[84] Why then, we might ask, should Cassirer have tried to incorporate everything within propositional understanding? Why could he not allow that there are legitimately non-propositional ways of engaging with the world: 'legitimate' in that they cannot be judged by the standards of science, or treated as if they must ultimately converge on scientific understanding? The answer is that Cassirer's is ultimately an Enlightenment project, more specifically a Kantian Enlightenment project: reason overcomes bigotry and prejudice, it secures cosmopolitan values, and in the form of science it provides the motor of civilization. To sacrifice this project would in effect have been to sacrifice civilization for Cassirer.

Before the Great War, such a view would have been widely shared in the European intellectual and cultural community, but in its wake, things had changed, and the viability of the Enlightenment project was being seriously questioned. Husserl is a case in point. Spiegelberg writes that, before and during the war,

a philosophy aiming at *Weltanschauung* seemed to Husserl incompatible with the objectives of philosophy as a rigorous science . . . . But this whole situation changed for Husserl after the First World War . . . During the War itself Husserl, who lost a brilliant son in action, had refrained deliberately from taking an active part by writing or speaking for the war effort. But in the aftermath he found it impossible to stay aloof from the questions of the day. Now the incapacity and unwillingness of science to face problems of value and meaning because of its confinement to mere positive facts seemed to him to be at the very root of the crisis of science and of mankind itself.[85]

In a 1935 lecture on 'Philosophy and the Crisis of European Humanity' for example, Husserl registers his disquiet at the way in which things were turning out in the post-war decades:

There are only two escapes from the crisis of European existence: the downfall of Europe in its estrangement from its own rational sense of life, its fall into hostility toward the spirit and into barbarity; or the rebirth of Europe from the spirit of philosophy through a heroism of reason that overcomes naturalism once and for all. If we struggle against this greatest of all dangers as 'good Europeans' with the sort of courage that does not fear even an infinite struggle, then out of the destructive blaze of lack of faith, the smoldering fire of despair over the West's mission for humanity, the ashes of great weariness, will rise up the phoenix of a new life-inwardness and spiritualization as the pledge of a great and distant future for man.[86]

[84] Joachim Ritter, 'Ernst Cassirers Philosophie der symbolischen Formen', *Neue Jahrbücher für Wissenschaft und Jugendbildung* 6 (1930), 593–605: 593.

[85] Herbert Spiegelberg, *The Phenomenological Movement: A Historical Introduction* (2nd edition, 2 vols, The Hague, 1971), i. 80. For a clear statement of Husserl's earlier view see his 'Philosophie als strenge Wissenschaft', *Logos* 1 (1910), 289–314.

[86] Edmund Husserl, *The Crisis of European Sciences and Transcendental Phenomenology: An Introduction to Phenomenological Philosophy* (Evanston, IL, 1970), 299. Husserl presumably had Heidegger in mind: in January 1931 he had written to Alexander Pfänder that Heidegger 'may be

The problem was that many considered that the Great War had revealed the values that Husserl is defending to be chimerical, that they are implicated in the horrors of its scientifically generated technology of war, and the chaos that it engendered and which continued in its wake, and that these had shown the aspirations of the Enlightenment to be ineffectual and empty.[87] This was reflected in a number of publications in the aftermath of the war with 'crisis' in the title.[88] Husserl was of course aware of this and reflected that Heidegger, whose mentor he had been, 'had been driven by the war and ensuing difficulties' into 'mysticism'.[89]

Independently of the charge of mysticism, we have here the context in which Cassirer's defence of the Enlightenment project at Davos was judged. The young German philosopher Franz Josef Brecht, for example, who was in the audience at Davos, highlighted the importance of the confrontation between Cassirer and Heidegger when he wrote that it 'was the meaningful high point of the conference; for here there stood two philosophers, as representatives not merely of two philosophical dispositions that ultimately were no longer susceptible to logical discussion, but rather, at the same time, as two philosophical generations'.[90]

For Heidegger, as for Cassirer and Husserl, the 'crisis of the European sciences' is fundamentally a philosophical crisis.[91] The values of science, as articulated through epistemology, are universal cosmopolitan values. For Heidegger these values have failed, science and technology having resulted in tools of war and unthinking consumerism. He rejects not only cosmopolitan values, which reflect only a thoroughly rationalized and inauthentic view of humanity, but also their associated notions of freedom and spontaneity. The latter were integral to the Neo-Kantian assimilation of sensibility to rationality, and it was precisely this assimilation that Heidegger refuses to accept. Our orientation in the world is not provided by reason on Heidegger's account, but by a very specific worldliness and temporality, one in which reason does not provide a means of escape, or a route to cosmopolitanism. Human existence can only be considered in all its worldly

involved in the formation of a philosophical system of the kind which I have always considered it my life's work to make forever impossible.' Quoted in Gordon, *Continental Divide*, 79.

[87] Criticisms of science as 'bankrupt' had in fact started to appear as early as 1885: see Roy Macleod, 'The "Bankruptcy of Science" Debate: The Creed of Science and its Critics', *Science, Technology, and Human Values* 7:4 (1982), 2–15.

[88] Among the more prominent, for example, were Rudolph Pannwitz, *Die Krisis der europäischen Kultur* (Nuremberg, 1917); Ernst Troeltsch, 'Die Krisis des Historismus', *Die neue Rundschau* 33 (1922), 572–90; Arthur Liebert, *Die geistige Krisis der Gegenwart* (Berlin, 1924). The 'crisis' was not just seen as philosophical but equally affected perceptions of mathematics and physics: Herman Weyl, 'Über die neue Grundlagenkrise der Mathematik', *Mathematische Zeitschrift* 10 (1921), 39–79; Richard von Mises, 'Über die gegenwärtige Krise der Mechanik', *Zeitschrift für angewandte Mathematik und Mechanik* 1 (1921), 425–31; Johannes Stark, *Die gegenwärtige Krise in der deutschen Physik* (Leipzig, 1922); Joseph Petzold, 'Zur Krisis des Kausalitätsbegriffs', *Naturwissenschaften* 19 (1922), 693–5; Albert Einstein, 'Über die gegenwärtige Krise der theoretischen Physik', *Kaizo* 4 (1922), 1–8.

[89] Quoted in Gordon, *Continental Divide*, 82.   [90] Quoted in ibid., 50.

[91] Cf. Karl Jaspers, *Die geistige Situation der Zeit* (Berlin, 1931).

forms, and in so considering it what we encounter is the sheer historical contingency of our human existence. What is at issue is a secular version of the religious experience of dependency, and it is worth noting that Heidegger, after a very brief spell in a Jesuit seminary, began his academic life studying theology at Freiburg, and wrote his habilitation thesis, heavily influenced by both Neo-Thomism and Neo-Kantianism, on Duns Scotus. Scotus' conception of Aristotle's metaphysics, which is in direct opposition to that of Aquinas, is mirrored in Heidegger's conception of Kantian metaphysics. Aquinas had seen metaphysics as a neutral discipline, dependent purely on reason, which was thereby able to judge conflicts between revelation and natural philosophy. Scotus, by contrast, took metaphysics to be 'the science of being-qua-being', so included all disciplines, including theology, giving rise to a highly theologized understanding of metaphysics.[92] In *Sein und Zeit*, Heidegger tells us that 'the positive outcome of Kant's *Kritik der reinen Vernunft* lies in its contribution to the working out of any Nature Whatever, not in a "theory" of knowledge'. [93] Two years later, in *Kant und das Problem der Metaphysik*, he writes that the purpose of the *Kritik* is 'completely misunderstood, therefore, if this work is interpreted as a "theory of experience" or perhaps a theory of the positive sciences'. It has nothing to do with knowledge, but is an account of 'ontology as *metaphysica generalis*'.[94]

In place of a rationality that enables us to create the world of scientific enquiry, we find ourselves 'thrown' into a world not of our making, and which we are unable to subject to rational control. It is a world characterized by mortality, care, anxiety, temporality, and our dependence on a history for which we can claim no responsibility. For Heidegger, the world we live in has become a science-driven world increasingly ruled by barbarians who are obsessed by consumerism and who have been able to deploy technology as a means of totalitarian control, a world in which culture and metaphysics have been banished. It is the 'final age' which brings Western culture to an end.[95] The only way to prevent this state of affairs coming to fruition, he believed, was total war. Perhaps early in his career, in his student days, he might have envisaged the result of this as being a return to a Christian medievalism, but by the 1920s he thinks through the question in what at least has the appearance of increasingly secular terms, namely in terms of a return to a primordial state, which he increasingly identified with the Pre-Socratics, although his conception of them serves the role of a secular Garden of Eden, with science a form of original sin.

What is of interest here is not Heidegger's apocalyptic scheme, with its bizarre return to origins, but the way in which, in the wake of the Great War, a challenge

---

[92] See Gaukroger, *The Emergence of a Scientific Culture*, 80–6.
[93] Martin Heidegger, *Sein und Zeit* (Tübingen, 1927), 10
[94] Martin Heidegger, *Kant and the Problem of Metaphysics* (Bloomington, 1962), 21
[95] This is the theme of a later writing, his *Nietzsche* (2 vols, Pfullingen, 1961), where we are told that the whole of Western thinking from the Greeks to Nietzsche is metaphysics, and that Nietzsche is the last metaphysician, heralding the modern epoch of nihilism (i. 479–80).

can be made to the idea that it is science and technology that carry civilization forward. Heidegger's was not the only such challenge, and Nietzsche had posed the question in his own inimical way in 1872 in *Die Geburt der Tragödie*, when he asked: 'What is the significance of science viewed as a symptom of life? . . . Is the resolve to be so scientific about everything perhaps a kind of fear of, and escape from, pessimism? A subtle last resort against—truth? and, morally speaking, a sort of cowardice and falseness?'[96] But the intensity of the debates following the First World War was of a different order, and they show clearly how philosophical discussions of the standing of science not only needed to reflect the transformations of the nature of scientific activity, but also how they could not simply fall back on pre-war understandings of science in the hope that they could be reworked to meet any challenges.

The Davos debate turned on extremes—science as the motor of civilization and culture, versus science and technology as the destroyers of civilization and culture—and there is no reason why we should accept that this is a fruitful way to formulate what are undoubtedly serious issues. In particular, we must face the question of how propositional understanding, in the form of rational enquiry manifested paradigmatically in science, can bear on non-propositional understanding: not simply opposing the one to the other as the only ultimate forms of understanding, as in the Davos debate. With this in mind, we shall now turn, in Part IV, to the question of the integration of propositional and non-propositional values, approaching it from two perspectives: technology and popular science. Technology, as I shall show, has a predominantly non-propositional character (if only because it produces not knowledge but material things), and its integration with science becomes so intimate that it is impossible to deal with science without taking this into account. We then turn to 'popular science', taken in a broad sense, which we shall see is the greatest transmitter of the values of civilization, so that, for the articulation of and promotion of the values of civilization, it is not science as such that we need to explore but, again, especially in genres such as utopian science fiction, something that integrates propositional and non-propositional ways of engaging with the world. One crucial upshot of our examination of these questions will be that the distinction between propositional and non-propositional ways of engaging with the world is far from absolute, and has a fluidity that we must capture if we are to understand changes in the standing of science.

---

[96] Friedrich Nietzsche, *Basic Writings of Nietzsche* (New York, 1968), 18.

PART IV

THE PURSUIT OF SCIENCE BY OTHER MEANS: 'APPLIED' AND 'POPULAR' SCIENCE

10

Technology and the Limits of Scientific Theorizing

In this real world of sweat and dirt, it seems to me that when a view of things is 'noble', that ought to count as a presumption against its truth, and as a philosophic disqualification. The prince of darkness may be a gentleman, as we are told he is, but whatever the God of earth and heaven is, he surely can be no gentleman.

William James, 'What Pragmatism Means' (1904)[1]

The successful aeroplane, like many other pieces of mechanism, is a huge mass of compromise.

Howard Wright, 'Aeroplanes from an Engineer's Point of View' (1912)[2]

It has been an unstated premise in the kind of understandings of science that we encountered in Part III that science is an essentially theoretical enterprise. Scientific theory is taken as a given, complete in itself and autonomous. On this conception, science, in the form of scientific theory, has two wholly dependent subsidiary exercises: 'applied science', offering applications to practical questions, and 'popular science', offering simplifications for a wider public. We are now about to see that technology, broadly conceived, cannot in fact be conceived as 'applied science', and indeed is not something deriving from science. In some respects, it is independent of science, in other respects it is an integral and inseparable part of science, practically and conceptually, but in no respect is it simply an offshoot of science. In Chapter 11 we shall see that cognate considerations hold for 'popular science': in some respects it is autonomous with respect to theoretical reasoning, whereas in others it forms an integral part of the scientific enterprise. In both cases, there are deep and complex conceptual issues that bear directly on the legitimation of the scientific enterprise. Accordingly, they are relevant to how we are to conceive of the association between science and civilization. The way in which the values of a 'pure' theoretical science bear on

---

[1] William James, 'What Pragmatism Means: Lecture Two', in William James, *Pragmatism: A New Name for an Old Way of Thinking* (Buffalo, NY, 1991), 34–5.

[2] Howard Theophilus Wright, 'Aeroplanes from an Engineer's Point of View', *Aero* 6 (1912), 374–80: 374.

Civilization and the Culture of Science: Science and the Shaping of Modernity, 1795–1935. Stephen Gaukroger, Oxford University Press (2020). © Stephen Gaukroger.
DOI: 10.1093/oso/9780198849070.001.0001

the understanding of modern civilization is not, and could not be, the same as the relation between civilization and a complex—and for all intents and purposes uncontrollable—mix of technology, scientific theory, and popularization. Yet it is precisely to this mix that we must look for what drives any sense of what modern civilization owes to science.

The upshot of this is that it is not possible to consider the issue of the unity of science from the middle of the nineteenth century without raising the question of the developing relation between science and technology, or without raising the question how we assess claims about the civilizing roles of science and technology. We shall see that we cannot simply collapse technology into science, but nor can we collapse science into technology. Either of these would efface crucial differences in modalities. Not only were the claims made for the civilizing effects of science generally different in kind from those for the civilizing effects of technology, for example, but the way in which they were promoted, despite areas of overlap, differs significantly. Science can be put into opposition with religion, for example, whereas technology cannot,[3] and this becomes a very distinctive feature of the promotion of science in the nineteenth century. At the same time, we have to reconsider how metascientific assessments are generated, in particular whether, rather than scientific achievements, it may (to the extent to which we can separate these) be that it is technological successes that are shaping them. Certainly, it would seem in some cases that there is a commitment to values that have come to be associated with science, such as objectivity, meritocracy, and freedom from superstition. But while some of these, such as freedom from superstition, may be the values of science, others, such as objectivity, while they seem like the values of science turn out on closer investigation to be values of practical and experimental procedures that can equally be those of engineering and technology.[4]

At the same time, there can be no doubt that when people considered the benefits of science in the period with which we are concerned, what they invariably had in mind were not any identifiable benefits of an increase in theoretical understanding of the natural world for example, but improvements in living and working conditions which they associated with technological innovation. Yet what was explicitly promoted was 'science', and it was absolutely crucial that it was science that was given ultimate responsibility for the manifest benefits that accrued, because it was science, not technology, that was considered to exercise civilizing effects on the population. It was science, not technology, that claimed to offer a coherent ordered view of the world and our place in it. But is this separation between science and technology, with a subsidiary role for the

---

[3] In the mid-nineteenth century, Pope Gregory XVI, as part of his opposition to science and modernity, had blocked the development of railways and the introduction of gas lighting in the streets of Rome: see Evans, *The Pursuit of Power*, 151–2. Here technology is treated as a manifestation of science, however: as a symbol of science, rather than something in opposition to religion in its own right.

[4] See Lorraine Daston and Peter Galison, *Objectivity* (New York, 2007).

latter, sustainable? Can we even imagine a world in which science was pursued completely independently of technology, a world in which there was no 'application' of science at all, just 'pure' science pursued for its own sake? Such a world would seem to be that envisaged by Sarton when he writes that 'the chief aim of scientific research is not to help mankind in the ordinary sense, but to make the contemplation of truth more easy and more complete';[5] or that of Henry Carhart, professor of physics at the University of Michigan, when he claimed that 'the quality of mind that discovers laws of nature is of a higher order than that which makes application of them';[6] or that envisaged by Charles Eliot, president of Harvard University, when he claimed that the goal of science had nothing to do with its practical applications, but the fact that science 'enables and purifies the mind'.[7] How, we may ask, would this 'pure' science differ essentially from philosophy?

With the technological transformations of scientific practice from the mid-nineteenth century onwards, accompanied by technology's integration into science, the latter loses the kind of theoretical purity that many of its advocates claimed for it. In the wake of the Great War, the idea of the unity of science gets caught up in deep and intractable questions about the extent to which it is possible to devise a notion of 'pure science' that can be separated from technological and other issues. The theoretical purity of science was seen, by many philosophers of science for example, as the timeless essence of science (even if they would not have put it in quite those terms), and giving it up was met not so much with resistance as with complete denial: with the result that the legitimation of science effectively becomes a matter of assimilation to philosophy, as if science were ultimately a form of conceptual analysis and derivation from basic conceptual truths. In this way, perhaps, it could avoid the stigma of being a tool of war, the image that it took on for many in the wake of the Great War.

Given this context, we might begin to wonder whether, as at first seems to be the case, the new post-war projects for the unity of science, such as those of the Logical Positivists, were simply ignoring the developments over the previous twenty or so years. Perhaps instead they were acutely aware of these developments (how could they not have been?), and were responding by trying to repurify science. In either case, the autonomy of science had now come into question in a way that these projects gave every appearance of refusing to address. Yet the very nature of science was at issue. As early as 1900, one of the pioneers of 'technical mechanics', August Föppl, was asking whether mathematics was auxiliary (*Hilfswissenschaft*) to technology or a foundation (*Grundwissenschaft*) for it, and urging

[5] Sarton, *The History of Science and the New Humanism*, 188.
[6] Henry S. Carhart, 'The Educational and Industrial Value of Science', *Science* n.s.1 (12 April 1895), 393–402: 399.
[7] Quoted in Kevles, *The Physicists*, 24.

the former.[8] The question became an especially pressing one in the disputes over aerodynamics in the first half of the century, as we shall see below. More recently, Peter Janich has summed up what is at stake in his claims that, 'in place of the musty ideology of the researcher who unravels nature's secrets, the physicist will understand himself to have just one task: *enabling technology*',[9] and that natural science 'is to be understood as a secondary consequence of technology rather than technology as an application of natural science'.[10]

It goes without saying that the issues we encounter in exploring the relationship between science and technology depend on what we assume as the model of science. I am taking the physical sciences as the model, because this has been very much a feature of the developments that we have been concerned with. But this model of scientific understanding is not universal, and the elevation of physics to this role, especially in the mathematical form that it had taken by the nineteenth century, is possibly quite unique. Medicine has certainly had a more prominent role in Islamic and Chinese cultures, for example. I have no doubt that consideration of medicine, whose aim is a practical one, namely that of the prevention and cure of illnesses, would have given us a far more straightforward route to understanding the relationship between theoretical, experimental, and technological practices. But following such a route would not have alerted us to those peculiarities of science in Western culture, such as the concern with the unity of science, that have gone to the core of its identity as something that is to be identified with civilization. It is these features that contribute to making Western science, as we have come to understand it in the wake of the eighteenth and nineteenth centuries, such a distinctive form of practice, and it is these, rather than some generic notion of science (of the kind one finds in theories of 'scientific method' for example), that we need to understand.

In what follows, I want to explore two questions central to understanding the nature and role of technology in the nineteenth and the early decades of the twentieth century. First, there is the problem of how technology engages with science. Here I shall be arguing that, to the extent to which science and technology can be integrated, what might once have been thought of as scientific developments should in fact be conceived in terms of a mixture of theory, experiment, and theory-free invention.[11] This unstable mixture inevitably confers

    [8] August Föppl, *Vorlesungen über technische Mechanik* (vol. 1, 2nd edition, Leipzig, 1900), ch. 1.
    [9] Peter Janich, *Zweck und Methode der Physik aus philosophischer Sicht* (Konstanz, 1973), 17.
    [10] Peter Janich, 'Physics—Natural Science or Technology', in W. Krohn, E. Layton Jr, and P. Weingart, eds, *The Dynamics of Science and Technology* (Dordrecht, 1978), 3–27: 13.
    [11] On invention, I simply draw attention to the comments of Elmer Sperry, one of the most successful and prolific inventors of the turn of the century, and the father of cybernetic, or feedback control, engineering. Distancing himself from any Eureka-style picture of the inventor, Sperry lays out his completely matter-of-fact approach: 'I would study the matter over; I would have my assistants bring before me everything that had been published about it, including the patent literature dealing with attempts to better the situation. When I had the facts before me, I simply did the obvious thing. I tried to discern the weakest point and strengthen it; often this involved

on 'science' a disunified, unruly, modular character. Second, I want to look at the different values of science and engineering: at the relative standing of scientists and engineers in the nineteenth and early twentieth centuries, and differences in approach to problem solving. In particular, we shall be looking at a case where science and engineering have a problematic fit: physics and engineering approaches to aerodynamics in the early decades of the twentieth century. Here we shall see how the separation of 'pure' scientific and 'practical' engineering concerns was unable to stem this unruliness, despite the claims of a rigorous foundational approach by those in a tradition of mathematical physics. I argue that the association of science and engineering can be so close that we must take seriously the non-discursive products of science such as machines. Once we consider not just the function and construction of machinery, but also the operation of machinery, we encounter questions very different from those that concern us in the study of 'pure' science, but to which we need to be attentive.

TECHNOLOGY AND THE 'UNRULINESS' OF SCIENCE

I referred earlier to an unruliness about scientific practice, something that is at odds not only with the epistemological reading of science but with the whole idea of the unity of science. In this latter respect, it is closely associated with the modular nature of the sciences, and with a pluralism in the types of explanations offered, most obviously, but certainly not exclusively, in the life sciences. This makes it especially important that we probe the sources of this unruliness, and we can use two issues to focus the questions. The first is the internal structure of modern science, an internal structure that was formed in the seventeenth century, when it comprised speculative systematic natural philosophy, experimental natural philosophy, and quantitative natural philosophy. The second is a nineteenth-century development, the merging of science and technology. In contrast to the idea that the connections between science and technology come via 'applied science', I will be arguing that the way to make sense of this merging is by exploring the connection between a largely autonomous experimental component in science and practices of 'invention' in a broadly technological context. Whereas the 'applied science' approach leaves science untouched by technology, exploring the connections between experiment and technology allows us to understand how technology can be integrated into scientific practice in such a way that the latter is shaped structurally by technology. Moreover, it allows us to do this without reducing the one to the other, and this is particularly important, as there remains a degree of incommensurability between the production of knowledge and the

alterations with many ramifications which immediately revealed the scope of the entire project.' Quoted in Hughes, *American Genesis*, 20.

production of useful objects and processes, even if there exists a significant grey area, and perhaps even some overlap, in the middle.

The tripartite composition of science that developed in the seventeenth century—comprising speculative natural philosophy, experimental natural philosophy, and quantitative programmes (mechanics, astronomy, optics, acoustics)—has been discussed in detail in earlier volumes,[12] and we touched on it in Chapter 4, but its development in subsequent centuries needs to be noted briefly in order to orientate our discussion. Confining our attention to the physical sciences, because these formed the model of science for almost all the figures with whom we have been dealing, the first point to note is that the three components were never fully integrated. Initially, for seventeenth-century mechanists, epitomized in Descartes, speculative natural philosophy and a quantitative programme were considered to go hand in hand. An account of the size, speed, and direction of motion of the micro-corpuscles constituting material things was considered to be all that was needed for a full explanation of the physical behaviour of macroscopic bodies. Experimental natural philosophy, initially in Boyle's pneumatics and Newton's work on the formation of the optical spectrum, was regarded as falling outside the scope of natural philosophy proper because the explanations offered were phenomenal and not causal. But in the eighteenth century, things began to take a new direction as the role of mechanics was rethought. The success of Newton's *Principia* encouraged the claims of rational mechanics: an axiomatic, analytical form of mechanics, which had aspirations ultimately to cover the whole of physics, but which did not operate in terms of a reduction of the macroscopic to the microscopic realm. Micro-corpuscularianism went rapidly into decline, and with it the kind of speculative natural philosophy that had provided seventeenth-century natural enquiry with a grand sense of purpose. Experimental natural philosophy, by contrast, provided a very productive route in physics, chemistry, and the life sciences in the eighteenth century, and by the beginning of the nineteenth century it was forming a fruitful union with analytical mechanics, although both retained a significant degree of autonomy. Systematic natural philosophy, revived briefly in Classical German Idealism, was largely discredited after the 1830s, but subsequently, in the work of the Neo-Kantians, as we have seen, the legitimatory programme was taken out of scientific enquiry as such and pursued at a metalevel, as it was in positivists such as Comte and Mill.

The numerous later experimental programmes that had evolved from experimental natural philosophy played a central structural role in the functioning of scientific practice. The crucial point about experimental practice is that it is inherently piecemeal: it is the epitome of modularity and pluralism. Its central role in scientific practice reinforces the latter's 'unruly' structure. We have already

---

[12] Gaukroger, *The Emergence of a Scientific Culture*, chs 8–11; *The Collapse of Mechanism*, chs 3–5, 8.

looked at the role of autonomous models, in the form of paper-tools, in chemistry. At the same time, Galison, as we saw earlier, shows how material tools, in the form of advanced technologies, shape accounts of the behaviour of micro-particles, and cannot be lifted out of their very different instrumental contexts when we come to the problem of reconciling these accounts: they remain in place as integral parts of the respective approaches. And just as Galison provides an illuminating account of the profound difficulties in distinguishing between science and technology in modern particle physics, likewise Hugh Aitken, in his study of the radio industry, shows how, in order to secure unimpeded information flow, engineering scientists are needed to devise an intermediary language to allow communication between scientific and technological concerns, so that information—which travels in both directions—takes a useable form.[13] More generally, Ronald Kline has noted that 'historians have shown that it is often difficult to distinguish between science and technology in industrial research laboratories; others have described an influence flowing in the opposite direction—from technology to science—in such areas as instrumentation, thermodynamics, electromagnetism, and semiconductor theory'.[14] Channell sums up the situation in these terms:

As historians began to examine the history of technology they found little evidence for a strong dependence upon science. A detailed historical analysis of such major technological inventions as movable type printing, the mechanical clock, guns and gunpowder, metallurgy, the steam engine, textile machines, machine tools, railroad, and the automobile led to the conclusion that such inventions depended little, if at all, on scientific knowledge, skill, or craftsmanship. Historians of technology also began to challenge the common assumption that the Scientific Revolution of the sixteenth and seventeenth centuries has been primarily responsible for the Industrial Revolution of the eighteenth and nineteenth centuries. Almost every important technological development that contributed to the Industrial Revolution, such as Abraham Darby's production of iron using coke, Richard Arkwright's textile machinery and Thomas Newcomen's steam engine, owed little to any scientific theory or discovery. Even when some connection between technology and science could be identified, the connection many times turned out to be either indirect or much more complex than the applied science model indicated.[15]

In the light of this, we need to address the second issue, the relation between experiment and technology: or, more specifically, the relation between scientific experiment and a category of technology that includes extra-theoretical forms of practice. Just as we should not think of experiment as simply an adjunct to scientific theory,[16] so we should not consider technology as simply applied science.

[13] Hugh G. J. Aitken, *Syntony and Spark: The Origins of Radio* (New York, 1976), 1–20.
[14] Ronald Kline, 'Construing "Technology" as "Applied Science": Public Rhetoric of Scientists and Engineers in the United States, 1880-1945', *Isis* 86 (1995), 194–221, 195.
[15] David F. Channell, *A History of Technoscience: Erasing the Boundaries between Science and Technology* (London, 2017), 10.
[16] This is clear as early as Galileo's treatment of falling bodies: see Gaukroger, *The Emergence of a Scientific Culture*, 413–20.

The category of 'applied science' is a nineteenth-century invention, and the distinction between 'pure' and 'applied' is a nineteenth-century one. The term first appears in English in Coleridge's 1817 *Treatise on Method,* an introduction to the *Encyclopedia Metropolitana.*[17] As with much in Coleridge, the idea can be traced back to Kant. His distinction between 'pure' and 'applied' science is one between synthetic a priori truths, which Coleridge links with the divine, and merely contingent truths based on empirical evidence.[18] Babbage subsequently draws on this distinction in his book-length entry on machinery for the *Encyclopedia Metropolitana,* defining the applied sciences as deriving 'their facts from experiment; but the reasonings, on which their chief utility depends, are the province of what is called abstract science'.[19] The idea of 'applied science'—which had been in competition with the slightly different ideas of 'practical science' and 'science applied to the arts' in the first half of the century—had become a standard one by mid-century.[20] Note however that 'pure' and 'applied' are context-dependent, and shifting, terms. While mathematics was treated as the model for physics in the early decades of the nineteenth century for example, optics, acoustics, and mechanics were deemed 'mixed mathematics', by contrast with 'pure mathematics', and of decidedly inferior standing as a result. Yet later in the century, for the defenders of the purity of science, these areas became archetypally pure mathematical physics, and they were contrasted favourably with applied science, without any change at all in the type of content of the disciplines.

Robert Bud has distinguished three components in the idea of applied science in the nineteenth century:

First, it became a category incorporating both machines and a kind of knowledge about them. Second, it entailed the claim of an historical relationship between carefully nurtured scientific knowledge and praxis which had already been the source of the nation's industrial success. Third, it expressed an expectation about the future, that in years to come, science would provide a key source of national wealth and prosperity.[21]

The second and third components, which are very distinctive of the second half of the nineteenth century, are particularly closely related. The second depends on a convenient swapping of the aims of science and technology. As Peter Dear puts it:

The authority of science in the modern world rests to a considerable extent on the idea that it is powerful, that it can do things. Artificial satellites or nuclear explosions can act as icons of

---

[17] Samuel Taylor Coleridge, *General Introduction; or, Preliminary Treatise on Method* (London, 1817).

[18] His 'applied science' is a translation of the term '*angewandte Wissenschaft*', coined by Kantian-inspired German scientists in the late 1780s and early 1790s. See Robert Bud, ' "Applied Science": A Phrase in Search of a Meaning', *Isis* 103 (2012), 537–45: 538–40.

[19] Charles Babbage, *On the Economy of Machinery and Manufactures* (London, 1832), 379–80.

[20] Bud, ' "Applied Science": A Phrase in Search of a Meaning', 541–5.

[21] Robert Bud, ' "Applied Science" in Nineteenth-Century Britain: Public Discourse and the Creation of Meaning, 1817–1876', *History and Technology* 30 (2014), 3–36: 4.

science because of the assumption that they legitimately represent what science really is: in such cases, the instrumentality of science stands for the *whole* of science. Conversely, when appeal is made to science as the authority for an account of how some phenomenon or object really is in nature . . . it then receives back from its presumed instrumental effectiveness an image of truthfulness that this instrumentality has already been accepted as confirming.[22]

The conflation of aims of science and technology paves the way for the third component: the idea that a science-led form of technology is what had delivered the industrial, modernizing benefits, and that it was in this that the future of civilization lay. As we have seen, Thomas Huxley was adamant about the civilizing power of science, its ability to save us from 'another flood of barbarous hordes', and its necessary role in securing 'physical and moral well-being'. In an 1880 lecture at Mason Science College (the future University of Birmingham) he pointed out that 'the practical man . . . may ask what all this talk about culture has to do with an institution, the object of which is defined to be "to promote the prosperity of the manufactures and the industry of the country"'. He continues:

I often wish that this phrase, 'applied science', had never been invented. For it suggests that there is a sort of scientific knowledge of direct practical use, which can be studied apart from another sort of scientific knowledge, which is of no practical utility, and which is termed 'Pure science'. But there is no more complete fallacy than this. What people call applied science is nothing but the application of pure science to particular classes of problems.[23]

This characterization of applied science, which as Gooday has noted became a touchstone in the ensuing debate about what kinds of science ought to exist and ought to be recognized as 'authoritative',[24] was nevertheless at odds with how many had seen the matter. The Victorians did not necessarily consider applied science as an applied form of pure science at all: for many it was an entirely autonomous domain of practical knowledge. For Huxley, by contrast, the purity of science was central if it was to play the civilizing role that he expected of it. For him, the crucial distinction was in fact not so much that between science and technology but that between science and barbarism. But Huxley's confidence in the dependence of technology and invention on science was manifestly misplaced. There are many cases—such as the magnetic compass needle, oxidizing agents, steam power, or penicillin—where the device or substance appeared before, or independently of, any scientific theory that might account for it.

---

[22] Peter Dear, 'What is the History of Science the History Of? Early Modern Roots of the Ideology of Modern Science, *Isis* 96 (2005), 390–406: 404.

[23] Thomas Huxley, *Science and Culture, and Other Essays* (London, 1881), 1–23: 19–21. Cf. Joseph Henry's comment in his 1850 retirement address to the American Association for the Advancement of Science: 'We leave to others with lower aims and different objects to apply our discoveries to what are called useful purposes.' Quoted in Nathan Reingold, 'Joseph Henry on the Scientific Life: An AAAS Presidential Address of 1850', in Nathan Reingold, ed., *Science, American Style* (New Brunswick, NJ, 1991), 156–68: 159.

[24] Graeme Gooday, ' "Vague and Artificial": The Historically Elusive Distinction Between Pure and Applied Science', *Isis* 103 (2012), 546–54: 550.

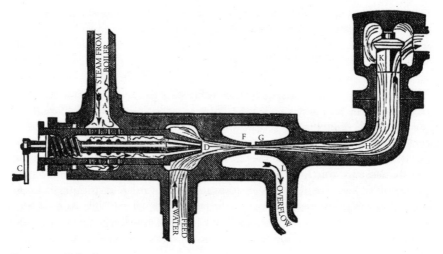

Fig. 10.1. Giffard's steam injector

A striking nineteenth-century example is the Giffard steam injector (Fig. 10.1), devised in 1858 by Henri Giffard, an engineer whose main interest was in the construction of steam-powered dirigibles. Giffard sought a feed apparatus for his dirigible that would not be subject to friction, by contrast with force pumps, which were hindered by friction, thereby absorbing power from the engine. It worked by delivering cold water to a boiler against its own pressure, using the boiler's own exhaust steam, and by the early 1860s it had completely replaced mechanical pumps. From the point of view of physics, however, the device presented a seemingly intractable problem: the process looked, *per impossibile*, to be a case of perpetual motion, and physicists struggled to understand how the injector worked.[25] The task was to reconcile scientific understanding with an established body of technological knowledge, but it was fifty years before a satisfactory thermodynamic explanation was offered.[26] Giffard's steam injector was a case where the technological development preceded the scientific under-standing. As Mokyr notes, the Industrial Revolution of the first half of the nineteenth century created a chemical industry without chemistry, an iron industry without metallurgy, and power machinery without thermodynamics.[27]

---

[25] See Eda Kranakis, 'The French Connection: Giffard's Injector and the Nature of Heat', *Technology and Culture* 23 (1982), 3–38.

[26] It came in Henri Poincaré, *Thermodynamique: Cours de physique mathématique de la faculté des sciences de Paris* (2nd edition, Paris, 1908), 323–34.

[27] Joel Mokyr, 'The Second Industrial Revolution, 1870-1914', in V. Castronovo, ed., *Storia dell'Economia Mondiale* (Rome, 1999), 219–45 (http://www.faculty.ecn.northwestern.edu/faculty/mokyr/castronovo.pdf).

A related kind of case is that where there are technological developments which, even when they did derive from scientific theories, turned out to be independent of the truth of the theory. A particularly striking example is Hertz's production of radio waves on the basis of Maxwell's electromagnetic theory, a central element in which was its assumption of an all-pervading aether. Hertz's account of radio waves was premised on the existence of an aether, but his subsequent development of the radio showed clearly that the production of radio waves was independent of this assumption, no matter how fundamental it may have been to the theory on which Hertz's discovery was based.[28] Moreover, even in cases where the successful technological development was explicitly based on a successful scientific one, the sense in which the latter can be said to have given rise to the former is questionable. Take the case of the steam turbine, a crucial turn-of-the-century innovation which displaced the wasteful steam engine. Its inventor Charles Parsons noted 'that the work was initially commenced because calculation showed that, from the known data, a successful steam turbine ought to be capable of construction. The practical development of this engine was thus commenced chiefly on the basis of the data of physics.'[29] The physics here doesn't tell us how we might go about constructing a machine of this kind, only that such a machine—which might or might not be able to be constructed—would not violate the laws of physics.

In short, technological developments do not just follow on from scientific ones. At the same time, scientific ones may be dependent on technological ones. In his 1950 autobiography, the physicist Robert Millikan suggests that results in the nineteenth- and twentieth-century physical sciences derived largely from developments in engineering, providing some revealing examples:

Historically, the thesis can be maintained that more fundamental advances have been made as a by-product of instrumental (i.e. engineering) improvement than in the direct and conscious search for new laws. Witness: (1) relativity and the Michelson-Morley experiment, the Michelson interferometer came first, not the reverse; (2) the spectroscope, a new instrument which created spectroscopy; (3) the three-electrode vacuum tube, the invention of which created a dozen new sciences; (4) the cyclotron, a gadget which with Lauritsen's linear accelerator spawned nuclear physics; (5) the Wilson cloud chamber, the parent of most of our knowledge of cosmic rays; (6) the Rowland work with gratings, which suggested the Bohr atom; (7) the magnetron, the progenitor of radar; (8) the counter-tube, the most fertile of all gadgets; (9) the spectroheliograph, the creator of astrophysics; (10) the relations of Carnot's reversible engine to the whole of thermodynamics.[30]

[28] See Jed Z. Buchwald, *The Creation of Scientific Effects: Heinrich Hertz and Electric Waves* (Chicago, 1994).

[29] Charles Parsons, *The Steam Turbine: The Rede Lecture 1911* (Cambridge, 1911), 1.

[30] Robert Millikan, *The Autobiography of Robert A. Millikan* (New York, 1950), 219.

It is clear that in these cases the idea of technology as 'applied science' simply gets things the wrong way round, and there are well-attested earlier examples of this phenomenon.[31]

The dependence of science on technology becomes clearer and more all-pervasive as we move into the era of post-First World War 'technoscience',[32] but there is no shortage of earlier examples. We saw in Chapter 4 how problems in machine design could be translated into forms of theoretical enquiry: in comparing Joule's original practical formulation of problems in thermodynamics with his later 'scientific' formulation, and in comparing Carnot's practical statement of the problem of heat engines with the later formulations of Clapeyron and William Thomson. The Joule and Thomson reformulations are cases in which science does not generate its own problems but starts from practical engineering and technological problems and reworks them, giving them a different kind of standing. This raises the question of the extent to which science generates its own subject matter. We need to come to terms with the fact that, particularly from the nineteenth century onwards, there are numerous cases where it is practical engineering and technological concerns that provide the subject matter for the physical sciences.[33]

The key question is not whether technology shapes the tasks of science, for the answer to that is clear; rather, it is *how* it shapes these tasks. The problem is not a new one, and as early as the middle of the nineteenth century there was already emerging a sophisticated grasp of the ways in which engineering and scientific considerations could merge. Consider Rankine's solution to problems in steam engine design in mid-century, a development that lay at the origins of what came to be designated engineering science.[34] One of the most pressing problems in the design of high-pressure steam engines centred on the question of the expansion of

---

[31] See, for example, Peter F. Drucker, 'The Technological Revolution: Notes on the Relationship of Science, Technology, and Culture', *Technology and Culture* 2 (1961), 342–51.

[32] The term is, I believe, due to Bruno Latour: see, for example, his *Science in Action: How to Follow Scientists and Engineers through Society* (Cambridge, MA, 1987). Latour's approach to technoscience, which construes preference for one theory over another in terms of authority and power in the institutions of science, is not plausible and is not an approach that I shall be following. On the other hand, he is certainly right to reject the philosophers' empty claim that 'truth' is the aim of science (on the general question of the aims of science in the formative early modern period see, for example, Gaukroger, *The Emergence of a Scientific Cult*ure, ch. 7). If the choice were simply between epistemological and sociological conceptions of truth we would indeed be in trouble, and I am offering something quite different from these here.

[33] As Thomson noted in an 1883 talk on all-important electrical units of measurement, 'resistance coils and ohms, and standard condensers and microfarads, had been for ten years familiar to the electricians of the submarine-cable factories and testing-stations, before anything that could be called electric measurement had come to be regularly practised in almost any of the scientific laboratories of the world.' William Thomson, *Popular Lectures and Addresses* (3 vols, London, 1891–4), i. 82–3.

[34] See David F. Channell, 'The Harmony of Theory and Practice: The Engineering Science of W. J. M. Rankine', *Technology and Culture* 23 (1982), 39–52, whose account I follow in my discussion here. See also Ben Marsden, 'Ranking Rankine: W. J. M. Rankine (1820–72) and the Making of "Engineering Science" Revisited', *History of Science* 51 (2013), 434–56.

steam, and there were two opposing views. The 'expansionists' advocated cutting off the supply of steam soon after it was introduced into the cylinder, letting the expanding steam do work on the piston. In this way, there was an economization of both steam and fuel. The 'antiexpansionists' by contrast questioned these savings, and advocated introducing steam into the cylinder throughout most of the stroke. What is of particular interest in the present context is the different kinds of rationales and defences offered by the two sides. By the 1850s, the use of a cut-off valve was generally accepted in Britain and in the United States, and the procedure seemed to be securely based on a basic scientific law, due to Boyle and Mariotte, that at constant temperature doubling the volume of a gas will be accompanied by a halving in pressure. But tests first performed in the naval dockyard in Brooklyn in 1859 cast doubt on the economy of using steam expansively because the procedure led to condensation, resulting in lower pressure, which neutralized the benefits of early cut-off. The expansionists argued that if there was a problem, it could hardly be due to the science: the failure must be owing to improperly constructed engines. It was a matter of making machines conform to fundamental scientific principles. As one of its advocates wrote: 'The nearer we make our practice conform to theory, by providing against the interference of circumstances unrecognized by our theory, the nearer will our practical results conform to our theoretical deductions.'[35] The antiexpansionists by contrast questioned the application of the Boyle–Mariotte law to steam engines. As the engineer in charge of the Brooklyn experiments put it: 'The whole theory is based on a pure and simple abstraction, that of the idea of perfect elasticity in gases and vapours unaffected by any conditions of matter or of the steam-engine.'[36] For the antiexpansionists the solution lay in relying on the practical experience of the builders of steam engines who, following rules of thumb and a long tradition in the design and construction of steam engines, knew what worked best and what did not work.

The issues underlying the dispute on the use of a cut-off valve ran deep, and were reflected at an institutional level. When the Regius Chair in Civil Engineering and Mechanics was instituted at Glasgow in 1840, there was staunch opposition to the subject from the professors of natural philosophy and mathematics, who argued that any theoretical questions were exclusively theirs, and the practical skills associated with engineering and mechanics should be taught outside the university through the apprenticeship system. The chemistry professor was particularly obstructive and managed, as a matter of principle, to prevent the teaching of engineering in any of the university rooms for the whole of the first year. This was the bifurcation of engineering that Rankine faced when he acceded to the

---

[35] R. H. Thurston, 'On the Economy Resulting from Expansion of Steam', *Journal of the Franklin Institute* 71 (1861), 193–5. Quoted in Channell, 'The Harmony of Theory and Practice', 41.
[36] Benjamin Franklin Isherwood, *Experimental Researches in Steam Engineering* (2 vols, Philadelphia, 1863), i. 140. Quoted in Channell, 'The Harmony of Theory and Practice', 42.

Chair in 1855. He combined serious interests in both science and technology, having been a natural philosophy student at Edinburgh in the late 1830s, where he was awarded the medal for his thesis on the relation between heat and light, and he subsequently spent the 1840s on engineering works: railways, river improvements, and harbours.[37] His aim at Glasgow, as set out in his inaugural address, 'Preliminary Dissertation on the Harmony of Theory and Practice in Mechanics', was to develop a genuinely integrated form of engineering science, something that cut across the institutional and professional distinctions.[38] Four years later, in the introduction to his treatise on steam engines and prime movers, he distinguished between two kinds of technological progress: 'the *empirical* and the *scientific*. Not the *practical* and the *theoretic*, for that distinction is fallacious: all real progress in mechanical art, whether theoretical or not, must be practical. The true distinction is this: that the empirical mode of progress is purely and simply practical; the scientific mode of progress is at once practical and theoretic.'[39]

'Empirical' progress is wholly practical and is characterized by gradual achievements in materials and workmanship, the result of 'individual ingenuity in matter of detail'. It is continuous and cumulative. 'Scientific' progress, by contrast, is discontinuous, and of a different nature:

When the results of experience and observation on the properties of the materials which are used, and on the laws of the actions which take place, in a class of mechanics, have been reduced to a science, then the improvement of such machines is no longer confined to amendments or enlargements in detail of previously existing examples; but from the principles of science practical rules are deduced, showing not only how to bring the machine to the condition of greatest efficiency consistent with the available materials and workmanship, but also how to adapt it to any combination of circumstances, how different soever from those which have previously occurred.[40]

Rankine's framework for engineering science was to analyse a machine into the essentials of materials and actions. The laws of action would rely more heavily on formal theoretical conceptions while the properties of materials would rely more heavily on practical experimental data, and in this way a route was open to accommodating both theoretical and practical considerations.[41] In terms of this model, what happens in the steam engine is that the combustion of fuel creates heat, which on being transmitted to the water transforms it into steam, which creates a state of mechanical energy in the piston, enabling it to perform useful work.[42] As Channell sums it up, 'the work of a steam engine depended on the properties of steam and the creation and disappearance of a state of heat in that

---

[37] Channell, 'The Harmony of Theory and Practice', 44.

[38] William John Macquorn Rankine, *A Manual of Applied Mechanics* (London, 1858), 1–11.

[39] William John Macquorn Rankine, *A Manual of the Steam Engine and other Prime Movers* (London, 1859), xix.

[40] Ibid., xx.   [41] Channell, 'The Harmony of Theory and Practice', 46.

[42] Rankine, *A Manual of the Steam Engine*, 223.

steam. Such a system combined theory and practice since the laws of heat relied more heavily on formal theoretical concepts, while an understanding of the properties of steam relied on experimental and practical data.'[43]

It was on this basis that Rankine tackled the controversy over the cut-off in the steam engine. Using data from the Brooklyn and other trials, he focused on how the laws of heat affected the properties of steam, showing that, although there is a degree of condensation of the steam, the reduction in the effective pressure that resulted was compensated for by the corresponding performance of work. Nevertheless, he argued, the data indicated that during the exhaust stroke the liquefied steam began to extract heat from the walls of the cylinder, the re-evaporated steam lowering the temperature of the cylinder, so that there was even more condensation in the next expansive stroke. Both expansionists and antiexpansionists, while assigning the blame differently, had accepted that an actual engine does not function as an ideal engine, but neither had identified the source of the lack of efficiency. Rankine was able to isolate it and as a result he was able to suggest remedies, such as showing how jacketing the cylinder would diminish condensation. Moreover, in his *Manual of Applied Mechanics*, he uses the resources of mechanics and technology in a similarly inventive way on questions of structural engineering, combining an account of the action of forces on a structure as a whole with the strength of individual material components of the structure, with a view to establishing the conditions of equilibrium of a structure such as a bridge.[44]

Rankine's distinction between 'empirical' and 'scientific' forms of technological progress is sharper than it needs to be for his establishment of engineering science to go through. There are cases of successful inventors, who, like Edison, were concerned with producing objects, such as an efficient incandescent light bulb,[45] and who had very limited success in offering scientific understanding,[46] but it is unhelpful to draw a sharp line between 'empirical' and 'scientific' technology, as this suggests a qualitative discontinuity, with everything clearly in one camp or the other. It should rather, at least provisionally, be seen in terms of a gradual transition from the one to the other, with inventions such as those of Edison, while not wholly devoid of a scientific element, being effectively theory-free, relying crucially on independent developments in devising sources of

---

[43] Channell, 'The Harmony of Theory and Practice', 47.   [44] Ibid., 49–50.

[45] Efficiency was the distinctive factor in Edison's light bulbs. As Ernest Freeberg notes: 'Edison could make no claim that he had invented the first working light bulb—the patent offices and newspapers provided ample evidence that others had accomplished this feat months and even years before. What he had done was create a complete lighting system that linked his powerful and more efficient dynamo, through a central main, feeders, and switches, to an incandescent bulb of superior design. . . . Alone among these rivals, his bulbs used a filament of high resistance, a crucial innovation that saved money by using a relatively small amount of current for each lamp.' *The Age of Edison*, 43.

[46] See W. Bernard Carlson, *Innovation as a Social Process: Eliuh Thomson and the Rise of General Electric, 1870–1900* (Cambridge, 1991); Paul B. Israel, *Edison: A Life of Invention* (New York, 1998); Ian Wills, 'Edison and Science: A Curious Result', *Studies in History and Philosophy of Science* 40 (2009), 157–66.

electricity—in this case the replacement of batteries with the new powerful type of dynamo developed in the 1870s—and teams of workers trying out various possibilities on what was often just a hit-and-miss basis.[47] As Thomas Hughes points out, scientists unfamiliar with inventions and development 'often denigrated this empirical approach, not realizing that to "hunt-and-try" was to hypothesize and experiment in the absence of theory. Thomas Midgley, the chemist and inventor responsible for the tetraethyl-lead additive for gasoline, remarked that the trick was to change a wild-goose chase into a fox hunt. Thomson, Sperry, and Edison, like other independent inventors, treasured their model builders, their chemists, their scientists, and their laboratories, because these facilitated experimentation, the life-blood of invention.'[48] At the same time, the experimental link between 'science' and 'engineering science' establishes a continuum between these, a continuum along which technological resources and highly theoretical ones can travel. It is this interaction of technology/engineering and experiment that plays such an important role in shaping a scientific theory.

Nevertheless, there is no doubt that talk of a continuum here can only be a first approximation. It suggests something linear, for example, but there is a case to be made that a linear continuum fails to capture the complexity of the relations between theoretical and non-theoretical forms of scientific and technological activity. Donald Stokes, in a study of Pasteur, has identified the problem with the linear model as lying in its implication that the closer some activity is to one end of the model the further it must be from the other. But Pasteur's work does not fit this description at all: he aimed at both a new theoretical understanding of the nature of diseases and at prevention of diseases such as rabies, and the spoilage of milk and wine. They were explicitly both part of the same exercise.[49] Moreover, Walter Vincenti, in a study of aeronautical engineering in the 1930s, identified a number of features of engineering practice which can be combined in different ways. He distinguishes 'descriptive' knowledge, that is, factual knowledge; 'prescriptive knowledge', knowledge of procedures or operations; and 'tacit knowledge', which is implicit, wordless, and pictureless. Different types of engineering activity involve different aspects of engineering knowledge. So, for example, he argues that learning to design requires descriptive and prescriptive knowledge, but (somewhat controversially) very little tacit knowledge, whereas learning to produce requires prescriptive and tacit knowledge but less descriptive knowledge.[50] It

---

[47] Edison said that out of a hundred experiments he did not expect more than one to be successful, and as to that one he was always suspicious until frequent repetition had verified the original results. Quoted in Frank Lewis Dyer and Thomas C. Martin, *Edison, His Life and Inventions* (2 vols, New York, 1910), ii. 612.

[48] Hughes, *American Genesis*, 52.

[49] Donald E. Stokes, *Pasteur's Quadrant: Basic Science and Technological Innovation* (Washington, DC, 1997), 12–18.

[50] Walter Vincenti, *What Engineers Know and How They Know it: Analytical Studies from Aeronautical History* (Baltimore, 1990), 197. For an attempt to identify the defining features of technology more generally, see John Staudenmaier, *Technology's Storytellers: Reweaving the Human*

is unlikely that there is a single model that fully captures the science–technology nexus in general terms. For our purposes, however, it is not necessary to pursue the complexities of this question here (we shall turn to some of these below when we look at aeronautics), but to register the real problems that these complexities pose, and to appreciate the inadquecy of the idea that science is, for example, 'knowing that' whereas technology is 'knowing how'.

If there is one area that has traditionally been seen to combine 'knowing that' and 'knowing how', it is experimental practice. It is possible for science to take on structural features from technology, above all its practical and piecemeal nature, because experiment, which is where the link with technology lies, already operates in a piecemeal way: and at the centre of scientific investigation, not at the periphery. It is this that helps foster an 'unruliness' in science, and a pluralism of explanation, something missed if one takes science to consist essentially in theories alone.

Note that practical, technological questions have always played a role in providing science with questions and goals. Ballistics, for example, provided the problems for, and effectively generated, mechanics in the late sixteenth century.[51] But there can be no doubt that the link between science and technology that comes to the fore in the middle of the nineteenth century means that understanding technology begins to have consequences for understanding science in a way, and on a scale, that it never did earlier. As Hughes has noted, for example, after 1880 'the British Admirality began specifying the performance characteristics it wanted in its engines, guns, ships, and other equipment, and challenged inventors, engineers, and industrialists to develop those designs. The military also began to pay at least a part of the costs for testing inventions. Invention of armaments for the major industrial and military powers then became a "command economy", or the supportive direction of innovation by government.'[52] And it is also perhaps worth remembering that from 1905 onwards the Wright brothers tried relentlessly to interest the US military in aeroplanes.[53]

These developments were consolidated with a wholly unprecedented government investment in technology in the Great War. Although it is true that there was a significant amount of catching up to do, and that the first three years of

---

Fabric (Cambridge, MA, 1985); Sven Ove Hansson, 'What is Technological Science?', *Studies in History and Philosophy of Science* 38 (2007), 523–7; and John D. Anderson, Jr, 'The Evolution of Aerodynamics in the Twentieth Century: Engineering or Science?', in Peter Galison and Alex Roland, eds, *Atmospheric Flight in the Twentieth Century* (Dordrecht, 2000), 207–22. See also Matthew Norton Wise, 'Mediations: Enlightenment Balancing Acts, or the Technologies of Rationalism', in Paul Horwich, ed., *World Changes, Thomas Kuhn and the Nature of Science* (Cambridge, MA, 1993), 207–56.

[51] See Klein, *Nützliches Wissen*, who sees the association of science and technology/engineering as a relatively continuous development since the sixteenth century. By contrast, I am arguing that, despite its sixteenth-century origins, matters only come to a head in a transformative way in the mid-nineteenth century.

[52] Hughes, *American Genesis*, 96–7.   [53] Ibid., 101–4.

the war were still fought with predominantly agrarian resources,[54] there was a determined move to mobilize science in everything from dyes for uniforms to armaments and aeroplanes right from the beginning. As the October 1914 editorial in *Nature* announced: 'This war, in contradistinction to all previous wars, is a war in which pure and applied science plays a conspicuous role'.[55] State-organized national cartels were established to manage innovation, institutionalizing invention and making mass production the norm.[56] As Agar notes, 'in general, the First World War intensified the organizational revolution—a gathering concern for scale, organization and efficiency that was discernible in the mid- to late-nineteenth century'.[57] This intensification did not come to an end in 1918, but was continuously built upon.[58] It shaped the subsequent history of scientific practice, as science was transformed into 'technoscience', characterizations of which differ, but the most appropriate for our purposes is that offered by Krige and Pestre, in describing the activities of Ernst Lawrence, founder of the Lawrence Radiation Laboratory at Berkeley: 'a profound symbiosis previously unknown in basic science; a fusion of "pure" science, technology and engineering. It was the emergence of a new practice, a new way of doing physics, the emergence of a new kind of researcher at once a physicist, in touch with the evolution of the discipline and its key theoretical and experimental issues, as conceiver of apparatus and engineer, knowledgeable and innovative in the most advanced techniques (like electronics at the time) and able to put them to good use, and entrepreneur, capable of raising large sums of money, of getting people with different expertise together, of mobilizing technical resources.'[59]

[54] Compare this to World War II, where, for the first time in history, a substantial rearming of combatants took place during the time-span of the conflict: see Alex Roland, 'Science, Technology, and War', in Mary Jo Nye, ed., *The Cambridge History of Science, volume 5: The Modern Physical and Mathematical Sciences* (Cambridge, 2003), 561–78: 565.

[55] Quoted in Agar, *Science in the Twentieth Century and Beyond*, 93. There were nevertheless precedents, e.g. in respect to the design and use of rifles, in the American Civil War and in the Boer War.

[56] See William H. McNeill, *The Pursuit of Power: Technology, Armed Force, and Society Since A. D. 1000* (Chicago, 1982), 318.

[57] Agar, *Science in the Twentieth Century and Beyond*, 92.

[58] Nevertheless, Vaclac Smil has argued in detail that 'the fundamental means to realize nearly all of the 20th-century accomplishments were put in place even before the [twentieth] century began, mostly during the three closing decades of the 19th century and in the years preceding WWI. That period ranks as history's most remarkable discontinuity not only because of the extensive sweep of its innovations but also because of the rapidity of fundamental advances that were achieved during that time': *Creating the Twentieth Century*, 5–6. Agar offers a similar assessment: *Science in the Twentieth Century and Beyond*, 63.

[59] John Krige and Dominique Pestre, 'Some Thoughts on the Early History of CERN', in Peter Galison and Bruce Helvy, eds, *Big Science: The Growth of Large-Scale Research* (Stanford, CA, 1992), 78–99: 93.

THE STANDING OF THE SCIENTIST

There is no more striking symbol of the scientific culture of the twentieth century than aviation, in both science fiction (see Plate 4) and in the promotion of its achievements (Plate 5). The development of aerodynamics in the early decades of the twentieth century presents us with an opportunity to look in more detail at why and how the different aims and procedures of mathematical physics and engineering science have clashed. In this section, which focuses on the claims of mathematical physics, we shall be looking at the 'why', in the next, which focuses on the claims of engineering, on the question of 'how'.

The 'why' is crucial, because we need to understand why technology has not had the standing, as it were, to lay claim to various developments in which it played a determining role. Conversely, what is it about the understanding of science that has required that any such developments must ultimately be counted as scientific ones? David Channell draws attention to a particularly egregious example when he writes:

Much of the confusion over what is science and what is technology originated during World War II. Vannevar Bush, an engineer who directed U.S. wartime research and headed the Office of Scientific Research and Development, said that when he came to discover that his British counterparts considered that 'the engineer was a kind of second-class citizen compared to the scientist,' he decided to designate all wartime researchers working in the Office of Scientific Research and Development as scientists. He noted that even after World War II the public was led to believe that such an achievement as the landing of the first astronauts on the moon was a great scientific achievement when in fact 'it was a marvelously skillful engineering job.'[60]

The 'why' question is a question about the public standing of science, and our first task must be to understand how this public standing arose and how it was consolidated: how there arose the deeply entrenched idea of mathematical physics as the ultimate model of understanding natural processes. The answer lies in developments in university education in the nineteenth century, particularly as regards mathematics teaching. The role of mathematics provides a good example of the problems facing attempts to make science part of a general cultural programme, and at the same time it throws light on the way in which this cultural programme offers an image of the *persona* of the scientist as someone uniquely equipped to provide a comprehensive understanding of the world.

---

[60] Channell, *A History of Technoscience*, 1. Cf. a detailed study for the Pentagon of the role of basic science in the study of weapons development in the 1960s, which concluded that basic science played 'an inconsequential role' in the development of major weapons systems: Glen R. Asner, 'The Linear Model, The U.S. Department of Defense, and the Golden Age of Industrial Research', in Karl Grandin, Nina Wormbs, and Sven Widmalm, eds, *The Science–Industry Nexus: History, Policy, Implications* (Sagamore Beach, MA, 2004), 3–30: 3. See also the discussion of industrial science in Steven Shapin, *The Scientific Life: A Moral History of a Late Modern Vocation* (Chicago, 2008), chs 4 and 5.

France provides a revealing starting point. In the immediate post-revolutionary era in France, there was a strong commitment to the unity of science ultimately founded on mathematics, the pursuit of which was considered to demonstrate a capacity for reason that was indicative of Enlightenment egalitarian values. The Écoles Centrales, set up in the wake of the Revolution for the teaching of mathematics as a core discipline, boasted that 'a few years will suffice for the son of an artisan to be more educated [through the study of science and mathematics] than the aristocrats once were after consuming their entire youth in numerous [classics] classes'.[61] But these institutions had effectively been abandoned by 1800.[62] While the importance of science and mathematics in the curriculum continued into the empire, mathematical and scientific education were gradually separated from the humanistic ideals of education.[63] Moreover, the approach to mathematics and the sciences was through the very formal path of analysis. The exclusion of Republicans such as Carnot and Monge from the Académie des Sciences meant that the practical programmes involving craftsmen and practical engineers for example were ignored, while analytical mechanics was unchallenged.[64]

Although there continued to be repeated calls for science education to replace classics at universities,[65] the attempt to establish mathematics as the basis for an education matching that in the humanities encountered numerous problems. There were limits to how much serious mathematics could be taught at a general level, and the autonomy of mathematics came to be stressed as a means of pursuing the discipline in a satisfactory manner. The promotion of their subject by mathematicians and scientists engendered conflicting interests within the disciplines, and the teaching of mathematics in universities in the early decades

[61] Quoted in L. Pearce Williams, 'Science, Education and the French Revolution', *Isis* 44 (1953), 311–30: 302. See also Ken Alder, 'French Engineers Become Professionals; or, How Meritocracy Made Knowledge Objective', in William Clark, Jan Golinski, and Simon Schaffer, eds, *The Sciences in Enlightened Europe* (Chicago, 1999), 94–125. Cf. Huxley, seventy years later: 'I weigh my words well when I assert, that the man who should know the true history of the bit of chalk which every carpenter carries about in his breeches-pocket, though ignorant of all other history, is likely, if he will think his knowledge out to its ultimate results, to have a truer, and therefore a better, conception of this wonderful universe, and man's relation to it, than the most learned student who is deep-read in the records of humanity and ignorant of those of Nature.' Thomas Huxley, *Lay Sermons, Addresses and Reviews* (London, 1877), 176–7.

[62] See L. Pearce Williams, 'Science, Education and Napoleon I', *Isis* 47 (1956), 369–82.

[63] See John Hubbel Weiss, *The Making of Technological Man: Social Origins of French Engineering Education* (Cambridge, MA, 1982).

[64] See Lorraine Daston, 'The Physicalist Tradition in Early Nineteenth Century French Geometry', *Studies in History and Philosophy of Science* 17 (1986), 269–95; and Ivor Grattan-Guiness, 'Work for the Workers: Advances in Engineering Mechanics and Instruction in France, 1800–1830', *Annals of Science* 41 (1984), 1–33. Note however that Carnot and Monge were key figures in the establishment of the École Polytechnique, which has remained hugely influential to the present day.

[65] See, for example, in the case of Britain, J. Graham Kerr, 'Biology and the Training of the Citizen', *Nature* 118 (1926), 102–12.

of the nineteenth century was caught between two different aims. In the case of Britain, for example, between the 1830s and the 1850s there was a fierce dispute between Whewell and the mathematician William Hopkins over whether mathematics had a purely pedagogical value or whether it was a practically useful and a morally worthwhile end in itself, revealing the mathematical principles of God's creation.[66] In the case of France, in the 1820s the mathematical community struggled hard with the opposing forces of 'exemplary' mathematics, where it was treated as the core of a humanistic culture, and 'separatist' mathematics—an abstract, analytical, autonomous form of advanced mathematics—and in the long run it was the separatists who prevailed.[67] In Germany 'exemplary' mathematics was pursued almost exclusively before the 1820s.[68] Mathematics was studied within the philosophy faculties which, as Joan Richards notes, 'laid the educational ground for students interested in going on to the more prestigious faculties of law, theology or medicine. Interest in the subject reflected this hierarchy of value. In 1800, the only major German mathematician was Carl Friedrich Gauss...who was a solitary researcher.'[69] This changed in the wake of the teaching and research reforms of the 1820s, and with the institution, at the University of Königsberg in 1834, of a joint mathematics-physics 'seminar' by Carl Jacobi and Franz Neumann, exemplary mathematics was decisively replaced by separatist mathematics.[70]

It would, however, be a mistake to see the shift from exemplary to separatist mathematics as an abandonment of a cultural role for mathematics. Mathematics continued to play a significant cultural role: it just shifted from a general cultural role to an elite one. This is particularly marked in Britain. Richards has drawn

---

[66] Andrew Warwick, *Masters of Theory: Cambridge and the Rise of Mathematical Physics* (Chicago, 2003), 94. Matters were complicated by other aspects of the question of the point of the exercise: while Cambridge was able to produce students with a very high level of technical competence, for example, this was largely a matter of technical skills for their own sake, and there was no discernible 'research school' in mathematics in Cambridge between the 1830s and the 1870s, when Maxwell's *Treatise* became the focus of collective research: see ibid., ch. 6.

[67] Joan L. Richards, 'Rigor and Clarity: Foundations of Mathematics in France and England, 1800–1840', *Science in Context* 4 (1991), 297–319.

[68] See, for example, the defence of the moralizing effect of mathematics in Christian Gottlieb Ferdinand Engel, *Welchen Einfluss äussert das Studium der Mathematischen Wissenschaften auf das Gemüth?* (Berlin, 1820).

[69] Joan L. Richards, 'The Geometrical Tradition: Mathematics, Space, and Reason in the Nineteenth Century', in Mary Jo Nye, ed., *The Cambridge History of Nineteenth and Twentieth Century Science* (Cambridge, 2003), 449–67: 458.

[70] See Gert Schubring, 'The Conception of Pure Mathematics as an Instrument in the Professionalisation of Mathematics', in H. Mehrtens, H. Bos, and T. Schneider, eds, *Social History of Nineteenth Century Mathematics* (Basel, 1981), 111–34; Gert Schubring, 'Die deutsche mathematische Gemeinde', in J. Fauvel, R. Flood, and R. Wilson, eds, *Möbius und sein Band. Der Aufstieg von Mathematik und Astronomie im Deutschland des 19. Jahrhunderts* (Basle, 1994), 394–406; Tom Archibald, 'Images of Mathematics in the German Mathematical Community', in U. Bottazzini and A. D. Dalmedico, eds, *Changing Images in Mathematics from the French Revolution to the New Millennium* (London, 2001), 49–67.

attention to the way in which mathematics began to replace classics as the vehicle for the cultivation of minds in Cambridge in the early decades of the century:

In a period of rising enrollments, examinations began to play an increasingly important role at both Cambridge and Oxford. At Oxford, the examinations were in classics, which focus was justified as a way of broadening young minds rather than as imparting specialized knowledge. The same kind of rationale was applied at Cambridge, where, however, the central examination was in mathematics. Until the middle of the century one needed to pass on this examination [the 'Tripos'] in order even to take the parallel examination in the classics. Even though the Tripos became more and more mathematically demanding, the justification for requiring that the students study for it continued to be broadly humanistic rather than specific or professional. Throughout the century the center of England's mathematical education pursued the subject as a way to help students become fully formed human beings.[71]

The mathematics Tripos had two parts, and it was the first, less advanced part, that featured as the general curriculum. But what was the 'mathematics' of the first part? The *Report of the Examination Board for 1849* sets out the course of study, which begins with basic geometry, algebra, plane trigonometry, and the geometry of conic sections. But as the list continues, we are in the realm of 'applied' mathematics: 'the elementary parts of Statics and Dynamics, treated without the Differential Calculus; the first three sections of Newton, the Propositions to be proved in Newton's manner; the elementary parts of Hydrostatics, without the Differential Calculus; the simpler propositions of Optics, treated geometrically; the parts of Astronomy required for the explanation of the more simple phenomena, without calculation'.[72] In other words, what is termed mathematics here is effectively a form of theoretical physics, and as we shall see, this is how its distinguished practitioners interpreted it.

The regime at Cambridge for those studying the second part of the Tripos—after 1848 open to those who had performed sufficiently well in the first part—was highly competitive, and advanced mathematics could not be learned from formal lectures, oral debate, and reading alone. As Andrew Warwick has pointed out, the vast majority of students 'needed an ordered and progressive plan of study, long periods of private rehearsal of problem solving, and regular face-to-face interaction with a tutor, someone who was not the professor or the college tutor, but a private tutor (usually a college fellow) who managed the student's studies for a fee'.[73] Tutors were crucial, because, as Weintraub notes, it was the Tripos examination, not those areas of mathematics—such as analysis—which were important in the eyes of the best European mathematicians, that 'defined the concerns of the students and the program. Indeed, the very best Cambridge

[71] Richards, 'Rigor and Clarity', 307–8.
[72] Quoted in Joan L. Richards, *Mathematical Visions: The Pursuit of Geometry in Victorian England* (San Diego, 1988), 40–1.
[73] Andrew Warwick, *Masters of Theory*, 37, and, for more detail, 227–85.

mathematicians, men like J. J. Sylvester, and Arthur Cayley, gave lectures that no students, or very few students, ever attended. Why should they have attended since that material was never going to appear on any examination?'[74] He continues:

Mathematics was thus defined, in England, by a set of tricks and details, based on Newton, which were linked to applied physics and mechanics, and which could be tested in a time-limited fashion. The function of the examination really was to provide a fixed ordering of the degree candidates. The top performer all the way down to the last found his place in the posted list of results. From Senior Wrangler . . . down to Wooden Spoon (last passing grade), the order of finish of the Tripos defined one's options in the world of scholarship at least.[75]

Moreover it was a crucial part of the ethos that intellectual endeavour had to be balanced against physical exercise, and the elite mathematics students were often elite athletes as well. As Warwick notes, by the 1840s the combined pursuits of mathematics and sport had become constitutive of the liberal education through which good undergraduate character was formed,[76] and an Indian student at the time observed with astonishment on arrival at Cambridge that his fellow students paid 'as much attention to their bodily as to their mental development'.[77] Here it is not so much that science has been integrated into English culture, but rather than it has begun to play a direct role in shaping that culture, through the educational formation of the country's elite: the names of the top mathematics students, the 'wranglers', for example, were published in the news-papers, particularly *The Times*, from the mid-1840s, occasionally accompanied by short biographies.[78]

On the question of the scientific shaping of university students, there was a link between the formation of a well-rounded character for the scientist and the idea that the scientist was participating in a unified understanding of the world. The dignity of science and the dignity of its practitioners were intimately tied. Taken in broad terms, this was not a new requirement. From the last decade of the

---

[74] E. Roy Weintraub, *How Economics Became a Mathematical Science* (Durham, NC, 2002), 13.

[75] Ibid., 14

[76] Warwick, *Masters of Theory*, 179. Note that exercise was largely collective and competitive: individual exercise was discouraged, except for long walks. There is a contrast with Germany here, where gymnastics was seen as a complement to intellectual endeavour: see Finkelstein, *Emile du Bois-Reymond*, 180–2. Du Bois-Reymond wrote a pamphlet defending gymnastics: *Über das Barrenturnen und über die sogenannte rationelle Gymnastik* (Berlin, 1862).

[77] Quoted in Warwick, *Masters of Theory*, 198. The exercise regime was also followed by mathematics coaches such as Routh, although here it was individual exercise (a two-hour walk every fine day at exactly the same time in Routh's case): ibid., 200. Routh was a brilliant applied mathematician, having pushed Maxwell into second place in the Tripos examination in 1854. His 1877 Adams' Prize essay, published as *A Treatise on the Stability of a Given State of Motion, Particularly Steady Motion* (London, 1877), became fundamental to the understanding in Britain of aerodynamic stability.

[78] Warwick, *Masters of Theory*, 202.

sixteenth century, a crucial part of Bacon's project for the reform of natural philosophy, for example, was a reform of its practitioners. One ingredient in this was the elaboration of a new image of the natural philosopher, an image that conveyed the fact that the natural philosopher was no longer an individual seeker after the arcane mysteries of the natural world, employing an esoteric language and protecting his discoveries from others, but a public figure in the service of the public good.[79] As we have seen, the need to establish the dignity of the scientific calling was far from secure in the seventeenth and early eighteenth centuries. By the nineteenth century, however, in the university environment science was a beneficiary of a more general rise of the standing of professorships.[80] The German university, the most admired institution of higher education in the Western world in the nineteenth century,[81] provides an instructive example. Leonore O'Boyle, writing on the idea of 'knowledge for its own sake' in German universities, notes the acceptance of the new academic profession, sparked by a change in the standing of philology, the field which, due to its radical reform of teaching methods from lectures to 'seminars', was in the vanguard of the changes, particularly affecting science teaching as it was transformed from passive attendance at lectures to hands-on laboratory training.[82] Philology itself was transformed from a remedial discipline to one that stood at the core of the Philosophy Faculty, something 'made evident in the public's willingness to confer money, autonomy, and status'.[83] How, O'Boyle asks, was this process served by the ideal of knowledge for its own sake?:

[79] See Gaukroger, *Francis Bacon*, ch. 2.

[80] See, for example, Avraham Zloczower, *Career Opportunities and the Growth of Scientific Discovery in Nineteenth Century Germany, with Special Reference to the Development of Physiology* (New York, 1981).

[81] See, for example, Edward Shils and John Roberts, 'The Diffusion of European Models Outside Europe', in Walter Rüegg, ed., *A History of the University in Europe, Volume III: Universities in the Nineteenth and Early Twentieth Centuries (1800–1945)* (Cambridge, 2004), 163–230; McClelland, *State, Society and University in Germany*; George Weisz, *The Emergence of Modern Universities in France* (Princeton, NJ, 1983); George Haines IV, *Essays on German Influence upon English Education and Science, 1850–1919* (Hamden, CT, 1969); Ronald L. Gougher, 'Comparison of English and American Views of the German University, 1840-1865: A Bibliography', *History of Education Quarterly* 9 (1969), 477–91.

[82] See Bas van Bommel, *Classical Humanism and the Challenge of Modernity: Debates on Classical Education in Nineteenth-Century Germany* (Berlin, 2015). Philology seminars were modelled on that instituted by August Böckh in Berlin in 1812. The statutes declared that the aim of the seminar was to 'sustain, propagate, and expand' the knowledge of antiquity. The natural science institutes usually started as service facilities for university 'cabinets' etc., but became part of the philosophy faculty's research objectives when they were transformed into *Praktika*—internships—where students started to be actively engaged in enquiry-based learning. The earliest example is Liebig's laboratory at Giessen. See Frederick Lawrence Holmes, 'The Complementarity of Teaching and Research in Leibig's Laboratory', *Osiris* (1989), 121–64.

[83] Leonore O'Boyle, 'Learning for its Own Sake: The German University as Nineteenth-Century Model', *Comparative Studies in Society and History* 25 (1983), 3–25: 8.

A summary answer is that the ideal helped those who professed it to move from a lowly class of origin to a position of alliance with a governing class of aristocrats and patricians. First, all suggestion of association with manual labor and the kinds of material reward satisfying to lesser men was eliminated. Second, the focus on abstract principles and forms in the philological disciplines was viewed as a concern with the general rather than the specific and concrete; all things were presumably seen by the scholars as parts of a whole rather than as isolated facts of the sort that distracted most men and indeed absorbed their life.[84]

Note two things here. The first is the importance of a shift in social status, crucial for the standing of the professional scientists who had almost completely replaced disinterested amateur practitioners by this stage. This shift is part of a long-term development in the *persona* of the natural philosopher/scientist which begins at the end of the sixteenth century with the removal of natural philosophy from the responsibility of clerical Aristotle commentators. There were significant national variations: in England, although Bacon had set out to establish the dignity of the collective research that he believed necessary for natural philosophy, he envisaged a solely supervisory role for a gentleman such as himself in this respect; in the Florentine court culture in which Galileo worked, the natural philosopher was a client of a patron and had to transform his social standing through demonstration of knowledge of the liberal arts for example; for Descartes, the kind of person best suited to natural philosophy was not someone who had been trained in the universities but the plain *honnête homme*, someone who embodied particular social, intellectual, and moral qualities; and so on.[85]

Second, note that the criterion for the elevation of the status of the scientist is a focus on seeing things as parts of a whole. This is reinforced by the idea that the standing of the scientist lies not just in his pursuing science but also in the general cultural standing that the *persona* of the scientist exhibits, and it in turn reinforces the conception of the *persona* of the scientist as someone who, through a commitment to the unity of science, manifests this quality in a paradigmatic way. It is the unity of science, something which indicates a comprehensive understanding of the world, that puts science in a controlling position. To this extent, the unity of science is really not so much about the nature of science as about the standing of the scientist: with the decline of religion, the scientist is the only person who can lay claim to a comprehensive understanding of the world.

I want to draw attention to two aspects of these developments. The first is the way in which the cultural standing of mathematics is able to meet the challenges of the shift from an 'exemplary' to a 'separatist' role. Indeed, the transition means that separatist mathematics has picked up a standing that mathematics did not have until it first took on an exemplary role, for it then inherited this standing from the latter. The second is the way in which, once its new separatist standing is

---

[84] Ibid., 8–9.
[85] Ch. 6 of Gaukroger, *The Emergence of a Scientific Culture* is devoted to these questions.

secure, it is immune to changes that would have compromised an exemplary form of mathematics. In the case of Britain for example, there was a political shift in the later decades of the nineteenth century that meant that the cultural value of science was independent of any contest with the humanities. As the century progressed, what civilization expected from science changed radically. As Frank Turner notes:

After approximately 1875 the spokesmen for British science shifted their rhetoric and the emphasis of their policy from the values of peace, cosmopolitanism, self-improvement, material comfort, social mobility, and intellectual progress toward the values of collectivism, nationalism, military preparedness, patriotism, political elitism, and social imperialism. Instead of being promoted as an instrument for improving the student morally and bringing greater physical security or personal benefit to humankind, science came to be portrayed as a means to create and educate better citizens for state service and stable politics, and to ensure the military security and economic efficiency of the nation.[86]

Did this mean that the image of mathematics, and the exact sciences more generally, shifted to some form of pragmatism? This was the view of some in engineering, such as Föppl. But the reaction among the Cambridge-trained mathematicians and scientists, for example, was the exact opposite. What Kevles has noted about the use of the term 'pure science' in the USA in the 1870s, namely that it 'was less purity of the subject as purity of motive',[87] captures perfectly the approach of those trained in the Cambridge Tripos. Moreover, the purity of the kind of mathematics pursued secured its absolute primacy over empirical concerns. George Bryan, whose 1911 *Stability in Aviation* set out the basic equations governing stability in aircraft,[88] and who was one of the most prominent representatives of the Tripos-educated mathematicians, made it clear that hydrodynamics in his view consisted in the study of partial differential equations and not 'town water supply, resistance of ships, screw propellers and aeroplanes', and if the mathematician were to adapt his work to the needs of engineers he might as well give up mathematics, relying instead 'on the introduction of constants and coefficients to save him from running his head against insoluble differential equations'. The hypothetical conditions postulated in the latter case do not reflect any empirical reality and, he writes, have 'pretty well done their duty when they have been made use of to write down differential equations'.[89] As David Bloor notes, for Bryan, 'it was not the empirical status of

[86] Frank M. Turner, 'Public Science in Britain, 1880-1919', *Isis* 71 (1980), 589–608: 592.

[87] Daniel J. Kevles, 'The Physics, Mathematics, and Chemistry Communities: A Comparative Analysis', in Alexandra Oleson and John Voss, eds, *The Organization of Knowledge in America* (Baltimore, 1979), 139–72: 141.

[88] George Bryan, *Stability in Aviation: An Introduction to Dynamical Stability as Applied to the Motion of Aeroplanes* (London, 1911).

[89] George Bryan, Review of R. de Villamil, *ABC of Hydrodynamics*, *Mathematical Gazette* 6 (1912) 379–80: 379.

these conditions (that is, their falsity) that counted, but their power to help the mathematician frame tractable equations'.[90]

SCIENCE AND MACHINES

> Every aeroplane is to be regarded as a collection of unsolved mathematical problems.
>
> George Henry Bryan, 'Researches in Aeronautical Mathematics' (1916)[91]

When the Wright brothers undertook the first sustained and controlled flights between 1903 and 1905, they had worked largely by trial and error, although they did occasionally experiment with rudimentary models. In the wake of their flights, those constructing aircraft continued in a largely trial-and-error fashion, building on the practical expertise of their predecessors. But at the same time there began attempts to develop a theoretical understanding of the action of the air on wings. The Wright brothers themselves, realizing that the lift data they had from the German aviation pioneer Otto Lillienthal were unreliable, had experimented with wind tunnels to produce theories of effective aerofoil shapes, although these were very crude. The aim of the new theories was to understand lift (the force on the wing that keeps it in the air), drag (the resistance of the air to motion), and stability (the ability to correct for pressure producing turning moments that would cause the wing to pitch).

The basic mathematical resources derived from hydrodynamics, the study of bodies moving through fluids, were developed in the eighteenth century by Daniel Bernoulli, d'Alembert, Euler, and Laplace among others. The area was mathematically challenging, and in order to make it tractable numerous simplifying assumptions had to be made. Euler's and Laplace's equations for the motion of a body through a fluid invoked a 'perfect fluid', and they provided a rigorous account of how a fluid would flow around a body if the body and the fluid were moving with respect to one another (it doesn't matter whether it is the body or the fluid that is moving, the effects are demonstrably identical). But perfect fluids are incompressible and inviscid (they have a viscosity of zero) and the flow exercises no resultant force on the body. The problem was that, despite the fact that it was well-developed mathematically, ideal-fluid theory was considered useless for the study of real fluids such as air: what was needed instead was an account of viscosity. Such an account had been provided by the Cambridge mathematician George Stokes mid-century in a series of equations called the Navier–Stokes

[90] David Bloor, *The Enigma of the Aerofoil: Rival Theories on Aerodynamics, 1909–1930* (Chicago, 2011), 95.

[91] George Henry Bryan, 'Researches in Aeronautical Mathematics', *Nature* 96 (1916), 509–11: 510.

equations (named after their co-discoverers). They extend Euler's equations into the area of viscous fluids, by applying Newton's second law to fluid motion. But the Navier–Stokes equations could only be solved in a few simple cases (it is not clear even now that, outside a handful of cases, solutions in three dimensions exist), so their application to the forces acting on wings was very limited.

In light of this, the task was to find a way of making ideal-fluid theory more realistic, and there were two basic approaches to this. The problem with perfect fluids arises from the fact that they are continuous and irrotational (they do not rotate around the body immersed in them). But as Bryan pointed out: 'As soon as you give up the equations of a medium derived from the assumed definitions of an ideal fluid, you obtain equations of which integrals cannot be found; at least mathematicians have tried over and over in vain to find them.'[92] Two different approaches attempted to solve the problem by introducing discontinuities in the one case, and circulating vortices around the moving body in the other. The approaches were associated with very different conceptions of what the understanding of physical phenomena consisted in. The first one maintained that any account must be anchored in—and ultimately be deducible from—mathematical physics, particularly as conceived in the Cambridge Tripos tradition, for as Groenewegen notes, the appreciation of mathematical knowledge as necessary and inevitable truth, derived axiomatically, 'was an aspect of Cambridge mathematical training which justified its pre-eminence in the university honours syllabus, combined as it was with methods by which such truths could be mastered. This was a point stressed by Whewell in his defence of the value of mathematical specialization. A high wrangler in particular would have been heavily imbued by this specialized feature of mathematical knowledge.'[93]

The alternative approach, which was in the tradition of 'practical mechanics', particularly as conceived in the tradition of the *Technische Hochschule*, rejected such foundational aspirations, and manipulated mathematical and theoretical resources in such a way as to achieve a particular engineering result. The situation becomes especially interesting when we consider the resistance that advocates of the first approach demonstrated to the success of the second. Such resistance has complex motivations, but I want to argue that one crucial motivation derived from the association of those in the Cambridge Tripos tradition with a model of science as something comprehensive and certain, which is accompanied by a conception of the *persona* of the scientist—in the form of the mathematical physicist—as the ultimate arbiter of what counts as a satisfactory outcome of scientific enquiry. To a large extent, the resistance arises from the fear that this conception would be compromised by abandoning the idea that the physical nature of the world can ultimately be derived from a unified set of fundamental, mathematically formulated laws.

[92] Bryan, Review of R. de Villamil, 379.
[93] Peter Groenewegen, *A Soaring Eagle: Alfred Marshall 1842–1924* (Aldershot, 1995), 116.

Consider first the theory of discontinuous flow. This was originally formulated by Helmholtz, and subsequently developed by Kirchhoff and Rayleigh.[94] Bloor summarizes the theory well:

Helmholtz argued that the result of zero resistance arose because an ideal fluid could wrap itself around an object and exert pressure from all sides in a way that cancelled out any resultant force. The discontinuous flow approach exploited the possibility that there could be discontinuities in the velocity of different bodies of ideal fluid that were in direct contact with one another. The flow was assumed to break away from the edges of an obstacle and create a wake behind it. The wake would be 'dead water' or 'dead air', and the main body of ideal fluid would flow past it. Such a flow pattern in an ideal fluid, with a wake of dead fluid, turned out to be compatible with the Euler equations. Furthermore, it could be established that, given such a discontinuous flow, the pressure on the front face of an object would be greater than the pressure of the dead fluid on the rear. . . . If the resultant force proved large enough, here was a theory that could, in principle, explain the lift of a wing as well as the resistance to motion, the 'drag'.[95]

However, by the second decade of the twentieth century, the prospects of the discontinuity theory were looking dire. Aerodynamic researchers identified a number of fundamental problems. It was a crucial assumption, for example, that the pressure in the dead-water region was the same as that of the undisturbed flow, but measurements showed it to be less than the latter. The young Cambridge mathematician Geoffrey Ingram Taylor noted in his 1914 PhD thesis that the ability to generate rigorous mathematical solutions in hydrodynamics was in inverse proportion to their experimental support, singling out the discontinuity theory, which he argued must now be relegated to the realm of mathematical curiosities.[96] At stake here was an issue that those involved in the practical aspects of aeroplane design felt strongly about. In a short 1916 piece, J. H. Ledeboer, the editor of *Aeronautics*, wrote:

I don't deny the infinitely valuable role of pure science, still less that of theory, but science should have some relation to practice, since it is its foster-mother. There is more than one aeroplane designer who knows just enough mathematics to make twice two work out at four, but he will turn out machines equal in performance to the best. We in this country

---

[94] Hermann von Helmholtz, 'Über discontinuirliche Flüssigkeits-Bewegungen', *Monatsbericht der königlich preussischen Akademie des Wissenschaften zu Berlin* (1868), 215–28. Gustav Kirchhoff, 'Zur Theorie freier Flüssigkeitsstrahlen', *Crelle's Journal für reine und angewandte Mathematik* 70 (1869), 289–98. Lord Rayleigh, 'On the Resistance of Fluids', *Philosophical Magazine* 2 (1876), 430–41; Lord Rayleigh, 'On the Irregular Flight of a Tennis Ball', *Messenger of Mathematics* 7 (1877), 14–16.

[95] Bloor, *The Enigma of the Aerofoil*, 77–8. I am indebted to Bloor's exemplary account in what follows.

[96] See ibid., 97. Taylor's thesis remained unpublished but he did draw on it for two subsequent 1916 reports: 'Pressure Distribution Around a Cylinder', and 'Pressure Distribution over the Wings of an Aeroplane in Flight', published in *Reports and Memoranda of the Advisory Committee for Aeronautics* 191 and 287 respectively.

know, as they do in the United States, of eminent designers who *see* a new type of machine rather than *design* it.[97]

The point is made in a less diplomatic way a year later by the founding editor of *The Aeroplane*, Charles Grey, who writes that much harm had been done 'both to the development of aeroplanes and to the good repute of genuine aeroplane designers by people who pose as "aeronautical experts" on the strength of being able to turn out strings of incomprehensible calculations resulting from empirical formulae based on debatable figures acquired from inconclusive experiments carried out by persons of doubtful reliability on instruments of problematic accuracy'.[98] Grey had no doubt that we should trust not the scientist but 'the man who guesses, and guesses right'.[99] In a 1922 short article on the development of aviation, W. H. Sayers, author of work on stability in aeroplane design, noted that developments before the Great War were the result of individual adventure, and 'individual designers worked, as artists worked, by a sort of inspiration as to what an aeroplane ought to be like, and built as nearly to their inspiration as the limited means, appliances and increasing knowledge they possessed would allow them'.[100] Brooke Hindle, in a detailed account of how invention typically proceeds by emulating established mechanical devices for new purposes, has described how visual imagination is central to invention, noting that many inventors were also artists.[101] As he remarks: 'Designing a machine requires good visual or spatial thinking. It requires mental arrangement, rearrangement, and manipulation of projected components and devices. It usually requires a trial construction of the machine, or at least a model of it, and then more mental manipulation of possible changes in order to bring it to an effective working condition.'[102]

The alternative to the discontinuity theory, which had now effectively become defunct, was circulation theory, which was more in tune with the concerns of engineers. The basic idea behind circulation theory is this.[103] It was known from Bernoulli onwards that, if the air behaves like an ideal fluid, then the faster the air flowing over a wing the less the pressure it exerts, and the slower the flow the greater the pressure. So for lift, the pressure of the moving air on the upper surface of the wing, which pushes downward, must be less than the pressure on the lower surface, which pushes upwards. It is this excess pressure that constitutes lift. If the

[97] John Henry Ledeboer, 'The Function of Literature', *Aeronautics* 11 (1916), 33.

[98] Charles Grey, Preface to Frank Barnwell, *Aeroplane Design* (London, 1917), 4.

[99] Charles Grey, 'Editorial Comment', *Aeroplane* 12 (1917), 1284.

[100] W. H. Sayers, 'The Arrest of Aerodynamic Development', *Aeroplane* 22 (1922) 138.

[101] Brooke Hindle, *Emulation and Invention* (New York, 1981).

[102] Brooke Hindle and Steven Lubar, *Engineers of Change: The American Industrial Revolution 1790–1860* (Washington, DC, 1986), 75.

[103] I draw here extensively on Bloor, *Enigma of the Aerofoil*, ch. 4. See also the collection of seminal papers collected by R. T. Jones, *Classical Aerodynamic Theory*, NASA Reference Publication 1050, 1979.

airflow around the wing follows the surface of the wing—i.e. is not discontinuous—then what needs to be explained is why the air immediately above the wing of an airborne aeroplane is moving more quickly than that immediately below the wing.

The circulation theory proceeds by distinguishing two distinct types of flow: a steady flow of air with constant speed and direction, and a swirling vortex that rotates around a central point, and whose speed drops off uniformly with distance from the centre. These are superimposed on one another (by a simple process of vector addition). Imagine—in two dimensions—the steady flow moving from left to right, and the vortex positioned to the right of it, the parts of the vortex moving in a clockwise direction. The motion of those parts of the vortex moving in the opposite direction to the flow will, on encountering the flow, be retarded because we will be subtracting one motion from the other, whereas the speed of those moving in the same direction will be increased, because we will be adding the motions. In the case envisaged, interaction of the airflow with the clockwise rotation would cause an increase in its speed over the wing and a decrease under the wing. Accordingly, if the wing generates a vortex, this would explain its lift.

The challenge for circulatory theories was to account for the origins of the assumed vortical motion. The first version of the theory was set out in Frederick Lanchester's 1907 *Aerodynamics*.[104] The argument starts with the observation that birds' wings, which have evolved into a shape that conforms to the pattern of airflow necessary for lift, have an arched profile with a slight downward inclination at the front edge. What must happen, Lanchester argues, is that the air must be moving upward as it approaches the leading edge of the wing and downward as it leaves the trailing edge. There is an exchange of momentum: the initial upward vertical component of the motion must be reduced to zero as the air passes over the wing, and then replaced with a downward vertical component.

Critics of the theory such as Taylor argued that it was lacking in physical content: it provided no mechanism by which circulation around the body could be created. Moreover, Taylor argued, Lanchester was assuming a perfect fluid, but in a stationary perfect fluid setting a body in motion could not create such a flow. As Bloor points out, Taylor's reaction to Lanchester assimilated Lanchester's analysis to the classical framework of perfect fluid theory, that is, to the equations of Euler and Laplace's equation. But Lanchester was well aware that rotation in an ideal fluid can neither be created nor destroyed: 'As an engineer working with real fluids, such as air and water, he hardly expected mathematical idealizations to be accurate. The important thing was to learn what one could from the idealized case but not be imposed upon by it.'[105]

---

[104] Frederick Lanchester, *Aerodynamics, constituting the First Volume of a Complete Work on Aerial Flight* (London, 1907).
[105] Bloor, *Enigma of the Aerofoil*, 147.

322 Technology and the Limits of Scientific Theorizing

One of the main complaints by critics of Lanchester was the narrow scope of his account. He focused on small angles of incidence, and defended the limited scope of his investigation, whereas in the Tripos tradition which his critics assumed, one might start with a single case but the point was then to generalize to all others. This was what genuine knowledge enabled one to do, and indeed this was the whole point of scientific investigation: one sought general explanations, not limited particular ones. There is nothing new in this, and the criticism largely mirrored criticisms of those working in experimental natural philosophy (in optics and pneumatics) in the seventeenth century. The general mathematics of flow accorded no special significance to the case of small angles of incidence that Lanchester examined. Accordingly, one objection that was made, for example by Cowley and Levy in their *Aeronautics in Theory and Experiment*,[106] was that Lanchester's choice was arbitrary, where the criterion whether something is arbitrary or not is a whether it is deducible from the basic equations of fluid dynamics. But Lanchester is not working to the criteria of fluid dynamics, which would have required a sharp separation between inviscid fluids (i.e. the perfect fluids of basic fluid mechanics) and viscous fluids. He includes both viscous and inviscid fluids in his account. This is not a lack of awareness of the standard mathematical procedures of aerodynamics on his part. Rather, it is the result of a flexibility in the handling of the mathematics.

In a memorandum written in 1936, well after the circulation theory had been accepted as the correct account of lift and drag, Lanchester wrote that the National Physical Laboratory had not taken his work seriously twenty years earlier because 'my methods did not appeal to them in view of their training. They mostly belonged to the Cambridge School, whereas I was a product of the Royal College of Science.'[107] What lies behind this difference is a deep divide between the view that the behaviour of aircraft wings is fundamentally a matter of a mathematically elaborated aerodynamics, and, by contrast, the bringing together of several different kinds of approach which may retain some degree of autonomy, but which can be manipulated with a view to providing a solution that makes sense in engineering terms.[108] The issues are best brought out by considering the way in which the German scientists/engineers of the *Technische Hochschule* dealt with the questions, and by focusing on the question of what the relation is between what the textbooks issuing from the Cambridge school had distinguished as ideal and real fluids.[109]

---

[106] William Lewis Cowley and Hyman Levy, *Aeronautics in Theory and Experiment* (London, 1918).

[107] Unpublished memorandum, quoted in Bloor, *Enigma of the Aerofoil*, 195. The Royal College subsequently became part of Imperial College.

[108] In this respect respect there are parallels with the establishment of atomism, which we looked at earlier, in Chapter 5.

[109] See, for example, Horace Lamb, *A Treatise on the Mathematical Theory of the Motion of Fluids* (Cambridge, 1879), and Horace Lamb, *Hydrodynamics* (5th edition, Cambridge, 1924).

Lanchester's circulation theory had been received enthusiastically right from the beginning by Ludwig Prandtl, who had been appointed professor of fluid mechanics in the Hannover *Technische Hochschule* in 1901. The *Technische Hochschulen* had been founded as what were in effect technical or engineering universities, and their future as research institutions was secured with the granting of the right to issue doctoral degrees in 1899.[110] Prandtl entered the *Technische Hochschule Munich* in 1894 and graduated with a doctorate six years later. He held seminars on Lanchester's *Aerodynamics* as soon as it appeared, immediately overseeing its German translation, and it was he who developed circulation theory in a particularly compelling way. Prandtl's student, Georg Fuhrmann, had carried out a number of experiments with model airships in a wind tunnel in 1910 to examine the predictions of ideal-fluid theory.[111] He distinguished two forms of drag: pressure drag, which acts normal to the surface, and is the kind of thing that can be caused by a perfect fluid; and friction drag, which acts at a tangent to the surface, and is caused by viscosity. He showed that the graph of the observed pressure distribution of the air flow was very close to that predicted by ideal-fluid theory except at the very tail of the airship models. Second, models with a rounded nose, slender body, and long tapered tail had extremely low resistance. Third, nearly all of the small residual drag could be accounted for by the frictional drag of the air on the surface. Even given the slight deviation at the tail and the effect of the air in immediate contact with the surface of the airship, what was surprising was the extent to which the air behaved like an ideal fluid.

In a report that he wrote up in 1923, Prandtl set out a survey of ideal-fluid theory followed by details of Fuhrmann's experiments, concluding that: 'The theoretical theorem that in the ideal fluid the resistance is zero, receives in this a brilliant confirmation by experiment.'[112] But as Bloor points out, for Prandtl's British counterparts, a 'theoretical theorem' could only be the result of deduction from the premises of the theory, and as such it would be something to be judged by logical, not experimental, criteria: it did not describe a real fluid but only an ideal one. Accordingly, Prandtl was claiming that the experiment provided 'brilliant confirmation' of what was essentially a mathematical theorem, which his critics considered ridiculous. By contrast, as Bloor notes, Prandtl and Fuhrmann 'found they could use the theory of ideal fluids to design airships that were very close approximations to the zero resistance entailed by the theory of perfect fluids. It helped them to identify the places where smooth flow was breaking down so that they could reduce it further. Their efforts were informed by an ideal

---

[110] On the development of technical education in Germany in the second half of the nineteenth century, see Kees Gispen, *New Profession, Old Order: Engineers and German Society, 1815–1914* (Cambridge, 1989), chs 3 and 4.

[111] Georg Fuhrmann, 'Theoretische und experimentelle Untersuchungen an Ballonmodellen', *Jahrbuch der Motorluftschiff-Studiengesellechaft* 5 (1911–12), 64–123.

[112] Ludwig Prandtl, 'Applications of Modern Hydrodynamics to Aeronautics', *National Advisory Committee for Aeronautics*, Report No. 116 (1923), 174.

that they were striving to attain. The ideal was not kept distinct from practice, or set in opposition to it, but was integral to it and gave practice its direction and purpose.'[113]

One thing that distinguishes the Tripos tradition of aerodynamics from that of the German engineers was set out clearly by Prandtl's thesis supervisor, August Föppl. The engineer, Föppl argued, needs to get answers, and this may mean using ideas that cannot be justified in terms of established physical knowledge. This is not unlike the difference I alluded to earlier, between the inventor and the scientist. For the inventor, it is an all-or-nothing exercise: one either produces the working object or one does not. Similarly the constraints on the engineer and the mathematician differ significantly. On Föppl's account, engineers need to develop their own concepts, which do not meet the logical demands of those forged in established theory. While the scientist can wait for inspiration, he writes, 'the engineer, by contrast, is subject to the force of necessity. He must, without delay, deal with the matter when some phenomenon interferes and interposes itself in his path. He must, in some way or another, arrive at a theoretical understanding of it as best he can.'[114] Accordingly, the engineer has a freedom to create new concepts that the scientist or mathematician does not have. Föppl doesn't deny that such concepts and procedures may be absorbed back into the main body of knowledge at some stage, but, by contrast with scientific concepts, this is not what their legitimacy consists in. None of the engineers directly asserted the literal truth of ideal-fluid theory, but they weren't worried about the fact that it was literally false because that didn't mean that one couldn't work with it.

It is in a 1923 paper by Hermann Glauert that we find the clearest statement of a rationale for the attempts of Prandtl and others to combine theoretical hydro-dynamics and aeronautical engineering.[115] Glauert identifies three steps in the solution of physical problems in aerodynamics. The first is the setting out of assumptions about what quantities—such as gravity, compressibility, and viscosity—can be neglected in dealing with the questions towards which the enquiry is directed. The second is the expression of the physical system in a mathematical form, that is, primarily, in terms of differential equations and boundary conditions. Third, the equations and the terms they contain must be manipulated until they yield numerical results that can be tested experimentally or used for some practical purpose. This may sound comparatively anodyne, but

---

[113] Bloor, *Enigma of the Aerofoil*, 178.

[114] Föppl, *Vorlesungen über technische Mechanik*, 11–12. See the discussion in Bloor, *Enigma of the Aerofoil*, 187–94.

[115] Hermann Glauert, 'Theoretical Relationships for the Lift and Drag of an Aerofoil Structure', *Journal of the Royal Aeronautical Society* 27 (1923), 512–18. This absolutely central paper has ramifications for all design professions, including architecture for example. Glauert came from an English germanophone family, and had gone through the Cambridge Tripos examinations (where he took a first class honours with distinction), but always had extensive connections with German engineers.

when Glauert comes to discuss the relationship between the first and the third stages, he captures something distinctive about the engineering approach. Because the mathematical problems may be insurmountable in the third stage, the engineer needs to return to the initial physical assumptions, simplifying them further or restricting attention to a smaller range of cases. There are, he notes, no absolutely rigid assumptions: any assumptions can be reworked to enable a solution to be yielded. In this way, the first and third steps become non-sequential, and this is the distinctive feature of the engineering approach, by contrast with the mathematical hydrodynamics approach. Problems encountered in the third step constantly feed back into the first step in an effort to make the problem tractable, and there is a constant movement between the first and third steps in a process of repeated problem redefinition. Applying this approach to the circulation theory, Glauert argues that while viscosity cannot be wholly neglected, this does not preclude the use of perfect fluid theory. We cannot ignore two facts about viscosity: the 'no-slip' condition, that is, at a solid boundary, a viscous fluid will have zero velocity relative to the boundary, by contrast with the behaviour of a perfect fluid, which has a finite slip; and second, the fact that viscous forces, which are proportional to the rate of change of velocity, are significant when close to the body but negligible at large distance. These are physical facts for which approximations must be found, but this cannot be done at Stage 1: the approximation is introduced after the fact, as it were, in Stage 3.[116]

For Glauert, under the right conditions, the equations of inviscid flow are legitimate approximations to the viscous equations. The mathematical hydrodynamics approach had separated out the two completely, arguing that ideal-flow theory was false and therefore could not be deployed in aerodynamics. But Glauert is arguing that there *is* a relationship between the two, because inviscid flow must be understood as a limiting case of viscous flow:

It is known that the solution obtained by ignoring the viscosity is unsatisfactory, but it is by no means obvious that the limiting solution obtained as the viscosity tends to zero is the same as the solution for zero viscosity. In particular, in the case of a body with a sharp edge, there is a region where the velocity gradient tends to infinity, and where the viscous forces will be of the same order of magnitude as the dynamic forces, however small the viscosity. On the other hand, the layer around the body in which viscosity is of importance can be conceived of as zero thickness in the limit, and this conception is equivalent to allowing slip on the surface of the body. It appears, therefore, that the non-viscous equations will be the same as the limit of the viscous equations, except in the region of sharp edges.[117]

In this context, consider a 1927 paper by Prandtl in which he asks how circulation arises (i.e. how an aircraft gets off the ground), and why perfect fluid theory,

---

[116] As noted in Bloor, *Enigma of the Aerofoil*, 329.
[117] Glauert, 'Theoretical Relationships', 514.

despite being strictly false, provides the solution.[118] On the first question, he argued that both his proposed circulatory flow and the irrotational flow of the mathematicians were consistent with the classical theorems of hydrodynamics. But only flow with circulation and the smooth confluence at the trailing edge is physically realizable. As Bloor notes, even if the process by which circulation was generated remained obscure, it was the logical possibility of circulation, and the logical right to postulate it, that really mattered for the perfect fluid approach.[119]

On the second question, Prandtl deploys his theory of the 'boundary layer', first developed in a paper published in 1905.[120] This paper is a classic instance of modelling achieving something that theorizing could not have done.[121] The problem to which it was directed was a discrepancy between the mathematical theory of fluid flow and experimental results: the former could give no account of why the very small frictional forces present in the flow of air or water around a body were able to create a no-slip condition at the solid boundary. To investigate this question, Prandtl built a small water tunnel in which fluid flowed past different kinds of bodies (Fig. 10.2). By these means, he was able to highlight different kinds of flows in regions of the fluid close to the body and in those more remote from it. He showed that in certain areas, frictional forces were insignificant, and in those where they were significant, approximations to the Navier–Stokes equations meant that they could be applied to the boundary layer. In this way he was able to provide a bridge that connected the behaviour of real viscous fluids and the unreal inviscid fluid of ideal-fluid theory. He showed how the field of flow could be divided into two areas separated by a boundary: a viscous flow inside the boundary layer, which was responsible for the drag experienced by the body; and one outside the boundary where viscosity was negligible, so that its behaviour was very close to that of an ideal fluid.

In a 1927 paper, he applies the boundary layer theory directly to the question of the success of perfect fluid theory in explaining lift. He notes that the boundary layer is the cause of the formation of vortices. The equations of classical hydrodynamics only apply to viscid flows, with the result that, in motion starting from rest, circulation will be zero (or close enough to zero) in fluid circuits that do not pass through the boundary layer. It is because of this that real fluids behave like perfect fluids at a distance from the solid body. But the boundary layer can be manipulated, as Prandtl demonstrated with photographs and films of experiments where it had been removed by suction, and this had a dramatic effect on the flow. This is Prandtl's justification for using perfect fluid theory. As he puts it: 'We thus get the unique characteristic that it is precisely these turbulent flows of low

[118] Ludwig Prandtl, 'The Generation of Vortices in Fluids of Small Viscosity', *Journal of the Royal Aeronautical Society* 31 (1927), 720–41.

[119] Bloor, *Enigma of the Aerofoil*, 387.

[120] Ludwig Prandtl, 'Über Flüssigkeitbewegung bei sehr kleiner Reibung, in A. Krazer, ed., *Verhandlungen des dritten Internationalen Mathematiker-Kongresses* (Leipzig, 1905).

[121] See the discussion in Morrison, 'Models as Autonomous Agents'.

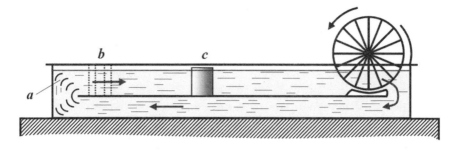

Fig. 10.2. Prandtl's hydraulic tunnel (1904)

resistance around bodies which can be so closely represented by the theory of the perfect liquid'.[122]

What we have in these debates in aerodynamics is a conflict between the procedures of mathematical physics, which in the Cambridge Tripos system for example are treated as models of physical explanation, and those of engineers whose procedures were taken by the Cambridge mathematicians as being at best pragmatic, but which I hope will now be clear should be seen rather as a different way of working, one which involves a degree of freedom or flexibility that is inappropriate in the mathematical tradition. The flexibility can involve mixing mathematical and experimental results, and regularly, and as a matter of course, going back to revisit premises when arguments based on these premises fail to deliver. Such constant revision and reworking is not compatible with the idea of deductively watertight inference that regulated the approach of the Cambridge mathematicians.

The issues that arise in such a stark form in the disputes over aerodynamics are not new ones. They mirror those in debates between mechanists and proponents of experimental natural philosophy in the seventeenth and eighteenth centuries.

---

[122] Prandtl, 'The Generation of Vortices', 739.

In considering the seventeenth-century cases in the first volume in this series,[123] I drew attention to the difference in how the two parties constructed the domain of investigation of physical enquiry. This difference was premised on two very different understandings of 'science' (or natural philosophy as it was then). For the mechanists, explanation was a matter of accounting for macroscopic behaviour in terms of the mechanically described behaviour of the corpuscules/atoms making up all material things. But in the 1660s, Boyle and Newton, though committed to this as a general model, found that in order to account for certain phenomena (pneumatic in Boyle's case, optical in Newton's) in a satisfactory way they had to suspend their commitment to corpuscularianism as the sole reference point for explaining physical behaviour. Since they also rejected what was in effect the only alternative comprehensive form of foundational explanation, namely an appeal to Aristotelian essences, they needed some way of organizing the phenomena under investigation other than in terms of underlying structure. Now underlying structure plays a crucial role in suggesting how the phenomena might be organized, not least in virtue of the clues it provides as to the connections between the phenomena. But, particularly in the case of the optical spectrum, these clues may turn out to be misguided, as Newton realized. One therefore has to proceed without the benefit of any supposed clues, but there is one possible source of guidance, the experimental apparatus itself. This produces a certain range of phenomena which defy explanation in 'fundamental' terms, but which cannot be dismissed because the results themselves cannot be faulted. The only way to proceed is to take the results at face value and start from them, but from a foundationalist perspective these results show no internal coherence, because in corpuscularian terms they are anomalous. The way in which they are generated is therefore crucial, not just because this is what legitimates them but also because, if they have any coherence at all, it has something to do with the way in which they are generated. It is the way in which they are generated, for example, that holds them together as connected phenomena, and that excludes what might otherwise seem—for example, on corpuscularian grounds—to be related phenomena. The way in which the results are generated is a function of the experimental apparatus, the way in which this apparatus is manipulated, and what one is able to do with it. The experiment or instruments bring a domain of investigation into focus, replacing the underlying structure that would traditionally have occupied this role. What occurs is a form of tailoring of the explanation to results produced by particular experiments or instruments, in short, a tailoring of *explanans* to *explanandum*, thereby reversing the normal direction of enquiry. But note that it is not a question of one side alone having to tailor things. The seventeenth-century mechanists always and as a matter of routine had to tailor the *explanandum* to

[123] Gaukroger, *The Emergence of a Scientific Culture*, ch. 10. I summarize material from this chapter here.

the *explanans*, and this tailoring—by means of the doctrine of primary and secondary qualities, for example—was in fact as problematic as it was radical.

The rejection—in Boyle and Newton in the seventeenth century, in Gray in the eighteenth, and in Faraday in the nineteenth—of the idea that a natural-philosophical system determines in advance what is to be explained and what is not, has obvious parallels with the engineers' rejection of the mathematicians' attempt to specify in advance the field of aeronautics and what resources one is and is not entitled to use in exploring it. In other words, science in the modern era has always been structurally complex. There is no doubt that the addition of engineering to experiment as a crucial part of the scientific enterprise has been very significant, but it has highlighted rather than transformed what might be termed the problem of the non-discursive products of science. The construction of ships in the eighteenth century, for example, has a number of parallels with the construction of aeroplanes in the twentieth. As Larry Ferreiro notes: 'sailing ships were the most complex engineering structures of the day. They combined heavy wooden construction of hull and masts with a dizzying array of standing rigging to support the masts, hundreds of lines and blocks to control the yardarms and sails, capstans for hauling up the anchors, tillers and wheels to turn the rudder, bilge pumps, and such, for which the constructors had overall responsibility to integrate into the ship.'[124] Even more striking is the fact that there were two opposing approaches to ship-building in the early decades of the eighteenth century: that of the French, who were trying to rationalize the various trades along the lines of a scientific method, and that of the (far more successful) British, whose constructors picked up their skills in shipyards through apprenticeships, and were often familiar with arithmetic and geometry in as far as it bore on their craft, but were otherwise illiterate.[125]

THE NON-DISCURSIVE PRODUCTS OF SCIENCE

These issues have a direct bearing on how we conceive of the non-discursive products of the science/engineering complex. In general terms, the question is: once we have abandoned the idea of technology as 'applied science'—that is to say, once we have abandoned the idea that science is essentially something purely theoretical, which may or may not then be applied—what are the consequences for our understanding of what science is and what it does? In most respects, asking this question is simply in the nature of an alert, for it goes beyond anything that we can attempt to answer here. But there are two issues that do require our immediate attention, however cursory. These are the nature of the products of

---

[124] Larrie D. Ferreiro, *Ships and Science: The Birth of Naval Architecture in the Scientific Revolution, 1600–1800* (Cambridge, MA, 2006), 24.

[125] Ibid., 24–6.

science/engineering, and the nature of the conditions under which they are produced and used. Moreover, we need to distinguish between the period between the middle of the nineteenth century and the First World War, when the products were predominantly machines, and the interwar period, when they became predominantly complex systems.[126]

Consider the period before the Great War first. What is at issue is not just the construction of these products, but their successful operation. From our discussion of aeronautics, one might be led to think in terms of the reconciliation of scientific knowledge with engineering knowledge. But that is misleading. Engineering produces a very different kind of product from science as traditionally conceived (i.e. knowledge). And if we are to understand the products of the integration of science and engineering, then, crudely put, it is a question of moving from a knowledge model to a working machine model. We might think of machines (devices that apply a power source other than human or animal) as embodied knowledge, but there is a complicating factor: the knowledge that they embody is not just that of the construction of the machine, but also that of its successful operation.

In asking for the conditions under which particular machines can actually be used, we can start by noting that there has often been a significant delay between the invention of a particular kind of machine and the use of the machine. Steam engines are a case in point. Newcomen's steam engine, first installed at a mine in 1712, has sometimes been lauded as heralding the Industrial Revolution in Britain, but steam engines were marginal in production in eighteenth-century Britain and, it would seem, from public awareness: there is no entry for them in Johnson's *Dictionary*, for example, and there is only one mention, in passing, of a 'fire engine' in Smith's *Wealth of Nations*. Early railways had been horse-drawn, working with an inclined plane over hilly ground, and as late as 1829, in the wake of the introduction of the Stephenson engine, there were still those who thought trucks pulled by a cable or by horses might be more efficient than steam engines. One issue was safety. If heavy machinery exploded the results were disastrous. British canon manufacturers, for example, were well aware that hidden hollows and impurities in the metal used in French canons in the Seven Years War had caused the canon to blow up at the point of firing with deadly consequences, and the pressure built up in steam engines was a significant cause of concern in this respect. James Watt was particularly opposed to high-pressure steam engines on safety grounds, regarding their use as irresponsible. Richard Trevithick, inventor of one of the most advanced forms of steam engine, wrote to a friend: 'I have been branded with folly and madness for attempting what the world calls impossibilities, and even from the great engineer, the late Mr James Watt, who said to an eminent scientific character still living, that I deserved hanging for bringing into

---

[126] Here the category of 'technoscience' is potentially problematic, for it tends to collapse these two into one another.

use the high-pressure engine.'[127] Another issue was the weight of the engine. Horses put no pressure on the rails at all, but steam engines were heavy and could crack and distort the rails, which meant that engines with full loads running on a regular basis could not run without wholly new, stronger sets of rails, but what were these to be made of?

Rails and roads provide a good example of the difficulties in getting technology to work in the requisite (economical and safe) way. The problem with rails was that wooden rails were too soft and perishable whereas iron ones were too brittle. The solution did eventually come, with the manufacture of wrought iron rails. The method of turning scrap iron into wrought iron, by melting it down with raw coal, in a process known as 'puddling and rolling', had been patented in 1783, but the crucial developments came only when the process was improved radically with the introduction of super-heated blast furnaces in 1828. The greater malleability and tensile strength of the wrought iron tracks, which from 1828 were rolled out in appropriate lengths, were as crucial for the sudden take off of interest in railways in the 1830s as were the engines. At the same time as the transformation of railway tracks, Britain witnessed a transformation of its roads. Significant attention had been paid to alleviating problems with road transport by improving the suspension of carriages in the eighteenth and early nineteenth centuries, to little avail. But in 1816 John Loudon McAdam, then sixty years old and with no scientific background or engineering experience, realized that the problem lay not in the suspension of the vehicle but in the surfaces of the roads on which it travelled. The solution was to provide a top layer of small stones which, when compressed by carriage wheels, formed a smooth well-drained surface.[128]

Steam engines, no matter how sophisticated, would not have been able to have anything like the impact they had without the wholly independent discovery of an economic process for the production of wrought iron. Nor could rail travel have been successful without the wholly independent invention of the electrical telegraph, which provided the basis for the long-distance signalling technology that allowed for the systematization and coordination of railway travel. And the same time, motor vehicles would have proved useless without prior innovations of the kind provided by McAdam for roads. Technological developments are dead ends without the right conditions for use. But it is not just a matter of material conditions here. Levels of expertise also play a necessary role in establishing the right conditions for use.

The question of machine expertise was paramount. The kind of large-scale machinery deployed in cotton and woollen mills, for example, was such that the manufacturer needed to dismantle the machinery and ship it in parts. This meant that a skilled engineer was needed at its destination to reassemble it. Consider the case of Benjamin Phillips' transfer of machines, including a version of Compton's

---

[127] Quoted in Weightman, *The Industrial Revolutionaries*, 49.   [128] Ibid., 123–5.

mule, to America in 1783. To avoid the customs restrictions caused by the Revolutionary Wars, he packed the machines in pottery casks, loading them on a ship at Liverpool and sending his son on to Philadelphia to receive them. Phillips himself was to accompany the machine parts, reconstructing them when he arrived in America, but he fell ill on the voyage and had to be sent home, where he died shortly afterwards, so it was left to his son to reassemble them on their arrival. But this was a formidable task, one the son was unable to accomplish, and ultimately he was forced to sell the parts on. The new purchaser found that he also was unable to assemble them, and ended up putting them in storage, eventually selling them on himself, to a new purchaser who realized he had no option but to send them back to England, where, now successfully reassembled, they stayed.[129]

Discussion of technological expertise has tended to focus on the role of expertise in the design of machines, but the role of expertise in the working of machines is equally important. Technology does not consist simply of machines, but is rather a union of machines and specific skills. This was perhaps most evident in the popular electrical machine exhibitions of the mid-nineteenth century. As Morus has pointed out, such exhibitions 'provided spaces where the machine-maker as showman could demonstrate his control over the machine and over the audience for his display. Those who worked and displayed the results of their labors at such places were skilled mechanics who needed a space where by demonstrating their control over their machines and experiments, they could also demonstrate their ownership of these technologies and the phenomena they produced. Invention was very much about controlling the products of labor.'[130]

Similarly in the more mundane context of commerce. Exporting machines typically meant sending an engineer who could set up the machine and keep it working. In some cases the skills required were not unlike those of the experimental scientist. This was particularly clear in the seventeenth century, when the two roles were performed by the one person. Hooke and Huygens noted that they were the only two people who were able to get their air pumps, particularly mechanically complex machines, to work satisfactorily. Here designing the machine and working it successfully are part of the same thing, and this, I suggest, should be our model for understanding how machines function. Machines and their successful manipulation go together. And just as the roles of invention of the machine and enabling it to work are part of the same enterprise, the physical object that constitutes the machine cannot be separated from the process by which it functions and the conditions under which this functioning is successful. But the non-discursive products of the science–engineering complex are not

---

[129] Ibid., 38–9. More generally see David Jeremy, *Artisans, Entrepreneurs and Machines: Essays on the Early Anglo-American Textile Industries* (Aldershot, 1998).
[130] Morus, *Frankenstein's Children*, 71.

confined to objects such as machines, and it is important that we register the transformations of the non-discursive products of the science–engineering complex that are typical of the interwar period.

Hughes draws attention to the problems in simply extrapolating from the prewar machine technology to that of the interwar period:

> To associate modern technology solely with individual machines and devices is to overlook deeper currents of modern technology that gathered strength and direction during the half-century after Thomas Edison established his invention factory at Menlo Park. Today machines such as the automobile and the airplane are omnipresent. Because they are mechanical and physical, they are not too difficult to comprehend. Machines like these, however, are usually merely components in highly organized and controlled technological systems. Such systems are difficult to comprehend, because they also include complex components, such as people and organizations, and because they often consist of physical components, such as the chemical and electrical, other than the mechanical. Large systems—energy, production, communication, and transportation—compose the essence of modern technology.[131]

As representative of the early developers of electrical technology, Hughes takes the electrical light and power systems of Samuel Insull, noting how different these were from the mechanical products of Ford and others. In particular, it was concepts of electrical circuitry rather than machinery that shaped the ways in which builders of electrical systems thought and acted: 'they manipulated interactions, not the simpler linear relations of cause and effect. The builders of electric-power and chemical-process plants also envisaged flow rather than the movement of batches of materials and mechanical parts. Instead of being the age of the machine, the interwar years emerged as the apogee of the age of electric power and chemical processes. The machine as a symbol of an age applies better to the British Industrial Revolution of more than a century earlier.'[132]

What is at issue here is not confined to the production of commercial products, but runs through the whole of the science–engineering complex, impacting on

---

[131] Hughes, *American Genesis*, 184–5. Hughes defines a system as being 'constituted of related parts or components. These components are connected by a network, or structure, which for the student of systems may be of more interest than the components. The interconnected components of technical systems are often centrally controlled, and usually the limits of the system are established by the extent of this control. Controls are exercised in order to optimise the system's performance and to direct the system toward the achievement of goals. The goal of an electric production system, for example, is to transform available energy supply, or input, into desired output, or demand. Because the components are related by the network of interconnections, the state, or activity, of one component influences the state, or activity, of other components in the system. The network provides a distinctive configuration for the system.' Thomas P. Hughes, *Networks of Power: Electrification in Western Society 1880–1930* (Baltimore, 1983), 5.

[132] Hughes, *American Genesis*, 186. Cf. Jon Agar's account, in *The Government Machine: A Revolutionary History of the Computer* (Cambridge, MA, 2003), of the uptake of technology in the British civil service, and the way in which the civil service, 'cast as a general-purpose universal machine, framed the language of what a computer was and could do.' (3) See also Agar, *Science in the Twentieth Century and Beyond*, ch.16, on information systems.

such 'basic' questions as the nature of experimental enquiry. To understand how the results of technologically complex experiments in nuclear physics are generated for example, it is often no longer a question of constructing the apparatus for oneself and checking that the results are as claimed. It is rather one of learning how the experimental set-up works in the course of prolonged visits to the laboratory where the experiment has been performed, learning the skills involved and the protocols used.[133] Laboratories have become large-scale enterprises, with all the organizational complexity this brings with it, and recognition of this has sometimes been slow. As Daniel Todes has noted, for example, in each of the four consecutive years that Pavlov was considered for the Nobel Prize, the committee were unable to decide how to credit the experimental work done in his large laboratory complex:

Guided by the image of the heroic lone investigator, the Nobel Prize Committee here confronted a different form of scientific production. Just as the workshop was yielding pride of place to the factory in goods production, so, as the nineteenth century wore on, were leading laboratory scientists increasingly likely to be the managers of large-scale enterprises, Justus von Liebig and Felix Hoppe-Seyler in chemistry, Karl Ludwig and Michael Foster in physiology, Robert Koch and Louis Pasteur in bacteriology, and Paul Ehrlich in immunology all presided over distinctively social enterprises involving substantial capital investment, a relatively large workforce, a division of labour, and a productive process that involved managerial decisions.[134]

The general lesson from the developments that we have examined in this chapter is that taking technological and engineering developments seriously—both in the production of explanations, and also in the generation of non-discursive products such as aeroplanes and electrical systems—means rethinking the nature of science. Instead of converging on an increasing unity, the modularity and pluralism of science become not simply reinforced but wholly transformed through the intimate connections between science and technology. As if that isn't enough, once we abandon the attempt to see science as driving modernity and focus instead on technology, we find that the idea that technology drives modernity is open to different kinds of objections, as we saw in Chapter 2. But if neither science nor technology, individually or collectively, can be considered to drive modernity, what then is their relation to it? To pose this question fruitfully we need to ask how science was promoted as carrying the values of progress and modernity in the nineteenth and twentieth centuries. The answer lies in what has been termed 'popular science', and I now want to show how popularized and

---

[133] See Collins, *Changing Order*; and Galison, *How Experiments End*.

[134] Daniel P. Todes, 'Pavlov's Physiology Factory', *Isis* 88 (1997), 205–46: 206. Cf. the remark of Edison's chief engineer, Francis Jehl: 'Edison is in reality a collective noun and means the work of many men': quoted in Richard Munson, *From Edison to Enron: The Business of Power and what it Means for the Future of Electricity* (Westport, CT, 2005), 14.

fictionalized forms of science played a formative role in establishing the standing of science. These have shaped the values that have been associated with science in the nineteenth and twentieth centuries to such an extent that it would not have been possible for it to have the standing it has had without them. In the end, it was not science as a model of truth that placed it at the centre of modern culture, but science as a model for the future.

11

Science For and By the Public

In examining the successful establishment, in the public realm, of the standing of science as a unique and exclusive source of understanding, we need to ask how science, typically promoted as a unified enterprise, was able to interact with a culture outside that of professional scientists and philosophers. In order to do this, clearly a crucial goal must be to understand how those outside the community of scientists, engineers, and philosophers would have encountered science. In particular, we need to identify how science was debated and promoted—consciously and unconsciously, successfully and unsuccessfully—in such a way that its public standing was established, both as a model for understanding the world and for understanding our place in it (the question of 'what it is to be human').

Although the responses to the newly extended standing of science by scientists and philosophers often had a direct bearing on a public agenda, they were in the first instance usually directed towards a narrow intellectual elite. During the eighteenth century, there had only been one institution in Britain devoted exclusively to the study of natural philosophy, the Royal Society. When the Royal Institution was founded in March 1799 at a meeting at the house of the president of the Royal Society, the plan was that the two would complement one another, as is evident from their titles. The full title of the Royal Society is 'The Royal Society of London for the Improving of Natural Knowledge', whereas what was to become The Royal Institution of Great Britain was designated as the 'Institution for Diffusing the Knowledge, and Facilitating the General Introduction, of Useful Mechanical Inventions and Improvements; and for teaching, by Courses of Philosophical Lectures and Experiments, the application of Science to the Common Purposes of Life'. One overriding aim in the formation of the Royal Institution was practical, especially the use of chemistry to improve agriculture, Britain having been effectively cut off from continental trade by the war with France. But with an annual subscription of five guineas, attendance at the Friday-night lectures was effectively limited to members and their guests. As the regulations governing the Friday-night meetings state: 'All the Members of the Institution have the privilege of introducing their Friends either personally or by proper tickets signed by themselves. No person, except Members of the Royal Institution or those especially invited by the Board of Managers can be admitted unless

Civilization and the Culture of Science: Science and the Shaping of Modernity, 1795–1935. Stephen Gaukroger, Oxford University Press (2020). © Stephen Gaukroger.
DOI: 10.1093/oso/9780198849070.001.0001

personally introduced by a Member or the production of a ticket signed by a Member.'[1]

But things were changing. The 1820s heralded a particularly productive era in England as far as the furtherance of science and technology was concerned.[2] The Astronomical Society and the Cambridge Philosophical Society were both formed in 1820, the former comprising both full-time scientists and 'gentlemen astronomers'. The Cambridge Philosophical Society was a forum for specialists, as were the London Electrical Society, founded in 1837, and the Chemical Society, founded in 1841. The Royal Society was predominantly (but not exclusively) for specialists.[3] By contrast, the British Association for the Advancement of Science, formed in 1831, was catholic in its membership, and could draw on very large audiences of scientists and gentlemen amateurs. Moreover, although it had originally aimed to exclude women from everything except public lectures, the measures were unsuccessful, as 'hordes' of scientifically curious 'elegant females' used their numbers to force their way into the early meetings, and thereafter remained active attendees.[4] By the time of the Great Exhibition in 1851, the attitude to science had been thoroughly transformed, as every kind of science education institution started to flourish, from the recently established mechanics institutes to the foundation of the medicine and science departments of the new University College and King's College, London.

From the middle of the nineteenth century, however, there were also responses that took place at the popular level and were concerned primarily with assessing scientific developments in terms of their social, political, and religious consequences. The debates over mesmerism in France in the 1780s and 1790s, for example, had questioned the basis on which members of scientific and medical establishment had a claim to expertise and authority in assessing new theories and cures.[5] A similar episode occurred in response to criticisms of Chambers' *Vestiges* in the 1840s and 1850s. Chambers responded to negative reviews by appealing to

---

[1] Managers' Minutes, in Frank Greenaway, ed., *Archives of the Royal Institution* (15 vols, London, 1971–6), 22 January 1827. Cf. Frank James, 'Running the Royal Institution: Faraday as an Administrator', in Frank James, ed., *'The Common Purposes of Life': Science and Society at the Royal Institution of Great Britain* (Aldershot, 2002), 119–46: 140.

[2] There were also developments in scientifically aware Edinburgh. Note in particular the Edinburgh Association, in which shopkeepers and small businessmen played a prominent role, which promoted its own radical brand of transformationist biology and phrenology from the 1820s. See Steven Shapin, 'Phrenological Knowledge and the Social Structure of Early Nineteenth-Century Edinburgh', *Annals of Science* 32 (1975), 219–43; Steven Shapin, ' "Nibbling at the Teats of Science": Edinburgh and the Diffusion of Science in the 1830s', in I. Inkster and J. Morell, eds, *Metropolis and Province: Science in British Culture, 1780–1850* (London, 1983), 151–78.

[3] In the 1830s and 1840s, the Royal Society experienced a transformation from what has aptly been described as from an absolute to a constitutional monarchy: Roy MacLeod, 'Whigs and Savants: Reflections on the Reform Movement in the Royal Society, 1830-1848', in I. Inkster and J. Morell, eds, *Metropolis and Province: Science in British Culture, 1780–1850* (London, 1983), 55–90: 56–7.

[4] Morrell and Thackray, *Gentlemen of Science*, 148–57.

[5] See Gaukroger, *The Natural and the Human*, 155–64

the public, challenging the claims of scientists to judge his work, and appealing to the rights of the layperson to speculate in scientific matters.[6] The evolutionary debates of the second half of the nineteenth century—first over *Vestiges* and then Darwin's *Origins of Species*—were genuinely public debates over science that reached across society as a whole, and presented a great challenge. The charge against evolution was that it turned life into an amoral chaos displaying no evidence of a divine authority or any sense of purpose or design. Because science had an established public standing by the nineteenth century, its advocates needed to be able to respond to such criticisms, in the process perhaps engaging questions that they would otherwise have been inclined to ignore. Such disputes reshaped science in crucial respects. They were far from merely ephemeral, and went to the core of the question of what values science brought to the understanding of civilization that it was fostering.

The forum in which debates took place changed somewhat in the course of the nineteenth century however. In the last three decades, both scientific and non-scientific disciplines became more specialized: scientists were now publicly accredited and separated from the general educated public. This marked a very significant shift in how the literate public engaged with science. Accepting the values of science meant more than just accepting particular scientific theories. With the emergence in the nineteenth century of what has come to be known as a knowledge society, science had to reform its esoteric image and to appeal to a broad spectrum of society, not just practising scientists. Our primary concern in this chapter is with the understanding of science and its social and cultural roles from the perspective of its reception and reworking for a wider public. The reworking of science comes as part of the reception package, for in asking how science was able to interact with a culture outside that of professional scientists and philosophers, we need to be alert to the reciprocal nature of this interaction. It is not just a question of exploring how those outside this culture encountered science. At the same time we need to understand how the image of science had to be reshaped, as scientific values became embedded in those of civilization, resulting in a new set of demands being made on science if it was to establish its legitimacy in such a role.

'Popular science' was crucial here, but we must be as circumspect about the category as we have been about that of 'applied science'. As I indicated in Chapter 10, on traditional conceptions of both, science is taken as complete in itself, with two wholly dependent subsidiary exercises, offering applications to practical questions in the one case, and simplifications for a wider public in the other. The second is as questionable as was the first. As Stephen Hilgartner notes:

---

⁶ [Robert Chambers], *Explanations, A Sequel* (London, 1845). See Richard Yeo, 'Science and Intellectual Authority in Mid-Nineteenth Century Britain: Robert Chambers and Vestiges of the Natural History of Creation', *Victorian Studies* 28 (1984), 5–31.

The culturally-dominant view of the popularization of science is rooted in the idealized notion of pure, genuine scientific knowledge against which popularized knowledge is contrasted. A two-stage model is assumed: first, scientists develop genuine scientific knowledge; subsequently, popularizers disseminate simplified accounts to the public. Moreover, the dominant view holds that any difference between genuine and popularized science must be caused by 'distortion' or 'degradation' of the original truths. Thus popularization is, at best, 'appropriate simplification'—a necessary (albeit low status) educational activity of simplifying science for non-specialists. At worst, popularization is 'pollution', the 'distortion' of science by such outsiders as journalists, and by a public that misunderstands much of what it reads.[7]

The parallels with 'applied science' are striking. Hilgartner continues:

A concept of purity requires one of contamination, and the notion of popularization shores up an idealized view of genuine, objective, scientifically-certified knowledge. Furthermore, the dominant view establishes genuine scientific knowledge, the epistemic 'gold standard', as the exclusive preserve of scientists; policy makers and the public can only grasp simplified representations. Finally, this view of popularization grants scientists broad authority to determine which simplifications are 'appropriate' (and therefore usable) and which are 'distortions' (and therefore useless—or worse).[8]

This kind of assessment was shared by many nineteenth-century writers. The anonymous author of an 1875 piece in the *Saturday Review* for example complains that 'real science' is being used through a 'counterfeit' to secure 'the patronage of the vulgar', who were attending the Royal Institution lectures either because they were simply followers of the latest fashion or because, despite the best intentions, they were 'incapacitated by lack of general education'. Popular science has been invented, the writer complained, 'to accommodate such feeble votaries'.[9] As H. G. Wells remarked twenty years later, 'popular science, it is to be feared, is a phrase that conveys a certain flavour of contempt to many a scientific worker.'[10]

In fact, popularizations serve a different role from science itself, not a diluted role, and, like technology, they cannot be treated merely as an added extra. They are the primary vehicle by which the standing of science as a cultural force is established. As Bernard Lightman notes, in criticizing the view that popularization consists simply in disseminating simplified accounts of science to a passive

---

[7] Stephen Hilgartner, 'The Dominant View of Popularization: Conceptual Problems, Political Uses', *Social Studies of Science* 20 (1990), 519–39: 519.

[8] Ibid., 520. Cf. Bernard Lightman, *Victorian Popularizers of Science: Designing Nature for a New Audience* (Chicago, 2007), ch. 1.

[9] Anon, 'Sensational Science', *Saturday Review* 40 (11 September 1875), 321–2. Quoted in Lightman, *Victorian Popularizers of Science*, vii.

[10] H. G. Wells, 'Popularising Science', *Nature* 50 (26 July 1894), 300–1: 300. Cf. the discussion of the stigma attached to professional scientists who wrote popular science material in Peter J. Bowler, *Science for All: The Popularization of Science in Early Twentieth-Century Britain* (Chicago, 2009), ch. 1.

readership, this model cannot be adopted as a heuristic guide to research 'because it uncritically assumes the existence of two independent, homogeneous cultures, elite and popular, and forces the latter into a purely passive role. Popular culture can actively produce its own indigenous science, or can transform the products of elite culture in the process of appropriating them, or can substantially affect the nature of elite science as the price of consuming the knowledge it is offered'.[11] It is crucial that we do not think of the difference in terms of science, considered as raw material, being 'packaged' for a popular audience. The research science literature packages science just as much as the popular science literature does, just differently. 'Packaging' is just a means of presentation, and there is no science that is not presented in some way. To the extent to which the research science literature is considered to give us 'raw' data, note that we rarely get accounts of contentious background assumptions, the choice and sifting of evidence, the dead ends that have been pursued, or how the procedures followed compare with competing programmes. Scientific research papers and books package the material presented in a way that matches the skills and expectations of its readers.[12] The different forms of packaging are designed to meet different ends. In the case of the popular science of the nineteenth century, particularly natural history and astronomy, the aims are similar to those of the metascientific accounts that we looked at in Chapter 9. Popular science shapes how science is presented and received at the metascientific level. Here, specific values, such as objectivity, meritocracy, scientific certainty, and the role of science in modern civilization, are shaped and promoted, no less than they are in the philosophical attempts to rethink the nature of science that we have looked at. And their impact has been significantly greater than the philosophical accounts. As with the metascientific accounts that we examined in Part III, which for these purposes could have been included under the broad rubric of 'popular science' along with the religious and other rationalizing accounts that we shall be looking at in what follows, these compete to shape the understanding of science, and it is these that tell the scientists what society

[11] Lightman, *Victorian Popularizers of Science*, 10. Cf. Richard Whitely, 'Knowledge Producers and Knowledge Acquirers: Popularisation as a Relation between Scientific Fields and Their Publics', in T. Shinn and R. Whitley, eds, *Expository Science: Forms and Functions of Popularisation* (Dordrecht, 1985), 3–28; Roger Cooter and Stephen Pumfrey, 'Separate Spheres and Public Places: Reflections on the History of Science Popularisation and Science in Popular Culture', *History of Science* 32 (1994), 237–67.

[12] Distinguishing science from popular science is not completely straightforward. One obvious way to distinguish science from popularizations is in terms of its audience. Morag Shiach suggests that 'popular just be taken to mean non-specialist, a group whose identity and constitution vary'. *Discourse on Popular Culture: Class, Gender and History in Cultural Analysis, 1730 to the Present* (Cambridge, 1989), 27–8. But there is significant difficulty identifying the difference between a specialist and a popular audience. Does a geneticist, reading an article in a general scientific journal about fundamental particle physics, count as a specialist or not? Does a physician who has never done any medical research count as a specialist in reading pharmaceutical research papers? Clearly, just being a scientist does not mean that one should be counted a specialist rather than a member of the educated public for these purposes.

expects of their work. Popular science is not research science (allowing for grey areas, one of which we will turn to in the last section), but the two are inseparable and mutually reinforcing in establishing the credentials of science as the core of modern culture.

Just as the philosophical metascientific accounts could have been included under popular science, popular science itself could in many respects have been included under metascience. But whereas the former had focused on epistemology, many of the developments that we shall be examining in this chapter worked at a religious level, and the question that came to the fore was that of wonder. Secular conceptions of science started at something of a disadvantage here, but the stakes could not have been higher. In order to fulfil the vastly expanded role that it had assumed from the end of the eighteenth century, one in which it was displacing religion as the key to understanding our place in the world, it was crucial that science be extended into the non-propositional realm—that is, engaging with the world in terms of desires, expectations, anxieties, fears, hopes, goals, raw beliefs, etc. This move can be seen in terms of a resurrection of a value that had looked to have been decisively abandoned in favour of intellectual curiosity in the sixteenth and seventeenth centuries: wonder. The shift from wonder to curiosity was a formative cultural development in early modern natural philosophy.[13] Wonder had traditionally been deemed pious, an unquestioning acceptance of divine authority and divinely ordained limits to enquiry, by contrast with curiosity, which had been dismissed as inquisitiveness for its own sake, a sin of Adamic proportions. In the early modern reversal of values, wonder was transformed into something unintelligent and stupefying, while curiosity became a cardinal virtue, an expression of a wholly legitimate desire to know, and indeed a route to self-understanding. Given its new roles, if nineteenth-century science was to enhance its ability to engage questions of non-propositional understanding, then it had to reinvent a concern with wonder.

But did this need to be a question of 'reinvention'? Wonder had never been completely abandoned in Christian culture, and it was able to be adapted to the new circumstances provided by the rise of disputes in natural history in the course of the nineteenth century. Here, in the popular science literature, there began to emerge something that looked like a perfect opportunity to provide a religious framework for the understanding and appreciation of science.[14] Lightman notes that female popularizers, in particular, 'developed an impressive array of narrative formats for communicating with their newly defined audience. Whatever genre they used, they all made aesthetic, moral, and religious themes central to their

---

[13] See, for example, Part III of Blumenberg, *The Legitimacy of the Modern Age*; and Lorraine Daston and Katherine Park, *Wonders and the Order of Nature, 1150–1750* (New York, 2001).

[14] As Mary Ward put it: 'no Entomologist can be an atheist.' Mary Ward and Lady Jane Mahon, *Entomology in Sport, and Entomology in Earnest* (London, [1859]).

work. Like the women of the earlier maternal generation, they presented themselves as guides to understanding the larger significance of scientific theories.'[15]

To the extent to which secular conceptions were able to counter these developments with a non-divine understanding of wonder, what they initially had to offer looked derivative: a secularized version of a religious notion with which they were distinctly uncomfortable. The problem was that they were fighting on a ground that was not one of their own choosing. I shall be arguing that the triumph of a secular view of wonder was fostered in the 'evolutionary epic' style of writing natural history, deriving originally from scientific 'spectacles', and above all from Chambers' *Vestiges*, as well as in a new secular genre of writing, science fiction, where secular and scientific values could be integrated in a way that not only did not look derivative but was entirely natural and unique. What was provided in all these cases was a way of conveying the mysteries of the universe that could completely bypass religious sensibilities. This was a new form of wonder, not just a revival of the old one. It is characteristically modern: above all, it is not circumscribed by authority, and it is future-orientated. And rather than being in competition with curiosity, it complements it, boosting its standing.

SCIENTIFIC LITERACY

Popular science was very successful in promoting a scientific culture in Britain, and it will be instructive to begin by providing some background to the rise of popular science by considering how it compared with the prevailing educational system, which, despite its organizational strengths, had very limited success in promoting a scientific culture. David Knight has noted that, by contrast with Scotland, which placed a high value on education from the eighteenth century onwards, in England in the first half of the nineteenth century education was seen as a privilege, and the English were concerned that too much education would produce overqualified and unemployable people.[16] This was reflected in literacy levels: as late as 1830 the numbers of literate and illiterate Britons was roughly equal.[17] Nevertheless, two decades into the eighteenth century, concerted attempts started to be made to imbue British culture with a scientific mentality, in mathematical education in the university, and as education in mechanics in various newly conceived institutions catering to the needs of relatively uneducated workmen flourished. For very different reasons, these programmes subsequently

[15] Lightman, *Victorian Popularizers of Science*, 135.
[16] David Knight, 'Scientists and Their Publics: Popularization of Science in the Nineteenth Century', in Mary Jo Nye, ed., *The Cambridge History of Nineteenth and Twentieth Century Science* (Cambridge, 2003), 72–90: 72.
[17] David Vincent, *Literacy and Popular Culture: England 1750–1914* (Cambridge, 1989), 22.

turned out to be largely failures as attempts to secure a popular scientific culture by traditional pedagogic means, in the university case because of the shift in the standing of mathematics from exemplary to separatist, as we saw in Chapter 10. A closer look at some of the non-university teaching institutions will enable us to probe more deeply into the problems facing this programme.[18]

For the Mechanics' Institutes, the problems of establishing a scientific culture by pedagogic means were especially challenging. The institutes, which flourished between the early 1820s and the mid-1840s, provided evening teaching for the working class. They were not sponsored by wealthy aristocrats, as in the case of the Royal Institution for example, but were voluntary associations. They derived from a wide variety of earlier institutions and events, including Adult Schools, Sunday Schools, Brotherly Societies, Schools of Arts, Temperance Societies, and itinerant science lectures, which had existed from the 1780s onwards.[19] The first to be established was the London Mechanics' Institute, proposed in November 1823. In the course of 1824 and 1825 Mechanics' Institutes had sprung up in Leeds, Newcastle, Lancaster, Aberdeen, Dundee, Alnwin, Manchester, Birmingham, Bolton, Halifax, Ashton, Devonport, Plymouth, and Lewes, and by 1850 there were 700 such institutions with around 120,000 members.[20] The Society for the Diffusion of Useful Knowledge, which was founded in 1825, and whose very popular *Penny Magazine* was the first magazine to offer scientific instruction for the working classes (see Fig. 11.1), provided an umbrella for some of the Mechanics' Institutes. There were competitors—in 1830s London, for example, by the Society for Promoting Practical Design, the St. Pancras Literary and Scientific Institute, and the Tower Street Mutual Instruction Society[21]—but this just went to demonstrate the huge demand for science education outside the educated elite. It represented a transformation in the standing of science and practical mechanics.

Mechanics was widely popular among men and boys of all classes, indicated by the proliferation of popular mechanics magazines, such as the weekly *Mechanics' Magazine*, which first appeared in 1823. A good example of the enthusiasm for mechanics is the 1829 prize competition for a steam engine to run the proposed Manchester to Liverpool line. Weightman notes that the railway company was inundated with suggestions, quoting Henry Booth, the prominent Liverpool backer of the railway and its treasurer and secretary at the time of the trials:

from professors of philosophy down to the humblest mechanic, all were zealous in their proffers of assistance. England, America and Continental Europe were alike all tributary ... The friction of the carriages was to be reduced so low that a silk thread would draw

[18] I treat 'pedagogic means' here as narrower than 'educational means', including in the latter museums, libraries, etc.

[19] See Ian Inkster, 'The Social Context of an Educational Movement: A Revisionist Approach to the English Mechanics' Institutes, 1820-1850', *Oxford Review of Education* 2 (1976), 277–307: 280.

[20] See ibid., 277. There is an extensive literature cited at 300 n.2   [21] See ibid., 288.

Fig. 11.1. *The Penny Magazine of the Society for the Diffusion of Useful Knowledge* (1832)

them, and the power to be applied was to be so vast as to rend a cable asunder . . . Every scheme which the restless ingenuity or prolific imagination of man could devise was liberally offered to the Company; the difficulty was to choose and decide.[22]

But, for all the enthusiasm behind mechanics, the Mechanics' Institutes were ill-equipped to deal with this general upsurge in interest. The problem that they faced has been summed up well by John and Barbara Hammond. Mechanics' Institutes, they point out, were established in the hope of popularizing scientific knowledge, and

incidentally making the workman better at his work. It was a time when there seemed no limit to the possibilities of scientific and mechanical discoveries, and it was hoped that the new institutions might benefit not only their members but science itself by 'uniting and concentrating the scattered rays of genius, which might otherwise be dissipated and lost to the scientific world.' Mechanics' Institutes had the difficult task of providing instruction for students on very different levels of book learning. Many members could not even read or write. Hence the institute had not only to spread scientific truth, but to act as glorified elementary evening schools as well, with classes for reading, writing and arithmetic.[23]

Unable to meet this impossible combination of demands, the Mechanics' Institutes went into decline in the late 1840s, despite a continued broad interest in mechanics. They did not revive, even though the 1850s and 1860s were a period of unparalleled prosperity in which popular science was a beneficiary of the significant increase in household income, particularly among the middle classes, who, in the wake of the Great Exhibition, threw themselves headlong into microscopy, and such activities as collecting specimens for the home aquarium and home fern collections,[24] with some models ingeniously combining the two (see Fig. 11.2). Rather, with the rapid increase in literacy in Britain in the second half of the century there emerged a market for cheap popular science books and magazines for the working classes, and with the introduction of mass production in printing in the 1870s (rotary printing, hot-metal typesetting, use of lithographic and photographic techniques, displacement of steam power by electricity),[25] not to mention the increased distribution of their products afforded by railway transport (and the bookshops that sprang up in the crowded railway stations), publishers started issuing large-imprint series such as the 'Science Primers', published by Macmillan from 1872; 'Manuals of Elementary Science', published by the Society for the Promotion of Christian Knowledge

---

[22] Weightman, *The Industrial Revolutionaries*, 133.

[23] John L. and Barbara Hammond, *The Bleak Age* (London, 1947), 163–4.

[24] See David Elliston Allen, *The Naturalist in Britain: A Social History* (London, 1976), 137. See also Lynn L. Merrill, *The Romance of Victorian Natural History* (Oxford, 1989), ch. 1. On aquaria, see the manual of their inventor: Philip Henry Gosse, *The Aquarium: An Unveiling of the Wonders of the Deep Sea* (London, 1856). On fern collecting and arranging, see David Elliston Allen, *The Victorian Fern Craze: A History of Pteridomania* (London, 1969).

[25] See Aileen Fyfe, *Science and Salvation: Evangelical Popular Science Publishing in Victorian Britain* (Chicago, 2004), 41–59.

Fig. 11.2. Combined aquarium/fernery

from 1873; 'Chambers's Elementary Science Manuals', published by Chambers from 1875; and 'Simple Lessons for Home Use', published by Stanford from 1877.[26]

FAIRIES AND MONSTERS

One of the most striking features of the scientific culture that emerged in the nineteenth century is the way in which science comes to be extended outside the confines of serious research, and to install itself at the centre of leisure, for both adults and children.[27] Scientific literacy becomes literacy per se. The commercial success of the new popular science book series gives a good indication of the progress of a scientific culture in Britain in the later nineteenth century, but in many respects the ground had been laid earlier, in children's literature.

The transformation of children's literature into a vehicle for science education was profound. In 1802 Charles Lamb had lamented the disappearance of children's classics from the shelves of bookstores in favour of pedagogical science books, singling out the children's author and essayist Anna Barboult:

Knowledge, insignificant & vapid as Mrs B's books convey, it seems, must come to the child in the shape of Science . . . Science has succeeded to Poetry no less in the little walks of Children than with Men.—Is there no possibility of averting this sore evil? Think what you would have been now, if instead of being fed with Tales and old wives fables, you had been crammed with Geography and Natural History? Damn them.[28]

Cartoons and commentary in *Punch* in 1848 tell the same story in pictorial terms (Fig. 11.3), and at the beginning of his 1854 novel *Hard Times*, Dickens has Mr Gradgrind declare that facts are what children need, and, the narrator tells us, Gradgrind's children had 'never known wonder.' But by the time of the publication of *Hard Times*, things were in fact changing radically. A concern with wonder was transforming children's activities—and here we can include, as well as literature, such things as table games, wall friezes, and toy sets—into vehicles for the teaching and appreciation of science, as well as creating a new focus on such children's daytime activities as rock collecting, shell collecting, the

[26] On popular science series, see Katy Ring, 'The Popularisation of Elementary Science through Popular Science Books, c. 1870–c. 1939', unpublished PhD dissertation, University of Kent at Canterbury, 1988, 69–90.

[27] I deal here with the British case. For parallel developments in France, see Sophie Lachapelle, *Conjuring Science: A History of Scientific Education and Stage Magic in Modern France* (London, 2015), ch. 2, on scientific home entertainments.

[28] Quoted in Tess Cosslett, *Talking Animals in British Children's Fiction, 1786–1914* (Aldershot, 2006), 27.

Nurseries will be turned into miniature laboratories, and we shall have the satisfaction of knowing that our children, as they grow out of their clothes, are becoming men, or rather hobbedehoys of science, every inch of them. A lesson will be contained in every toy; our

THE OLD.   THE NEW.

lamb's-wool dogs will be taught to bark chemistry; our speaking-dolls be made to talk ten languages, and the most abstruse sciences be made easy to the smallest understanding by the aid of a plaything.

THE NEW ROCKING-HORSE.

Fig. 11.3. Proposal for new children's toys, *Punch* (1848)

identification of plants, and the night-time activity of star-gazing. With this transformation, wonder was introduced into the core of scientific education.[29]

The development is well illustrated in a review of recent English editions of Grimm's fairy tales in the *Athenæum* of 25 February 1863, where the reviewer notes the way in which the progress of science has been mirrored by a renewed interest in fairy tales:

Coincident with the world of Fact, in nearly all ages and among all nations, and lying by the side of that world like a fantastic shadow, has been the world of fairy Fiction:—a

---

[29] See Melanie Keene, *Science in Wonderland: The Scientific Fairy Tales of Victorian Britain* (Oxford, 2015), to which I am particularly indebted in this section; and Laurence Talairach-Vielmas, *Fairy-Tales, Natural History and Victorian Culture* (Basingstoke, 2014).

domain almost as rich and various, if not so majestic, as the actual universe by which we are surrounded . . . That there must have been some necessity for the existence of such an intellectual creation—that it arose to answer some want felt by the universal human heart—is sufficiently proved by the general diffusion of stories of enchantment over all races of men, civilized or savage . . . . It is a singular fact, the philosophy of which might be worthwhile, when the occasion shall serve, to seek, that the present age, which has made more extraordinary advances in science and practical invention than any other, should not only have returned to a more mystical order of poetry, but should be distinguished by a renewed love for those old-fashioned fairy tales which had almost died out from our literature . . . . [T]he dreams of old-world faith have again arisen, and kept pace with the steam engine and the electric telegraph. It may be well perhaps for mankind that this balance has been maintained; since the analytical tendency of Science, which views the Universe simply in its details, might lead us into a morbidly-exclusive perception of the mechanical anatomy of things, were it not for Imagination, which feels and enjoys results by means of the instincts of the heart.[30]

The challenge in the new children's scientifically aware literature was to combine didacticism, for it remained decidedly didactic, with wonder. Eighteenth-century science books for children had lacked an element of wonder. The Introduction to Robert Davidson's very popular *Elements of Geography* (1787), for example, while insisting that knowledge should be made 'a pleasure not a task', attempts to engage children (specifically boys) with the importance of travel, exploration, and imperial expansion in a patronizing way.[31] In the most popular science for children series, the 'Tom Telescope' books, which first appeared in 1761 and were being republished well into the nineteenth century, the precocious Tom lectures his friends on Newtonian natural philosophy. These are a little more imaginative, as in the illustration of inertial forces (Fig. 11.4), but despite using nursery objects to demonstrate points (Fig. 11.5) the treatment is baldly didactic. This instructional literature must be placed in the context of the (Lockean-inspired)[32] reform of domestic and institutional pedagogy, and it was not confined to books: jigsaws (dissected puzzles), paper dolls, card games, table games, alphabets, and wall friezes were all deployed. In the *Rational and Moral Game*, the French émigré the Abbé Gaultier set out the benefits of the use of games in instruction:

1. By rendering instruction amusing it *prevents discouragement*, the natural consequence which attends the dryness of subjects . . .
2. It prevents [children] associating the idea of study with that of fatigue

[30] [Edmund Ollier], Review of '*Yule-Tide Stories* . . . and *Household Stories*', *The Athenæum* 1322 (25 February 1863), 247–8: 247.

[31] Robert Davidson, *The Elements of Geography, Short and Plain* (London, 1787). See Jill Shrefin, ' "Make it a Pleasure not a Task": Educational Games for Children in Georgian England', *Princeton University Library Chronicle* 60 (1999), 251–75.

[32] See John H. Plumb, 'The New World of Children in Eighteenth-Century England', *Past and Present* 67 (1975), 64–95.

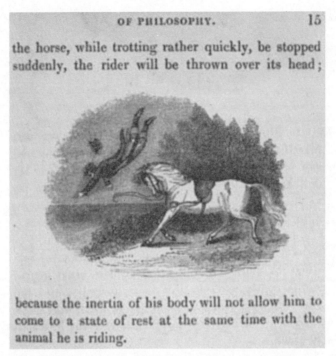

OF PHILOSOPHY.   15

the horse, while trotting rather quickly, be stopped
suddenly, the rider will be thrown over its head;

because the inertia of his body will not allow him to
come to a state of rest at the same time with the
animal he is riding.

Fig. 11.4. Illustration of inertial forces from 'Tom Telescope', *The Newtonian Philosophy, and Natural Philosophy in General; Explained and Illustrated by Familiar Objects in a Series of Entertaining Lectures* (1838)

3. The method of Games *is most analogous to the feeble organization of children.*
    At that age, what can we expect from an intense application of the mind?
    A forced application, in some measure, obstructs the progress of reason, and
    is prejudicial to the unfolding of the moral and physical faculties.[33]

As the nineteenth century progressed, children's stories became both far more
imaginative and more concerned with scientific questions, and a number of new
fictional devices were introduced. The 1861 *Boy's Own Book of Natural History* by
the best-selling author John Wood promoted the use of anecdotes as valid forms
of evidence,[34] and his publisher, Routledge, was willing to invest huge sums of
money for the lavish illustrations in the books.[35] Many children's science books in

---

[33] Abbé A. E. C. Gaultier, *A Rational and Moral Game* (London, c.1805), 5; quoted in Shrefin,
'"Make it a Pleasure not a Task"', 254.
[34] John George Wood, *Boy's Own Book of Natural History* (London, 1861), iv.
[35] See Livingstone, *Victorian Popularizers of Science*, 182.

Fig. 11.5. Illustration of the relationship between the sun, the earth, and the planets from 'Tom Telescope', *The Newtonian System of Philosophy Adapted to the Capacities of Young Gentlemen and Ladies* (1770)

the period between 1780 and 1840 were the work of women authors, and adopted the 'familiar format', a fictional literary device that used letters, dialogues, and conversations. They presented women who were both knowledgeable about science and in positions of authority in the home as religious and moral educators of the young.[36]

At the same time, traditional children's themes were often adapted to scientific educational use. This is best illustrated by two of the most popular themes: fairies and monsters. Fairies were the antithesis of the values of modernity,[37] and fairy tales were attacked by some writers as deceitful in early nineteenth-century England. In W. F. Sullivan's 1817 *The Young Liar!!! A Tale of Truth and Caution; for the Benefit of the Rising Generations*, for example, the young Wilfred Storey, having been brought up on fairy tales by his nursemaid, Fibwell, dies tragically as a result.[38] But there arose a counter-movement in Victorian England, a revival of

[36] The tradition started in the eighteenth century with Anna Letitia Barbauld, *Lessons for Children* (4 Parts, London, 1778–9), and continued in Sarah Trimmer, *An Easy Introduction to the Knowledge of Nature and the Holy Scriptures* (London, 1780); Priscilla Wakefield, *An Introduction to Botany, in a Series of Familiar Letters with Illustrative Engravings* (London, 1796); Priscilla Wakefield, *Mental Improvement; or the Beauties and Wonders of Nature and Art. In a Series of Instructive Conversations* (New Bedford, MA, 1799); [Maria Jacson], *Botanical Dialogues, between Hortensia and her four Children . . .* (London, 1799).

[37] In France, where modernity was strongly associated with anti-clericalism, belief in angels was assimilated to belief in fairies. See, for example, André Lefèvre's entry on 'Fées' in Abel Hovelacque et al., *Dictionnaire des sciences anthropologique* (Paris, 1889), 475.

[38] Keene, *Science in Wonderland*, 12.

interest in fairies in painting and literature, which has been seen as a form of relief from contemporary anxieties about the effects of industrialization, the remoteness of the past, the value of culture, and the way in which science threatened to undermine religion and spirituality.[39] In children's instructional literature, rather than fairies being summarily dismissed, they became popular as a teaching device, the most successful being Arabella Buckley's *The Fairy-Land of Science* (1879).[40] Their effect was occasionally striking, as in the science fairy in the library instructing the children (Fig. 11.6), and in the charming representations of molecules in Lucy Meyer's *Real Fairy Folks*, where atoms take the form of fairies, whose behaviour, dress, and limbs—holding hands to indicate atomic bonds (Fig. 11.7)—reflect their chemical properties, and their grouping into families and cousins represents similarity of chemical properties.[41] These images are possible through a Victorian transformation of fairies. Unlike the traditional mischievous and malevolent fairies of folklore, Victorian fairies are wholly benign, take on wings, and are shrunken to the size of butterflies or dragonflies, with whom they come to be associated in the children's instructional literature on entomology, above all in the three-volume *Episodes of Insect Life* (1849–51) by 'Acheta Domestica', *nom de plume* of Louise M. Budgen, where anthropomorphized insects display their particular skills.[42]

As well as fairies, children's scientific instructional literature also used monsters, but unlike fairies, these were translated into a non-fictional form. This was not wholly unprecedented, and children's stories had traditionally made reference to fabulous beasts such as dragons and griffins that were described as having lived in an earlier era. In books such as John Cargill's 1859 Christmas book, *Fairy-Tales of Science*, in the first chapter, 'The Age of Monsters', the reader is told:

Fortunately truth is stranger than fiction; the revelations of modern science transcend the wildest dreams of the old poets; and in exchange for a few shadowy griffins and dragons, we are presented with a whole host of monsters, real and tangible monsters too, who in the early days of the world's history were the monarchs of all they surveyed, and had no troublesome Seven Champions to dispute their sway. We are on the shores of the Ancient Ocean. We search in vain for any sign of Man's handiwork; no iron steam-ship, no vessel of war, no rude canoe even, has yet been launched upon its bosom, though the tides ebb and flow, and the waves chant their eternal hymn, according to those immutable laws which the Creator ordained at the beginning. The ocean teems with life, but it contains no single creature which has its exact likeness in modern seas.[43]

The results of palaeontology here become brought into line with the creatures of legends and myths. Some works, such as Hutchinson's *Extinct Monsters*, came close

[39] See Nicola Bown, *Fairies in Nineteenth-Century Art and Literature* (Cambridge, 2001).
[40] Arabella Buckley, *The Fairy-Land of Science* (London, 1879).
[41] See the discussion in Keene, *Science in Wonderland*, 74–81.   [42] See ibid., 56–64.
[43] John Cargill, *The Fairy-Tales of Science: A Book for Youth* (London, 1859), 4.

Fig. 11.6. The Library Fairy, frontispiece to *Fairy Know-a-Bit: A Nutshell of Knowledge* by A.L.O.E. [Charlotte Maria Tucker] (1868)

to identifying the two,[44] but Cargill and other authors simply filled the vacuum left by the dismissal of mythical creatures with scientifically discovered and described specimens. *Fairy-Tales of Science* is, as Keene notes, not a fanciful work; rather, 'its generalized sense of wonder and curiosity was often deemed the essential starting point for further scientific investigation', and the fairy tale framing 'was able to

[44] Henry Neville Hutchinson, *Extinct Monsters: A Popular Account of Some of the Larger Forms of Ancient Animal Life* (London, 1897).

HYDRO-CHLORIC ACID.

Fig. 11.7. A water molecule and a hydrochloric acid molecule, from Lucy Ryder Meyer, *Real Fairy Folks, Or, the Fairyland of Chemistry: Explorations of the World of Atoms* (1887)

emphasize this much more readily than less narrative works'.[45] The concern with 'ancient monsters' was widespread mid-century. When the Crystal Palace building was dismantled and reconstructed in East London in 1854, for example, it was expanded, and it now included a sequence of twenty-one full-size three-dimensional antediluvian creatures in its grounds (Fig. 11.8).[46]

Virtually every scientific subject was treated in the new Victorian children's literature, and generally speaking the material was uncontentious, but there was one exception. In the wake of the publication of Darwin's *Origin of Species* in 1859, there was a flood of books on evolutionary theory, and children's literature was as engaged in the issues as any other literary genre. There were works both for evolutionary theory—such as Arabella Buckley's 1880 *Life and her Children*—and against it—such as (with qualifications) Albert and George Gresswell's 1884 *The Wonderland of Evolution*—as well as enduring classics such as Charles Kingsley's 1863 *Water-Babies*.[47] Some works of children's literature began to take a

[45] Keene, *Science in Wonderland*, 26.

[46] See ibid, 38–53. See also James Secord, 'Monsters at Crystal Palace', in Soraya de Chadarevian and Nick Hopwood, eds, *Models: The Third Dimension of Science* (Stanford, CA, 2004), 138–69.

[47] See Keene, *Science in Wonderland*, 110–38. More specifically on Kingsley's engagement with evolutionary theory, see Piers J. Hale, 'Monkeys into Men and Men into Monkeys: Chance and Contingency in the Evolution of Man, Mind and Morals in Charles Kingsley's *Water-Babies*', *Journal of the History of Biology* 46 (2013), 551–95. Note that *Water-Babies* was subsequently reissued in an abridged version in 1928, leaving out much of the scientific material, presumably to make it fit more closely into the genre of children's literature, as that was conceived at the time. It is the 1928 abridgement that many modern editions use.

Fig. 11.8. Artists' Working Shed, Crystal Palace: from *Illustrated London News*, 31 December 1853

resolutely secular tone: Edward Clodd's *Childhood of the World* (1873),[48] for example, sets out to explain Bible stories in scientific terms, drawing on new developments in anthropology. Here the interests of children's literature and those of the adult natural history popular literature converge.

NATURAL HISTORY AND THE SECULARIZATION OF NATURE

Natural history had been an area of potential conflict from the early decades of the century, but in the mid-1840s Chambers' *Vestiges* had sharpened the issues considerably, and the task of providing a religious framework for science became ever more pressing among Christian apologists. Natural history books for the general reader were the most prolific type of instructional literature from the middle decades of the century onwards.[49] In 1857, Routledge launched its best-selling 'Common Objects of the Seashore', and it was followed by a number of equally popular series: 'Series of Natural History for Beginners' (1866), 'Natural History Rambles' (1879), 'Fur, Feather, and Fin Series' (1893), and 'Naturalist's Library' (1894). The natural history literature was unlike other popular science books and magazines, in that it came to engage with metascientific issues in a way that would have been quite alien to the mechanics literature (outside astronomy) for example. The terrain of the metascientific disputes was not that of philosophy but that of religion: science was rationalized in terms of its religious and social consequences. The natural history authors were predominantly Anglican clergy-men and laymen, nonconformists, and women writers on the one side, and scientific naturalists such as Huxley and Tyndall on the other. From the 1840s onwards, the role of clergymen in scientific societies such as the British Association had been diminishing, to be replaced by a professional membership.[50] But as Lightman notes, although Huxley and his allies may have driven Anglican clergymen out of the institutions and societies that they controlled, 'their power did not extend to the periodical press and the great publishing houses. The growth of a reading public eager to learn about natural history created an opportunity for members of the Anglican clergy to pursue their scientific interests as popular-izers'.[51] Women, to whom the domestic religious and moral education of their children fell, also had a stake in natural history as it became suffused into the

[48] Edward Clodd, *Childhood of the World* (London, 1873).

[49] I am particularly indebted in this section to Lightman, *Victorian Popularizers of Science*.

[50] See Frank M. Turner, *Contesting Cultural Authority: Essays in Victorian Intellectual Life* (Cambridge, 1993), 179–87. Women were also a target: for Huxley, for example, removing women from scientific societies allowed the societies to upgrade their professional status. See Evelleen Richards, 'Huxley and Woman's Place in Science', in J. Moore, ed., *History, Humanity, and Evolution* (Cambridge, 1989), 253–84.

[51] Lightman, *Victorian Popularizers of Science*, 41.

culture of middle-class families in the second half of the century. Women writers formed a distinctive category along with male Anglicans and nonconformists, and they were able to establish themselves as authorities in a range of areas of natural history. The three groups had common interests in integrating religious and moral values with scientific ones, though their audiences initially differed. The intended audience for women writers prior to 1840 was mothers and their children, and they worked within what Lightman has termed the 'maternal' tradition, in which a female narrator figures in a mother role. After 1840 their audience was expanded into a general one, and women writers, at least in the case of adult literature, now abandoned the image of the mother narrator.

As I have indicated, religious conceptions of science were able to import a conception of wonder into their accounts of the natural realm which put them, at least initially, at a significant advantage over secular ones. Early religious natural history writings, such as William Martin's 1830s volumes on quadrupeds and birds,[52] were in effect natural history companions to scripture, but by 1842, in his account of reptiles,[53] such considerations have almost completely disappeared,[54] and it is the wonder of nature in its evidence of design that now comes to the fore. Similarly with the works of Charles Johns, whose *Flora Sacra* of 1840 matches descriptions of plants with quotations drawn from the Bible,[55] but in his short 1853 *First Steps to Botany*, designed with working-class readers in mind, he stresses the wonder of design in leaves and fruit, and in *Birds' Nests*, published the following year, 'the wisdom of God in creation' is a recurring theme.[56] There was a general move among Christian authors against including scriptural passages as the century progressed, and one of the most successful natural history authors, John Wood, was evidently infuriated when editors of religious magazines inserted illustrative scriptural quotations into his articles.[57] Wood preferred a natural-theological approach to a scriptural one, and one device that he exploited extensively was analogical reasoning. This was a tool that William Paley had forged in his formulation of an argument from design which drew on the analogy between the systematic complexity of a watch and the systematic complexity of nature.[58] Wood does not use analogy to explicitly promote arguments from design: rather,

---

[52] [William Martin], *Popular Introduction to the Study of Quadrupeds; with a particular notice of those mentioned in Scripture* (London, 1833); [William Martin], *Introduction to the Study of Birds, with a Particular Notice of the Birds Mentioned in Scripture* (London, 1835).

[53] [William Martin], *Popular History of Reptiles, or Introduction to the Study of Class Reptilia, on Scientific Principles* (London, 1842).

[54] This does not mean that the genre itself ceased to exist. George Henslow's *Plants of the Bible* (London, 1896) was in some ways an updating of Martin's early work.

[55] Charles Alexander Johns, *Flora Sacra; or, The Knowledge of the Works of Nature Conducive to the Knowledge of the God of Nature* (London, 1840).

[56] Charles Alexander Johns, *First Steps to Botany* (London, 1853); *Birds' Nests* (London, 1854).

[57] See Lightman, *Victorian Popularizers of Science*, 187. H. G. Wells cited Woods as the most important influence on his intellectual development as a child: see ibid., 216–17.

[58] William Paley, *Natural Theology: or, Evidences of the Existence and Attributes of the Deity, collected from the Appearances of Nature* (London, 1802). On the extensive use of analogy in natural

in setting out to show how all human constructions mirror natural ones in their design, he leaves it to the reader to draw the analogy between the role of humans in the construction of human artefacts and the design of natural structures.

Charles Kingsley was one of the earliest natural history writers to make the whole exercise turn on the question of wonder. His 1846 lecture 'How to Study Natural History', was devoted to the study of a pebble. The human imagination would find 'inexhaustible wonders, and fancy a fairy-land' not just in the pebble, he announces, but in 'the tiniest piece of mould or decayed fruit, the tiniest animalcule'.[59] His most popular natural history book, *Glaucus, or the Wonders of the Shore*, published eleven years later, is suffused with the idea that natural history invokes wonder, drawing the enquirer 'out of the narrow sphere of self-interest and self-pleasing, into a pure and wholesome region of solemn joy and wonder'.[60] Similarly, 'delight' and 'wonder' are key themes of the natural history books of William Houghton, like Kingsley an Anglican minister, including works on the microscope (1871), insects (1875), and fish (1879).[61] As Lightman points out: 'All the language of wonder reminiscent of natural theology is in these books but there is no attempt to demonstrate that design in nature proves the existence of God. To enhance his appeal to the reader's sensibilities, Houghton included a large number of visual images.'[62] Wood's *Common Objects of the Country* and Houghton's *Sketches of British Insects* give a good idea of striking nature of some of these illustrations (Plates 6 and 7), while George Lewes offers a *tour de force* of 'wonder at nature' prose, where 'Life' (capitalized) occupies the place of wonder which Christian authors had reserved for God:

Come with me, and lovingly study nature, as she breathes, palpitates, and works under myriad forms of Life—forms unseen, unsuspected, or unheeded by the mass of ordinary men. Our course may be through park and meadow, garden and lane, over swelling hills and spacious heaths, beside running and sequestered streams, along the tawny coast, out on the dark and dangerous reefs, or under dripping caves and slippery ledges. It matters little where we go: everywhere—in the air above, in the earth beneath, and waters under the earth—we are surrounded with Life.[63]

This sense of wonder is not just a rhetorical device intended to engage the reader. Alfred Russel Wallace, for example, who made a living collecting specimens, but

theology, see John Hedley Brooke and Geoffrey Cantor, *Reconstructing Nature: The Emergence of Science and Religion* (Edinburgh, 1998), 190.

[59] Charles Kingsley, *Scientific Lectures and Essays* (London, 1890), 304. The natural history tradition of beginning the discussion with an apparently insignificant object has continued. See, for example, Stephen Jay Gould, *Hen's Teeth and Horse Toes: Further Reflections on Natural History* (New York, 1983).

[60] Charles Kingsley, *The Water-Babies and Glaucus* (London, 1908), 224.

[61] William Houghton, *The Microscope and Some of the Wonders it Reveals* (London, [1871]); William Houghton, *Sketches of British Insects: A Handbook for Beginners in the Study of Entomology* (London, 1875); William Houghton, *British Fresh-Water Fishes* (London, 1879).

[62] Lightman, *Victorian Popularizers of Science*, 84.

[63] George Henry Lewes, *Studies in Animal Life* (London, 1862), 1.

was as ingenuous a man as one could imagine, remarks of a butterfly he has just netted:

The beauty and brilliancy of this insect are indescribable, and none but a naturalist can understand the intense excitement I experienced when I at length captured it. On taking it out of my net and opening the glorious wings, my heart began to beat violently, the blood rushed to my head, and I felt much more like fainting than I have done when in apprehension of immediate death. I had a headache the rest of the day, so great was the excitement produced by what will appear to most people a very inadequate cause.[64]

One significant issue in natural history books after Chambers' *Vestiges* was evolution, and a good part of the success of *Vestiges* had come from its use of 'the evolutionary epic', a structuring of the narrative in terms of a journey from the colasescence of planets from swirling nebulae, through the 'chemico-electric' generation of the first living globules and the subsequent ascent of life, to 'the perfection of man'. In fact, as Ralph O'Connor has pointed out, the 'epic' or 'pageant' form first emerges around 1830 in geological writing, where 'these panoramic visions of the ancient earth blended human, sacred, and natural history to point towards a single cosmic pagent, dimly discerned behind the visible universe. They represent a major literary achievement, adapting ancient narrative structures to enact the supreme authority of modern science.'[65] 'Cosmic history' proved extremely popular for subsequent natural history writers, endowing evolution with a strong element of wonder missing in Darwin's writings for example. As Secord points out, *Vestiges* provided 'a template for the evolutionary epic-book-length works that covered all the sciences in a progressive synthesis'.[66] Here finally was a device inducing a form of wonder that could be deployed as a secular counter to an intrinsically religious form of wonder at a divinely inspired beauty and harmony of nature. The evolutionary epic did not necessarily have to be deployed for secular purposes, and it was also taken up by a number of Christian natural history writers, but the point is that it offered a very different form of wonder from the traditional one, with which it was in competition. To see how this works, consider four examples of the evolutionary epic in popular natural history books from the second half of the nineteenth century.[67]

Our first example is David Page, author of *The Past and Present Life of the Globe* (1861) and *The Earth's Crust* (1868),[68] the most successful popular accounts of

[64] Alfred Russel Wallace, *The Malay Archipelago: The Land of the Orang-utan and the Bird of Paradise, A Narrative of Travel with Studies of Man and Nature* (London, 1869), 351.
[65] Ralph O'Connor, *The Earth on Show: Fossils and the Poetics of Popular Science, 1802–1856* (Chicago, 2007), 27.
[66] Secord, *Victorian Sensation*, 461.
[67] See Lightman, *Victorian Popularizers of Science*, ch. 5, to which I am particularly indebted here.
[68] David Page, *The Past and Present Life of the Globe: Being a Sketch in Outline of the World's Life-System* (London, 1861); David Page, *The Earth's Crust: A Handy Outline of Geology* (Edinburgh, 1868).

geology of their time. He uses a number of tried and tested literary devices, including the focus on commonplace objects to reveal evolutionary origins: evolution encompassed everything, and was thereby able to provide an overall bridging structure for the account not just of organic things, but of such phenomena as volcanic activity and geological stratification. Page uses his evolutionary epic to refute what he saw as a move against design in accounts of evolution such as that of Darwin, and in particular to refute materialist tendencies which he detected in Lamarck, Chambers, and Darwin. 'The highest aim of our science', he writes, 'is to discover the Creative Plan which binds the whole into one unbroken and harmonious life-system'.[69]

A second proponent of the evolutionary epic was Arabella Buckley, committed to a broadly Darwinian perspective except on the question of the evolution of human beings, where she and Wallace developed what is usually described as 'spiritualism'.[70] Buckley's epic of evolution accordingly issues in a separate and distinctive form of evolution for the human mind, giving it purpose and meaning. Note that Christianity and spiritualism have often been antithetical. Spiritualism promotes some form of existence after death for humans, but does not consider this in a religious context, so that such afterlife does not necessarily have any bearing on the existence of a deity. Many spiritualists, including the leading utilitarian philosopher Henry Sidgwick for example, tended to treat the existence of an afterlife as an empirical matter,[71] wholly distinct from any Christian advocacy of personal immortality.[72] Here was an avowedly secular reading of evolution, and the fact that more orthodox Darwinians such as Huxley and Tyndall might have found it credulous does not diminish its thoroughly secular potential, one replete with wonder.

The third writer, Grant Allen, began his writing career with two research monographs devoted to extending evolutionary theory to new areas, aesthetics and colour sensation,[73] but soon turned his attention to the short essay, being a prolific contributor to magazines.[74] But again, Allen was no orthodox Darwinian, and was principally indebted to Spencer. In his *Charles Darwin*, he argues that it was Darwin who established the empirical credentials of evolution, but that it was

---

[69] Page, *The Past and Present Life of the Globe*, 21.

[70] Arabella Buckley, 'Soul, and the Theory of Evolution', *University Magazine* 93 (January 1879), 1–20.

[71] Correlatively, the existence of the soul after death was also sometimes treated in empirical terms. In 1907, Dr Duncan MacDougall, a Massachusetts physician, weighed six patients who were in the process of dying from tuberculosis just before and just after death, with a view to determining the weight of the departing soul. He came up with an average mass of 21 grams.

[72] See Bart Schultz, *Henry Sidgwick, Eye of the Universe: An Intellectual Biography* (Cambridge, 2004), ch. 5; and more generally Janet Oppenheim, *The Other World: Spiritualism and Psychical Research in England, 1850–1914* (Cambridge, 1985), ch. 4.

[73] Grant Allen, *Physiological Aesthetics* (London, 1877), and Grant Allen, *The Colour-Sense: Its Origin and Development. An Essay in Comparative Psychology* (London, 1879).

[74] See Peter Morton, *'The Busiest Man in England': Grant Allen and the Writing Trade, 1875–1900* (New York, 2005).

Spencer who showed how evolution extended beyond the events described in *The Origin of Species*. It was, as he put it, Spencer who established 'the total esoteric philosophic conception of evolution as a cosmical process, one continuous from nebula to man, from star to soul, from atom to society'.[75] Spencer's formative works, such as *Social Statics*, were written well before the publication of *The Origin of Species*, so it is important for Grant to stress that evolutionary principles, in the requisite sense, predated the appearance of Darwin's version of evolution, having been recognized in astronomy in the nebular hypothesis by Kant and Laplace, in geology by Lyell, in biology by Lamarck, and, most importantly for Allen, in psychology by Spencer.[76] Allen's particular skill lay in starting with commonplace objects, such as a feather, and drawing a whole evolutionary epic out of them.[77] In doing so he was presenting evolution not just in a non-Darwinian way, but in a way that Darwin would have rejected, even though Darwin enthusiastically supported Allen's popularizing work in other respects. It cannot be emphasized enough that a commitment to evolutionary theory in the wake of Darwin is not a commitment to the Darwinian version of evolution, and this is something particularly marked in the 'evolutionary epic' popular science writers. Nevertheless, there were lines beyond which one could not go, even though to modern eyes they seem somewhat arbitrarily placed. Among those who worked in the evolutionary epic genre but whose commitment to Darwinism was deemed too loose, was Samuel Butler, who attempted to revise Darwin's account in the direction of Lamarckism. Darwin had accepted Lamarckism in conjunction with natural selection, but he sought to distance himself from the Lamarckian radicalism that was emerging in Britain. Lamarck was increasingly viewed with suspicion, on the grounds of inheritance of acquired characteristics but also on the grounds of the rapidity of Lamarckian evolution compared to natural selection. In the wake of the publication of his *Life and Habit* in 1878, Butler was ostracized in Darwinian circles, particularly after Darwin's death in 1882, when the opposition between Darwinian and Lamarckian evolution became polarized.[78]

Of the popular evolutionary epic writers, it was Edward Clodd who was closest to orthodoxy, but it is instructive that even he departed from Darwin on crucial points, and in a way that Darwin would certainly have rejected. Whereas writers such as Huxley presented themselves as agnostics, a new term devised at the time to describe their views, Clodd, one of the most successful Darwinian popular science writers, while paying lip-serve to agnosticism, clearly saw atheism as the natural consequence of a scientific attitude to the world, and evolutionary theory in particular, and later in his career he closely associated himself with the Rationalist Society. For Clodd, it was his reading of Huxley's *Man's Place in Nature* (1863) and Edward Tylor's *Primitive Culture* (1871) that shaped his

[75] Grant Allen, *Charles Darwin* (London, 1885), 191.
[76] Grant Allen, 'Evolution', *Cornhill Magazine* 57 (January 1888), 34–47.
[77] See Lightman, *Victorian Popularizers of Science*, 272–5.   [78] Ibid., 289–94.

thinking about evolution. The first extended evolutionary theory to human beings, the second to every branch of knowledge. As well as popular accounts of evolution interpreted from a staunchly secular perspective—*The Story of Creation* (1888), *The Story of Primitive Man* (1895), *A Primer of Evolution* (1895), and *Pioneers of Evolution* (1897)—in 1880 he published his most radical book, *Jesus of Nazareth*, a debunking of the idea of God which unsurprisingly provoked an outburst of criticism. Like Chambers and unlike Darwin, Clodd saw evolution in cosmic, not just purely biological, terms, and the *Story of Creation* for example combines the nebular hypothesis and evolutionary theory to generate a totalizing form of evolution, whose impact is on a par with that of Chambers, but on a slightly sounder basis. Here again, a secular form of wonder completely bypasses anything on which the religious tradition could draw. Its impact was widespread. In an 1897 letter to Clodd, for example, Thomas Hardy praised the way in which he had managed to present the evolutionary story in *Pioneers of Evolution*, writing that he had 'a breadth of grasp, and a power of condensing the stupendous ideas scattered over time in fragments, which you have never before equaled. It is just as when one sees a landscape of miles length reproduced in a charming miniature picture inside a camera.'[79]

It is through the popular natural history books in the genre of the evolutionary epic that evolutionary theory becomes established as part of the broader culture in Britain in the last decades of the nineteenth century. In the hands of writers like Allen and Clodd for example, the wonders of the evolutionary epic completely replace those promoted by natural theology, establishing evolution at the centre of the culture of scientific wonder. Even what might properly be considered scientific textbooks in natural history, such as Forbes' *History of British Starfishes* and Gosse's *Actinologia Britannica*, employ the evocative language of wonder of the popular natural history literature.[80] By contrast, Thomas Huxley, although he was one of the most popular speakers of the age,[81] always found it difficult to 'put the truths learned in the field, the laboratory and the museum, into language which, without bating a jot of scientific accuracy shall be generally intelligible, taxed such scientific and literary faculty as I possessed to the uppermost'.[82] Huxley was inclined to be far less speculative than the writers that we have just looked at on

[79] Quoted in ibid., 261.
[80] Edward Forbes, *History of British Starfishes, and Other Animals of the Class Echinodermata* (London, 1841); Philip Henry Gosse, *Actinologia Britannica: A History of British Sea-Anemones and Corals* (London, 1860). See the discussion in Merrill, *The Romance of Victorian Natural History*, ch. 3.
[81] See, for example, David Knight, 'Getting Science Across', *British Journal of the History of Science* 29 (1997), 129–38.
[82] Thomas Huxley, *Discourses Biological and Geological* (New York, 1897), v. An American Darwinian philosopher and historian, on meeting Huxley for the first time wrote that there was no alternative to seeing some people: 'Reading their books don't give you the flesh-and-blood idea of them. But once to see such a man as Huxley is never the forget him': quoted in Adrian Desmond, *Huxley: The Devil's Disciple* (London, 1994), xiv. On Huxley's character see also Paul White, *Thomas Huxley: Making the 'Man of Science'* (Cambridge, 2003).

the question of cosmic evolution, and accordingly the genre of the evolutionary epic had less appeal for him, although he had no hesitation in speculating on the benefits of science and technology for the development of civilization. In line with some of the evolutionary epic treatments, his account is future-orientated. It is not so much that evolution in its own right points to a particular future, as Spencer and others had argued, but that evolution is a core part of the scientific enterprise, which does so. What was of primary concern to Huxley was establishing the authority of the Darwinian theory of natural selection: he saw himself above all as a research scientist rather than a popularizer, and indeed he was the leading vertebrate palaeontologist of his time. He was concerned not just about the authority of natural selection within the life sciences, but also with its commanding extra-scientific standing. In particular, he was convinced that science locks on to culture, redirecting it along a favourable path, but at the same time he was clear that the advocacy of natural selection rules out teleology, so that this redirection did not require any grand cosmic schemes.

What it did require in Huxley's view was education.[83] The 1867 Reform Act had given working-class men the vote, and there was a widespread view that they had to be educated in order to participate in elections in a responsible way. An indicative statement of the approach of the government is given by the Liberal politician Robert Lowe, chancellor of the exchequer from 1868 to 1873, and subsequently home secretary, in a speech given in Edinburgh in 1867:

Is it not better that gentlemen should know the things which the working men know, only know them infinitely better in their details, so that they may be able, in their intercourse and their commerce with them, to assert their superiority over them which greater intelligence and leisure is sure to give, and to conquer back by means of a wider and more enlightened cultivation some of the influences which they have lost by political change? . . . The lower classes ought to be educated to discharge the duties cast upon them. They should be educated so that they may appreciate and defer to higher cultivation when they meet it; and the higher classes ought to be educated in a very different manner, in order that they may exhibit to the lower classes that higher education to which, if it were shown to them, they would bow down and defer.[84]

The 1870 Education Act was designed to meet the educational needs created by the Reform Act, and we have in Lowe's statement an indication of how the condescending view of many politicians might have conflicted with that of the aims of those promoting the study of science. The stakes had been raised considerably, and Huxley and his circle began to see gaining control over the

[83] See the essays collected in his *Science and Education* (New York, 1898), especially 'A Liberal Education: And Where to Find it [1868]', 76–110.

[84] Quoted in Simon Heffer, *High Minds: The Victorians and the Birth of Modern Britain* (London, 2013), 431.

organs of education—including the popular science literature—as a higher priority than original research.[85]

'THE SCIENTIFIC AMUSEMENTS OF LONDON'

Between 1851 and 1914, 187 international fairs promoting commerce and technology were held across the world. There had been a few such fairs prior to this, but nothing on the scale of the 1851 Great Exhibition of the Works of Industry of all Nations, held in London (Figs 11.9 and 11.10). Lynn Merrill has remarked that the 'Crystal Palace', which housed this exhibition, 'was a cathedral of sorts. Its arching glass and iron architecture emphasized light, verticality, and spaciousness (and even glorified nature, since the building had been designed to accommodate several tall trees within it), and in this vast chamber, material objects of commerce were arrayed like objets d'art or sacred relics. The Palace represented that peculiarly Victorian conflation of sacred and profane, spiritual and material satisfaction.'[86] In this last respect, it is striking that, on opening the exhibition, precedent was put to one side and instead of a Bible, the Queen held in her hand a catalogue of the exhibition.

Fig 11.9. The Great Exhibition: Crystal Palace exterior

[85] See Lightman, *Victorian Popularizers of Science*, 364–421.
[86] Lynn Merrill, *The Romance of Victorian Natural History*, 110

Fig 11.10. The Great Exhibition: Crystal Palace interior

In Britain, the decades prior to the Great Exhibition were politically fraught, from the Peterloo Massacre of 1819 onwards. The Reform Bill of 1832, abolishing corrupt boroughs and extending the vote to more householders, was preceded by mass marches, riots, and fire-bombings, and was bitterly opposed by Tories. Hopes that the radicals newly elected to parliament in the wake of the Reform Bill would see working men finally get the vote were dashed, leading to the drawing up of the People's Charter in 1838, which demanded universal male suffrage with no property qualification, annual (as opposed to indefinite) parliaments, equal size constituencies, payments for members of parliament, and secret ballots. The demands terrified the political establishment, as did the huge Chartist rallies, the 1839 rally being a particularly bloody event as a result of the military being called in. There was widespread fear that the Great Exhibition, with its huge numbers in confined spaces, would foment revolution, and the King of Prussia wrote to Prince Albert, its main sponsor, of his fear that 'countless hordes of desperate proletarians, well organised and under the leadership of blood-red criminals, are on their way to London now'.[87] But by the 1850s a thriving economy was soaking up excess capital and labour in Britain, and the concerns about machinery displacing labour that had been prevalent in the early decades were largely assuaged. The Great Exhibition turned the celebration of machinery

---

[87] Letter of 8 April 1851, quoted Heffer, *High Minds*, 305.

(particularly British machinery) into a spectacle, and the Machinery Court (Fig. 11.11) was, despite being the noisiest, evidently the most popular spectacle.[88] Subsequent major international exhibitions followed suit: notably, in the nineteenth century, the Exposition Universelle in Paris in 1867, the Centennial Exposition in Philadelphia in 1876, the Exposition Universelle in Paris in 1878, the Colonial and Indian Exhibition in London in 1886, the Exposition Universelle in Paris in 1889, and the World's Columbian Exhibition in Chicago in 1893.

The Great Exhibition was a cultural and commercial success—it was attended by six million people, equivalent to a third of the population of Britain—and the surplus revenue that it generated was used to found a number of new museums: the Victoria and Albert Museum (1852), the Science Museum (1857), and the Natural History Museum (1881). At the same time, a sister institution, the Imperial Institute, was established in 1888 (without government funding), to house industrial and commercial products and to carry out scientific research. There were also various private spin-offs, the most notable being John Gould's display of 320 species of hummingbird (stuffed, not live) in the Zoological

Fig. 11.11. The Machinery Court, the Great Exhibition, from *The London Illustrated News* (1851)

[88] See Asa Briggs, *Victorian People* (London, 1954), 38.

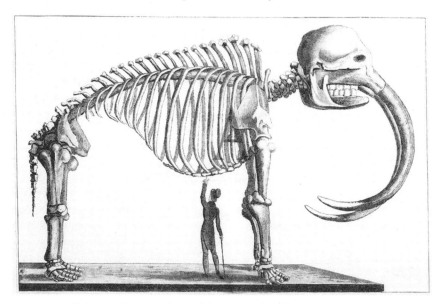

Fig. 11.12. Skeleton of a Mastodon, exhibited in Pall Mall in 1802.

Gardens at Regent's Park, which had only been open to the public for four years. His display attracted a remarkable 75,000 paying visitors.

Spectacular scientific displays for the public can in fact be dated back to the beginning of the century. In 1802 the first reasonably complete mammoth skeleton went on display for the paying public in London's Pall Mall (Fig. 11.12). What was distinctive about this exhibition was not the presence of mammoth bones as such: the Hunterian Museum, for example, had an extensive collection of fossils, but they were just lined up in cabinets, whereas the Pall Mall exhibition was an independent exhibit, in which spectacle played the major role.[89] Similarly with the exhibition of antiquities of William Bullock, naturalist and antiquarian.[90] In the late 1790s he opened a Museum of Natural Curiosities in Sheffield, moving it to Liverpool around 1801, at which point he issued a catalogue,[91] and in 1812 he started to exhibit his collection of 'curiosities' in the newly built 'Egyptian Hall' in Piccadilly. The collection, which by the time it

[89] See O'Connor, *The Earth on Show*, ch. 1.

[90] There were earlier large private collections, such as those of John Hunter and Hans Sloane. On Sloane's collection see James Delbourgo, *Collecting the World: Hans Sloane and the Origins of the British Museum* (Cambridge, MA, 2017).

[91] William Bullock, *A Companion to the Liverpool Museum, containing a Description of . . . Natural and Foreign Curiosities, Antiquities & Productions of the Fine Arts, Open for public inspection . . . at the house of William Bullock, Church Street* (Liverpool, *c.*1801).

moved to Piccadilly comprised 32,000 artefacts, was very mixed, but a prominent place was devoted to stuffed animals and Bullock issued a guide to preservation of natural specimens.[92] As Rachel Poliquin notes, 'Bullock was among the first curators to present his creatures in theatrical, almost atmospheric displays, with an eye to both the scientific interest and the spectacular entertainment value of exotic specimens.... Down the center of his museum, he arranged his large exotics, including a giraffe, an elephant, a lion, and a rhinoceros surrounded by models of tropical plants, in order to produce a "panoramic effect of distance... affording a beautiful illustration of the luxuriance of a torrid clime" '[93] (see Fig. 11.13).

Natural history displays were exclusively of skeletons and stuffed animals. With the emergence of zoos, the public had access to live exotic animals, but under very restrictive viewing constraints. Animal rides provided rare exceptions (see Fig. 11.14), but generally animals were caged or enclosed in some way. The

Fig. 11.13. Display of exotic animals in Bullock's Museum, *c.*1812

[92] William Bullock, *A Concise and Easy Method of Preserving Objects of Natural History: Intended for the use of sportsmen, travellers, and others; to enable them to prepare and preserve such curious and rare articles* (London, 1818).

[93] Rachel Poliquin, *The Breathless Zoo: Taxidermy and the Cultures of Longing* (State University, PA, 2012), 44.

Fig. 11.14. *Monday Afternoon at the Zoological Society's Gardens*, by J. C. Staniland (1871)

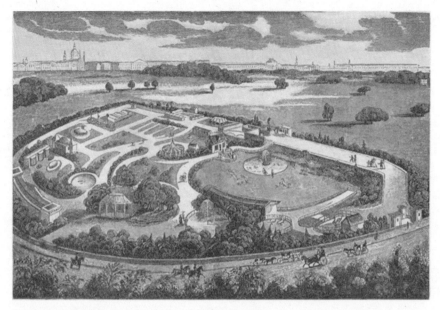

Fig. 11.15. Zoological Gardens, Regents Park (1829)

idea of the 'Zoological Society of London' was the initiative of Humphry Davy and Thomas Stamford Raffles. Davy had thought of it as a vocational club for sporting noblemen and zoological specialists,[94] but Raffles thought of it as a display of the exotica of the empire, arguing the need for a new Society because the Linnean Society focused too narrowly on botany. The initial aim was to promote the classification and description of animals, but also to domesticate new animals with a view to agriculture and recreation. This did not fit well with Davy's original idea of locating it at a farm in Surrey however, and the zoo was opened in Regent's Park in 1828 (Fig. 11.15). Between 1830 and 1835 a huge sum was devoted to developing a landscaped Zoological Gardens and stocking it with exotic animals. The stated goal was the 'formation of a collection of living animals; a museum of preserved animals, with a collection of comparative anatomy; and a library connected with the subject'.[95] However, as Nigel Rothfels points out, with the founding of the gardens, 'the collection of animals swiftly assumed an added

[94] See John Bastin, 'The First Prospectus of the Zoological Society of London: New Light on the Society's Origins', *Journal of the Society for the Bibliography of Natural History* 5 (1970), 369–88; John Bastin, 'A Further Note on the Origins of the Zoological Society of London', *Journal of the Society for the Bibliography of Natural History* 6 (1973), 236–41; and Desmond, *The Politics of Evolution*, 134–44.

[95] Henry Scherren, *The Zoological Society of London: A Sketch of its Foundations and Development* (London, 1905), 20.

character that was anything but scientific. Raffles' early proposals included the idea that the zoological collection should also both "interest and amuse the public." In the end, this latter quality would become perhaps the most important mandate behind the development of the gardens throughout the century.... By the end of the 1840s... the public was admitted Monday through Saturday, paying sixpence on Monday and a shilling the rest of the week'.[96] In short, the Zoological Gardens had become essentially an amusement park, with Sundays devoted to military bands, exotic entertainments, and garden teas.

Clearly, the rationale for the collections of museums and zoos varied radically. The British Museum, which held the large natural history collections of Hans Sloane that seeded the institution, was particularly hostile to displaying its specimens. The trustees, predominantly members of the aristocracy and senior Anglican clergy, believed that the museum should house national treasures, not advance knowledge. In the report of all-party Select Committee set up in 1835 to examine the running of the museum, Robert Grant—Fullerian Professor at the Royal Institution and later dean of the University College medical faculty—noted that comparison of the British Museum with the Paris Jardin des plantes was ludicrous. In the Jardin des plantes, he writes:

all the collections are extensive, well preserved, well exhibited, classified, and named by the first authorities in Europe; they have been increasing for more than a century; they are supported by large annual grants from Government, and they are directed and superin-tended by many of the most eminent zoologists living, who have each their particular departments, with numerous assistants under them . . . The Zoological department of the British Museum is miserable in funds, miserable in science, miserable in materials, and its collections are for the most part without either classification or nomenclature; so that the Museum more resembles a store-house than a school of zoology.[97]

One of the defenders of the arrangements as they stood was Sir Robert Inglis, MP for Oxford University, who insisted that wealth and rank were essential for trustees if they were to solicit patronage for a public museum. He defended the titled trustees, their efficiency and record, and disputed the advantage of putting men of science on the board.[98] Much political struggle continued, and it was ten years before the first display gallery, the 'Coral Room', was opened.

But things were slowly changing. Rachel Poliquin notes that between 1887 and 1891 the London Natural History Museum 'installed an introductory series of cases on evolutionary theory variously demonstrating species variation under domestication, protective mimicry, albinism and excess pigmentation malenism, and seasonal colour variation. Prominently positioned in the main entrance hall, the cases were meant to offer a clear-eyed view of potentially abstract theory with

[96] Nigel Rothfels, *Savages and Beasts: The Birth of the Modern Zoo* (Baltimore, 2002), 32.
[97] Report from the Select Committee on the British Museum (Parliamentary Papers, 14 July 1836), 133. Quoted in Desmond, *The Politics of Evolution*, 149.
[98] See Desmond, *The Politics of Evolution*, 146.

the animals themselves.'[99] But it was far from clear that the one institution could accomplish both instruction for the general public and scientific research. John Wood, one of the most successful popularizers of science, in an 1887 article entitled 'The Dulness of Museums', complained that while the big public museums may be of interest to the expert, they are 'intolerably dull', and he suggested that three types of museum were in fact needed: one type for purely scientific purposes, one for those trying to learn the rudiments of science, and one for the general public.[100] This was not the first complaint of this kind. Nineteen years earlier, Darwin, Huxley, and others had issued a similar plea, petitioning the government to radically rework the display, arrangement, and care of natural history collections. Without this, they argued, the public would simply be dazzled and confused by unexplained objects crammed in together, while specialists would be unable to remove specimens for examination. To make matters worse, the public inevitably brought dust and dirt into the rooms (the British Museum refused tickets to those whose clothes and appearance did not conform to strict standards of hygiene),[101] but hermetically sealing the cabinets was not a solution because it would prevent specialists from examining the specimens.[102] The problems facing natural history collections in some ways mirror those facing early nineteenth-century university educators in deciding between mathematics as an exemplary subject and as a research programme, though they are perhaps not as severe.

Whether viewing natural history displays of stuffed animals, or viewing caged animals in zoos, the spectator was wholly passive. But natural history by no means exhausted scientific displays in nineteenth-century Britain, and there was nothing passive about popular electrical displays. As Morus notes, machines were becoming increasingly ubiquitous as objects of display by the 1830s, and 'electricity in these exhibitions therefore also fitted into the world of machinery. Experiments and apparatus took their place surrounded by displays of the products and adjuncts of labor.'[103] Electrical experiments competed with natural history displays for public attention in the middle decades of the nineteenth century. Prominent among the popular science institutions that sprang up during the 1830s were those where experiments were publicly performed. The most popular were the Adelaide Gallery and the Royal Polytechnic Institution. They are described by a contemporary in an article on 'The Scientific Amusements of London':

The places in fact which furnish the great mass of the public with demonstrations of science. The other institutions are in a manner exclusive: membership or the introduction of a member is, with few exceptions, imperative on those who would be present. The

99   Poliquin, *The Breathless Zoo*, 131.
100 John George Wood, 'The Dulness of Museums', *Nineteenth Century* 21 (1887), 384–96.
101 O'Connor, *The Earth on Show*, 218.   102 Darwin, *Correspondence*, vii. 525–7.
103 Morus, *Frankenstein's Children*, 71.

exceptions are in the institutions where single tickets are sold for single lectures. But the exhibitions now before us . . . are accessible to all who proffer their shilling at the door; and we venture to say that these two institutions have played an important part in diffusing among the middle, and even the working classes, that better knowledge of natural phenomena which is fast spreading among us, and that taste for physical philosophy which is becoming so general.[104]

The National Gallery of Practical Science, called the Adelaide Gallery after the street on which it was located, just off the Strand, was established in London in 1832 by the American inventor and entrepreneur Jacob Perkins. The display was dominated by his own inventions, one correspondent complaining that the idea of erecting a hall 'to exhibit two or three inventions of a particular individual—not all original and some of them mere abortions—was, to say the least, exceedingly preposterous'.[105] Nevertheless, the exhibition gradually acquired more exhibits, including an oxyhydrogen microscope (a particularly effective microscope illuminated by burning lime swept by a current of a mixture of oxygen and hydrogen), a Jacquard loom in action, demonstrations of chemical processes such as bleaching, and electrical apparatuses: the latter perhaps not always too safe, as it was reported that the Duke of Wellington, on a visit to the gallery, had accidently received a severe electric shock from one of the batteries.

In 1838, the Polytechnic Institution (granted a Royal charter the following year) was opened in Cavendish Square (see Fig. 11.16), modelled in many ways on the Adelaide Gallery. Its most popular feature was a diving bell, which accommodated four or five paying customers as it submerged into a tank. The original manager had been Charles Payne, who had been superintendent of the Adelaide Gallery, but in 1854 the chemist John Henry Pepper became the manager, and he realized that the institution needed reforming to meet the expectations that had been raised by visits to the Great Exhibition. Pepper supplied books of reduced-price tickets to factories so that workers would be able to afford to attend, as well as publishing popular science books based on his lectures and demonstrations, books which were in part designed to promote the Polytechnic Institution.[106] At the same time, he exploited the relationship between the polytechnic and the extensive London popular entertainment scene by bringing in music and spectacle.[107] Spectacular displays were to be a feature of electricity throughout the century. In 1879, Edison announced his luminescent bulb by illuminating the outside of his laboratory in Menlo Park, a spectacular display of the first electrically illuminated building that attracted thousands. Five

[104] 'Zeta', 'The Scientific Amusements of London', *London Polytechnic Magazine and Journal* 1 (1844), 225–34: 229–30.

[105] Anon., 'Exhibition of Works of Popular Science', *Mechanics' Magazine* 17 (1832), 378.

[106] Among the most popular were his *The Boy's Playbook of Science* (London, 1860) and *Playbook of Metals* (London, 1861).

[107] On Pepper, see Lightman, *Victorian Popularizers of Science*, 200–16.

Fig. 11.16. Main Hall of the Royal Polytechnic Institution

years later, he lined up all his employees in a march along a city street, described at the time as a 'presidential parade'.[108] The marchers formed a phalanx around a horse-drawn dynamo, steam engine, tankstand, and boiler, the electricity being conveyed to bulbs on their heads through wires tucked up their sleeves (Fig. 11.17).

SCIENCE AND THE FUTURE

By the early decades of the twentieth century, there was a transformation of popular literature on science. The number of popular expositions of science had diminished somewhat,[109] and what was being produced tended to be written by scientists, who offered less speculative accounts than those that had predominated in the Victorian era. But at the same time, there emerged a new very popular genre, science fiction, in the form of books, stories, and films, which took over the promotion of the wonders of science from the nineteenth-century literature in such a way that speculation on the future development of science and technology was the core theme. It was in the potentialities of future development that the

[108] Reported in *Scientific American*, 15 November 1884, p. 310, from which the illustration is taken.
[109] See, for example, Roy MacLeod, 'Evolutionism, Internationalism, and Commercial Enterprise in Science: The International Science Series, 1871-1910', in A. J. Matthews, ed., *Development of Science Publishing in Europe* (Amsterdam, 1980), 63–93.

Fig. 11.17. Edison Company's 'Electric Torchlight Procession' (1884). The electricity is conveyed to bulbs on the heads of the marchers through wires up their sleeves, as illustrated in the top left-hand panel

wonder of science and technology now lay, and these fell within the ambit of science fiction rather than science itself.

While they did not engage religion in the way that their Victorian forebears had done, the popular science books written by early twentieth-century scientists tended to display their religious beliefs, or lack of religious beliefs, openly. Arthur Eddington, James Jeans, Oliver Lodge, and J. Arthur Thomson for example,[110] the first two of whom were also early contributors to the radio, were explicit in their writings in their espousal of the need for a Christian understanding of the world. Eddington's *The Nature of the Physical World* is particularly instructive in its argument for the incompleteness of science, an incompleteness that, he argued, requires it to be supplemented by Christianity. It is instructive because it is in the

[110] For example, Arthur Stanley Eddington, *The Nature of the Physical World* (Cambridge, 1928); James Jeans, *The Mysterious Universe* (Cambridge, 1930); Oliver Lodge, *Man and the Universe: A Study of the Influence of the Advance in Scientific Knowledge upon our Understanding of Christianity* (London, 1908); J. Arthur Thomson, *Science and Religion* (London, 1924).

questions where Eddington finds science to be incomplete that wonder reappears. On the other hand, Joseph McCabe, E. Ray Lankester, and Arthur Keith all stressed that science is complete without religion, and they promoted forms of rationalism.[111] In both cases it is of interest how cosmology, as Bowler has noted, 'replaced biological evolution as the most exciting area for creating a grand narrative defining humankind's place in nature'.[112]

Wonder is certainly less in evidence in the books of these writers than it had been in the Victorian literature, and this reflects a shift in the social role of science. The issues now turned on legitimating science by incorporating it into government programmes, less with promoting public awareness of science (which would in any case have been counterproductive given the mechanized slaughter and chemical attacks of the Great War). The main question for scientists was their influence on government thinking, with Lankester spearheading a campaign in 1916 to highlight the 'Neglect of Science'.[113] The British government had in fact been building up a programme of science and technology from the end of the nineteenth century, accelerated during the Great War, when most of Britain's leading scientists worked on military research.[114] Moreover in the years before the Great War popular science books for boys and adults routinely included the military applications of technology. In 1917 Richard Gregory, a writer for *Nature*, published a condemnation of narrow specialized research and its applications in schemes for world domination, and defended the idea of science as the pursuit of truth driven by sheer curiosity.[115] But the idea of a pure science guided by a disinterested pursuit of truth was not what the popular literature promoted. On the contrary, the popular science literature mixed science, technology, and engineering, with no sharp lines drawn. As Bowler notes, while theoretical science was not absent from the new popular science literature, 'it is very much subordinated to practical applications. The reader is told just enough to get an idea of the scientific background to the new technologies. There will only be a hint of the disturbing conceptual implications of theoretical developments in physics. There is little or no emphasis on the methodology of science—here invention is more important than discovery: indeed, discovery emerges very much as a consequence of practical men trying to solve technical problems'.[116]

The era of the grand synthesizing works of the 'evolutionary epic' genre of Victorian popular science literature had not completely come to an end however. There was still the odd major work in the genre to come. H. G. Wells'

---

111 For example, Joseph McCabe, *Evolution: A General Sketch from Nebula to Man* (London, 1910), and *The Existence of God* (London, 1934); E. Ray Lankester, *Science from an Easy Chair* (London, 1910); Arthur Keith, *Darwinism and What it Implies* (London, 1928).

112 Bowler, *Science for All*, 97.   113 See ibid., 19.

114 See Peter Alter, *The Reluctant Patron: Science and the State in Britain, 1850–1920* (Oxford, 1987); David Edgerton, *Science, Technology, and the British Industrial 'Decline'* (Basingstoke, 1991).

115 Richard Gregory, *Discovery; or, The Spirit and Service of Science* (London, 1917).

116 Bowler, *Science for All*, 25.

extraordinarily successful *The Outline of History*, published in 1920,[117] was a 'cosmic history' of the kind promoted by late nineteenth-century writers, but written with the facility of a great novelist. Wells (with help from his friend Lankester) began with a 'rationalist', religion-free account of the natural world from which humans emerged. Reflecting on the success of the book, Wells later wrote that this 'was as great a discovery for his publishers as for himself' and that what it taught was that 'there existed an immense reading public in the world which was profoundly dissatisfied with the history it had learnt at school, and which was eager for just what the *Outline* promised to be, a readable, explicit summary of the human adventure'.[118]

That it was Wells, one of the greatest science fiction writers of the century, who came closest to continuing the nineteenth-century tradition of wonder is not surprising, for it was in science fiction that scientific wonder reappeared. Yet just what counts as science fiction is not always clear. Jules Vernes' 1870 *Twenty Thousand Leagues under the Sea*,[119] for example, is generally taken as a science fiction novel, and indeed was a formative influence on later science fiction writers. But it is unlike its twentieth-century successors, in that it uses science not so much to promote or reflect upon the potentialities of technology, but rather in a very Victorian matter of fact way. As Merrill points out, it exhibits a passion for quantification: 'Statistics abound: exact figures for latitude, longitude, ocean temperatures, dates, distance, lengths of objects, land volumes . . . . The assistant to Professor Aronnax, his servant Conseil, is a walking compendium of taxonomy, always ready to categorize any sea creature by order, family, genus, and species, in a post-Linnean frenzy.'[120] It is perhaps not so surprising that when, in 1924, Léon Groc writes his reworking of Verne, *Two Thousand Years under the Sea*, he could do little more than update the science, focusing on Darwinian evolution.[121] By contrast, Albert Robida, author of an 1880s trilogy of futuristic novels about the twentieth century,[122] extrapolated from contemporary everyday life. Robida was also a prolific illustrator, and his charming but wholly unrealistic illustrations of an imagined late twentieth-century Paris (Fig. 11.18) illustrate the dangers of such extrapolation. Similarly with Edward Bellamy's very popular utopian novel, *Looking Backwards, 2000–1887*, which pictures a future in which invention is no longer driven by commercial gain but simply to serve mankind.[123]

---

[117] H. G. Wells, *The Outline of History: Being a Plain History of Life and Mankind* (2 vols, London, 1920). See Matthew Skelton, 'The Paratext of Everything: Constructing and Marketing H. G. Wells's *The Outline of History*', *Book History* 4 (2001), 237–75.

[118] H. G. Wells, *The Work, Wealth and Happiness of Mankind* (London, 1932), 10.

[119] Jules Verne, *Vingt mille lieues sous les meres* (Paris, 1870).

[120] Merrill, *The Romance of Victorian Natural History*, 132.

[121] Léon Groc, *Deux mille ans sous la mer* (Paris, 1924).

[122] Albert Robida, *Le Vingtième Siècle* (Paris, 1883); *La Guerre au vingtième siècle* (Paris, 1887); *Le Vingtième Siècle. La vie électrique* (Paris, 1890).

[123] Robert Bellamy, *Looking Backwards, 2000–1887* (Boston, MA, 1888).

Fig. 11.18. Robida, Paris Skyline, *c.*1955, from *La Vingtième Siècle* (1883)

In what might be regarded as the fictional parallel to Wells' *Outline*, Olaf Stapledon's remarkable *Last and First Men* of 1930[124] describes the history of humanity from the present to two billion years in the future. The journey takes us across eighteen distinct human species, each of which proceeds through a fall into savagery followed by a rebirth in which the new species always rises to greater heights than the previous one. The emergence of telepathy plays a crucial role, and the story anticipates genetic engineering. Seven years later, Stapledon expanded the timescale to include the whole history of the universe in his *Star Maker*,[125] where the traveller discovers that his own cosmos is only of many (Fig. 11.19). A pervading theme is the unity of living things, humans and aliens. A similar theme had been pursued in 1920, in David Lindsay's *Voyage to Arcturus*,[126] which describes an interstellar voyage in which the protagonist encounters different states of mind, seeking the meaning of life. Similarly, Thea von Harbou's 1928 *Die Frau im Mond*[127] can be interpreted as a feminist voyage of discovery,

[124] Olaf Stapledon, *Last and First Men: A Story of the Near and Far Future* (London, 1930).
[125] Olaf Stapledon, *Star Maker* (London, 1937).
[126] David Lindsay, *Voyage to Arcturus* (London, 1920).
[127] Thea von Harbou, *Die Frau im Mond* (Berlin, 1928), made into a film in the following year.

Fig. 11.19. Covers of Olaf Stapledon's *Last and First Men* (1930) and *Star Maker* (1937)

demonstrating the female protagonist's skill, and her bravery in remaining behind on the moon.

Such works manifest the spirit of wonder most clearly, for they comprise a genre of self-discovery, in which we are invited to imagine our powers expanded and to reflect on the consequences. But not all such explorations have an optimistic outlook. In some cases, it was science that acted to put a check on utopian ideas. William Morris' utopian *News from Nowhere* (1890) was immensely influential for a time, but as followers such as George Bernard Shaw and H. G. Wells began to wonder what evolution meant for socialism, they abandoned Morris' vision of the future as incompatible with what Darwinism told us about human development.[128] By contrast, in the wake of the Great War, confidence in the ability of science to solve the problems of humanity was called into question. In a 1925 novel, Ernest Pérochon imagined a world a thousand years after the twentieth century, when humanity, who have survived a devastating war, seems to live in an ideal world based on science. But tensions arise as advanced technology is used to overcome an opponent.[129] A similar theme is

[128] See Hale, *Political Descent*, esp. ch. 6.
[129] Ernest Pérochon, *Les Hommes frénétiques* (Paris, 1925).

found in H. G. Wells' 1933 *The Shape of Things to Come* (made into a film in 1936),[130] although here the aviators who have contributed to the destruction in a decades-long war ultimately come to the rescue of civilization, only for their association of technology and civilization to come into question. In this genre, we might also mention Alfred Döblin's 1924 science fiction novel *Berge Meere und Giganten*,[131] which follows the development of human beings from the Great War up to the twenty-seventh century, focusing on the conflicts between technologies and natural forces, with the characters at one point losing their human individuality and growing into the earth. Again, the reaction to technology is ambivalent, as it is in Thea von Harbou's *Metropolis*,[132] where the benefits of technology are separated from those who generate its power. *Metropolis* was made into a film (it probably existed as a film script before it was written up as a novel) and released in 1927. It became one of the most famous science fiction films ever made, but the theme was not a new one in German cinema, and in 1920, in Hans Werckmeister's *Algol*, aliens give a human a machine that would enable him to rule the world, but it creates massive economic upheaval, and he finally realizes he has been corrupted by its power and destroys the machine. The Great War, with its mechanization of warfare, seems to underpin the worries driving the narrative here.

Encounters with aliens, which represent a major genre of the science fiction literature, work on a different level. Various forms of space travel were described in science fiction novels of the era, but it has a longer history. As early as 1643 a fictional work by Kepler, posthumously edited by his son, appeared, giving a detailed account of how the earth would look when viewed from the moon.[133] Jules Verne's 1865 *From the Earth to the Moon*[134] was wildly successful, and prompted some of the great early amateur investigations of the mechanics of space flight.[135] But in the nineteenth and early twentieth centuries, attention shifted to Mars, primarily as a result of the work of three astronomers who raised the issue whether it showed signs of life. The first was the Italian Giovanni Schiaparelli, who made extensive telescopic observations of Mars at the Breda Observatory in Milan, and in 1877 observed a dense network of linear structures which he called *canali*, meaning channels. Schiaparelli explored what the conditions of life on Mars might be,[136] but he did not subscribe to the idea of Martian 'canals' (a

[130] H. G. Wells, *The Shape of Things to Come* (London, 1933).

[131] Alfred Döblin, *Berge Meere und Giganten* (Berlin, 1924).

[132] Thea von Harbou, *Metropolis* (Berlin, 1926).

[133] Johannes Kepler, *Somnium, seu Opus Postvmvm de Astronomia Lvnari, Divulgatum à M. Ludovici Kepplero Filio* (Frankfurt, 1634). The protagonists are propelled up to the moon by demons.

[134] Jules Verne, *De la terre à la lune* (Paris, 1865).

[135] Namely Robert Goddard in the USA, Oberth in Germany, and Tsiolkovsky in Russia. See Dennis Piszkiewicz, *The Nazi Rocketeers: Dreams of Space and Crimes of War* (Mechanicsberg, PA, 2007), 6.

[136] Giovanni Schiaparelli, *La vita sul pianeta Marte* (Milan, 1893).

misleading translation of *canali*), which suggested to some writers construction by intelligent life. The French astronomer Camille Flammarion, for example, author of works on psychical research and science fiction novels as well as astronomical works, pursued such an interpretation. In his first book, published in 1862, he argued that there were many inhabited planets in the cosmos,[137] and during the 1880s and 1890s, he researched the Martian 'canals'. In his book on Mars of 1892, he argued that the canals were rectifications of old rivers whose purpose was to direct water to the Martian surface continents, constructed by an advanced civilization struggling to survive in a dying world.[138] The most tenacious defender of Martian canals was the American Percival Lowell, who became determined to devote himself to astronomy and the observation of Mars as a result of reading Flammarion. In 1894, he built an observatory at Flagstaff, Arizona, the first observatory built in a high remote location conducive to observation of the night sky. He made detailed drawings of the 'canals' (Fig. 11.20), and in a series of books he set out to explore what he argued were non-natural features of the planet's surface.[139]

Mars provided a setting for a number of widely read science fiction novels, starting with Kurd Lasswitz's 1897 Martian novel in which humans encounter a

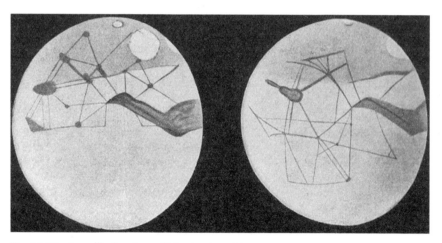

Fig. 11.20. Lowell's drawings of Martian 'canals'

[137] Camille Flammarion, *La pluralité des mondes habités: Étude ou l'on expose les conditions d'habitude des terres célestes discutées au point de vue de l'astronomie, de la physiologie at de la philosophie naturelle* (Paris, 1862).

[138] Camille Flammarion, *La planète Mars et ses conditions d'habitabilité: Encyclopédie générale des observations martiennes* (Paris, 1892), 586–9.

[139] Percival Lowell, *Mars* (New York, 1895); *Mars and its Canals* (New York, 1906); *Mars as an Abode of Life* (New York, 1908). See William C. Heffernan, 'Percival Lowell and the Debate over Extraterrestrial Life', *Journal of the History of Ideas* 42 (1981), 527–30.

Martian civilization which is older and more advanced than ours but which is running out of water.[140] In Aleksey Tolstoy's Russian science fiction novel *Aelita*[141]—made into a wonderfully stylish film in 1924, a year after the publication of the novel—an advanced civilization is discovered on Mars, but the huge gap between the rulers and the workers causes political problems when the Martians experience life on earth through a telescopic apparatus. In *Les Navigateurs de l'infini*, by J. H. Rosny, explorers on Mars discover an extraterrestrial race—*Tripèdes*—with three limbs, six eyes, and a supernatural beauty. These Martians turn out to be representatives of an ancient species which is slowly giving way to a new form of life, the *Zoomorphes*, mineral and less intelligent.[142] In many respects the comparisons with aliens, Martians or otherwise, are similar to the more traditional devices that used an outsider to comment on aspects of human culture from a critical perspective. Voltaire had used this device in his 1752 novella *Micromégas*, where giant inhabitants from an outer planet of the solar system are used to reflect on how features such as the (comparatively short) length of human lives affect our mentality. More mundane versions of the device figure in his 1747 novella *Zadig ou la destinée*, as well as in works that use Eastern civilizations as a comparison, such as Montesquieu's *Lettres persanes* and Voltaire's *Les Moeurs*, which we looked at earlier.

SPACE: SCIENCE FICTION AND ENGINEERING

Science fiction offered a novel way of articulating the values and dangers of science, prompting questions about the nature of modernity that a theoretical understanding of the world was unable to address. At the same time, in areas such as the early development of space flight what it had to offer was in competition with the claims of a theoretical understanding. Space flight represents the most spectacular achievement of twentieth-century technoscience, and it is its greatest feat of engineering, yet the history of its development and makeup is far richer than that of technology or science alone. Space travel has always figured prominently in science fiction, and as such it brings together central themes from this and from the last chapter in a striking way. The theoretical research characteristic of 'pure science' played a subsidiary role. Indeed, two of the most notable early developments in rocket science, in Germany and in Russia, were the work of amateurs and science fiction writers.

Fritz Lang's 1929 film of Harbou's *Frau im Mond* featured a rocket tearing across space to the moon. The rocket ship that it featured was designed by Hermann Oberth, a Romanian school teacher. Although Oberth and his

[140] Kurd Lasswitz, *Auf zwei Planeten* (2 vols, Leipzig, 1897).
[141] Aleksey Tolstoy, *Aelita* (Moscow, 1923).
[142] J. H. Rosny, *Les Navigateurs de l'infini* (Paris, 1925).

assistant, the popular science writer Willy Ley, had a real rocket in mind, what was depicted in the film was of course not a functioning rocket, but in the wake of the success of the film Oberth tried to talk the film company into funding a working model. Oberth had trained as a doctor, but moved towards physics and mathematics and in 1922 submitted his PhD in space flight at the University of Heidelberg, introducing problems of weightlessness in space and the importance of using liquid propellants.[143] The PhD was promptly failed, but Oberth reworked it as a book, which appeared in 1923.[144] Although the science and engineering establishment showed some limited interest in the more technical aspects of his account, they completely rejected the idea of space travel. But the reading public did not, and the book became a bestseller, establishing Oberth as the person to go to if one were interested in space flight. This is what led to the approach from Lang, and there was a spike in the popularity of the idea of space flight in the wake of *Frau im Mond*. Nevertheless, his project to obtain funding from the film studio to build a working version of a rocket floundered. His only hope lay with the amateur space travel societies that were beginning to be set up. He was invited to join the first of these, the Verein für Raumschiffahrt, founded in 1927, and he quickly became its most prominent figure. Through these connections, in 1930, an offer of funding came from a government institute, the Chemische-Technische Reichsanstalt, and in July of that year he and his colleagues succeeded in launching a rocket on a ninety-second journey. This completed, the project effectively came to an end for Oberth who, having a wife and family to support, needed a regular salary, and returned to teaching in Romania. One of his assistants, Wernher von Braun, was approached by the German military in August 1932 however, and agreed to develop the work on rockets in the direction of missile research.

There is a strong case to be made that spaceflight become possible through an unlikely merging of military engineering and utopian imagination, and this is even more evident in Russia than it was in Germany. The American geneticist Leslie Dunn, reflecting on a visit to what was to become the Institute of Experimental Biology in Moscow in 1927, noted that whereas 'Westerners were inclined to go in through the traditional front door, our Soviet colleagues seemed at times to break in through the back door or even to come up through the floor'.[145] Coming up through the floor is perhaps the best way to characterize the distinctive science fiction route to space flight in Russia from the late nineteenth century onwards.

In his study of the Russian space programmes from the late nineteenth century, culminating in *Sputnik*, the spaceflight historian Asif Siddiqi has argued that the

---

[143] On Oberth's career see Helen B. Walters, *Hermann Oberth: Father of Space Travel* (New York, 1962).
[144] Hermann Oberth, *Die Rakete zu den Planetenräumen* (Munich, 1923).
[145] Leslie Dunn, 'Science in the USSR: Soviet Biology', *Science* 99 (1944), 65–7: 66.

developments occurred by bridging 'imagination and engineering', and he argues that they should be seen not as discrete and sequential,

but as mutable, intertwined, and concurrent. Both imagination and engineering were necessary to attain the reality of space exploration. Russian imagining of the cosmos dated back to the late nineteenth century, a time when the first seeds of cosmic enthusiasm were sown in a broader literate public. This curiosity percolated into a burst of utopian fascination in the 1920s that inspired and then intersected with the practical realities of rocket engineering a decade later. Imagination and engineering not only fed each other but were coproduced by key actors who maintained a delicate line between secret work on rockets (which interested the military) and public prognostications on the cosmos (which captivated the populace). *Sputnik* was the outcome of both large-scale state imperatives to harness science and technology and populist phenomena that frequently owed little to the whims and needs of the state apparatus.[146]

The primary state institution typically associated with the advancement and sponsorship of Soviet science, the Academy of Sciences, was, Siddiqi argues, 'only marginally involved in the genesis and creation of one of the greatest public advertisements for Soviet science, their space program. Instead, the century-long origin of the Soviet space program provides evidence of a kind of "science from below", which later intersected with the military imperative to build rockets in giving birth to the space program.'[147]

The Russian translations of the space-themed novels of Jules Verne appeared in the 1880s, and those of Flammarion in the 1890s, and these were a formative influence on the Russian literature,[148] alongside the home-grown utopian works of V. F. Odoevskii, which had been censored but were circulated in manuscript form from the 1840s. Although Aleksandr Bogdanov's *Red Star* (*Krasnaya Zvezda*, 1908), which envisages a communist utopia on Mars, provided the template for revolutionary science fiction over the next two decades, the most influential figure from the end of the nineteenth century to the early decades of the twentieth was Konstantin Tsiolkovskii, an impoverished rural schoolteacher who did not move in scientific circles but whose writings provide, in an unmistakable form, a complex mix of utopianism, science fiction, engineering, and technology. In Tsiolkovskii's three science fiction novels, published in the 1890s and in 1920 (see Fig. 11.21), a host of arcane ideas are presented in stilted language, with the plot subsumed under scientific detail. He introduces a great number of ideas—space stations, multi-stage rockets, space-suits, life-support systems—that were only later to become the basics of space travel. As Siddiqi notes, 'the odd tension

[146] Asif. A. Siddiqi, *The Red Rockets' Glare: Spaceflight and the Soviet Imagination, 1857–1957* (Cambridge, 2010), 8.

[147] Ibid.

[148] See Darko Suvin, 'The Utopian Tradition in Russian Science Fiction', *Modern Languages Review* 66 (1971), 139–59; and Richard Stites, *Revolutionary Dreams: Utopian Vision and Experimental Life in the Russian Revolution* (Oxford, 1989).

Fig. 11.21. Illustration from Konstantin Tsiolkovskii, *Na Lune* (Moscow, 1893)

between his vision of the future—fantastic, unbelievable, and utopian—and the language that he used to communicate this vision—torpid, turgid, and inelegant—gave his fiction a strange tenor, one that firmly linked his fiction to popular science writing rather than creative literature'.[149] Tsiolkovskii was an avid experimenter to the extent to which his very limited resources would allow, setting out detailed plans for building a metallic dirigible, and carrying out aerodynamic

---

[149] Siddiqi, *The Red Rockets' Glare*, 22.

experiments. He was the first to propose the use of liquid propellants for rockets, and explored the idea of how to provide enough acceleration to escape the Earth's velocity through the 'reaction principle', whereby a body is powered by ejecting matter; and in 1903 he set out an equation showing the exponential relation between the weight and fuel supply of a rocket and its maximum velocity. In the later reception of these ideas, popular science journalism played a crucial role. At the same time, the evolution of the many amateur 'cosmic societies' that sprang up in the 1920s and 1930s, from voluntary astronomical societies into communities of spaceflight enthusiasts, was effected to a large extent by the heavily illustrated self-education magazines, starting with the very popular weekly, *Priroda i Liudi*, which appeared from 1889, and whose publisher, Petr Soikin, also issued a stream of English and French science fiction works in Russian translation.

The vast bulk of Tsiolkovskii's writings were unpublished, and the majority of those that were published were self-published. Consequently, they could not have had the great effect they subsequently did on the 'cosmic societies', without being taken up by popular science writers, the most influential of which was Iakov Perel'man. His 1913 book on 'physics for entertainment' was singularly successful and seeded a series of 'for entertainment' books on arithmetic, algebra, astronomy, geometry, and mechanics. In his writings, Perel'man was a staunch promoter of the view that reason and science were unique sources of truth, by contrast with the superstition, mysticism, and religion of traditional Russian society. In translating Tsiolkovskii's work into a more readable form and promoting it in widely read articles and books, Perel'man established the standing of this work. As Siddiqi notes: 'The notion that Tsiolkovskii was doing legitimate science, even as he was both ignored by the West and unappreciated by the Russian public, ran through the pages of probably the most important popular work on space exploration in the imperial era, Perel'man's first complete monograph on the topic, *Mezhplanetnye puteshestviia* (*Interplanetary Journeys*), issued by Soikin publishers in 1915.'[150] The combination of Tsiolkovskii's mechanics and mathematical computations, the popular science journalism of Perel'man, and the widespread availability of spaceflight in science fiction literature, together shaped a culture in which men and women joined to gather in urban centres across the Soviet Union to establish an informal and dynamic network for the exchange of ideas on the possibility of space exploration, a network that, in spite of the lack of official support, lay at the basis of the future space programme. Some of the groups in this network, such as the Association of Self-Educated Naturalists, asserted their independence from the scientific establishment, but other existed on small state subsidies and membership fees. Although official recognition did come to Tsiolkovskii in the early 1930s, it is notable that his citation mentioned 'conquering the stratosphere', saying nothing about the space exploration that Tsiolkovskii

---

[150] Ibid., 39.

himself, and the many cosmic societies, were concerned with. These were simply not of any official interest.

The enthusiasts of the cosmic societies often had no formal education in the natural sciences, and many were autodidacts, relying on the popular science literature. They embraced not only technological utopianism, but were in many cases motivated by a form of cosmic (religious and non-religious) mysticism. At the same time, their use of popular forms of communication distanced them from the orthodox scientific community. They shared a commitment to technological determinism with the latter, and this strengthened the hand of those promoting space travel not just as an idea for the remote future but one with immediate prospects, for it could be seen as a natural and perhaps inevitable consequence of the development of technology. Throughout the 1920s and 1930s, the Russian media was full of discussion of space travel, but in the early 1930s it was the cosmic societies that organized the first practical research organizations dedicated to developing rockets. There was no significant state support, with enthusiasts often selling personal belongings to keep projects going. But by the end of the decade, these enthusiastic amateurs were able to start promoting their cause to the state, whose exclusively military interests finally began to overlap with theirs. Siddiqi notes that, even as late as the 1950s, 'defence industry designers on the "inside" formed an effective alliance with journalists on the "outside" to mobilize public opinion in support of a space program, once again despite a lack of interest from the government'. Whether in the 1930s, 1940s, or 1950s, 'the nature of the discourse generated or the type of rocket produced was neither predetermined nor inevitable but resulted from a complex play of social, political, and military factors'.[151]

What the Russian case shows clearly is that 'science' does not have control over the products of popular science, even though at the same time it needs its resources. The association of science and civilization, crucial to the standing of science in the modern era, would simply not be possible without popular science—natural history books, children's science books, museums, science fiction, amateur voluntary associations of enthusiasts—because science doesn't have the cultural resources to establish its standing in its own right. More specifically, as I have indicated, in order to meet the expectations of the vastly expanded role that it had assumed from the end of the eighteenth century, one in which it was displacing religion as the key to understanding our place in the world for example, it was crucial that science be extended into the non-propositional realm, engaging with the world in terms of desires, expectations, anxieties, fears, hopes, goals, raw beliefs, etc. Only popular science could do this, but it could never just have been a promotional exercise, however much some scientists may have wished for this. The burgeoning science fiction literature in the wake of the Great War reflects a

---

[151] Ibid., 10.

growing ambivalence about the standing of science and technology as the guar-antors of civilization. It captures both aspirations and misgivings that neither the scientific community nor respective governments could articulate. It opened up deep questions about the direction in which science might proceed, questions that had a direct bearing on the responsibilities of scientists, on the relation between science and technology, and above all on the social and political standing of science.

PART V

SCIENCE AND THE
CIVILIZING PROCESS

12

The Modernization of the Population

Accommodating the Human to the Scientific Image

The public was not merely a passive recipient of science, as we have just seen. But nor was science something that the public could control, assimilating it as it saw fit. Transition to a science-driven modernity was accompanied by a profound cultural transformation, and having looked in Part IV at how the public encountered and responded to the sciences, we now turn to consideration of how a scientific culture encountered and responded to the public, looking specifically at how the population was shaped in such a way that it became amenable to scientific investigation, at how it was reconfigured and redirected accordingly. This process has a distinctive feature. Just as the project of the unity of the sciences was concerned to include certain disciplines and exclude others, and to tier and rank those that it included, so the resources that the shaping of the population drew upon followed this model in crucial ways. Questions of inclusion and exclusion are now pursued in terms of the establishment of norms. Ranking within the population comes to turn on success in various forms of testing and assessment, not least intelligence testing, which, at least notionally, replaced ranking by class, gender, and race with ranking by intelligence, for as one commentator has noted, 'underlying all concepts of intelligence is the conservative assumption that one cannot have order without hierarchy, either in society or nature'.[1]

In certain respects, attempts at shaping the public—which we can think of in the present case in terms of a 'modernization' of the population—were not unprecedented. In the Christian West in the fifteenth and sixteenth centuries, a concerted attempt had been made to transform the behaviour and attitudes of the general population by introducing a new set of moral and social values. Two developments are of particular interest here. The first was the shift from a very heterogeneous collection of practices which varied from region to region and person to person, to a relatively uniform religious sensibility. The second was a general 'internalization' of religion, a shift from something public to something

---

[1] Christopher F. Goodey, *A History of Intelligence and 'Intellectual Disability': The Shaping of Psychology in Early Modern Europe* (Farnham, 2011), 74.

Civilization and the Culture of Science: Science and the Shaping of Modernity, 1795–1935. Stephen Gaukroger, Oxford University Press (2020). © Stephen Gaukroger.
DOI: 10.1093/oso/9780198849070.001.0001

personal. These were features of both the Reformation and the Counter-Reformation, for in spite of their differences they shared a sense that Christianity needed to be renewed and given a new focus. As Jean Delumeau has shown, the key ingredient in this process is the fostering of a *contemptus mundi*, something originally developed and refined in monasteries, and later transmitted to the whole of society, in the first instance through the mendicant orders.[2] We can identify three ingredients in this 'contempt for the world': hatred of the body and the world, the pervasiveness of sin, and a sharp sense of the fleetingness of time. There was a fundamental concern with self-reform, motivated by feelings of guilt and repentance, evident not just in the ecclesiastical literature but also in secular literature from Montaigne onwards. These developments were accompanied by massively increased vigilance on the part of the various churches, which instituted a determined campaign against 'paganism'.[3] What was at issue in this campaign was not so much those non-Christian practices that had traditionally been deemed in competition with Christianity, and accordingly had been treated as a threat, but practices that had traditionally been considered as simply falling outside and hence being irrelevant to Christian culture. These were now treated as antithetical, as the understanding of what Christianity amounted to became both more all-encompassing and more exhaustively defined. Systematic attempts were now made to purge ancient rites and beliefs that had previously been tolerated, for example, on the assumption that they were harmless and would simply die out. At the same time, major programmes of reform were instituted by the Catholic Church in the sixteenth century, countering or matching those instituted by Protestantism, and the way in which this reform was instituted was the same for the Catholic and Protestant Churches in that there was a concerted programme of 'Christianization' of the population.[4] As Delumeau puts it, the problem was 'how to persuade hundreds of millions of people to embrace a severe spiritual and moral discipline of the sort which had never actually been demanded of their forebears, and how to make them accept that even the most secret aspects of their daily lives should thenceforth be saturated by a constant preoccupation with things eternal'.[5]

We can refine our understanding of this programme of 'Christianization' if we turn to a pathbreaking book first published in 1939, Norbert Elias' *Über den*

---

[2] See Jean Delumeau's magnificent tetralogy: *La Peur en occident (XIVe–XVIIIe siècles): Une cité assiégée* (Paris, 1978); *Le Péché et la peur: La culpabilisation en occident, XIIIe–XVIIIe siècles* (Paris, 1983); *Rassurer et protéger: Le sentiment de sécurité dans l'occident d'autrefois* (Paris, 1989); *L'Aveu et le pardon* (Paris, 1992).

[3] See Phillipe Ariès, *Religion populaire et réforme liturgique* (Paris, 1975).

[4] This had been a programme of the Church since at least Clement of Alexandria in the second century—see Peter Brown, *The Body and Society: Men, Women and Sexual Renunciation in Early Christianity* (London, 1988)—but there it was a matter of the regulation of a small sect, whereas by the sixteenth century it was the European population in general.

[5] Jean Delumeau, 'Prescription and Reality', in Edmund Leites, ed., *Conscience and Casuistry in Early Modern Europe* (Cambridge, 1988), 134–58: 147–8.

Prozess der Zivilization, in which Elias showed how the question of the civilizing process needed to be at the centre of our understanding of the transformation of values in the transition from feudal to early modern culture.[6] In a detailed discussion of table manners, natural functions, blowing one's nose, spitting, bedroom behaviour, behaviour of men and women towards one another, aggressiveness, and chivalrous behaviour, he transformed the standing of enquiry into questions of the fundamental reform and homogenization of behaviour by looking at the reform of 'manners'. One of Elias' distinctive theses was that the civilizing process that we find so marked from the early sixteenth century onwards provides the prototype for what he terms the conversion of 'external to internal compulsion', whereby the values of civilization are secured through a form of internalization.

The phenomenon of early modern Christianization can be thought of fruitfully in terms of a transformation of external to internal compulsion. Elias offers a way of thinking of the civilizing process in terms of the internalization of certain values, and this captures the process of Christianization perfectly. Moreover, as a form of internalization of values, Christianization has much in common with the later programmes of the 'age of empire', which were rationalized in terms of spreading civilization. Here, the civilizing process took on an international dimension. Bayly has remarked that, as world events became more interconnected and interdependent in the nineteenth century, so forms of human action adjusted to each other and came to resemble each other across the world. He points out that we can trace 'the rise of global *uniformities* in the state, religion, political ideologies, and economic life as they developed through the nineteenth century. This growth of uniformity was visible not only in great institutions such as churches, royal courts, or systems of justice. It was also apparent in . . . "bodily practices": the way in which people dressed, spoke, ate, and managed relations within families.'[7]

But for uniformities to have spread, it was important that a form of self-awareness consonant with the cultural norms of the dominant powers emerged in populations that did not initially share these cultural norms. Accordingly, as Bayly notes, 'exposure to global changes could encourage literati, politicians, and ordinary people to stress difference rather than similarity. By the 1880s, the impact of Christian missionaries and Western goods, for example, had made Indians, Arabs, and Chinese more aware of their distinctive religious practices, forms of physical deportment, and the excellence of their local artisans.'[8] In other

---

[6] Norbert Elias, *Über den Prozess der Zivilization: Soziogenetische und psychogenetische Untersuchungen* (2 vols, Basle, 1939).

[7] Bayly, *The Birth of the Modern World,* 1.

[8] Ibid., 4. On the way in which different 'religions' come to be distinguished in colonial societies as part of the colonizing process, see Brent Nongbri, *Before Religion: A History of a Modern Concept* (New Haven, CT, 2013).

words, it was not a question of a forced homogenization of values, so much as one of the provision of a common framework within which values could be identified as such, marshalled, and compared and assessed accordingly. Nevertheless, this common framework worked in terms of norms, and it is indicative of these developments that Cesare Lombroso, in his immensely influential study of the 'criminal type', *L'uomo delinquente* (1889), could contrast the savage state, not with the civilized one, but with *normali*, normal people.[9]

The wholescale, international nature of this exercise is as distinctive of modernity as any of the other factors that we have considered. As might be expected from a project that consists in classification, differentiation, and assessment, the quantitative methods of science were called into play. But these methods did not simply function as an aid. They went to the heart of the exercise, acting to radically reshape the subject matter, because they were intimately tied up with a drive towards a quantitative understanding of human behaviour, and towards the idea that human capacities and behaviour are indeed in essence something that is quantitatively discrete.

In this chapter I want to bring these kinds of issues under the general rubric of the modification of behaviour, with an emphasis on what Elias identified as the shift from external to internal compulsion. This is not just something facilitated by the resources of science. Above all, it acts to accommodate the human to a scientific image, to fit the human to a scientific template, because this is now what understanding the human being is considered to consist in. In the early modern Christianization of Europe, an exhaustive range of qualities were identified as being those on which the character of the individual rested. At the same time, procedures of self-examination were introduced for the general population which had previously been the preserve of monastic culture, and more recently of the clergy generally. As a result, the responsibilities of the general population were effectively assimilated to those of the clergy. With varying degrees of success, there was now introduced a strict template against which both a traditionally loosely regulated group, the secular clergy, and a traditionally unregulated group, the population in general, could be examined, and the primary mechanism of examination was self-examination. One thing that unites Christianization with the later developments with which we are concerned is that in both cases they work via an internalization of values. What distinguishes them from one another is that the template that Christianization provides is one of what collectively might be described as virtues, whereas what develops in the nineteenth and twentieth centuries is a template of norms: the degree to which one matched the requirements of the 'type' into which one was classified. It enables various

---

[9] Cesare Lombroso, *L'uomo delinquente in rapporto all'antropologia, alla giurisprudenza ed alle discipline carcerarie* (Turin, 1889), 168.

standards of normality to be defined, and deviance—economic, sexual, criminal, political, intellectual—to be identified and analysed.[10]

The earlier template had judged against a standard of moral excellence, broadly conceived: the new one judges against a standard of norms, often taking the form of averages. The norms template reflects the shift in understanding human behaviour in collective rather than individual terms. As we saw earlier, early modern writers of 'philosophical' histories (as opposed to histories of events such as wars or reigns) such as Montesquieu and Hume realized that the driving forces behind constitutional, economic, and cultural history were best discerned at a collective rather than an individual level. But it was the shift in thinking about practical reasoning in morality that precipitated the most significant transformation. Without the guidance provided by Christianity, which its nineteenth-century critics considered dogmatically conservative, a number of more 'neutral' criteria of the moral worth of behaviour were explored. In the present context, the most significant of these is that of utilitarianism, where a criterion is devised for assessing the moral worth of behaviour that worked for collective actions, such as commercial behaviour, because it did not rely on character or intentions, and it was then transformed into a general criterion that could also be applied to individual behaviour. In the case of morality, this meant that any prior classification into the virtuous and the non-virtuous was irrelevant: all that mattered was an ability to appreciate the consequences of one's action in terms of the pleasure, or pain and suffering, they caused. In the case of economics, it meant that any prior classification into landlords, capitalists, and labourers for example was rejected: it was simply a question of maximizing individual utility.

THE CONSTRUCTION OF NORMS

What is ultimately at issue in the modern concern with norms is the shaping of the population into the kinds of person who can occupy a scientifically modelled form of civilization, and this requires elaborate testing and grading. We should note from the outset that there are powerful social and political imperatives behind the establishment of norms and the testing procedures that accompany them, imperatives that have often led to a results-orientated use of norms that are devoid of statistical significance. Lombroso, for example, in his study of 'the criminal type', simply took the 'normal' with which the criminal type was contrasted to be something clear to all, and accordingly it was left unexamined.[11]

---

[10] Michel Foucault offered a deservedly influential investigation of sexual norms in his *Histoire de la sexualité I: La Volonté de savoir* (Paris, 1976), which has acted as a model for much of the subsequent literature on norms.

[11] By contrast, Durkheim, in his *Les Règles de la Méthode Sociologique* (Paris, 1895), insisted that crime had to be considered something normal, inevitable, and indeed in some respects useful to society, although he qualifies this by arguing that it does not follow that 'the criminal is a person

Attention was focused on the non-normal, for it was this class of persons who posed a grave threat to society, and hence these were the object of study.[12] Although statistics were occasionally deployed, they were not used as an exploratory instrument but merely as a way of confirming what one already knew, or at least believed one knew. Similarly with anthropometric testing in the United States in the early decades of the twentieth century, dominated by the work of James Cattell. Although he subsequently came to be greatly influenced by Galton's project of measuring the psychological differences between people, Cattell exhibited no interest at all in what, for Galton, were the scientific foundations of his work, the statistical law of normal distribution. The historian of anthropometric testing, Michael Sokal, notes that Cattell's programme in the last decade of the nineteenth century was almost wholly devoid of scientific content. His major enterprise of anthropometric mental testing, 'to which he devoted more than a dozen years, failed miserably. But despite the obsolescence of his laboratory work, the failure of his tests, and especially the shallowness of his ideas, Cattell is possibly the nineteenth-century psychologist with the greatest influence on the twentieth century, particularly in the United States. The man who coined the term "mental test" and who sold "science" to the public did much to shape modern culture.'[13] How could this have happened? The answer is that Cattell's work met a perceived need, and despite the collapse of its empirical credentials, anthropometric testing continued into the first years of the twentieth century, because

America at that time was engaged in what has been called 'The Search for Order'. Millions of new immigrants—most with cultural backgrounds totally different from those of the early twentieth century—were flocking to the New World. The rapid industrialization of the period and the rise of the new professions placed a heavy premium on a standardized work style and on the development of formalized criteria for judging applicants for universities and jobs. Many citizens looked to education as an ordering, and Americanizing, process, and compulsory education laws were enacted by 1900. The rising concern for welfare, and evil influence, of the delinquent, dependent, and defective classes led to the rapid growth of institutions to serve their needs and to protect the public from them.[14]

We can trace the concerns that drive such programmes back to the nineteenth-century attempts to move from the establishment of biometric norms, where notions of health are assimilated to the idea of the norm, and illness conceived as

---

normally constituted from the biological and psychological viewpoints.' See Fournier, *Émile Durkheim*, 181–2.

[12] See Peter Cryle and Elizabeth Stephens, *Normality: A Critical Genealogy* (Chicago, 2017), ch. 5.

[13] Michael M. Sokal, 'James McKeen Cattell and the Failure of Anthropometric Mental Testing, 1890–1901', in William R. Woodward and Mitchell G. Ash, eds, *The Problematic Science: Psychology in Nineteenth-Century Thought* (New York, 1982), 322–45: 322.

[14] Ibid., 338–9.

deviation from the norm which is established on statistical grounds,[15] to the idea of statistical averages acting as social norms, where a notion of health of the general population goes beyond anything specifically medical. As early as 1835, Quetelet, in his *Sur l'homme*, explicitly links the progress of civilization to adherence to norms, announcing that:

one of the principal acts of civilization is to increasingly compress the limits within which the different elements relative to man oscillate. The more that enlightenment is pursued, the more will deviations from the mean diminish, and as a consequence we will tend to associate ourselves with what is beautiful and what is good. The perfectibility of the human species is derived as a necessary consequence of all our investigations. Defects and monstrosities disappear more and more from the body; the frequency and the gravity of maladies are combatted with greater effectiveness through the progress of medical science; our moral qualities will meet with improvements no less tangible; and the more we advance, the less we need to fear the effects and the consequences of great political upheavals and wars, the plagues of humanity.[16]

For Quetelet, the faculties of the average man are 'in a proper state of equilibrium, in a perfect harmony, equally distant from excesses and defects of every kind, in such a way that . . . one must consider him as the type of all that is good', and that 'an individual who epitomized in himself, at a given time, all the qualities of the average man, would represent at once all the greatness, beauty, and goodness of that being'.[17] Porter rightly notes in Quetelet a will to mediocrity, a tendency for the 'enlightened' to resist the influences of external circumstances and, like the sick body, seeking always to return to a normal and balanced state.[18]

Later in the century, and particularly in the early decades of the twentieth century, there emerged scientifically inspired programmes of research along these lines in which definite goals were adumbrated, and which can be seen as being in many respects detailed updatings of Quetelet's programme. The 1933 'Science of Man' prospectus of the Rockefeller Foundation, one of the largest and most influential sources of funding for the sciences in the USA at the time, sets out one explicit version of such a programme:

Can man gain an intelligent control of his own powers? Can we develop so sound and extensive a genetics that we can hope to breed, in the future, superior men? Can we obtain enough knowledge of the physiology and psychobiology of sex that man can bring this pervasive, highly important, and dangerous aspect of life under rational control? Can we unravel the tangled problem of the endocrine glands, and develop, before it is too late, a therapy for the whole range of mental and physical disorders which result from glandular disturbances? Can we resolve the mysteries of the various vitamins so that we can nurture a race sufficiently healthy and resistant? Can we release psychology from its present

---

[15] See Georges Canguilhem, *The Normal and the Pathological* (New York, 1991), 115–229.

[16] Lambert Adolphe Jacques Quetelet, *Sur l'homme et le développment de ses facultés, ou essai de physique sociale* (2 vols, Paris, 1835), ii. 342.

[17] Ibid., ii. 287, 289.   [18] Porter, *The Rise of Statistical Thinking*, 103.

confusion and ineffectiveness and shape it into a tool which every man can use every day? Can man acquire enough knowledge of his own vital processes so that we can hope to rationalize human behaviour? Can we, in short, create a new science of man?[19]

This approach was not just an interwar phenomenon. In 1951, for example, Hans Reichenbach's *The Rise of Scientific Philosophy*, the first primer of Logical Positivism for American audiences, set out a programme of Logical Positivism in terms of science taking over a range of moral and political issues. David Hollinger sums up Reichenbach's programme in these terms:

Once the preference for democracy over other forms of government was accepted... virtually every other issue of social concern within democracy was cognitive, not strictly emotive, was a matter for resolution by rational assessment of cause-and-effect relationships in the real world. Should, for example, private property be abolished? This was a matter for the infant science of sociology to decide on an empirical basis. It sounded like a moral issue, Reichenbach explained, but actually it was a scientific one. Reichenbach confidently listed, as examples of aspects of life that can be organized on a scientific basis, 'education, health, sex life, the civil law, the criminal code, and the punishment of criminals.' Individual wills may differ on such matters, but these wills are to be 'harmonized' through group interactions that are guided by 'cognitive relations', facts about the relationship of means to ends.[20]

The 'new science of man' is not just a descriptive project, but one that involves a fundamental reform of the population. The reform project can be traced back at least to the beginning of the nineteenth century, to Bentham's elaboration of a utilitarian ethics in terms of a quantitative assessment of outcomes of actions. By the beginning of the twentieth century, various new accounts of human behaviour, from psychoanalysis to behaviourism, were going beyond traditional questions of intentionality and probing the sources of human behaviour with a view to modifying it accordingly. These developments were certainly not confined to a domestic context. In his study of the uses of psychology in managing the British Empire, Erik Linstrum notes how new techniques in psychology around the turn of the twentieth century came to be accepted as reliable and useful sources of knowledge about the inner lives of people living in what was then the world's biggest and most diverse political system. These innovations, he writes,

marked a departure from late Victorian ideas about the supposedly inscrutable and alien 'native mind', and they captured the imaginations of critics and rulers of empire alike. Laboratory experiments, psychoanalysis, and mental testing offered ways to challenge racial hierarchies and expose pathologies at the root of the relationship between colonizer and colonized. But they also furnished methods for governing populations: making factories and armies run more smoothly; recruiting the most talented subjects for government jobs and scarce school places; combating anticolonial rebellions; and remolding

---

[19] Quoted (from the Rockefeller Foundation's Archives) in Kay, *The Molecular Vision of Life*, 45.
[20] Hollinger, 'Science as a Weapon in *Kulturkämpfe*', 447.

families, economies, and societies. From anthropologists and missionaries to bureaucrats, schoolteachers, and industrialists, the science of mind had wide appeal in the imperial context because it promised technical solutions to political problems.[21]

The rise of global uniformities in the nineteenth and twentieth centuries involved a reform of mentalities, but it also involved what Bayly called 'bodily practices': the way in which people dressed, spoke, ate, and managed relations within families and within the workplace. A good example of this is the late nineteenth-century establishment of norms of bodily proportions. In the last decades of the nineteenth century, for example, Dudley Allen Sargent, the director of physical education at Harvard, calculated the average physical dimensions and proportions of college students. As Cryle and Stephens note, Sargent's aim

was to develop a 'scientific' approach to physical education, and he saw anthropometrics as the best way to achieve this. During the 1870s and 1880s he measured the bodily size and strength of thousands of (young, white, male) college students. He then compiled these individual results into a single chart, using the statistical mean—the apex of the bell curve—as the most typical for each measurement. He took measurements of forty individual parts of the body, comparing left and right sides to assess their symmetry. He also tested for strength and stamina, using new devices he spoke of as his own inventions, calling them such names as the spirometer, the manometer, and the dynamometer.[22]

Sargent commissioned composite statues based on mean bodily proportions for the 1893 World's Fair in Chicago, labelling them 'The Typical American: Male and Female'. The statistical charts had an explicitly normalizing function, in that they allowed individuals to chart their position in relation to statistical norms and to try to move closer to those norms, and Galton was later to take up the idea of an anthropometric laboratory and make it a standard feature of a publicly available programme of self-examination (see Fig. 12.1). Cryle and Stephens make the crucial point that, while normalization imposes homogeneity, it is also what constitutes the modern subject as an individual, noting that this is exactly how normal standards were conceptualized in Sargent's anthropometrics:

For those whose measurements fell below the fiftieth percentile, the charts could be used to encourage self-improvement and correction. For those with measurements above the fiftieth percentile, Sargent's charts could provide a means by which to optimize their health and fitness and to measure their superiority in quantified ways. Sargent actively promoted this use of his charts at the 1893 World's Fair, where he ran a participatory Anthropometric Laboratory at which members of the public were measured and their anthropometric data recorded . . . . However, Sargent's charts and the statues he modelled from them were intended to serve a new purpose. Previously, anthropometric data had been collected for the anthropologist or other professional to interpret—primarily so that

[21] Erik Linstrum, *Ruling Minds: Psychology in the British Empire* (Cambridge, MA, 2016), 1.
[22] Cryle and Stephens, *Normality*, 294.

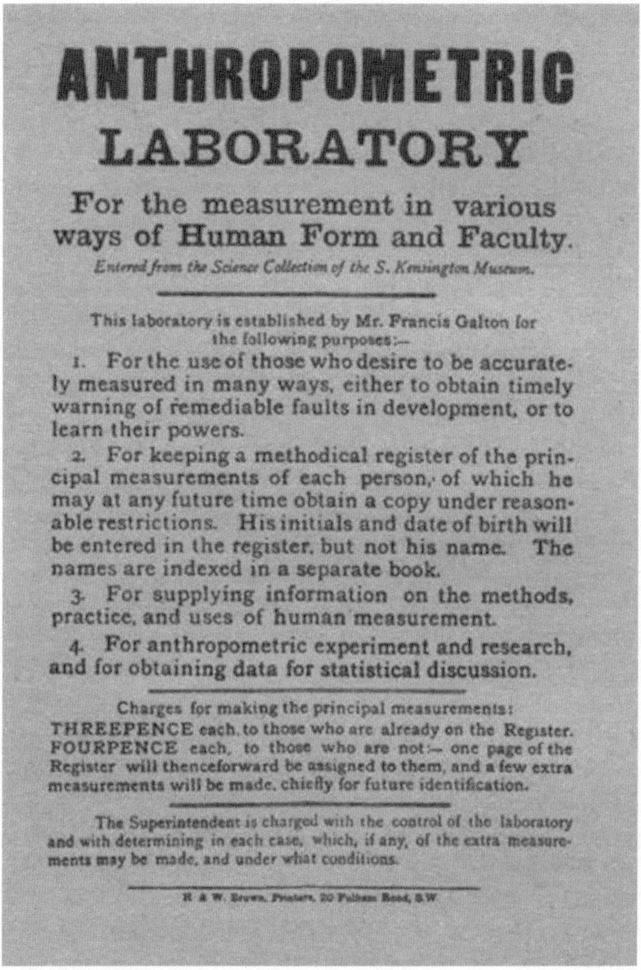

ANTHROPOMETRIC

LABORATORY

For the measurement in various
ways of Human Form and Faculty.

Entered from the Science Collection of the S. Kensington Museum.

This laboratory is established by Mr. Francis Galton for
the following purposes:—

1. For the use of those who desire to be accurately measured in many ways, either to obtain timely
warning of remediable faults in development, or to
learn their powers.

2. For keeping a methodical register of the principal measurements of each person, of which he
may at any future time obtain a copy under reasonable restrictions. His initials and date of birth will
be entered in the register, but not his name. The
names are indexed in a separate book.

3. For supplying information on the methods,
practice, and uses of human measurement.

4. For anthropometric experiment and research,
and for obtaining data for statistical discussion.

Charges for making the principal measurements:
THREEPENCE each, to those who are already on the Register.
FOURPENCE each, to those who are not:— one page of the
Register will thenceforward be assigned to them, and a few extra
measurements will be made, chiefly for future identification.

The Superintendent is charged with the control of the laboratory
and with determining in each case, which, if any, of the extra measurements may be made, and under what conditions.

R & W. Brown, Printers, 20 Fulham Road, S.W.

Fig. 12.1. Galton, anthropometric testing notice

individual subjects could be identified according to their particular type. Sargent, however,
wanted to train the general public to apply these anthropometric charts to programs of self-
improvement that they monitored and managed themselves.[23]

---

[23] Ibid., 295–6. See also Sarah E. Igo, *The Averaged American: Surveys, Citizens, and the Making of
a Mass Public* (Cambridge, MA, 2007); Jürgen Link, *Versuch über den Normalismus: Wie Normalität
produziert wird* (Göttingen, 2013); and Fenneke Sysling, 'Science and Self-Assessment: Phrenological
Charts 1840–1940', *British Journal for the History of Science* 51 (2018), 261–80.

Sargent realized that there is unlikely to exist anyone with fully 'normal' proportions, because so many measurements are involved, and because it was only a tiny proportion of the population that had symmetrical physiques. The norm is an ideal body, and this is why it acts as a goal, for everybody needs to engage in acts of normalization—self-measurement, self-monitoring, and self-management—to try to meet it.[24]

The issues go beyond the question of normalization of bodily proportions however. There was, for example, also the question of how human bodies were able to function most efficiently in manufacturing processes. Rabinbach summarizes the developments in these terms:

In the mental life of the nineteenth century, work was at the center not only of society but of the universe itself. Social modernity, the project of superseding class conflict and social disorganization through the rationalization of the body, emerged at the intersection of two broad developments: the thermodynamic 'model' of nature as labour power, and the concentration of human labor power and technology of the second industrial revolution. The metaphor of human motor united these developments in the single idea that the working body is a productive force capable of transforming universal natural energy into mechanical work and integrating the human organism into highly specialized and technical work processes.[25]

The developments had in fact begun in the eighteenth century. Wedgewood had introduced a division of labour in his factory at Etruria, and this was a formative moment in the development of modern manufacturing. But at the end of the nineteenth century a new element was introduced, which was explicitly promoted as 'scientific', as a number of projects emerged, in Europe and in the USA, to rework the division of labour on a scientific basis, measuring the human body on analogy with a motor. The thermodynamic 'model' in play here rests on the idea of the conservation of energy, but also on the second law of thermodynamics, the decrease in total available energy. This is mirrored, Rabinbach has argued, in the replacement of idleness as the paramount cause of resistance to work by fatigue. European scientists, he writes,

devised sophisticated techniques to measure the expenditure of mental and physical energy during mechanical work—not only of the worker but also of the student, and even of the philosopher. If the working body was a motor, some scientists reasoned, it might even be possible to eliminate the stubborn resistance to perpetual work that distinguished the human body from a machine. If fatigue, the endemic disorder of industrial society, could be analysed and overcome, the last obstacle to progress would be eliminated.[26]

From the 1890s, the language of disorderly and dissolute workers on the one side, competing with that of exploited and suffering workers on the other, was

---

[24] Cryle and Stephens, *Normality*, 298. Cf. Daniela Döring, *Zeugende Zahlen: Mittelmaß und Durchschnittstypen in Proportion, Statistik und Konfektion* (Berlin, 2011).
[25] Rabinbach, *The Human Motor*, 289.   [26] Ibid., 2.

beginning to be replaced by scientific study, as 'experts in fatigue, nutrition, and the physiology of the "human motor" sought to provide a neutral, objective solution to economic and political conflicts arising from labor'.[27] The solution lay in the scientific investigation of bodily movements and rhythms, providing new techniques of measurement and photographic study.

In the USA, Frederick Taylor was advocating a system of 'scientific management', the principles of which were set out in his book of that title in 1911.[28] Here the labour force was conceived in purely functional terms, and an inefficient worker was like a poorly designed machine part. The production process was conceived in terms of a machine in which the mechanical and human parts were for all intents and purposes indistinguishable. Although Taylor's principal efforts were directed towards improvement in plant layout and tools and machinery—which were redesigned as standardized models—the use of instruction cards and reports in this enterprise meant reorganizing the labour process in a comprehensive way, as indeed did the linking of wages to output.[29] From the early 1880s, he used timing of labour processes in a novel fashion, breaking down what he considered to be complex processes into elementary ones, and timing these latter in the cases of those workers whom he considered most efficient. Once he had done this, he resynthesized these efficiently performed component motions into a new complex set. The result, as Hughes notes, 'was a detailed set of instructions for the worker and a determination of time required for the work to be efficiently performed. This determined the piecework rate; bonuses were to be paid for faster work, penalties for slower. He thus denied the individual worker the freedom to use his body and his tools as he chose.'[30] As far as Taylor was concerned, the

---

[27] Ibid., 6.

[28] Frederick W. Taylor, *The Principles of Scientific Management* (New York, 1911). The extent to which what Taylor was offering was either 'scientific' or 'management' are questionable. As one commentator has pointed out: 'A jack-of-all-trades in industrial affairs, he could hardly engage in the most casual diversion without concocting some new method or piece of equipment for doing it better. He had little feeling for the ambiguities of human motivation and no patience for the subtleties of human psychology, having nothing whatever to contribute to personnel recruitment, welfare work, promotion and career strategies, or "industrial psychology" generally.' Gavin Wright, 'The Truth about Scientific Management', *Reviews in American History* 9 (1981), 88–92: 91.

[29] The great divide between European proponents of a science of work such as Jean-Marie Lahy, and advocates of Taylorism, was that the fundamental premise of the approach of the former was that job satisfaction and greater productivity were linked and that both could lead to social justice and greater happiness: by contrast with Taylorism, which worked on the assumption, as Rabinbach puts it, 'that by offering higher wages as premiums for productivity, the "natural" unhappiness of workers could simply be compensated by nonwork-related material rewards.' *The Human Motor*, 242. See Jean-Marie Lahy, *Le Système Taylor et la physiologie du travail professionnel* (Paris, 1916) for the reaction in France, and Wilhelm Kochmann, 'Das Taylorsystem und seine volkswirtschaftliche Bedeutung', *Archiv für Sozialwissenschaft und Sozialpolitik* 38 (1914), 391–424, for a German reaction.

[30] Hughes, *American Genesis*, 191. In the light of this alone, it is hard to credit Daniel Nelson's claim that 'scientific management in practice had little direct impact on the character of work or the activities of production workers': *Frederick W. Taylor and the Rise of Scientific Management* (Madison, WI, 1980), 137. In his *Labor and Monopoly Capitalism: The Degradation of Work in the Twentieth*

merits of different approaches to work processes must be subject to scientific criteria. How novel was this appeal to science? So far as I can tell, no one ever claimed that there was anything 'scientific' about Wedgwood's division of labour at Etruria. It was simply the result of ingenious attempts to improve efficiency. Taylor, by contrast, sees what he is doing in grander terms, as effecting a conceptual revolution, in which the measurement of efficiency plays a pivotal role.[31] At the same time, the new scientific approach set out to transform traditional social and political questions. It was promoted as a means of escaping from class conflict. As Agar points out, by insisting that attention be devoted to efficiency rather than division of surplus, 'Taylor and his followers argued that the wealth of increased production would benefit both workers and owners. In language that anticipates late twentieth-century theories of the knowledge society, industrial strife would wane as knowledge was gathered and wielded by neutral experts.'[32]

Sargent and Taylor, two contemporary American pioneers of human measurement, both had aspirations to being harbingers of new sciences. The important point, however, is not whether what they were proposing was actually scientific, but rather the fact of the claim itself, which hinged on the idea that the quantification procedures that they used secured scientific standing for their enterprises. The aim was to put them on a new footing, for any criticism could now in effect only be backward looking. Quantifying something that hadn't been quantified before was assumed to generate scientific standing.

HEREDITY AND THE IMPROVEMENT
OF THE POPULATION

What is striking about the attempts to establish scientifically characterized norms of behaviour, is the way in which they mirror the religious reforms of the early modern period, in particular the way in which anthropometrics mirrors and replaces Christianization. Just as with the Christianization of the population 500 years earlier, normalization not only imposes homogeneity but also constitutes the subject as an individual. Moreover, failure to meet the standards of orthodox social/biological characteristics had consequences, just as did failure to

Century (New York, 1974), Harry Braveman identified the features of Taylorism as: the dissociation of the labour process from the skills of workers; the separation of concept and execution; and the use of a monopoly on knowledge to control each step of the labour process and the means of its execution.

[31] Taylorism in the USA was single-mindedly a question of efficiency and productivity, but in Europe it was applied more selectively: see Charles S. Maier, 'Between Taylorism and Technocracy: European Ideology and the Vision of Industrial Productivity in the 1920s', *Journal of Contemporary History* 5.2 (1970), 27–61.

[32] Agar, *Science in the Twentieth Century and Beyond*, 181–2.

meet the standards of orthodoxy in Christian beliefs in the earlier case. Sargent's norm of the ideal body, as represented in the Chicago World's Congress statues, was being compared with contemporary body shapes by eugenicists in the 1920s to demonstrate a degeneration in American masculinity as a result of mixing with inferior European stock.[33] In 1920, Harry Laughlin, assistant director at the Eugenics Records Office in Cold Spring Harbor, in response to charges of the unconstitutional character of available eugenics proposals, drafted a constitutionally sound proposal targeting anyone who, 'regardless of etiology or prognosis, fails chronically in comparison with normal persons, to maintain himself or herself as a useful member of the organized social life of the state'.[34] In the wake of the 1921 Second International Congress of Eugenics, voluntary sterilization on these grounds was passed, first in Virginia in 1924, and then subsequently in thirty states.[35]

These developments were reinforced by socio-political readings of natural selection. Very soon after the publication of *The Origin of Species*, natural selection had begun to play a crucial role in thinking about the future of humanity. As early as 1865 Friedrich Lange argued that the core social question was whether the struggle for life 'remains the only way to the perfection of man, or whether, with the strengthening of human reason, a new factor enters, and with it a turning point in the struggle for life'.[36] One way of dealing with the 'struggle for life' was culling some of the population and encouraging others to breed. The idea goes back at least as far as Plato, and the related questions of overpopulation and the role of inherited characteristics in shaping human behaviour had a significant history before the late nineteenth-century eugenics movement: Malthus had identified solutions to overpopulation in the eighteenth century, and Carlyle and Mill were engaging in a public controversy over whether human behaviour and social relations were fixed by nature (Carlyle) or were malleable (Mill).[37] Moreover, there was already an experimental community in 1848 founded in Oneida, New York, where eugenic principles regulated the choice of reproductive partners and the rearing of children.[38]

[33] See Mary Coffrey, 'The American Adonis: A Natural History of the Average American (Man), 1921–1932', in Susan Currell and Christina Cogdell, eds, *Popular Eugenics: National Efficiency and American Mass Culture in the 1930s* (Athens, OH, 2006), 185–216.

[34] Harry H. Laughlin, *Eugenical Sterilization in the United States* (Chicago, 1922), 446.

[35] Cryle and Stephens, *Normality*, 302. Not all sterilization was voluntary: see Mark A. Largent, *Breeding Contempt: The History of Coerced Sterilization in the United States* (New Brunswick, NJ, 2008). More generally see Mark H. Haller, *Eugenics: Hereditarian Attitudes in American Thought* (New Brunswick, NJ, 2008).

[36] Friedrich Albert Lange, *Die Arbeitfrage: Ihre Bedeutung für Gegenwart und Zukunft* (Leipzig, 1910), 38.

[37] See Diane B. Paul and Ben Day, 'John Stuart Mill, Innate Differences, and the Regulation of Reproduction', *Studies in History and Philosophy of the Biological and Biomedical Sciences* 39 (2008), 222–31.

[38] See Garland E. Allen, 'The Misuse of Biological Hierarchies: The American Eugenics Movement, 1900–1940', *History and Philosophy of the Life Sciences* 5 (1983), 105–28; Edwin

Among the first to draw the social and political consequences of natural selection were the economist Hugo Thiel and the psychologist William Preyer, who worked closely together.[39] Thiel used Darwin in an attempt to show that free competition in the economy is a natural necessity. Mutual struggle is a crucial element in political and social strivings, he argued, because it allows the best to triumph more quickly, with fewer victims.[40] Preyer developed the view that 'the inequality of man is a necessity of nature which must become more pronounced the more culture advances',[41] announcing that: 'Competition is not only the soul of industry and trade but also the most important lever of scientific and artistic progress, the most important driving power for labour, self-development, the development of the dispositions of character and all talents and virtues, for the perfection of the material and intellectual well-being of the individual and the entire nation.'[42] Translating this reading of natural selection into developmental physiology and developmental psychology, Preyer set out, in two influential works on the 'mind of the child' and the 'mental development of the child',[43] a programme of 'psychogenesis' combining research in comparative anatomy and physiology with observations of the instincts of animals and the intellectual development of schoolchildren.[44] Inherited traits play a crucial role in Preyer's account, as indeed they were doing in Galton's understanding of heredity.[45] Development is seen as the unfolding of a potentiality, completely distinct from any external, social, or environmental factors, but since, he argues, no mind could develop simply on the basis of one's own experiences, 'each must develop and revive the hereditary dispositions and the traces of the experiences and activities of his forefathers'.[46] However, this move to bypass any question of a social and environmental contribution to development, by making inherited traits fill the

Black, *War Against the Weak: Eugenics and America's Campaign to Create a Master Race* (Washington, DC, 2012), ch. 3.

[39] See Siegfried Jaeger, 'Origins of Child Psychology: William Preyer', in William R. Woodward and Mitchell G. Ash, eds, *The Problematic Science: Psychology in Nineteenth-Century Thought* (New York, 1982), 300–21.

[40] Hugo Thiel, *Über eigene Formen der landwirtschaftliche Genossenschaften* ([Poppelsdorf], 1868). Thiel sent Darwin a copy of the pamphlet.

[41] William Preyer, *Darwin. Sein Leben und Wirken* (Berlin, 1869), 169.

[42] William Preyer, *Naturwissenschaftliche Tatsachen und Probleme* (Berlin, 1880), 93. Quoted in Jaeger, 'Origins of Child Psychology', 310.

[43] William Preyer, *Die Seele des Kindes, Beobachtungen über die geistige Entwicklung des Menschen in den ersten Lebensjahren* (Leipzig, 1882); *Die geistige Entwicklung in der ersten Kindheit nebst Anweisungen für die Eltern, dieselbe zu beobachten* (Stuttgart, 1893).

[44] See Jaeger, 'Origins of Child Psychology', 314–17, whose account I draw upon here.

[45] This motivated Galton's studies of fraternal and identical twins in his attempts to fix similar features. The latter played a crucial role in his investigations, for the degree of resemblance in their behaviour suggested to him—or perhaps confirmed what he already believed—that even the most complex abilities were largely determined by heredity. See Francis Galton, 'The History of Twins, as a Criterion of the Relative Powers of Nature and Nurture', *Journal of the Anthropological Institute* 5 (1875), 324–9.

[46] Preyer, *Die Seele des Kindes*, ix.

role that the former would have played, does not mean that social and environmental factors play no role. Preyer makes it clear that 'education and instruction must master hereditary defects, and increase hereditary advantages'.[47]

Projects for the improvement of the human stock were not restricted to education and instruction, but, as we have already seen, formed part of broader eugenics programmes. In his *Memoirs*, Galton set out the rationale for eugenics in these terms:

Its first object is to check the birth-rate of the Unfit, instead of allowing them to come into being, though doomed in large numbers to perish prematurely. The second object is the improvement of the race by furthering the productivity of the Fit by early marriages and healthful rearing of their children. Natural selection rests upon excessive production and wholesale destruction; Eugenics on bringing no more individuals into the world than can be properly cared for, and those only of the best stock.[48]

Galton distinguished two kinds of eugenics. Positive eugenics took the form of increasing the proportion of the population with desirable traits, consisting in 'watching for the indications of superior strains or races, and in favouring them that their progeny shall outnumber and gradually replace that of the old one'.[49] In his first eugenics publication, in 1865, for example, he offers a model based on animal stock-breeding,[50] imagining a process whereby those receiving the highest marks in state-administered competitive exams would be married in Westminster Abbey and given a generous endowment allowing them to start a family immediately, remarking that if only 5 per cent of what is spent on the improvement of breeds of horses and cattle were devoted to the enhancement of the human race, then 'what a galaxy of genius might we not create!'[51] The idea of a eugenic choice of marriage partner continued well into the twentieth century, promoted in the popular press (see Fig. 12.2).[52] Negative eugenics, by contrast, consisted in

---

[47] William Preyer, *Naturforschung und Schule* (Stuttgart, 1887), 37.

[48] Francis Galton, *Memoirs of My Life* (London, 1908), 323.

[49] Francis Galton, *Inquiries into Human Faculty and Its Development* (London, 1883), 307.

[50] The idea was in the air. In the same year there appeared three essays by Vasilii Florinskii on 'human perfection and degeneration'. Like Galton, its author drew parallels with animal breeding: 'Much attention is paid to, and whole doctrines exist about, the betterment of stocks in cattle, sheep, dogs, even chickens, pigeons, and so on, and the goal is actually being achieved. Systematically cultivated breeds of animals astonish us by their perfection; whilst man in the successive generations breeds diseases and physical weakness rather than perfection.' Quoted in Nikolai Krementsov, *With and Without Galton: Vasilii Florinskii and the Fate of Eugenics in Russia* (Cambridge, 2018), 7 (http://dx.doi.org/10.11647/OBP.0144).

[51] Francis Galton, 'Hereditary Talent and Character Part I', *Macmillan's Magazine* 12 (1865), 157–66: 165. Cf. Francis Crick, in the early 1980s: 'I do not suggest that only the very rich or the very intellectual should have children (what a thought!) but roughly that upper and upper-middle class families be encouraged to have say 3 or 4 on average and manual labourers and obviously dim and disturbed people have 0 or 1.' Quoted in Matt Ridley, *Francis Crick: Discoverer of the Genetic Code* (New York, 2009), 161.

[52] Note the appreciation of Darwin's demonstration in *The Descent of Man* that it was females, not males, who are the agents of sexual selection.

decreasing the proportion of those with undesirable traits, with contraception and sterilization emerging as the preferred procedures.

Paul and Moore write that what was distinctive about Galton's approach was that it was the first framing of the issue to be inspired by the *Origin of Species*, the first to make an evolutionary argument about human nature linking questions of human breeding to the anxieties about biological decline that the reception of Darwin had provoked.[53] But what exactly was its debt to the *Origin*? To what extent can his account be said to rest on evolutionary theory, and to what extent is it more a question of his putting an evolutionary gloss on a something rather more prosaic? Galton took his initial bearings from the discussion of artificial selection in first chapter of *The Origin of Species*. But the intellectual thrust and novelty of the first chapter had been to use a well-known and uncontentious form of selective breeding to question the distinction between varieties and species, arguing that some products of artificial selection were as legitimately characterized as new species as they were new varieties, thus preparing the way for the main argument to emerge in the rest of the book, namely that, just as in artificial selection, so natural selection could result in new species. But Galton's concern was to improve the human species, not to produce a new species, and there is nothing new or especially Darwinian about selective breeding.

This leads to two questions: how does Galton's eugenics engage with evolutionary theory; and what scientific credentials does it lay claim to? On the first question, we might begin by noting that one of the great hopes of many evolutionists was to give human natural selection a beneficial direction. The problem was not just that Darwinian natural selection was blind, but, more disturbingly, that allowing natural selection to run its course seemed to mean disallowing those measures that attempted to check it—public charities, vaccinations, public health measures, and the like. If measures to protect less able members of society were used, the fear was that this would mean that they would not be selected out and would eventually compete with able members, possibly reversing an assumed beneficial 'direction' of evolution that natural selection had secured.

As we have seen, natural selection, while it was by far the primary mechanism of evolution for Darwin, was only one of a number of evolutionary mechanisms that he himself was prepared to accept, and these included Lamarckian-style forms of selection: indeed, Lamarckism takes on increasing importance in consecutive editions of the *Origins*. Lamarckian inheritance of acquired characteristics would solve the problem that Galton raises, but there was in Lamarck no mechanism by which characteristics are acquired, and Galton rejected it on these grounds. Instead, he explores the theory of 'pangenesis' proposed by Darwin in *The*

[53] Diane B. Paul and James Moore, 'The Darwinian Context: Evolution and Inheritance', in Alison Bashford and Philippa Levine, eds, *The Oxford Handbook of the History of Eugenics* (Oxford, 2010), 27–42: 29.

Dr. Ira S. Wile, Psychiatrist,
Expresses the Opinion That the State
May Control All Mating a Century
Hence, With Eugenics as the
Chief Thought in Mind.

Fig. 12.2. 'Marriage 100 Years From Now', *Oakland Tribune*, 25 June 1933

Variation of Animals and Plants under Domestication (1868), which offered a mechanism for both regeneration and the inheritance of acquired characteristics. Pangenesis postulates that every part of the body emits tiny 'gemmules' that contribute heritable information to the gametes. Galton reasoned that such gemmules should be carried in the blood, and that accordingly in the transfusion of blood from one breed of animal to another, the latter should begin to exhibit traits of the former. A number of experiments carried out by Galton on rabbits failed to pick up any such transfer however, and although Darwin himself subsequently said all that was shown was that gemmules were not carried in the blood, a commitment to pangenesis went into rapid decline.[54]

The only remaining solution, Galton considered, lay in eugenics. Since most modern understandings of eugenics differ from those of its originators and early supporters, however, it is important that we are not misled into associating it exclusively with the genocidal practices of ethnic and racial cleansing that came to characterize it from the 1930s.[55] For example, Dirk Moses and Dan Stone, in a

---

[54] See the discussion in Nicholas Wright Gillham, *A Life of Sir Francis Galton: From African Exploration to the Birth of Eugenics* (Oxford, 2001), ch. 13.

[55] This is now less common than it was. Consider for example the definition given in the editors' introduction to the *Oxford Handbook the History of Eugenics*: 'Eugenic practice sometimes aimed to prevent life (sterilization, contraception, segregation, abortion in some instances); it aimed to bring about fitter life (environmental reforms; *puériculture* focused on the rearing of children, public

study of the relation between eugenics and genocide, note that eugenics was typically conceptualized and practised with respect to the same group. It was not normally directed against other groups, let alone intended to bring about their destruction.[56] The original understanding of eugenics, while it often had racial and class undertones, was formulated very much in terms of improvement, and its advocates considered it as bringing together a wide range of scientific and humanistic disciplines (see Fig. 12.3). As Nils Roll-Hansen points out:

> Eugenics started as a science-based movement to combat threatening degeneration. It was initiated by idealistic scientists and was inspired by a humanistic Enlightenment ideal of science as the servant of human welfare, in which the general goal was to improve the biological heredity of human populations. In the abstract this appeared as a good and unobjectionable aim—provided the means were acceptable. Before the 1930s and the traumatic experiences of Nazi population policies, the word 'eugenics' had mostly positive connotations . . . . There was broad acceptance that the knowledge of genetics and other biological science should inform social policy.[57]

Eugenics as an alternative to Lamarckism and pangenesis can be thought of, as we have just seen, along the lines of what Darwin had discussed under the rubric of artificial selection in the first chapter of *The Origin of Species*. Indeed, artificial selection had provided the template for his understanding of variation, and although his discussion of breeding had been confined to plants and to animals such as pigeons, it was clear that it could in theory also be used as a means of controlling the breeding of humans themselves. He welcomed Galton's attempts to develop such a programme, writing to him after reading his *Hereditary Genius* that: 'I must exhale myself, else something will go wrong in my inside. I do not think I ever in all my life read anything more interesting and original.'[58] In *The Descent of Man*, which cites *Hereditary Genius* frequently, Darwin is insistent on the need for control of human breeding, and even though he is generally only in favour of voluntary measures, he writes that, if checks are not put in place to 'prevent the reckless, the vicious, and the otherwise inferior members of society from increasing at a quicker rate than the better class of men, the nation will retrograde, as has occurred too often in the history of the world'.[59]

---

health); it aimed to generate more life (pronatalist interventions, treatment of infertility, "eutelegenesis"). And at its most extreme, it ended life (the so-called euthanasia of the disabled, the non-treatment of neonates).' Philippa Levine and Alison Bashford, 'Introduction: Eugenics and the Modern World', in Alison Bashford and Philippa Levine, eds, *The Oxford Handbook of the History of Eugenics* (Oxford, 2010), 5–24: 5.

[56] A. Dirk Moses and Dan Stone, 'Eugenics and Genocide', in Bashford and Levine, eds, *The Oxford Handbook of the History of Eugenics*, 192–209: 194.

[57] Nils Roll-Hansen, 'Eugenics and the Science of Genetics', in Bashford and Levine, eds, *The Oxford Handbook of the History of Eugenics*, 80–97: 81.

[58] Francis Darwin and A. C. Seward, eds, *More Letters of Charles Darwin* (London, 1903), 41.

[59] Darwin, *The Descent of Man*, 166.

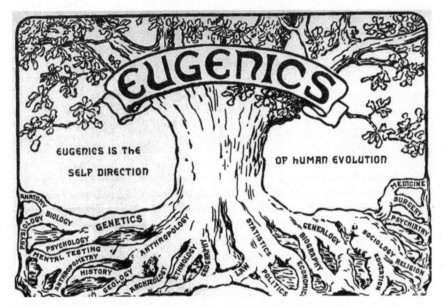

Fig 12.3. Chart of eugenics from the Cold Spring Harbour Lab.

The benefit of a eugenics modelled on artificial selection was that, unlike natural selection, which was blind, it enabled human evolution to be given a direction, to be steered as it were. For Galton, eugenics provided a way of circumventing the adverse effects of natural selection, while working in tandem with it. The alternative—'leaving things to nature'—amounts on his account to leaving things to natural selection, for there is no third way: the only choice is between natural selection and artificial selection. Galton and Darwin were nevertheless the first to admit that they had no understanding of the mechanism of transmission via heredity. Galton's eugenics was completely speculative, and could draw no support from evolutionary theory as such. Its claim was that it offered the best answer to what was considered to be a pressing social consequence of Darwin's theory of natural selection. It was an answer to a problem generated by a scientific theory. This did not make the answer—even if it were a satisfactory one—itself a scientific answer, but Galton did certainly claim scientific standing for his answer, and he pursued this scientific standing through a statistically based anthropometrics.

There are a number of aspects of his approach that are of particular interest in this respect. The first concerns his use of the error law, a formula for the distribution of errors originally introduced to deal with the existence of variations among astronomical observations, allowing the application of probability theory to an indefinite number of independent events. Quetelet had interpreted the law as showing that variation could be neglected in favour of the study of mean values. But Galton used it in a very different way, as a precision tool to study the nature and effects of natural variation. He rejected Quetelet's deployment of the statistical techniques of astronomy as unsuitable because they were designed to eliminate errors and deviations, whereas it was exactly these latter that he was interested in. As Porter points out: 'Galton admitted Quetelet's use of the error law, perhaps precisely because he paid no attention to Quetelet's interpretation of it. Galton understood the error law as an invaluable means for taking account of natural variation, and he was accordingly critical of contemporary statists who, because of their infatuation with averages, failed to make use of it.'[60]

It was the combination of this grasp of statistical techniques with a concern to counter the effects of natural selection through a scientifically directed programme of controlled breeding that is the most distinctive feature of his advocacy of eugenics. At the same time, Galton brought some highly imaginative modelling techniques to his investigation of natural variation. His ingenious representations of a defined range of features in composite portraits is particularly striking, and makes his point in a way that tables of numbers or even graphs could not. Layering a number of partially exposed individual portraits onto a single photographic plate (Plate 8), he set out to 'portray an imaginary figure possessing the average features of any group of men'.[61] The idea behind the portraits was that deviations from any given type must be distributed in the same way as errors of observation, so superimposing the images would result in averaging these out, producing an image very close to the type itself.[62] His use of the statistical techniques was also innovative, and it was appropriately flexible. In dealing with attributes that cannot be measured but can be ranked for example, he devised a method of 'intercomparison' as 'a method for obtaining simple statistical results which has the merit of being applicable to a multitude of objects lying outside the present limits of statistical enquiry'.[63] The core of the method was the substitution of the median for the mean, which he realized would introduce only a very small degree of inaccuracy,[64] and meant that it was simply a question of ordering by rank and identifying the middle value.

Galton did not have the ability to follow up the advanced mathematics needed to develop his programme fully, but Karl Pearson, who did, took up the reins

[60] Porter, *The Rise of Statistical Thinking*, 129.

[61] Francis Galton, 'Composite Portraits Made by Combining Those of Many Different Persons into a Single Figure', *Nature* 18 (1878), 97–100: 97. See also Francis Galton, 'On the Application of Composite Portraiture to Anthropological Purposes', *Report of the British Association for the Advancement of Science* 51 (1881), 690–91; 'An Inquiry into the Physiognomy of Phthisis by the Method of "Composite Portraiture"', *Guy's Hospital Reports* 25 (1882), 475–93.

[62] Galton, 'Composite Portraits', 98.

[63] Francis Galton, 'Statistics by Intercomparison, with Remarks on the Law of Frequency of Error', *Philosophical Magazine* 49 (1875), 33–46: 37.

[64] He learned the technique of substituting medians for means from Gustav Theodore Fechner, *Elemente der Psychophysik* (Leipzig, 1860): see Porter, *The Rise of Statistical Thinking*, 144.

from the 1890s onwards. Pearson, by his own account, took his starting point from Galton's development of the idea of correlation, which he introduced as 'a category broader than causation . . . of which causation was only the limit, and that this new concept of correlation brought psychology, anthropology, medicine and sociology in large parts into the fields of mathematical treatment. It was Galton who freed me from the false prejudice that sound mathematics could only be applied to natural phenomena under the category of causation. Here for the first time was the possibility—I will not say a certainty of reaching knowledge—as valid as physical knowledge was then thought to be—in the field of living forms and above all in the field of human conduct.'[65] Pearson's larger programme was reminiscent of Quetelet's advocacy of the idea that the faculties of the average man are 'in a proper state of equilibrium, in a perfect harmony, equally distant from excesses and defects of every kind'. In his *The Grammar of Science* (1892), Pearson argues that, for the 'scientific man' it is affective states that must be avoided to achieve such equilibrium.[66]

INTELLIGENCE, SOCIAL STANDING, AND RATIONALITY

Galton, reflecting on the demands of the modern era in his 'Hereditary Character', wrote:

The average culture of mankind is become so much higher than it was, and the branches of knowledge and history so various and extended, that few are capable even of comprehending the exigencies of our modern civilization; much less of fulfilling them. We are living in a sort of intellectual anarchy, for want of master minds. The general intellectual capacity of our leaders requires to be raised, and also to be differentiated.[67]

What was needed in response is what Pearson referred to as the 'scientific man', someone who is completely rational. But this is not just associated with the ability to rise above one's personal feelings and beliefs, and ridding oneself completely of everything not based on evidence. It is also associated with forms of cognitive ability that came to be classified under the name of 'intelligence'. As Agar notes, in the early nineteenth century, intelligence 'meant a faculty, quickness of understanding or communication, more often a relation rather than a thing, and very rarely something to be measured. The mid-century phrenologists had offered one means of measuring faculties, including intelligence. But, from the late nineteenth

[65] Karl Pearson, *Speeches Delivered at a Dinner Held in University College, London, in Honour of Professor Karl Pearson, 23 April, 1934* (Cambridge, 1934), 22–3. See also Theodore M. Porter, *Karl Pearson: The Scientific Life in a Statistical Age* (Princeton, NJ, 2004), ch. 9.

[66] Karl Pearson, *The Grammar of Science* (London, 1892), 7.

[67] Galton, 'Hereditary Talent and Character Part I', 166.

century, measurement of intelligence came together with statistical techniques to offer a powerful hybrid.'[68]

The rise of 'intelligence' should be seen in the context of the naturalization of reason, and a corresponding shift from reason to rationality, in Weberian terms from *Wertrationalität* to *Zweckrationalität*. In *The Natural and the Human*, I argued that anthropological medicine pitted reason against nervous sensibility, that philosophical anthropology pitted reason against sensation, and that what was important about nervous sensibility and sensation, and what distinguished them from reason as traditionally conceived, was that they were amenable to empirical enquiry. Both anthropological medicine and philosophical anthropology were naturalization projects, and amenability to empirical enquiry was crucial. Reason, it was considered, could not be naturalized (at least not without proceeding through a naturalized sensibility, as in Herder). It was utilitarianism that changed this. Bentham had naturalized moral reasoning, arguing that the morality of behaviour was to be assessed in terms of observable consequences. Mill had subsequently extended the naturalization programme to all forms of reasoning, construing logical inference in terms of a combination of empirical psychology and empirical generalization. With Galton, we encounter a different treatment of reasoning, one which subjects it to empirical testing and ranks reasoning abilities on a scale. This development became an integral part of a eugenics programme. Unprecedented though its procedures were, however, it was really just the latest move in a social stratification exercise presented as a form of scientific enquiry, one whose claim to legitimacy was that the ranking that it offered was based wholly on merit rather than class, wealth, or race.

The quantification of skills and abilities in the population plays a crucial role in thinking of civilization in scientific terms and, in particular, 'intelligence' comes to the fore as a form of social standing. Crucial here is what might be termed a depersonalization of judgement: a person's ability to make considered judgements is replaced by something that can be measured and allocated a place on a scale. Correlatively, we witness the idea of 'reason' as the highest faculty becoming transformed into 'rationality', a set of impersonal, context-free, potentially algorithmic devices that enable decision making, and the same game-theoretical model comes to regulate everything from missile strategy to moral judgements. As Erikson et al. put it, 'reason' and 'rationality' had previously:

either been used as rough synonyms or had each been assigned its own domain: reason referred to the highest intellectual faculty with the most general applications, from physics to politics to ethics; rationality referred more narrowly to the fitting of means to ends (sometimes called instrumental reason) and was especially associated with economics and engineering. What was distinctive about Cold War rationality was the expansion of the domain of rationality at the expense of that of reason, asserting its claims in the loftiest

---

[68] Agar, *Science in the Twentieth Century and Beyond*, 72.

realms of political decision making and scientific method—and sometimes not only in competition with but in downright opposition to reason, reasonableness, and common sense.[69]

In fact, the rise of rationality can be traced back to Bentham's consequentialism and it has pervaded much political theory since Bentham's time. The idea that political theory is a matter of securing rational rules, for example, has been summed up well by John Gray when he writes that philosophers

no longer aim to improve the political reasoning of ordinary citizens. Instead, they see their task as providing a body of principles which can best be interpreted by a court. In this view, best represented in John Rawls' deeply meditated and intricately reasoned, yet supremely parochial and unhistorical book, *A Theory of Justice* (1971), politics is redundant. Philosophers supply principles dictating the scope of individual liberty and the distribution of social goods; judges apply them. Little if anything of importance remains for political decision. Rawls calls his theory 'political liberalism', but it is more accurately described as a species of anti-political legalism.[70]

In the present context, however, our concern is not with political or moral reasoning but with the association of rationality and intelligence. Here we face a core question, that of where to locate 'intelligence'. It would be wholly misleading to place it under the rubric of reason, where it had in fact never been located, as if it told us something about sound or considered judgement for example. Rather, it must be located within the realm of rationality. It is the measurable manifestation of rationality.

Galton's approach to intelligence testing shows the influence of his time at Cambridge studying for the mathematics Tripos, and it provides some insight into just what was at issue. Suffering two nervous breakdowns, he did not perform particularly well in the first part of the Tripos, and was discouraged from continuing by his mathematics coach, so that he only graduated (like the majority of students) with an ordinary degree. The Tripos examination was, unlike earlier forms of assessment in Oxford and Cambridge, a time-sensitive examination. The questions started from straightforward problems and increased in difficulty. The final questions were so difficult that even the best students, those who were able to proceed that far in the test, were only able to solve a fraction of them.[71] But the

---

[69] Paul Erikson et al., *How Reason Almost Lost its Mind: The Strange Career of Cold War Rationality* (Chicago, 2013), 2.

[70] John Gray, review of Alex Callinicos, *Equality*, *Times Literary Supplement* no. 5116 (20 April 2001), 3. Helen Irving has pointed out to me that the emergence of human rights legislation in the post-war period is a crucial factor here. The point of the legislation was precisely to take human rights questions out of the realm of political decisions, in the light of the experience of the war, and to make them take the form of basic inalienable principles. The problems arise because what has happened is that the protection of human rights has been taken as a general model, pre-empting what would otherwise be political decision making.

[71] This is reflected in the marks of the top-performing students. These were never officially released, but Galton provides figures for three years in the 1860s. Out of a total of 17,000 possible

crucial thing about the examination was not just being able to solve difficult problems, but the rate at which one solved them. Unlike other forms of testing, where it was simply a question of the ability to solve problems, in this case no one taking the test had enough time to solve all the problems. One raced against the clock: the ability tested was one that depended on speed of response. It was as if the examinations, with their highly competitive nature, had been based on tests of athletic ability rather than intellectual ability, perhaps not so surprising in Galton's case given the association in the Cambridge system between building up mathematical skills and engaging in sporting activities, as we saw in Chapter 10.

This is what became the model for Galton's intelligence testing, and it received a 'scientific' vindication when Cattell, who had trained at Leipzig between 1883 and 1886 under Wundt, performing reaction-time experiments, introduced Galton to Wundt's experimental programme.[72] Wundt's work at his Leipzig Institute in the 1880s had played a crucial role in the development of empirical psychology. Building on the experiments of Helmholtz, his teacher, on the measurement of the speed of reflex transmission, Wundt set out to measure the rate at which the brain responded to nervous impulses by measuring reaction times.[73] Here was a cognitive operation that was measureable, and the thought was that the sensory operation that it measured could be mirrored in the measurement of an intellectual operation. The common feature was the variable rate at which the operation could be performed.

There were many refinements and revisions to intelligence tests. Independently of Galton, Alfred Binet and Théodore Simon devised a (non-time-sensitive) system of testing for children with learning difficulties.[74] Binet's view was that intelligence testing was subject to variability, was not simply a matter of genetics, and was affected by various environmental influences. But in 1908 the Binet–Simon test was taken over wholesale by the eugenicist Henry Goddard and converted into an intelligence test for the general population. Goddard's aim was to use intelligence testing to show the superiority of the white race, and he was condemned by Binet on a number of grounds, above all in that it illegitimately

---

marks, the score of the senior wrangler was 7,634 in the first year, between 5,500 and 6,000 in the second, and 9,422 (considered exceptionally high) in the third. The lowest-ranked candidates received scores in the low hundreds. Francis Galton, *Hereditary Genius: An Inquiry Into its Laws and Consequences* (London, 1869), 19–20.

[72] Cattell himself was evidently not interested in differences in reaction times between different subjects, being concerned instead solely with reaction times under different conditions of attention, practice, and fatigue. Moreover, despite Wundt's urging, he was not at all interested in the meaning of these results, but solely with collecting mathematical data. See Sokal, 'James McKeen Cattell', 326

[73] Robert S. Harper, 'The First Psychological Laboratory', *Isis* 41 (1950), 158–61; Woodward, 'Wundt's Program for the New Psychology'; David K. Robinson, 'Reaction-Time Experiments in Wundt's Institute and Beyond', in R. W. Rieber and D. K. Robinson, eds, *Wilhem Wundt in History* (Boston, 2001), 160–204.

[74] Alfred Binet and Theodore Simon, *Les Enfants anormaux: Guide pour l'admission des Enfants anormaux dans les classes de Perfectionnement* (Paris, 1907).

construed intelligence as a single, unitary construct.[75] At the same time, in developments in England, Charles Spearman and Cyril Burt set out to identify the strictly intellectual component in Galton's talk of mental powers, and devised intelligence tests for children to detect 'sub-normal' intellectual abilities, which became general IQ tests. Later, when cognitive ability came to be associated with information processing, speed began to be considered a measure of the efficiency of the brain.[76]

The universal feature of these generalized IQ tests was that it was the speed at which one solved increasingly difficult problems that was the underlying criterion of ability. Questions were raised as early as the 1920s as to why it was speed that was being tested, rather than accuracy for example, and it was noted that tests in the USA on American Indians indicated that they routinely aimed at accuracy rather than speed.[77] At the same time, there was much dispute over whether the 'intelligence' being measured in the tests was a single coherent thing. 'Intelligence', like the rationality of which it was the measureable indicator, was in reality something purely artificial which had been generated in the tests. As Goodey has noted, the procedure for establishing intelligence as a scientific concept 'consists *first* in conjuring up the notion of a mean purely as such. Subsequently, and only subsequently, this mean becomes something concrete.'[78] In other words, it is not as if there is something 'there in nature' that is being measured; rather, it is something that is constructed via measurement in the tests and has no identity outside them.[79]

This is not to deny that the quality referred to as *ingenium* or 'wit' had traditionally been valued in the modern era. But assessments of what it amounted to varied. Goodey, for example, notes that, over time, *ingenium*

often came to describe not just the operation (one among several) of one faculty, but a whole faculty or ability in itself. Speed became entangled with the notion of a single overarching *ingenium*, especially among medical writers.... The humanistic *ingenium*, with its connotations of speed, gained prominence through this reduction of complexities. Scholastic philosophers had allocated speed no special value in any of the various operations

---

[75] See Sheldon White, 'Conceptual Foundations of IQ Testing', *Psychology, Public Policy, and Law* 6 (2000), 33–43.

[76] See Goodey, *A History of Intelligence*, 39–47.

[77] See, for example, Otto Klineberg, *An Experimental Study of Speed and Other Factors in 'Racial' Differences* (New York, 1928).

[78] Goodey, *A History of Intelligence*, 71.

[79] This in itself does not mean that intelligence is merely an arbitrary invention, or that the tests lack any objectivity. While they certainly do not, and could not, have the kind of standing that their advocates often claimed for them, IQ tests do offer a broad-grained test of certain cognitive abilities, and I have argued elsewhere that if we think of the tests in terms not of intelligence but of the suitability for achieving particular educational goals, they may have some value when combined with other forms of assessment: see Gaukroger, *Objectivity*, ch. 2. And indeed, as I read him, this is how Binet saw his original tests, which eugenicists had subsequently transformed into something designed to register a single inheritable feature.

of the reasoning faculty such as *discursus* (the relating of premises to conclusions), *contemplatio* ('study' or meditation of these) or even *discretio* (their subsequent application), let along *ingenium* as they conceived it. On the rare occasions when operations are described in terms of speed, they are *so* quick as to take place in zero time, and then only in angels, 'prophets' or exceptional humans . . . . But in the Renaissance this territory was invaded by the more everyday quick-wittedness of *ingenium*, as the humanists conceived it.[80]

But at the same time there was resistance among many early modern writers to the association of speed and *ingenium*, stressing the importance of depth and penetration instead.[81] Speed was at best irrelevant to reason considered as the highest faculty. Considered as the source of sound and considered judgement, it is inimical to it, and speed in coming to decisions has often traditionally been condemned in terms of hasty judgement.

If we stand back from the details of the debates over intelligence testing, and examine its association with rationality, we can ask two questions. The first is what role rationality and the associated notion of intelligence have played. The second is what their limits are, and whether they engender any misunderstandings, particularly on the problem of the relation between reason and rationality.

On the question of the role of intelligence, what is at issue is not just its social standing in the modern era, but how this social standing is formed and operates. Here Goodey has offered a helpful account of the difference in the way in which concepts embodying social status distinctions are articulated. Status is usually seen as a two-tiered structure: there is a level comprising an abstraction of social goals, and a level of concrete evidence or collateral that one might have for claiming such status, the latter varying with the values of a given historical or cultural context. This two-tiered structure, Goodey argues, is inadequate as an account of how social status is established. We need to introduce another level mediating between those of abstract status and collateral. Prior to the modern period, the ideas of honour and grace played this role. Honour and grace are not externally verifiable, and are examples of what Goodey calls bidding modes: 'they connect status at its highest level, as an abstraction of values and goals, to its lower level, as concrete collateral to be used in support of a bid. That is why such modes are not susceptible to objectivist definition, why there is endless controversy about what constitutes them and why people claiming status will talk about their honour or state of grace as if it were self-evident when actually the terms are purely self-referential.' Intelligence tends to be thought of either as something objective and independently testable, or in sociological terms, but, like honour and grace, it should rather be seen as a mode of bidding for status, for, like them, it is located midway between status as a sum of general goals and status as the array of concrete items upon which one calls when making a bid.[82]

---

[80] Goodey, *A History of Intelligence*, 55.   [81] See ibid., 56–8.   [82] Ibid., 64.

How then, more precisely, do these concepts function? All three, Goodey argues, entail personal destiny and collective perfectibility. The motto 'honour before death', for example, illustrates the way in which honour 'lifts one so far above the nonentity of the mass that withdrawal of status is the difference between life and death. Similarly, with grace come salvation and life everlasting among the company of the elect, thereby marking one off from the reprobate and doomed. Likewise intelligence is the individual possession of the intelligent society.' At the same time,

All three modes sanctify the person. Each confirms the legitimacy of an individual's behaviour by referring it to an external authority. In a form appropriate to the mode (the king disburses honourable titles, God dispenses grace, the psychologist allocates IQ scores), selected individuals are invested with some of the superior's sacred authority. Although this authority is in fact arbitrary, in receiving its blessings we abnegate our right to it, thereby binding ourselves to accept practices which a different generation, in different historical circumstances, might regard as utterly wrong....We find too an anxiety about *self-authentification*, directed inwards at one's self-esteem and personal autonomy. A gentleman's ancestry is the permanent guarantee of his honour, yet depends also on an unattainable certainty as to who his real father was; if his pure bloodline falls in doubt, so does his autonomy and freedom from having to bow to the will of others, and perhaps he must join the servile out-group which by definition has no such autonomy. People otherwise certain of their own salvation worry about their status with God whenever they suffer a personal setback, however clearly the Bible says that grace is conferred once and for all. And one's intelligence, that which endows modern personhood with its sense of permanence, may be denied by the formal entrance requirements of educational and social institutions.[83]

Finally, note how intelligence takes on scientific credentials on this account. In the shift in ideological balance from honour and grace to intelligence, we find that 'as scientific knowledge began to be thought of as the genius's *possession*, the branch of it known as psychology gave rise to scientific knowledge *about* the genius (and, of course, its opposite). A status concept was thereby transformed into an objective intuited intellect (*nous*, for which "intelligence" was the preferred translation).'[84]

If the modern notion of intelligence is a marker of social standing, where does this leave the associated notion of rationality? Examination of the notion of intelligence brings out the vast gulf between reason and rationality. The fact that they are such different kinds of things means that we need to exercise great caution in what exactly we are committing ourselves to in extolling the virtues of reason, by contrast with the virtues of rationality, even though the same word may be being used for both. A particularly problematic case is that of the defence of 'Enlightenment rationality', where its defenders tend to take what is at issue as 'reason', in the sense that I have considered it, whereas its critics, such as the

[83] Ibid., 65–6.   [84] Ibid., 69.

Frankfurt School, take what is at issue as something purely instrumental, namely rationality proper.

Once we turn to explicit full-blooded defenders of rationality proper, the dire consequences of its dissociation from reason become evident. One of the most egregious examples of the phenomenon of 'rational decision making' is the recommendations of Hermann Kahn, senior game theorist at the RAND corporation, which provided research and analysis for the American armed forces in the Cold War. In his 1960 book *On Thermonuclear War*, Kahn used game theory to think through the advantages and disadvantages of nuclear war for the USA. He offered a game-theoretical analysis in which he argued for the rationality of a first-strike nuclear attack on the USSR. In spite of the millions who would die in such a war, the wholescale genetic mutation that would result, and the decimation of agriculture and industry, he argued that life would go on, just as it had in the wake of the Black Death, and his cost–benefit analysis concluded that the defeat of communism and the eventual worldwide establishment of an American-style society that it would bring about was on balance the most favourable outcome.[85] There are parallels here with Heidegger's conclusion that 'total war' was the only solution to the ills of society, and the parallels run deeper than might immediately be evident. It might seem, for example, that in Heidegger's case what is distinctive is the rejection of reason in favour of a metaphysical engagement with the world. But, although Kahn is certainly not advocating the latter, he *is* rejecting reason: in favour of rationality, for nowhere is the conflict between the algorithmic decision-making procedures of game-theoretical rationality and the judgements of reason more evident than in Kahn's calculations, where game theory is exclusively about 'winning'.

Although some game theorists admired Kahn's commitment to following through the consequences wherever they might lead,[86] many were horrified at the use to which he had put the procedures. But the point is that, in rejecting Kahn's analysis, it is not rationality to which one appeals, since that was the source of the problem in the first place. As one of the formative figures in game theory, Anatol Rapoport, admitted: 'For the most part, decisions depend on the ethical orientations of the decision-makers themselves. The rationales of choices may be obvious to those with similar ethical orientations but may appear to be only rationalizations to others. Therefore, in most contexts, decisions cannot be

[85] Hermann Kahn, *On Thermonuclear War* (Princeton, NJ, 1960). More generally on game theory and Cold War rationality, and their development in more recent thinking, see Sonia M. Amadae, *Rationalizing Capitalist Democracy: The Cold War Origins of Rational Choice Liberalism* (Chicago, 2003); and Sonia M. Amadae, *Prisoners of Reason: Game Theory and Neoliberal Political Economy* (Cambridge, 2016).

[86] These include Hugh Everett III, the inventor of the many worlds interpretation of quantum mechanics, who subsequently took up a career in game theory, employed by the US military. In his biography of Everett, Peter Byrne notes that he admired Kahn's 'cold-blooded, cost-benefit approach to "thinking the unthinkable"': *The Many Worlds of Hugh Everett III: Multiple Universes, Mutual Assured Destruction, and the Meltdown of the Nuclear Family* (Oxford, 2010), 64.

defended on purely rational grounds. A normative theory of decision which claims to be "realistic", i.e. purports to derive its prescripts exclusively from "objective reality", is likely to lead to delusion.'[87] In short, a theory of rational action cannot be used to get itself out of trouble. Once it reaches its limits, or descends into a moral quagmire, we are forced to fall back on reason, which offers a sense of proportion, and a sense of moral integrity, wholly lacking in game theory, with its ideology of 'winning', and the rationality—in social terms, who is and is not 'smart'—to which it is committed.

It is inevitable that the shift to a scientific mentality characteristic of modernity has had as its highest aspiration the scientific understanding of human behaviour. Of the many attempts to apply scientific criteria to the understanding of human behaviour—to model the population on the scientific image—by far the most egregious has been that of abandoning reason in favour of rationality. There is nothing inevitable about the move to rationality, which has been encouraged by the idea that there must be some unified scientific understanding of the world, and by the associated idea of science as the master discipline. But the move illuminates, better than any other phenomenon of which I am aware, the pitfalls of a blind commitment to a 'scientific understanding'.

---

[87] Anatol Rapoport, *Strategy and Conscience* (New York, 1964), 75. This is a major concession from the author of the 1957 'Scientific Approach to Ethics', q.v.

CONCLUSION

13

Science and the Shaping of Modernity

With this volume, we have reached the final point in our exploration of how all knowledge claims came to be assimilated to scientific ones—how science, through the provision of cognitive norms, came to serve as a model for all forms of purposive behaviour. The story has been a long and complex one, and a number of the developments that we have followed are intrinsically labyrinthine. Now is the appropriate point to pull back on Ariadne's thread, leaving the labyrinth with what we have learned, offering an overview in which some main lines of the development of a scientific culture in the West can be highlighted.

The emergence of a scientific culture has often been treated as a simple success story. It is not. As the cultural critic H. L. Mencken once pointed out, for every complex problem there is always a simple solution, and it is always wrong. Trying to establish a clearer grasp of the issues, we must see complex questions for what they are. It is only by negotiating our way through this complexity that we are able to face up to the sheer contingency of scientific development and consolidation in modern times. An important starting point is the recognition that, in crucial respects, we are concerned with *Western* science, something that developed in a context of concerns that were not mirrored elsewhere. In particular, the emergence of a scientific culture in the West was not a realization of something common to all scientific cultures, something at which they had all aimed, but rather, from the point of view of cultures outside the West, at least before the end of the nineteenth century, it was an anomaly. Rather than trying to understand why other scientific programmes hadn't accomplished what was achieved in the West in the Scientific Revolution, the question is rather why the development of science in the West took a path quite different from every other earlier and contemporary culture. The model in the other cases was a boom/bust one, in which interest in scientific questions was very much related to solving particular questions, and when these were dealt with, there was a decline of interest in science as such. By contrast, the development of science from the late sixteenth century onwards manifests a rate of growth that is pathological by the standards of earlier cultures, but this is legitimated by the cognitive standing that science takes on.

To begin to understand the origins of this development, we need to distinguish the emergence of theoretical and experimental scientific programmes from the cultural consolidation of these programmes. These are not the same thing. Many

Civilization and the Culture of Science: Science and the Shaping of Modernity, 1795–1935. Stephen Gaukroger, Oxford University Press (2020). © Stephen Gaukroger.
DOI: 10.1093/oso/9780198849070.001.0001

attempts to explore the success of science in the West have assumed that what is unique about the Scientific Revolution is that its practitioners hit upon the only really successful way of pursuing science, and that the scientific practice that was produced in the wake of the Scientific Revolution represents the only way in which scientific practice could have developed with any long-term viability. In this view, science in the early modern era was so spectacularly successful that it not only displaced competing accounts, but it was also able to export the method by which it achieved these fundamental results to all cognitive domains. In short, there is an implicit assumption that a good account of emergence is a good account of consolidation. But a very different kind of explanation is needed for consolidation, and the crucial point is that none of the earlier successful scientific cultures engaged in the legitimatory venture of consolidating scientific gains in such a way as to establish science as a model for all forms of understanding. Large-scale consolidation of this type was never part of the programmes of Alexandrian, Arab-Islamic, or Chinese science, for example. Quite the contrary, the solution of a defined range of specific problems was the rule, and success in this enterprise usually brought an end to significant attention to theoretical scientific problems. The idea of large-scale consolidation is not something inherent in the scientific enterprise as such, but it *is* inherent in Western scientific enterprise after the Scientific Revolution. Without this kind of consolidation, we would simply not have had the Scientific Revolution: we would have had a development on a par with what happened in the vibrant, but shorter-term, scientific cultures of early medieval Baghdad or Andalusia, or in Sung and Ming dynasty China. Successful consolidation, of a kind that aims to promote the cognitive claims of science and build a legitimate scientific culture around them, is the characteristic feature of Western science in the wake of the Scientific Revolution. But such consolidation isn't simply a question of success; it's a question of success in achieving an aim—an aim absent from earlier Western scientific cultures and from those outside the West.

The origins of Western scientific culture lay in fact not in scientific developments as such but in a particular set of political and religious problems, and it was thought of and defended in the context of a Christian understanding of the world until the middle of the eighteenth century. By the nineteenth century, science was becoming dissociated from Christianity, but taking its cue from Christianity it began to be presented as an autonomous enterprise representing truly universal values, by contrast with those of Christianity, now increasingly considered dogmatic and parochial. The technical details of particular scientific programmes had always been largely autonomous, especially in the case of mathematical disciplines such as mechanics, and indeed were shared with scientific developments outside the West, but once one moves outside this area, into the broader ambit of early modern natural philosophy, all kinds of considerations come into play, not least compatibility with different conclusions arrived at on competing philosophical, religious, or even empirical grounds. This is obscured by the fact that the success of technical areas such as mechanics, now incorporated into natural philosophy in the seventeenth century, encouraged the view that the latter had freed itself from contextual constraints. What actually happened was far more complex, and in understanding it, it is crucial to distinguish as far as possible between the consolidation of scientific enquiry and the consolidation of a scientific culture, especially since it is the latter—not the former—that marks out what happened in Europe as quite different from the development of science elsewhere. The questions become more complex as what it is that one seeks to understand becomes even broader, as questions of what scientific understanding consists in begin to be explored, as problems about its scope and limits are raised, and as its benefits over other ways of understanding the world are probed.

I have taken it as not requiring much argument to establish that one cannot extrapolate, in a sociological fashion, from the broad features of efforts to promote and legitimate science, which are manifestly highly contextualized, to the technical details of particular enquiries. Similarly with obverse attempts to extrapolate from the autonomous nature of areas such as mechanics to science in general, despite the continued prevalence of these in the literature, particularly the philosophical and popular science literature. Conceptual investigation will not throw much light on the complex political and cultural factors shaping the differences in the uptake of Newton's theory of gravitation in England and France in the early eighteenth century for example, or on the promotion of the role and standing of intelligence testing; any more than social and institutional investigation will illuminate the problems caused by identifying inertia with equilibrium in statically modelled accounts of dynamics in the seventeenth century, or why it was organic chemistry, not inorganic chemistry, that guided research on chemical valency in the nineteenth century.

The chief difficulty lies rather in the fact that there is no line stretching from the one to the other, marking increasing degrees of autonomy as one approaches apparently self-contained technical areas. At the same time, the problem is not so much the reductionist temptation to treat all issues in either conceptual or social terms, but rather that there is a danger that these kinds of approaches can be taken to exhaust the alternatives. It is crucial for the kind of project with which the volumes have been concerned that this is not assumed. New tools occasionally need to be forged. To take just one example, I have devoted considerable attention to developing an account of the shaping of the *persona* of the natural philosopher/ scientist. This has played a significant role in thinking through a number of questions. The changes to the standing, aspirations, perceived expertise, and sense of identity of the natural philosopher are not something that can be captured straightforwardly in social or institutional terms, and they bear directly on how the legitimacy of natural philosophy/science can be presented and defended. The skill is to devise an account of this realm, and in the process to identify the resources that one needs in order to explore it, without reducing these questions to psychology, politics, or sociology, while at the same time avoiding the reduction of the content of theories or experimental practices to effects of a particular type of *persona*. One develops one's explanatory resources as one proceeds: deciding in advance whether one is an 'internalist' or an 'externalist', for example, or whether one is going to pursue an epistemological or sociological approach because this is where one perceives one's (seemingly unalterable) expertise to lie, would get us nowhere in probing the kinds of issues with which these volumes have been concerned. In short, one does not start by reflecting on one's tools and asking what one can do with them. Rather, one must start by identifying a significant, pressing issue and try to devise or identify the resources to deal with it. If this means that it is no longer clear whether one is writing as an historian or as a philosopher for example, this is because it is the issues that are dictating the resources, not the resources dictating the issues.

*

Our starting point has been the second decade of the thirteenth century. Since I have argued that the major developments in disciplines making up natural philosophy did not occur until the middle to late decades of the sixteenth century at the earliest, one might ask what was so significant about what happened three and a half centuries earlier. The answer is that we can trace back to this earlier period a number of connected developments that shape the emergence of a scientific culture. These resulted from the abandonment of an integrated conception of the world that was formulated in the early Church Fathers, and which received its canonical statement in the writings of Augustine of Hippo at the end of the fourth century. On this integrated conception, philosophy was a tool of theology, and there was no division into autonomous secular and ecclesiastical realms. It was with the Investiture Controversy (1050–1122) that this conception first came apart in a politico-theological context. The Controversy hinged on the appointment of clergy by monarchs and nobles, which was condemned by Pope Gregory VII, whose first target was the removal of the monarch's power to appoint popes. Asserting papal supremacy over the entire Western church, he declared its independence from secular control. The resolution of the Controversy resulted in the church achieving a legal identity independent of emperors, kings, and feudal lords. But the relevant upshot of this for our purposes was a separation of the ecclesiastical and secular realms.

The bifurcation into *ecclesia* and *mundus* gradually covered every aspect of political and intellectual life, and its first manifestation was the legal division into autonomous ecclesiastical and secular legal systems, initiating a tradition of reconciling the two legal systems so that they did not conflict. These procedures provided a model for the various forms of bifurcation that followed, not least in the realm of understanding, and this brings us to a formative development in the establishment of a scientific culture: the revision of the Aristotelian conception of metaphysics, one that enabled it to play a mediating role between what became two independent sources of knowledge, natural philosophy on the one hand, and revelation and Christian teaching on the other. Aristotelianism made sense perception the sole source of knowledge of the natural world, and allowing an understanding of the natural world that deployed purely natural resources was a fundamental concession. It was necessitated by the realization that Aristotelianism provided far more powerful systematic philosophical resources than anything else available for resolving fundamental theological dilemmas such as transubstantiation and the trinity. As a consequence, it was now metaphysics, not theology, that tempered claims about the natural world.

Despite much resistance, Aristotelianism became the pre-eminent means for understanding natural philosophy and metaphysics by the end of the thirteenth century. In contrast to the Platonism that had dominated theological and philosophical thought up to that time, the core of the new understanding of the world and one's place in it lay in natural philosophy. Here, for the first time, there emerges a culture in which scientific values take centre stage. Early modern science was not continuous with medieval science, as some commentators have claimed. But neither was it formed via a dissociation of science and theology, as others have maintained. As part of an attempt to bolster the resources on which theology could draw, metaphysics becomes detheologized in the hands of Aquinas: in contrast to earlier (and some contemporary) conceptions, on which metaphysics is a science of 'being-qua-being', that is, something that covers both the philosophical and the theological realms, Aquinas construed metaphysics as a neutral discourse which, governed solely by reason, can decide conflicts between natural philosophy, reliant on sense perception, and theology, reliant on revelation. The latter was considered to have a degree of certainty lacking in natural philosophy, which gave it a more secure standing, but nevertheless it was metaphysics that was called upon to resolve contradictions between the two, on the immortality of the soul, for example, and on the eternity of matter.

In short, theological resources were no longer deployed in our understanding of the natural world, which had now become exclusively tied to natural philosophy. I stress that this was a theologically motivated move: the adoption of Aristotelian metaphysics provided resources for dealing with pressing theological issues in a far more satisfactory way than the alternatives. But Aristotelian metaphysics came as part of a package in which natural philosophy provided exclusive access to the natural world. In adopting it, Christianity took on a unique feature: a fundamental engagement with science, which set it apart from all other religions, transforming it radically and giving it a distinctive identity. Islam was the only other religion which had had a full-scale engagement with science, and Aristotelianism had also provided the route for this engagement: indeed the scholars of thirteenth-century Paris were deeply indebted to the Arabic Aristotle tradition. But the commitment to theocracy in medieval Islamic culture (something that the division of responsibilities that emerged in the wake of the Investiture Controversy had effectively made impossible in Christian Europe) resulted in the Arab engagement coming to an end just as the Christian one was beginning. The upshot was that, with Christianity from the thirteenth century onwards, science became an integral part of the understanding of the world. This had very significant consequences for Christianity, and at the same time it resulted in the gradual transformation of the Christian West into a scientific culture. These developments have occasionally been resisted: nineteenth-century Catholicism retreated into a staunchly anti-science form of dogmatism, struggling vainly to institute a form of theocracy, and twentieth-century evangelical Protestantism attempted to subordinate science to a biblical literalism. But the benefits of a scientific culture—particularly when associated with technological and other developments that have transformed daily life—together with the intellectual standing of science, have been so overwhelming that there was never any chance of these developments being considered as anything other than marginal and misguided.

Nevertheless, the developments were problematic. The move, with the development of Scholasticism, to treat natural philosophy as autonomous relied on the assumption that metaphysics would be able to play a decisive role in adjudicating between natural philosophy and theology. This is where the problems began, for by the beginning of the sixteenth century it had become clear that a Thomist-style metaphysics was unable to provide the necessary guidance on the most pressing philosophical-theological issue of the day: the doctrine of the personal immortality of the soul. These considerations arose in the context of a conflict between Aristotelian natural philosophy and theology. This is the arena into which the new natural philosophies of mechanics and micro-corpuscularianism were born, and by the early decades of the seventeenth century Aristotelian natural philosophy had been replaced, largely by micro-corpuscularianism, and two new forms of scientific enquiry appeared: a rejuvenated form of mechanics, initially in astronomy, optics, and statics; and later in the century we witness the emergence of experimental natural philosophy.

There is a widely held view that the success of Western science—initially characterized by mechanics and micro-corpuscularianism—lay, at least in part, in its ability to dissociate itself from religion. It is certainly true that the relations between religion and natural philosophy shifted quite radically in the sixteenth and seventeenth centuries, but these shifts were by no means straightforward, and the outcome was by no means a turn away from religion, but rather in many respects a turn towards it. The sixteenth and seventeenth centuries were the most intensely religious centuries Europe has known. A range of exacting moral standards, accompanied by demands for self-vigilance, which had been the preserve of monastic culture throughout the Middle Ages, were transferred wholesale to the general populace in the course of the Reformation and Counter-Reformation. Religious sensibilities in the secular population were deep and intense in the early modern era, and these religious sensibilities motivated much natural-philosophical inquiry well into the nineteenth century.

A good part of the distinctive success of the consolidation of the scientific enterprise derived not from a separation of religion and natural philosophy, but rather from the fact that natural philosophy could be accommodated to projects in natural theology: what made natural philosophy attractive to so many in the seventeenth and eighteenth centuries were the prospects for the renewal of natural theology that it offered. Far from science breaking free of religion in the early modern era, its consolidation depended crucially on religion being in the driving seat: Christianity took over natural philosophy in the seventeenth century, setting its agenda and projecting it forward in a way quite different from that of any other scientific culture, and in the end establishing science as in part constructed in the image of religion. The union of natural philosophy and Christian theology meant that the former took on the aspirations of the latter, as it became part of a project not just of understanding the world but also of understanding our place in it. This legitimation of science, in which it was effectively installed at the heart of the culture, was unique to the early modern West. It provided science with a role that its technical successes, no matter how significant, could by themselves never secure for it.

But just when science was assuming this new role, the features that gave it the appearance of being such a comprehensive success, and which enabled it to be a partner with theology, were coming under threat. Mechanism—the reduction of all physical phenomena to the mechanically characterized behaviour of microscopic corpuscles—was a model of systematic natural philosophy, offering an exhaustive account of the natural world that made it the only serious competitor to the Aristotelian natural philosophy. Its most developed general form was that offered by Descartes in the 1640s, but in many areas mechanism was little more than a promissory note. And the promises turned out to be empty in crucial areas such as chemistry, electricity, and physiology, which were at the cutting edge of research in the eighteenth century. Those working in these areas gradually turned away from mechanism and pursued experimental natural philosophy, which had no systematic aspirations; they rejected 'speculative natural philosophy', which had very explicit systematic aspirations.

Speculative natural philosophy was in many respects a continuation of an ancient conception of the subject, seeking explanations for macroscopic events in microscopic ones. Proponents of experimental natural philosophy—the models were Boyle's account of air pressure and Newton's account of the production of spectral colours—adopted forms of explanation that rejected the idea that one had to look beyond the phenomena to what caused them, where the cause was always to be located at a more fundamental level. By the middle decades of the eighteenth century this experimental approach had notable successes in chemistry (Geoffroy) and electricity (Franklin and then in the next century with Faraday), and there emerged a strong view that trying to incorporate experimentally guided explanations into a systematic account was fruitless. Indeed, there was a widespread rejection of appeals to systems as intellectually dishonest and an obstacle to explanation, right across the spectrum, from conservative Jesuits to radical Parisian *philosophes*.

The seventeenth-century union of natural philosophy and Christian theology meant that natural philosophy became part of a broad ambitious programme in which the aim was an understanding of the natural world and our place in it. But by the middle of the eighteenth century, natural philosophy had lost much of the systematic coherence that had enabled it to serve as a model for knowledge and understanding generally. Even more significantly, the kind of comprehensive, systematic aspirations fostered in theology and metaphysics came to be rejected, most influentially in the work of Hume. There was plenty of scientific innovation in the second half of the eighteenth century, but the kind of legitimation of science that mechanism had provided in the seventeenth century, and which in turn had enabled it to act as a model for understanding generally, was now lost. As natural philosophy was no longer considered a single system, but was rather coming to be thought of as a collection of theories and experimental results that lacked any intrinsic unity, it could no longer play the role of cognitive model in quite the same way.

With the decline of the kind of guaranteed legitimacy that the union of natural theology and natural philosophy had provided to mechanism, and with nothing to replace it, we have to ask why a scientific culture—above all a belief in science as an intrinsically worthwhile enterprise, that is, something that has a value independently of the success of particular scientific discoveries or advances—did not collapse around the middle of the eighteenth century. Certainly the reasons why such a conception had been successful in the seventeenth century no longer had the force they had possessed then. Natural philosophy had taken on a greater standing, but could it continue to meet the demands that standing brought with it?

To the extent to which it could meet these demands in the second half of the eighteenth century, it was not because of a return to comprehensive systems. Its legitimacy was no longer tested primarily against the physical sciences, but in a new domain that it effectively created for itself, the human sciences. From the 1750s on, there were three sets of questions that dominated debate about the relations among science, religion, and metaphysics. The first was the assessment of the merits and drawbacks of systematic and non-systematic understanding. The second was the contrast between propositional understanding—the kind of understanding traditionally associated with scientific inquiry, where all under-standing is construed in terms of knowing *that* something is the case—and non-propositional understanding, that is, approaching the world in terms of aspir-ations, fears, anxieties, plans, raw beliefs, desires, etc. The third concerned the respective roles of reason and sensibility in understanding the world and our place in it. System, propositional understanding, and reason were, as one might expect, defended together, as were anti-system arguments, non-propositional understand-ing, and sensibility. Sensibility was particularly important here, for a rethinking of the relation between reason and sensibility led to a discrediting of the model of knowledge offered by seventeenth-century mechanism. By the middle of the century in France, for example, it was being argued, most notably by Condillac, Bonnet, Diderot, and Buffon, that it was sensibility that underlay our understanding of the world, not reason, because only sensibility could provide us with a realization that our sensory states were caused by something external to us.

The fundamental role that sensibility took on had the consequence that understanding our place in the world now came to be seen on a par with—and in some cases conceptually prior to—understanding the world. It is in this context that the emergence of the human sciences, from the second half of the eighteenth century onward, should be considered. The human sciences were the outcome of a move to naturalization: the opening up to empirical inquiry of questions that had previously been treated as purely conceptual matters. In particular, questions that had been thought unique to a religious and philosophical domain were transformed into questions subject to empirical probing. The naturalization of the human brought into play a new stage in the consolidation of a scientific culture, and with the inclusion of the human in the scientific ethos the totalizing aspirations of a scientific understanding were realized, the whole exercise now finally taking on a distinctively modern aspect. But the routes to the naturalization of the human were fraught and were routinely at variance with one another, each of them offering its own distinctive key to naturalization, whether through one form or another of physical anthropology, through a form of social arithmetic, or through one of the programmes for naturalizing the human that used sensibility as the route to naturalization.

The latter offered particularly profound challenges to the understanding of human behaviour. Anthropological medicine, primarily a French phenomenon, built on the advances in the physiology of nervous sensitivity in order to physiologize, and then medicalize, the human being as a whole, holding that earlier theological and metaphysical accounts of human nature were misguided. In particular, the medical thinkers criticized the treatment of the mind by philosophers, pointing out that the latter only dealt with healthy minds, whereas attention to these alone was insufficient: unhealthy minds needed to be included. And once one included these, and considered them in a medical context, then one was alerted to a range of psychosomatic and somatopsychic phenomena which philosophers had missed, but which were crucial to understanding the relation between mind and body. At the same time, philosophical anthropology, primarily a German phenomenon, was also using sensibility as the route to naturalization, albeit in a very different fashion. Both Kant and Herder had a naturalizing approach to anthropology, but whereas for Kant this effectively made it a purely descriptive exercise, Herder set out to use naturalization to shape anthropology into a genuinely powerful scientific account of the human. He begins by translating reason, something abstract, into thought, something concrete and embodied. Next, he approaches thought via what the Wolffian tradition had characterized as 'empirical psychology', a naturalized set of techniques for studying one's own psychological states and those of others. However, whereas Wolf saw empirical psychology merely as providing the raw materials for a much more traditional 'rational psychology', Herder completely reworks it. Rejecting Wolff's introspective and behavioural tools, he instead uses language as the means of analysis, treating it as an empirically analysable manifestation of thought. Importantly, language so construed has an essential historical dimension, and the comparative history of languages becomes a history of thought, which in turn is a history of reason. In this way, naturalized sensibility provides, via language, a route to mental states and a model for approaching reason.

For eighteenth-century thinkers, reason was considered unnaturalizable in its own right, which was why the route to the naturalization of cognitive and affective states was through sensibility. But at the end of the eighteenth century, utilitarianism rejected all talk of sensibility and offered what looked like a direct form of naturalization of reason. Bentham for example naturalized moral reasoning, arguing that the morality of behaviour was to be assessed in terms of observable consequences. Mill subsequently extended the naturalization programme to all forms of reasoning, construing logical inference in terms of a combination of empirical psychology and empirical generalization. I say that this *looked like* a naturalization of reason because the naturalization process actually translated reason, taken as the highest faculty and the source of considered judgement, into something that was very different: a set of algorithmic decision-making procedures. The problem lies not with these processes as such, since they are valuable in some contexts (such as those concerned with uncertainty and risk), but with attempts to use them to replace or dislodge reason as the source of judgement about human behaviour. The quantitative character of rationality stimulated a number of projects to measure the abilities of population and their suitability for the tasks required of a technologically advanced modern society. Just as others devised norms for bodily shape and size, or workplace norms, Galton for example, preoccupied with identifying new norms of behaviour, subjected reasoning to empirical testing and ranked reasoning abilities on a scale centred on a statistical norm of intelligence.

*

These last developments were premised on the idea that there must be some unified scientific account of the world, and that the understanding it embodies is theoretical (as opposed to practical or instrumental for example) in nature. Challenging this assumption has been one of the main tasks, and one of the *leitmotifs*, of the volumes, coming to a head in this final volume.

A core issue is the evolution of science. By this I don't mean the evolution of scientific theories, a topic that has been the almost exclusive concern of historians and philosophers of science, but the evolution of 'science' itself, that is, what comprises science. In recent years, historians of science have spoken of 'technoscience', which gives a sense of one kind of development at issue, but in itself leaves open the crucial question of what the relationship between science and technology is. It also bypasses the important question whether concurrent developments, such as the emergence of new genres of popular science, are in any way a reflection of a deeper transformation.

Separating the emergence of science and the emergence of a scientific culture is a condition of understanding the development and consolidation of scientific programmes, and when dealing with the early modern period, this separation is relatively straightforward. But it becomes far more difficult with the emergence of technologically orientated science and with the emergence of a popular science that critically engages with the issues of the day. The importance of the distinction for my account in the early volumes lay in countering attempts to extrapolate from the success of scientific theory to the completely different issue of the success of a scientific culture more generally. But the difficulties that arise when considering nineteenth-century developments reveal that the problems lie far deeper than such attempts at extrapolation, and at the same time they prompt us to question the idea that science, or at least its core identity, lies in scientific theorizing. On the one hand, innovations and developments that have been assumed to be the result of scientific theorizing turn out to have been quite independent of it: to have been due to technological developments, or matters of free invention, which are independent of theoretical considerations. At the same time, when theoretical considerations have been brought to bear on these, they have sometimes offered illumination, but they have occasionally acted as obstacles to understanding. It is in this context that I have questioned the idea of technology and engineering as 'applied science'. On the other hand, I have also questioned the idea that 'popular science', as it emerged in the middle decades of the nineteenth century, was simply a promotional exercise, a means of disseminating simplified accounts of science to a passive readership. In the course of the early twentieth century, popular science reflected a growing ambivalence about the standing of science and technology as the guarantors of civilization. It captured both aspirations and misgivings that neither the scientific community nor respective governments could articulate, opening up deep questions about the direction in which science might proceed, and above all on its social and political standing.

What emerges from this is a conception of science as a mixture of theory, experiment, and theory-free invention, articulated through a number of channels, including popular science, and resulting in both discursive products (theoretical understanding) and non-discursive products (machines and complex systems). These factors interact in ways that are highly dependent on context, and the outcome of the interaction cannot be discerned in advance. 'Science' emerges from this as an unstable mixture, and this reinforces its modular character, and the need for pluralism in its explanations. As is indicated in the discussion in earlier volumes, of the non-reductive and phenomenal explanatory programme of the experimental natural philosophy tradition, this is something that dates back to the emergence of modern science in the seventeenth century. In those early stages it looks as if, despite the best efforts of those seeking a unity of science, we have two parallel programmes which feed off one another to some extent. But by the late nineteenth century, the modular pluralistic factors have entered the core of scientific programmes, leaving behind the idea of the unity of science, inducing a unification panic among many scientists and philosophers, and giving science as a whole an unruly character if viewed from the perspective of 'pure theory'. At the same time, the consequences of the integration of technology and engineering into the core of scientific practice are reinforced by popular science. These go to the very heart of the question of just what science does, what its role in the modern world is.

The grand natural-philosophical schemes of the seventeenth century were seen by their proponents as partners in a project of uncovering the nature of the world. Of necessity, they shared a theoretical character with the theological conceptions they complemented or replaced, and in this respect they drew on the philosophical tradition, one that had been shaped by theologians, not the scientific traditions—in mechanics, optics, or astronomy for example. Understanding the world was theoretical understanding: at its most refined, something that takes the form of basic principles, virtually indistinguishable from philosophical abstractions. The assimilation of science to a philosophical model meant that, in turn, science itself provided philosophy with a model of truth, a conception that proved resilient. It received a renewed formulation in the writings of the Logical Positivists in the early twentieth century, where its connection with the idea of the unity of science was explicit. But the conception of science as being essentially scientific theory is exactly what had to be jettisoned if the technological and engineering developments were to be understood. It had in fact never been satisfactory, as the long story of experimental natural philosophy testifies, but it had seemed possible to proceed as if it were only a matter of time before the latter could be assimilated to the theoretical model. In this sense, the late nineteenth- and early twentieth-century developments did not so much create a new idea of science as open up fissures that had always been latent in scientific practice.

It is not possible to make sense of these developments without abandoning the idea of the unity of science, an idea that has been premised on two things: a secularized religious model of nature in which nature is the result of a single coherent act of creation; and that of science as an essentially abstract enterprise, a body of theory open to application and popularization, neither of which have a bearing on its essential characteristics. In the circumstances, it is unsurprising that science took on an aura of religious authority from the late eighteenth century onwards. But the integration of science, technology, invention, and engineering, and the need to articulate scientific values in a wide variety of ways, including in terms of popular science if they are to be effective, have resulted in something that the traditional understanding of science wholly fails to capture. In particular, the idea that science could take on the mantle of a unity of understanding, that it could replace Christianity in a role for which the latter had been found wholly inadequate, without asking why Christianity encountered the problems that it did in the first place, shows an egregious lack of will to confront the issues. Christianity, with its notion of a created purposive natural realm, was actually better equipped to offer a unified understanding. It was the nature of the exercise that was wrong, not whether science had the credentials that Christianity lacked.

Since the middle of the nineteenth century, science has shed the last remnants of its purity and come of age, its inherent unruliness and pluralism brought to the surface for all to see. As a result, its standing as the bearer of the values of civilization has now begun to look mysterious at best. Accordingly, the question of what exactly we want out of science becomes an urgent one. There is not some perennially valid response available here. An answer that addresses itself to current issues cannot be the same as that which was appropriate in the Scientific Revolution, or in the Enlightenment, or indeed in any period prior to the middle of the nineteenth century. Above all, it cannot lie in some quasi-religious, quasi-metaphysical quest for the ultimate truth about the world. To the extent to which a general model can be illuminating, I admit to being tempted by medicine, by contrast with physics, as a way of thinking through how the sciences operate. Human anatomy, physiology, pathology, etc. are not theories which happen to find application in the prevention and treatment of illness. This is to get things the wrong way around. These theoretical and experimental disciplines have been brought into existence and shaped by a medical agenda, and their standing is subservient to the social and other goods that medicine sets out to achieve. The physical sciences are not quite like this, but it might perhaps be possible, with care (with a lot more care than has been exercised to date), to model the physical sciences on this basis, in terms of providing the means of transforming living and working conditions. But this is the task of another project, which would be different from that which I have been pursuing, not least in that comparative questions would be far more central. But however much value there may be in the exploration of such a medical model, it would not reflect how a scientific culture has in fact been promoted and developed in the West, and it is this that I have been concerned to elucidate. The historical development of a scientific culture has been problematic in many respects, but aware as we need to be of the limits of the values that have been attributed to science, and to a science-based civilization, these values cannot simply be dismissed. They form the basis of our own culture, and we cannot simply step outside this culture.

Bibliography of Works Cited

'Acheta Domestica' [Louise M. Budgen]. *Episodes of Insect Life* (3 vols, London, 1849–51).

Ackermann, Jacob Fidelis. *Versuch einer physikalischen Darstellung der Lebenskräfte organischer Körper* (2 vols, Frankfurt-on-Main, 1797–1800).

Ackernecht, Erwin. *Medicine at the Paris Hospital, 1794–1848* (Baltimore, 1967).

Adas, Michael. *Machines as the Measure of Man: Science, Technology, and Ideologies of Western Dominance* (Ithaca, NY, 1989).

Adelmann, Dieter. *Einheit des Bewusstseins als Grundproblem der Philosophie Hermann Cohens* (Heidelberg, 1968).

Agar, Jon. *The Government Machine: A Revolutionary History of the Computer* (Cambridge, MA, 2003).

Agar, Jon. *Science in the Twentieth Century and Beyond* (Cambridge, 2012).

Agassiz, Louis. *Recherches sur les poissons fossiles* (5 vols, Neuchâtel, 1833–45).

Aitken, Hugh G. J. *Syntony and Spark: The Origins of Radio* (New York, 1976).

Alborn, Timothy L. 'Negotiating Notation: Chemical Symbols and the British Society, 1831–1835', *Annals of Science* 46 (1989), 437–60.

Alder, Ken. 'French Engineers Become Professionals; or, How Meritocracy Made Knowledge Objective', in William Clark, Jan Golinski, and Simon Schaffer, eds, *The Sciences in Enlightened Europe* (Chicago, 1999), 94–125.

Alder, Ken. *The Measure of All Things: The Seven-Year Odyssey that Transformed the World* (London, 2002).

Allen, David Elliston. *The Victorian Fern Craze: A History of Pteridomania* (London, 1969).

Allen, David Elliston. *The Naturalist in Britain: A Social History* (London, 1976).

Allen, Garland E. 'The Misuse of Biological Hierarchies: The American Eugenics Movement, 1900–1940', *History and Philosophy of the Life Sciences* 5 (1983), 105–2.

Allen, Grant. *Physiological Aesthetics* (London, 1877).

Allen, Grant. *The Colour-Sense: Its Origin and Development. An Essay in Comparative Psychology* (London, 1879).

Allen, Grant. *Charles Darwin* (London, 1885).

Allen, Grant. 'Evolution', *Cornhill Magazine* 57 (January 1888), 34–47.

Alter, Peter. *The Reluctant Patron: Science and the State in Britain, 1850–1920* (Oxford, 1987).

Amadae, Sonia M. *Rationalizing Capitalist Democracy: The Cold War Origins of Rational Choice Liberalism* (Chicago, 2003).

Amadae, Sonia M. *Prisoners of Reason: Game Theory and Neoliberal Political Economy* (Cambridge, 2016).

Anderson, Elizabeth. 'The Epistemology of Democracy', *Episteme* 3 (2006), 8–22.

Anderson, John D. Jr. 'The Evolution of Aerodynamics in the Twentieth Century: Engineering or Science?', in Peter Galison and Alex Roland, eds, *Atmospheric Flight in the Twentieth Century* (Dordrecht, 2000), 207–22.

Anderton, Keith. 'The Limits of Science: A Social, Political, and Moral Agenda for Epistemology in Nineteenth Century Germany', unpublished PhD dissertation, Harvard University, 1993.

Anon. 'Exhibition of Works of Popular Science', *Mechanics' Magazine* 17 (1832), 378.

Anon. 'The British Association for the Advancement of Science', *Oxford University Magazine* 1 (1834), 401–12.

Anon. 'Sensational Science', *Saturday Review* 40 (11 September 1875), 321–2.

Anstey, Peter. 'Experimental versus Speculative Natural Philosophy', in Peter Anstey and John Schuster, eds, *The Science of Nature in the Seventeenth Century: Patterns of Change in Early Modern Philosophy* (Dordrecht, 2005), 215–42.

App, Urs. 'William Jones's Ancient Theology', *Sino-Platonic Papers* 191 (2009), 1–125.

App, Urs. *The Birth of Orientalism* (Philadelphia, 2010).

Appel, Toby. *The Cuvier-Geoffroy Debate: French Biology in the Decades before Darwin* (Oxford, 1987).

Archibald, Tom. 'Images of Mathematics in the German Mathematical Community', in U. Bottazzini and A. D. Dalmedico, eds, *Changing Images in Mathematics from the French Revolution to the New Millennium* (London, 2001), 49–67.

Ariès, Phillipe. *Religion populaire et réforme liturgique* (Paris, 1975).

Arnold, Matthew. *Schools and Universities on the Continent* (London, 1868).

Arrow, Kenneth J. 'Rationality of Self and Others in an Economic System', *Journal of Business* 59.4 (1986), S385–99.

Ash, John. *A New and Complete Dictionary of the English Language* (2 vols, London, 1775).

Asner, Glen R. 'The Linear Model, The U.S. Department of Defense, and the Golden Age of Industrial Research', in Karl Grandin, Nina Wormbs, and Sven Widmalm, eds, *The Science–Industry Nexus: History, Policy, Implications* (Sagamore Beach, MA, 2004), 3–30.

Avogadro, Amedeo. 'Essai d'une manière de déterminer les masses relatives des molécules élémentaires des corps, et les proportions selon lesquelles elles entrent dans ces combinaisons', *Journal de Physique* 73 (1810), 58–76.

Babbage, Charles. *Reflections on the Decline of Science in England and on Some of its Causes* (London, 1830).

Babbage, Charles. *On the Economy of Machinery and Manufactures* (London, 1832).

Baer, Karl Ernst von. *Über Entwickelungsgeschichte der Thiere: Beobachtung und Reflexion* (2 vols, Königsberg, 1828–37).

Bain, Alexander. *The Senses and the Intellect* (London, 1855).

Bain, Alexander. *John Stuart Mill: A Criticism with Personal Recollections* (London, 1882).

Baird, Davis. *Thing Knowledge: A Philosophy of Scientific Instruments* (Berkeley, 2004).

Bakhtin, Mikhail. *Rabelais and his World* (Cambridge, MA, 1968).

Baldwin, Melinda. *Making Nature: The History of a Scientific Journal* (Chicago, 2015).

Balfour, Arthur J. 'Address: Reflections Suggested by the New Theory of Matter', *Report of the Seventy-Fourth Meeting of the British Association for the Advancement of Science* (London, 1905), 3–14.

Balfour, Francis M. *A Treatise on Comparative Embryology* (2 vols, London, 1880).

Banks, Erik C. *Ernst Mach's World Elements: A Study in Natural Philosophy* (Berlin, 2003).

Barbauld, Anna Letitia. *Lessons for Children* (4 Parts, London, 1778–9).

Barnwell, Frank. *Aeroplane Design* (London, 1917).

Barrow, John D. and Frank J. Tipler. *The Anthropic Cosmological Principle* (Oxford, 1986).

Bastin, John. 'The First Prospectus of the Zoological Society of London: New Light on the Society's Origins', *Journal of the Society for the Bibliography of Natural History* 5 (1970), 369–88.

Bastin, John. 'A Further Note on the Origins of the Zoological Society of London', *Journal of the Society for the Bibliography of Natural History* 6 (1973), 236–41.

Bayertz, Kurt. 'Darwinismus und Freiheit der Wissenschaft. Politische Aspekete der Darwinismus-Rezeption in Deutschland 1863–1878', *Scientia* 118 (1983), 267–81.

Bayertz, Kurt. 'Dépasser la philosophie par la science: Le matérialisme naturaliste en Allemagne au XIXe siècle', in Charlotte Morel, ed., *L'Allemagne et la querelle du matérialisme (1848–1866)* (Paris, 2017), 67–82.

Bayly, Christopher A. *The Birth of the Modern World, 1780–1914: Global Connections and Comparisons* (Oxford, 2004).

Beiser, Frederick C. *The German Historicist Tradition* (Oxford, 2010).

Beiser, Frederick C. *Late German Idealism: Trendelenburg and Lotze* (Oxford, 2013).

Beiser, Frederick C. *The Genesis of Neo-Kantianism, 1796–1880* (Oxford, 2014).

Beiser, Frederick C. *Weltschmerz: Pessimism in German Philosophy 1860–1900* (Oxford, 2016).

Bellamy, Robert. *Looking Backwards, 2000–1887* (Boston, MA, 1888).

Belloc, Hilaire. 'Science as the Enemy of Truth', in Hilaire Belloc, *Essays of a Catholic Layman in England* (London, 1933), 195–236.

Berdoe, Marmaduke. *An Enquiry into the Influence of the Electric Fluid in the Structure and Formation of Animated Beings* (Bath, 1771).

Beretta, Marco. 'Between the Workshop and the Laboratory: Lavoisier's Network of Instrument Makers', *Osiris* 29 (2014), 197–214.

Bernal, J. D. *The Social Function of Science* (London, 1939).

Bernard, Claude. *Introduction à l'étude de la médecine expérimentale* (Paris, 1865).

Bernard, Claude. *Leçons sur les phénomènes de la vie communs aux animaux et aux vegetaux* (2 vols, Paris, 1878–9).

Bernard, Claude. *Leçons de physiologie opératoire* (Paris, 1879).

Bernard, Claude. *An Introduction to the Study of Experimental Medicine* (New York, 1957).

Bernard, Claude. *Cahiers de notes, 1850–1860*, ed. M. D. Grmek (Paris, 1965).

Bernoulli, Daniel. *Hydrodinamica sive de viribus et motibus fluidorum commentarii* (Strassburg, 1738).

Berthollet, Claude Louis. *Essai de statique chimique* (2 vols, Paris, 1803).

Berzelius, J. Jacob. 'An Address to those Chemists Who Wish to Examine the Laws of Chemical Proportions, and the Theory of Chemistry in General', *Annals of Philosophy* 5 (1815), 122–31.

Bichat, Xavier. *Anatomie Générale, appliquée à la Physiologie et à la Médicine, Première Partie. Tome Premier* (Paris, 1801).

Bigg, Charlotte. 'Evident Atoms: Visuality in Jean Perrin's Brownian Motion Research', *Studies in History and Philosophy of Science* 39 (2008), 312–22.

Bigg, Charlotte. 'A Visual History of Jean Perrin's Brownian Motion Curves', in Lorraine Daston and Elizabeth Lunbeck, eds, *Histories of Scientific Observation* (Chicago, 2011), 156–79.

Binet, Alfred and Theodore Simon. *Les Enfants anormaux: Guide pour l'admission des Enfants anormaux dans les classes de Perfectionnement* (Paris, 1907).

Binoche, Bertrand. 'Civilisation: Le mot, le schème et le maître-mot', in Bertrand Binoche, ed., *Les Équivoques de la civilisation* (Seyssel, 2005), 9–30.

Biot, Jean Baptiste. *Traité de physique expérimentale et mathématique* (4 vols, Paris, 1816).

Biot, Jean-Baptiste and François Arago. 'Mémoires sur les affinités des corps pour la lumière, et particulièrement sur les forces réfringentes des différens gaz', *Mémoires de la Classe des Science Mathématiques et Physiques de l'Institut National de la France* 7 (1806), 301–87.

Black, Edwin. *War Against the Weak: Eugenics and America's Campaign to Create a Master Race* (Washington, DC, 2012).

Bloor, David. *The Enigma of the Aerofoil: Rival Theories on Aerodynamics, 1909–1930* (Chicago, 2011).

Blue, Gregory. 'Scientific Humanism and the Founding of UNESCO', *Comparative Criticism* 23 (2001), 173–200.

Blumenberg, Hans. *The Legitimacy of the Modern Age* (Cambridge, MA, 1983).

Boantza, Victor. *Matter and Method in the Long Chemical Revolution: Laws of Another Order* (Farnham, 2013).

Bodin, Jean. *Methodus ad facilem historiarum cognitionem* (Paris, 1566).

Bogdanov, Aleksandr. *Krasnaya Zvezda* (St. Petersburg, 1908).

Bölsche, Wilhelm. *Das Liebesleben in der Natur: Eine Entwickelungsgeschichte der Liebe* (3 vols, Leipzig, 1898–1903).

Boltzmann, Ludwig. 'Über der mechanische Bedeutung des Zweiten Hauptsatzes der Wärmetheorie', *Wiener Berichte* 53 (1866), 195–220.

Boltzmann, Ludwig. 'Studien über das Gleichgewicht der lebendigen Kraft zwischen bewegten materiellen Punkten', *Wiener Bericht* 58 (1868), 517–60.

Boltzmann, Ludwig. 'Weitere Studien über das Wärmegleichgewicht unter Gasmolekülen', *Wiener Bericht* 66 (1872), 275–370.

Boltzmann, Ludwig. 'Über die beziehung dem zweiten Haubtsatze der mechanischen Wärmetheorie und der Wahrscheinlichkeitsrechnung respektive den Sätzen über das Wärmegleichgewicht', *Wiener Berichte* 76 (1877), 373–435.

Boltzmann, Ludwig. 'Ein Wort der Mathematik an die Energetik', *Annalen der Physik* 57 (1896), 39–71.

Boltzmann, Ludwig. 'Entgegnung an die wärmetheoretischen Betrachtungen des Hrn. E. Zermelo', *Annalen der Physik* 57 (1896), 772–84.

Boltzmann, Ludwig. *Populäre Schriften* (Leipzig, 1905).

Bommel, Bas van. *Classical Humanism and the Challenge of Modernity: Debates on Classical Education in Nineteenth-Century Germany* (Berlin, 2015).

Bossuet, Jacques-Bénigne. *Discours sur l'histoire universelle a monseigneur le dauphin pour expliquer la fuite de la religion et les changemens des empires* (Paris, 1681).

Bowler, Peter J. *The Eclipse of Darwinism: Anti-Darwinian Evolution Theories in the Decades Around 1900* (Baltimore, 1983).

Bowler, Peter J. *Reconciling Science and Religion: The Debate in Early Twentieth-Century Britain* (Chicago, 2001).

Bowler, Peter J. *Science for All: The Popularization of Science in Early Twentieth-Century Britain* (Chicago, 2009).

Bown, Nicola. *Fairies in Nineteenth-Century Art and Literature* (Cambridge, 2001).

Brandis, Joachim Dietrich. *Versuch über die Lebenscraft* (Hannover, 1795).

Braveman, Harry. *Labour and Monopoly Capitalism: The Degradation of Work in the Twentiethy Century* (New York, 1974).

Braveman, Harry. *Frederick W. Taylor and the Rise of Scientific Management* (Madison, WI, 1980).

Brazill, William J. *The Young Hegelians* (New Haven, CT, 1970).

Breidbach, Olaf. 'The Culture of Science and Experiments in Jena Around 1800', in R. M. Brain, R. S. Cohen, and O. Knudsen, eds, *Hans Christian Ørsted and the Romantic Legacy in Science* (Dordrecht, 2007), 177–216.

Bridgewater Treatises, on the Power, Wisdom, and Goodness of God as Manifested in the Creation, The (8 vols, London, 1834–7).

Briggs, Asa. *Victorian People* (London, 1954).

Brisseau-Mirbel, Charles-François. *Exposition et défense de ma théorie de l'organisation végétale* (The Hague, 1808).

British Association for the Advancement of Science. *Report of the Third Meeting of the British Association for the Advancement of Science, held in Cambridge in 1833* (London, 1834).

Brock, William H., and Knight, David M. 'The Atomic Debates: "Memorable Evenings in the Life of the Chemical Society"', *Isis* 56 (1965), 5–25.

Brockway, Lucile H. *Science and Colonial Expansion: The Role of the British Royal Botanic Gardens* (New Haven, CT, 2002).

Brodie, Benjamin Collins. 'The Croonian Lecture, on some Physiological Researches, respecting the Influence of the Brain on the Action of the Heart, and on the Generation of Animal Heat', *Philosophical Transactions of the Royal Society of London* 101.1 (1811), 36–48.

Brodie, Benjamin Collins. 'Further Experiments and Observations on the Influence of the Brain on the Generation of Animal Heat', *Philosophical Transactions of the Royal Society of London* 102.2 (1812), 378–93.

Broman, Thomas H. *The Transformation of German Academic Medicine, 1750–1820* (Cambridge, 2002).

Brongniart, Adolphe. *Mémoire sur la génération et le développement de embryon dans les végétaux phanérogames* (Paris, 1827).

Brongniart, Adolphe. 'Nouvelles recherches sur le pollen et les granules spermatiques des végétaux', *Annales des Sciences Naturelles* 15 (1828), 381–90.

Brooke, John Hedley. 'Wöhler's Urea and its Vital Force?—A Verdict from the Chemists', *Ambix* 15 (1968), 84–114.

Brooke, John Hedley and Geoffrey Cantor. *Reconstructing Nature: The Emergence of Science and Religion* (Edinburgh, 1998).

Brougham, Henry. *A Discourse of Objects, Advantages, and Pleasures of Science* (London, 1827).

Brown, C. Mackenzie. 'Western Roots of Avataric Evolutionism in Colonial India', *Zygon* 42 (2007), 423–48.

Brown, Peter. *The Body and Society: Men, Women and Sexual Renunciation in Early Christianity* (London, 1988).

Brown, Robert. 'A Brief Account of Microscopical Observations Made in the Months of June, July, August, 1827, on the Particles Contained in the Pollen of Plants; and on the

General Existence of Active Molecules in Organic and Inorganic Bodies', *Philosophical Magazine* 4 (1828), 161–73.

Brown, Robert. 'Additional Remarks on Active Molecules', *Philosophical Magazine* 6 (1829), 161–6.

Browne, Janet. *Charles Darwin: Voyaging* (London, 1995).

Browne, Janet. *Charles Darwin: The Power of Place* (London, 2002).

Browne, Janet. *Darwin's Origin of Species: A Biography* (London, 2006).

Bruford, Walter H. *The German Tradition of Self Cultivation: Bildung from Humboldt to Thomas Mann* (Cambridge, 1975).

Brumfitt, J. H. *Voltaire Historian* (Oxford, 1958).

Brush, Stephen G. 'Mach and Atomism', *Synthese* 18 (1968), 192–215.

Bryan, George Henry. *Stability in Aviation: An Introduction to Dynamical Stability as Applied to the Motion of Aeroplanes* (London, 1911).

Bryan, George Henry. Review of R. de Villamil, *ABC of Hydrodynamics*, *Mathematical Gazette* 6 (1912) 379–80.

Bryan, George Henry. 'Researches in Aeronautical Mathematics', *Nature* 96 (1916), 509–11.

Buchanan, Roderick and James Bradley. '"Darwin's Delay": A Reassessment of Evidence', *Isis* 108 (2017), 529–52.

Büchner, Ludwig. *Kraft und Stoff. Empirische-naturphilosophische Studien. In allgemein verständlicher Darstellung* (Frankfurt am Main, 1855).

Buchwald, Jed Z. *The Creation of Scientific Effects: Heinrich Hertz and Electric Waves* (Chicago, 1994).

Buckland, William. *Reliquiae diluvianae: or, observations on the organic remains contained in caves, fissures, and diluvial gravel, and on other geological phenomena, attesting the action of an universal deluge* (London, 1823).

Buckle, Henry Thomas. *History of Civilization in England* (2 vols, London, 1857–61).

Buckle, Henry Thomas. 'On the Influence of Women on the Progress of Knowledge', *Fraser's Magazine* 57 (1858), 395–407.

Buckle, Henry Thomas. 'Mill on Liberty', *Fraser's Magazine* 59 (May 1859), 509–42.

Buckley, Arabella. *The Fairy-Land of Science* (London, 1879).

Buckley, Arabella. 'Soul, and the Theory of Evolution', *University Magazine* 93 (January 1879), 1–20.

Buckley, Arabella. *Life and her Children: Glimpses of Animal Life from the Amoeba to the Insects* (London, 1880).

Bud, Robert. '"Applied Science": A Phrase in Search of a Meaning', *Isis* 103 (2012), 537–45.

Bud, Robert. '"Applied Science" in Nineteenth-Century Britain: Public Discourse and the Creation of Meaning, 1817–1876', *History and Technology* 30 (2014), 3–36.

Buffon, George Louis Leclerc, Comte de. *Histoire naturelle, générale et particulière* (15 vols, Paris, 1749–67).

Bullock, William. *A Companion to the Liverpool Museum, containing a Description of... Natural and Foreign Curiosities, Antiquities & Productions of the Fine Arts, Open for public inspection... at the house of William Bullock, Church Street* (Liverpool, c.1801).

Bullock, William. *A Concise and Easy Method of Preserving Objects of Natural History: Intended for the use of sportsmen, travellers, and others; to enable them to prepare and preserve such curious and rare articles* (London, 1818).

Burdach, Karl Friedrich. *Über die Aufgabe der Morphologie* (Leipzig, 1817).

Burnet, Thomas. *Archaelogiae philosophiae: sive doctrina antiqua de rerum originibus* (London, 1692).

Bush, Vannevar. *Science, the Endless Frontier: A Report to the President on a Program for Postwar Scientific Research* (Washington, DC, 1946).

Butler, Samuel. *Life and Habit* (London, 1878).

Byrne, Peter. *The Many Worlds of Hugh Everett III: Multiple Universes, Mutual Assured Destruction, and the Meltdown of the Nuclear Family* (Oxford, 2010).

Cahan, David. *An Institute for an Empire: The Physikalisch-Technische Reichsanstalt 1871–1918* (Cambridge, 1989).

Cahan, David, ed. *Hermann von Helmholtz, Science and Culture* (Chicago, 1995).

Cairnes, John Elliott. *Character and Logical Method of Political Economy* (New York, 1875).

Calmet, Augustine. *Histoire universelle sacrée et profane, depuis le commencement du monde jusqu'à nos jours* (Strasbourg, 1735).

Campbell, Lewis and William Garnett. *The Life of James Clerk Maxwell. With a Selection from his Correspondence and Occasional Writings and a Sketch of his Contribution to Science* (London, 1882).

Caneva, Kenneth L. *Robert Mayer and the Conservation of Energy* (Princeton, NJ, 1993).

Canguilhem, Georges. *Etudes d'histoire et de philosophie des sciences* (Paris, 1975).

Canguilhem, Georges. *The Normal and the Pathological* (New York, 1991).

Cannon, Susan Faye. *Science in Culture: The Early Victorian Period* (New York, 1978).

Cardano, Girolamo. *De subtilitate rerum* (Nuremberg, 1550).

Cardwell, D. S. L. *The Organisation of Science in England* (London, 1972).

Cargill, John. *The Fairy-Tales of Science: A Book for Youth* (London, 1859).

Carhart, Henry S. 'The Educational and Industrial Value of Science', *Science* n.s.1 (12 April 1895), 393–402.

Carhart, Michael C. *Leibniz Discovers Asia: Social Networking in the Republic of Letters* (Baltimore, 2019).

Carlson, W. Bernard. *Innovation as a Social Process: Eliuh Thomson and the Rise of General Electric, 1870–1900* (Cambridge, 1991).

Carnap, Rudolph. 'Die physikalischer Sprache als Universalsprache der Wissenschaft', *Erkenntnis* 2 (1932), 12–26.

Carnap, Rudolph. 'Psychologie in physikalischer Sprache', *Erkenntnis* 3 (1932), 107–42.

Carnap, Rudolph. 'Carnap's Intellectual Biography', in P. A. Schilpp, ed., *The Philosophy of Rudolph Carnap* (La Salle, IL, 1963), 3–84.

Carnot, Sadi. *Réflexions sur la puissance motrice du feu et sur les machines propres à développer cette puissance* (Paris, 1824).

Carpenter, William. *Principles of General and Comparative Physiology* (London, 1841).

Carroll, Sean B. *Endless Forms Most Beautiful: The New Science of Evo Devo and the Making of the Animal Kingdom* (London, 2005).

Cartwright, Nancy. *How the Laws of Physics Lie* (Oxford, 1983).

Cartwright, Nancy. *The Dappled World: A Study of the Boundaries of Science* (Cambridge, 1999).

Carty, John J. 'The Relation of Pure Science to Industrial Research', *Science* 44 (1916), 511–18.

Cassini, Jean Dominique. 'Reflexions sur la Chronologie Chinoise par Monsieur Cassini', in Simon de la Loubere, *Description du royaume de Siam* (2 vols, Amsterdam, 1691), ii. 304–21.

Cassirer, Ernst. 'Hermann Cohen und die Erneuerung der Kantischen Philosophie', *Kant-Studien* 17 (1912), 252–73.

Cassirer, Ernst. 'Der Begriff der Symbolischen Form im Aufbau der Geisteswissenschaften', *Vorträge der Bibliothek Warburg 1921/22* (1923), 11–39.

Cassirer, Ernst. *Philosophie der symbolischen Formen. Erster Teil: Die Spracht* (Berlin, 1923).

Cassirer, Ernst. *Philosophie der symbolischen Formen. Zweiter Teil: Das mythische Denken* (Berlin, 1925).

Cassirer, Ernst. *Philosophie der symbolischen Formen. Dritter Teil: Phänomenologie der Erkenntnis* (Berlin, 1929).

Cassirer, Ernst. *The Philosophy of Symbolic Forms* (3 vols, New Haven, CT, 1955).

Cat, Jordi. 'On Understanding: Maxwell on the Methods of Illustration and Scientific Metaphor', *Studies in History and Philosophy of Modern Physics* 32 (2001), 395–441.

Cat, Jordi, Nancy Cartwright, and Hasok Chang. 'Otto Neurath: Politics and the Unity of Science', in Peter Galison and David J. Stump, eds, *The Disunity of Science* (Stanford, 1996), 347–69.

Cesalpino, Andrea. *De plantiis libri XVI* (Florence, 1583).

Chalmers, Alan. *The Scientist's Atom and the Philosopher's Stone: How Science Succeeded and Philosophy Failed to Gain Knowledge of Atoms* (Dordrecht, 2011).

[Chambers, Robert]. *The Vestiges of the Natural History of Creation* (London, 1844).

[Chambers, Robert]. *Explanations, A Sequel* (London, 1845).

Channell, David F. 'The Harmony of Theory and Practice: The Engineering Science of W. J. M. Rankine', *Technology and Culture* 23 (1982), 39–52.

Channell, David F. *A History of Technoscience: Erasing the Boundaries between Science and Technology* (London, 2017).

Chayut, Michael. 'J. J. Thomson: The Discovery of the Electron and the Chemists', *Annals of Science* 48 (1991), 527–44.

Churchill, Frederick B. *August Weismann: Development, Heredity, and Evolution* (Cambridge, MA, 2015).

Clapeyron, [Benoît Paul] Émile. 'Mémoire sur la puissance motrice de la chaleur', *Journal de l'Ecole Royale Polytechnique* 22 (1834), 153–91.

Clarke, John and Thomas Hughes. *The Life and Letters of the Reverend Adam Sedgwick* (2 vols, Cambridge, 1890).

Clausius, Rudolf. 'Über die bewegende Kraft der Wärme und die Gesetze, welche sich daraus für die Wärmelehre selbst ableiten lassen', *Poggendorffs Annalen* 79 (1850), 368–97, 500–24.

Clausius, Rudolf. 'Ueber die Art der Bewegung, welche wir Wärme nennen', *Annalen der Physik* 100 (1857), 353–79.

Clodd, Edward. *Childhood of the World* (London, 1873).

Clodd, Edward. *Jesus of Nazareth* (London, 1880).

Clodd, Edward. *The Story of Creation: A Plain Account of Evolution* (London, 1888).

Clodd, Edward. *A Primer of Evolution* (London, 1895).

Clodd, Edward. *The Story of Primitive Man* (London, 1895).

Clodd, Edward. *Pioneers of Evolution: From Thales to Huxley* (London, 1897).

Coffrey, Mary. 'The American Adonis: A Natural History of the Average American (Man), 1921–1932', in Susan Currell and Christina Cogdell, eds, *Popular Eugenics: National Efficiency and American Mass Culture in the 1930s* (Athens, OH, 2006), 185–216.

Cohen, H. Floris. *The Scientific Revolution: A Historiographical Inquiry* (Chicago, 1994).

Cohen, Hermann. *Kants Theorie der Erfahrung* (Hildesheim, 1871).

Cohen, Hermann. 'Zur Kontroverse zwischen Trendelenburg und Kuno Fischer', *Zeitschrift für Völkerpsychologie und Sprachtwissenschaft* 7 (1871), 249–96.

Cohen, Hermann. *Das Prinzip der Infinitesimal-Methode und seine Geschichte: Ein Kapitel zur Grundlegung der Erkenntniskritik* (Berlin, 1883).

Cohen, Hermann. *System der Philosopie, Erster Teil: Logik der reinen Erkenntnis* (Berlin, 1902).

Cohen, Hermann. *Einleitung mit kritischem Nachtrag, zur neunten Auflage von Langes Geschichte des Materialismus* (Leipzig, 1914).

Cohen, Hermann. *Hermann Cohens Schriften zur Philosophie und Zeitgeschichte* (2 vols, Berlin, 1928).

Cole, Henry. *Popular Geology Subversive of Divine Revelation! A Letter to the Reverend Adam Sedgwick . . . being a Scriptural Refutation of the Geological Positions and Doctrines Promulgated in his Lately Published Commencement Sermon, Preached in the University of Cambridge, 1832* (London, 1834).

Coleman, William. *Georges Cuvier Zoologist: A Study in the History of Evolution Theory* (Cambridge, MA, 1962).

Coleman, William. *Biology in the Nineteenth Century: Problems of Form, Function, and Transformation* (New York, 1971).

Coleridge, Samuel Taylor. *General Introduction; or, Preliminary Treatise on Method* (London, 1817).

Coleridge, Samuel Taylor. *On the Constitution of the Church and State, According to the Idea of Each* (3rd edition, London, 1839).

Collingwood, R. G. *An Essay on Metaphysics* (Oxford, 1940).

Collins, Harry M. *Changing Order: Replication and Induction in Scientific Practice* (London, 1985).

Colquhoun, Patrick. *A Treatise on Indigence; exhibiting a general view of the national resources for productive labour; with propositions for ameliorating the condition of the poor, and improving the moral habits and increasing the comforts of the labouring people* (London, 1806).

Combe, George. *The Constitution of Man Considered in Relation to External Objects* (London, 1828).

Combe, George. *A System of Phrenology* (5th edition, 2 vols, Edinburgh, 1843).

Comte, Auguste. *Cours de Philosophie Positive* (6 vols, Paris, 1830–42).

Comte, Auguste. *Système de politique positive, ou Traité de sociologie, Instituant la Religion de l'Humanité* (4 vols, Paris, 1851–4).

Comte, Auguste. *The Positive Philosophy of Auguste Comte, Freely Translated and Condensed by Harriet Martineau* (2 vols, London, 1853).

Comte, Auguste. *Early Political Writings*, ed. and trans. H. S. Jones (Cambridge, 1998).

Condorcet, Marquis de. *Esquisse d'un tableau historique des progrès de l'esprit humain* (Paris, 1795).

Condorcet, Marquis de. *Outlines of an Historical View of the Progress of the Mind* (London, 1795).

Cooter, Roger and Stephen Pumfrey. 'Separate Spheres and Public Places: Reflections on the History of Science Popularisation and Science in Popular Culture', *History of Science* 32 (1994), 237–67.

Cosslett, Tess. *Talking Animals in British Children's Fiction, 1786–1914* (Aldershot, 2006).

Costard, George. 'A Letter from the Rev G. Costard to the Rev. Thomas Shaw, D.D. F.R. S. and Principal of St. Edmund-Hall concerning the Chinese Chronology and Astronomy', *Philosophical Transactions* 44 (1747), 475–92.

'Country Gentleman, A'. *The Consequences of a Scientific Education to the Working Classes of This Country Pointed out; and the Theories of Mr Brougham on That Subject, Confuted; in a Letter to the Marquess of Lansdown* (London, 1826).

Cowley, William Lewis, and Hyman Levy. *Aeronautics in Theory and Experiment* (London, 1918).

Creath, Richard. 'The Unity of Science: Carnap, Neurath, and Beyond', in Peter Galison and David J. Stump, eds, *The Disunity of Science* (Stanford, 1996), 158–69.

Crosland, Maurice and Crosbie Smith. 'The Transmission of Physics from France to Britain, 1800–1840', *Historical Studies in the Physical Sciences* 9 (1978), 1–61.

Crosse, Andrew. 'On the Production of Insects by Voltaic Electricity', *Annals of Electricity* 1(1836–7), 242–44.

Croze, Mathurin Veyssière de la. *Histoire du Christianisme des Indes* (The Hague, 1724).

Cryle, Peter and Elizabeth Stephens, *Normality: A Critical Genealogy* (Chicago, 2017).

Culotta, Charles A. 'Tissue Oxidation and Theoretical Physiology: Bernard, Ludwig, and Pflüger', *Bulletin of the History of Medicine* 44 (1970), 109–40.

Cunningham, Andrew. 'The Pen and the Sword: Recovering the Disciplinary Identity of Physiology and Anatomy Before 1800. I: Old Physiology—the Pen', *Studies in History and Philosophy of Biology and Biomedical Sciences* 33 (2002), 631–55.

Cuvier, Georges, Baron. 'Mémoire sur les espèces d'éléphants vivantes et fossiles. Lu le premier pluvoîse an IV', *Mémoires de l'Institut national des sciences et des arts. Sciences mathématiques et physiques* 2 (1799), 1–22.

Cuvier, Georges, Baron. *Recherches sur les ossemens fossiles de quadrupèdes, où l'on rétablit les caractères de plusieurs espèces d'animaux que les révolutions du globe paroissent avoir détruites* (4 vols, Paris, 1812).

Cuvier, Georges, Baron. *Le régne animal distribué d'après son organisation pour servir de base à l'histoire naturelle des animaux et l'introduction à l'anatomie comparée* (5 vols, Paris, 1829–30).

Dalton, John. 'Absorption of Gases by Water and Other Liquids', *Memoirs of the Literary and Philosophical Society of Manchester* 2nd series, 1 (1805), 271–87.

Dalton, John. *A New System of Chemical Philosophy* (3 vols, Manchester, 1808–27).

Dalton, John. 'Dr. Bostock's Review of the Atomic Principles of Philosophy', *Journal of Natural Philosophy, Chemistry and the Arts* 29 (1811), 143–51.

Darlu, Alfonse. 'Réflexions d'un philosophe sur la question du jour. Science, morale et religion', *Revue de métaphysique et de morale* 3 (1895), 239–51.

Darrigol, Olivier. 'Henri Poincaré's Criticism of Fin-de-Siècle Electrodynamics', *Studies in History and Philosophy of Modern Physics* 26 (1995), 1–44.

Darrigol, Olivier. 'The Electromagnetic Origins of Relativity Theory', *Historical Studies in the Physical and Biological Sciences* 26 (1996), 241–312.

Darrigol, Olivier. *Worlds of Flow: A History of Hydrodynamics from the Bernoullis to Prandtl* (Oxford, 2005).

Darwin, Charles. *The Zoology of the Voyage of H.M.S. Beagle* (5 vols, London, 1839–43).

Darwin, Charles. *The Structure and Distribution of Coral Reefs. Being the first part of the geology of the voyage of the* Beagle, *under the command of Capt. Fitzroy, R.N. during the years 1832 to 1836* (London, 1842).

Darwin, Charles. *Geological Observations on the Volcanic Islands visited during the voyage of H.M.S. Beagle, together with some brief notices of the geology of Australia and the Cape of Good Hope. Being the second part of the geology of the voyage of the Beagle, under the command of Capt. Fitzroy, R.N. during the years 1832 to 1836* (London, 1844).

Darwin, Charles. *Geological Observations on South America. Being the third part of the geology of the voyage of the* Beagle, *under the command of Capt. Fitzroy, R.N. during the years 1832 to 1836* (London, 1846).

Darwin, Charles. *On the Origin of Species by means of Natural Selection, or the Preservation of Favoured Races in the Struggle for Life* (London, 1859).

Darwin, Charles. *The Descent of Man, and Selection in Relation to Sex* (London, 1871).

Darwin, Charles. *The Origin of Species by Means of Natural Selection, or the Preservation of Favoured Races* (6th edition, London, 1872).

Darwin, Charles. *The Foundations of the Origin of Species: Two Essays written in 1842 and 1844 by Charles Darwin*, ed. Francis Darwin (Cambridge, 1909).

Darwin, Charles. *The Correspondence of Charles Darwin*, ed. F. Burkhardt and S. Smith (9 vols, Cambridge, 1985–94).

Darwin, Charles. *The Works of Charles Darwin*, ed. Paul H. Barrett and R. B. Freeman (29 vols, London, 1986).

Darwin, Francis and A. C. Seward, eds, *More Letter of Charles Darwin* (London, 1903).

Daston, Lorraine. 'The Physicalist Tradition in Early Nineteenth Century French Geometry', *Studies in History and Philosophy of Science* 17 (1986), 269–95.

Daston, Lorraine. 'The Empire of Observation, 1600–1800', in Lorraine Daston and Elizabeth Lunbeck, eds, *Histories of Scientific Observation* (Chicago, 2011), 81–113.

Daston, Lorraine and Peter Galison. *Objectivity* (New York, 2007).

Daum, Andreas. *Wissenschaftspopularisierung im 19. Jahrhundert: Bürgerlich Kultur, naturwissenschaftliche Bildung und die deutsche Öffentlichkeit 1848–1914* (Munich, 2002).

Davidson, Robert. *The Elements of Geography, Short and Plain* (London, 1787).

Davy, Humphry. *Collected Works of Sir Humphry Davy* (9 vols, London, 1840).

de Bellaigue, Christopher. *The Islamic Enlightenment: The Modern Stuggle Between Faith and Reason* (London, 2017).

de Vries, Jan. 'The Industrial Revolution and the Industrious Revolution', *Journal of Economic History* 54 (1994), 240–70.

de Vries, Jan. *The Industrious Revolution: Consumer Behavior and the Household Economy, 1650 to the Present* (Cambridge, 2008).

de Vries, Jan and Adriaan van de Woude, *The First Modern Economy* (Cambridge, 1997).

Dear, Peter. 'What is the History of Science the History *Of?* Early Modern Roots of the Ideology of Modern Science, *Isis* 96 (2005), 390–406.

Delbourgo, James. *Collecting the World: Hans Sloane and the Origins of the British Museum* (Cambridge, MA, 2017).

Deltete, Robert. 'Helm and Boltzmann: Energetics at the Lübeck Naturforschversammlung', *Synthese* 119 (1999), 45–68.

Deltete, Robert. 'Wilhelm Ostwald's Energetics 1: Origins and Motivations', *Foundations of Chemistry* 9 (2007), 3–56.

Deltete, Robert. 'Wilhelm Ostwald's Energetics 2: Energetic Theory and Applications, Part I', *Foundations of Chemistry* 9 (2007), 265–316.

Deltete, Robert. 'Wilhelm Ostwald's Energetics 3: Energetic Theory and Applications, Part II', *Foundations of Chemistry* 10 (2008), 187–221.

Delumeau, Jean. *La Peur en occident (XIV*ᵉ*–XVIII*ᵉ *siècles): Une cité assiégée* (Paris, 1978).

Delumeau, Jean. *Le Péché et la peur: La culpabilisation en occident, XIII*ᵉ*–XVIII*ᵉ *siècles* (Paris, 1983).

Delumeau, Jean. 'Prescription and Reality', in Edmund Leites, ed., *Conscience and Casuistry in Early Modern Europe* (Cambridge, 1988), 134–58.

Delumeau, Jean. *Rassurer et protéger: Le sentiment de sécuritée dans l'occident d'autrefois* (Paris, 1989).

Delumeau, Jean. *L'Aveu et le pardon* (Paris, 1992).

Dennett, Daniel C. *Brainchildren: Essays on Designing Minds* (Cambridge, MA, 1998).

Descola, Philippe. *Beyond Nature and Culture* (Chicago, 2013).

Desmond, Adrian. *The Politics of Evolution: Morphology, Medicine, and Reform in Radical London* (Chicago, 1989).

Desmond, Adrian. *Huxley: The Devil's Disciple* (London, 1994).

Despretz, César-Mansuète. 'Recherches expérimentales sur les Causes de la chaleur animale', *Annales de chimie et de physique* 26 (1824), 337–64.

Dewey, John. 'Unity of Science as a Social Problem', in Otto Neurath, Rudolph Carnap, and Charles Morris, eds, *Foundations of the Unity of Science* (2 vols, Chicago, 1970), i. 29–38.

Dibdin, William. *Public Lighting by Gas and Electricity* (London, 1902).

Diderot, Denis and Jean le Rond d'Alembert, *Encyclopédie ou Dictionnaire raisonné des sciences, des arts et des métiers par une société des gens de Lettres, mis en ordre et publié par Diderot et quant à la Partie mathématique par d'Alembert* (2nd edition, 40 vols, Geneva, 1777–9).

Dietrich, Michael. 'Paradox and Persuasion: Negotiating the Place of Molecular Evolution Within Evolutionary Biology', *Journal of the History of Biology* 31 (1998), 87–111.

Dikötter, Frank. *The Discourse of Race in Modern China* (London, 1992).

Dilthey, Wilhelm. *Einleitung in die Geisteswissenschaften: Versuch einer Grundlegung für das Studium der Gesellschaft und der Geschichte* (Leipzig, 1883).

Döblin, Alfred. *Berge Meere und Giganten* (Berlin, 1924).

Dobzhansky, Theodosius. *Genetics and the Origin of Species* (New York, 1937).

Döllinger, Ignaz. 'Über den jetzigen Zustand der Physiologie', *Jahrbücher der Medicin als Wissenschaft* 1 (1805), 119–42.

Döring, Daniela. *Zeugende Zahlen: Mittelmaß und Durchschnittstypen in Proportion, Statistik und Konfektion* (Berlin, 2011).

Dougherty, Frank. 'Über den Einfluß Johann Friedrich Blumenbachs auf Kielmeyers feierliche Rede von 1793: Mit einer Anhang über Kielmeyers Göttinger Lektüre', in K. T. Kanz, ed., *Philosophie des Organischen in der Goethezeit* (Stuttgart, 1944), 50–80.

Douglas, Janet Mary. *The Life and Selections from the Correspondence of William Whewell* (London, 1881).

Drake, Stillman. *Galileo Studies* (Ann Arbour, 1970).

Drayton, Richard. *Nature's Government: Science, Imperial Britain, and the 'Improvement' of the World* (New Haven, CT, 2000).

Drobisch, Moritz Wilhelm. *Philologie und Mathematik als Gegenstände des Gymnasialunterrichts betrachtet, mit besonder Beziehung auf Sachsens Gelehrtenschulen* (Leipzig, 1832).

Droysen, Johann Gustav. *Grundriss der Historik* (Leipzig, 1868).

Drucker, Peter F. 'The Technological Revolution: Notes on the Relationship of Science, Technology, and Culture', *Technology and Culture* 2 (1961), 342–51.

Du Bois-Reymond, Emil. *Untersuchungen über thierische Elektricitricität. Erster Band* (Berlin, 1848).

Du Bois-Reymond, Emil. *Über das Barrenturnen und über die sogenannte rationelle Gymnastik* (Berlin, 1862).

Du Bois-Reymond, Emil. *Reden von Emil du-Bois Reymond* (2 vols, Leipzig, 1912).

Du Bois-Reymond, Emil. *Über die Grenzen des Naturerkennens: Die Sieben Welträtseln* (Leipzig, 1916).

Duchesneau, François. *Genèse de la théorie cellulaire* (Montreal and Paris, 1987).

Dugas, René. *La Théorie Physique au Sense de Boltzmann et ses Prolongements Modernes* (Paris, 1959).

Dulong, Pierre-Louis. 'Mémoire sur la chaleur animale', *Annales de chimie et de physique*, 3rd series, 1 (1841), 440–55.

Dumas, Jean-Baptiste. 'Memoire sur quelques Points de la Théorie atomistique', *Journal de Chimie Physique* 33 (1826), 337–91.

Dumas, Jean-Baptiste. 'Dissertation sur la Densité de la Vapeur de quelques corps simples', *Journal de Chimie Physique* 50 (1832), 170–8.

Dumas, Jean-Baptiste. *Leçons sur la philosophie chimique* (Paris, 1837).

Dumas, Jean-Baptiste. 'Premier mémoire sur les types chimiques', *Annales de Chimie et de Physique* 73 (1840), 73–103.

Dumas, Jean-Baptiste and Justus Liebig. 'Note sur l'état actuel de la Chimie', *Comptes Rendus des Séances de l'Académie des Sciences* 5 (1837), 567–72.

Dunn, Leslie. 'Science in the USSR: Soviet Biology', *Science* 99 (1944), 65–7.

Dupré, John. *The Disorder of Things* (Harvard, MA, 1995).

Dutrochet, René Joachim Henri. *Mémoires pour servir à l'histoire anatomique et physiologique des végétaux et des animaux* (Paris, 1824).

Dyer, Frank Lewis and Thomas C. Martin. *Edison, His Life and Inventions* (2 vols, New York, 1910).

East, Edward M. 'The Nucleus-Plasma Problem', *American Naturalist* 68 (1934), 289–303, 402–39.

Eddington, Arthur Stanley. *The Nature of the Physical World* (Cambridge, 1928).

Edgerton, David. *Science, Technology, and the British Industrial 'Decline'* (Basingstoke, 1991).

Edgerton, David. 'British Scientific Intellectuals and the Relations of Science, Technology, and War', in P. Forman and J. M. Sánchez, eds, *National Military Establishments and the Advancement of Science and Technology: Studies in Twentieth Century History* (Dordrecht, 1996), 1–35.

Edgerton, David. *The Rise and Fall of the British Nation: A Twentieth Century History* (London, 2018).

Edgeworth, Francis Ysidro. *Mathematical Psychics: An Essay on the Application of Mathematics to the Moral Sciences* (London, 1881).

Einstein, Albert. 'Über einen die Erzeugung und Verwandlung des Lichtes betreffenden heuristischen Gesichtspunkt', *Annalen der Physik* 17 (1905), 132–48.

Einstein, Albert. 'Über die molekularkinetischen Theorie der Wärme geforderte Bewegung von in ruhenden Flüssigkeiten suspendierten Teilchen', *Annalen der Physik* 17 (1905), 549–60.

Einstein, Albert. 'Über die gegenwärtige Krise der theoretischen Physik', *Kaizo* 4 (1922), 1–8.

Einstein, Albert. 'Autobiographical Notes', in P. A. Schilpp, ed., *Albert Einstein, Philosopher-Scientist* (2 vols, New York, 1959), i. 1–96.

Eisenstadt, Shmuel N. *Modernisation, Protest and Change* (Englewood Cliffs, NJ, 1966).

Eisenstadt, Shmuel N. 'The Civilizational Dimension of Modernity: Modernity as a Distinct Civilization', *International Sociology* 16 (2001), 320–40.

Elias, Norbert. *Über den Prozess der Zivilization: Soziogenetische und psychogenetische Untersuchungen* (2 vols, Basle, 1939).

Elias, Norbert *The Civilizing Process: Sociogenic and Phylogenetic Investigations* (Oxford, 2000).

Elkana, Yehuda. 'The Conservation of Energy: A Case of Simultaneous Discovery?', *Archives internationales d'histoire des sciences* 90.1 (1970), 31–60.

Elman, Benjamin. *On Their Own Terms: Science in China, 1550–1900* (Cambridge, MA, 2005).

Elshakry, Marwa. *Reading Darwin in Arabic, 1860–1950* (Chicago, 2013).

Elshakry, Marwa. 'Spencer's Arabic Readers', in B. Lightman, ed., *Global Spencerism: The Communication and Appropriation of a British Evolutionist* (Leiden, 2015), 35–55.

Engel, Christian Gottlieb Ferdinand. *Welchen Einfluss äussert das Studium der Mathematischen Wissenschaften auf das Gemüth?* (Berlin, 1820).

Engelhardt, Dietrich von. *Historisches Bewußtsein in der Naturwissenschaft von der Aufklärung bis zum Positivismus* (Freiburg im Breisgau, 1979).

Engels, Eve-Marie. 'Die Lebenskraft—metaphysiche Konstrukt oder methodologisches Instrument? Überlegungen zum Status von Lebenskräften in Biologie und Medizin im Deutschland des 18. Jahrhunderts', in K. T. Kanz, ed., *Philosophie des Organischen in der Goethezeit* (Stuttgart, 1994), 127–52.

Engemann, Walter. *Voltaire und China: ein Beitrag zur Geschichte der Völkerkunde und zur Geschichte der Geschichtsschreibung sowie zu ihren gegenseitigen Beziehungen* (Leipzig, 1933).

Enros, Philip C. 'The Analytical Society (1812–1813): Precursor of the Renewal of Cambridge Mathematics', *Historia Mathematica* 19 (1983), 24–47.

Erikson, Paul, Judy Klein, Lorraine Daston, Rebecca Lemov, Thomas Sturm, and Michael Gordin. *How Reason Almost Lost its Mind: The Strange Career of Cold War Rationality* (Chicago, 2013).

Erxleben, Johann Christian Polycarp. *Anfangsgründe der Naturgeschichte* (Göttingen, 1768).

Erxleben, Johann Christian Polycarp. *Anfangsgründe der Naturlehre* (Göttingen, 1772).

Evans, Richard J. *The Pursuit of Power: Europe 1815–1914* (London, 2016).

Faraday, Michael. *Experimental Researches in Electricity* (3 vols, London, 1839–55).

Farber, Paul Lawrence. *Finding Order in Nature: The Naturalist Tradition from Linnaeus to E. O. Wilson* (Baltimore, MD, 2000).

Febvre, Lucien. 'Civilisation. Evolution d'un mot et d'un groupe d'idées', in L. Febvre et al., *Civilisation. Le mot et l'idée* (Paris, 1930), 1–59.

Febvre, Lucien. *The Problem of Unbelief in the Sixteenth Century: The Religion of Rabelais* (Cambridge, MA, 1982).

Fechner, Gustav Theodore. *Elemente der Psychophysik* (Leipzig, 1860).

Ferguson, Eugene. 'The Mind's Eye: Non-Verbal Thought in Technology', *Science* 197 (1977), 827–36.

Ferreiro, Larrie D. *Ships and Science: The Birth of Naval Architecture in the Scientific Revolution, 1600–1800* (Cambridge, MA, 2006).

Finkelstein, Gabriel. *Emil du Bois-Reymond: Neuroscience, Self, and Society in Nineteenth-Century Germany* (Cambridge, MA, 2013).

Fischer, Joachim. *Philosophische Anthropologie: Eine Denkrichtung des 20. Jahrhunderts* (Freiburg, 2008).

Fischer, Kuno. *Anti-Trendelenburg* (Jena, 1870).

Fisher, R. A. *The Genetical Theory of Natural Selection* (Oxford, 1930).

Flammarion, Camille. *La pluralité des mondes habités: Étude ou l'on expose les conditions d'habitude des terres célestes discutées au point de vue de l'astronomie, de la physiologie at de la philosophie naturelle* (Paris, 1862).

Flammarion, Camille. *La planète Mars et ses conditions d'habitabilité: Encyclopédie générale des observations martiennes* (Paris, 1892).

Fontana, Michela. *Matteo Ricci: A Jesuit in the Ming Court* (Lanham, MD, 2011).

Föppl, August. *Vorlesungen über technische Mechanik* (vol. 1, 2nd edition, Leipzig, 1900).

Forbes, Edward. *History of British Starfishes, and Other Animals of the Class Echinodermata* (London, 1841).

Forman, Paul. 'Weimar Culture, Causality, and Quantum Theory, 1918–1927: Adaptation by German Physicists and Mathematicians to a Hostile Intellectual Environment', *Historical Studies in the Physical Sciences* 3 (1971), 1–116.

Forrester, John. 'Chemistry and the Conservation of Energy: The Work of James Prescott Joule', *Studies in History and Philosophy of Science* 6 (1975), 273–313.

Foucault, Michel. *Histoire de la sexualité I: La Volonté de savoir* (Paris, 1976).

Fourier, Joseph. *Recherches statistiques sur la ville de Paris et le Département de la Seine* (2 vols, Paris, 1821–3).

Fourier, Joseph. *Théorie analytique de la chaleur* (Paris, 1822).

Fournier, Marcel. *Émile Durkheim: A Biography* (Cambridge, 2013).

Fox, Robert. 'Dalton's Caloric Theory', in D. S. L. Cardwell, ed., *John Dalton and the Progress of Science* (Manchester, 1968), 187–201.

Fox, Robert. 'The Rise and Fall of Laplacian Physics', *Historical Studies in the Physical Sciences* 4 (1974), 89–136.

Frank Jr, Robert G. *Harvey and the Oxford Physiologists* (Berkeley, 1980).

Frankopan, Peter. *The Silk Roads: A New History of the World* (London, 2015).

Freeberg, Ernest. *The Age of Edison: Electric Light and the Invention of Modern America* (New York, 2013).

Freud, Sigmund. *The Standard Edition of the Complete Psychological Works of Sigmund Freud* (24 vols, London, 1953–74).

Friedman, Michael. *A Parting of Ways: Carnap, Cassirer, and Heidegger* (Chicago, 2000).

Friedman, Milton. 'The Methodology of Positive Economics', in Milton Friedman, *Essays in Positive Economics* (Chicago, 1953), 3–43.

Fuhrmann, Georg. 'Theoretische und experimentelle Untersuchungen an Ballonmodel-len', *Jahrbuch der Motorluftschiff-Studiengesellschaft* 5 (1911–12), 64–123.

Furkawa, Yasu. 'Macromolecules: Their Structures and Functions', in Mary Jo Nye, ed., *The Cambridge History of Science, volume 5: The Modern Physical and Mathematical Sciences* (Cambridge, 2003), 429–45.

Fyfe, Aileen. *Science and Salvation: Evangelical Popular Science Publishing in Victorian Britain* (Chicago, 2004).

Gadamer, Hans-Georg. *Truth and Method* (New York, 1982).

Gadamer, Hans-Georg. *Reason in the Age of Science* (Cambridge, MA, 1983).

Galison, Peter. *How Experiments End* (Chicago, 1987).

Galison, Peter. 'Introduction: The Context of Disunity', in Peter Galison and David J. Stump, eds, *The Disunity of Science* (Stanford, 1996), 1–33.

Galison, Peter. *Image and Logic: A Material Culture of Microphysics* (Chicago, 1997).

Galison, Peter. *Einstein's Clocks, Poincaré's Maps* (London, 2004).

Galton, Francis. 'Hereditary Talent and Character Part I', *Macmillan's Magazine* 12 (1865), 157–66.

Galton, Francis. *Hereditary Genius: An Inquiry Into its Laws and Consequences* (London, 1869).

Galton, Francis. 'The History of Twins, as a Criterion of the Relative Powers of Nature and Nurture', *Journal of the Anthropological Institute* 5 (1875), 324–9.

Galton, Francis. 'Statistics by Intercomparison, with Remarks on the Law of Frequency of Error', *Philosophical Magazine* 49 (1875), 33–46.

Galton, Francis. 'Composite Portraits Made by Combining Those of Many Different Persons into a Single Figure', *Nature* 18 (1878), 97–100.

Galton, Francis. 'On the Application of Composite Portraiture to Anthropological Pur-poses', *Report of the British Association for the Advancement of Science* 51 (1881), 690–91.

Galton, Francis. 'An Inquiry into the Physiognomy of Phthisis by the Method of "Composite Portraiture"', *Guy's Hospital Reports* 25 (1882), 475–93.

Galton, Francis. *Inquiries into Human Faculty and Its Development* (London, 1883).

Galton, Francis. *Memoirs of My Life* (London, 1908).

Gambarotto, Andrea. 'Vital Forces and Organization: Philosophy of Nature and Biology in Karl Friedrich Kielmeyer', *Studies in History and Philosophy of Biological and Biomedical Sciences* 48 (2014), 12–20.

Gane, Mike. *Auguste Comte* (London, 2006).

Garber, Elisabeth Wolfe. 'Clausius and Maxwell's Kinetic Theory of Gases', *Historical Studies in the Physical Sciences* 2 (1970), 299–319.

Garstang, Walter. 'The Theory of Recapitulation: A Critical Restatement of the Biogenic Law', *Zoological Journal of the Linnean Society* 35 (1922), 81–101.

Gaukroger, Stephen. *Cartesian Logic: Descartes' Conception of Inference* (Oxford, 1989).

Gaukroger, Stephen. *Francis Bacon and the Transformation of Early-Modern Philosophy* (Cambridge, 2001).

Gaukroger, Stephen. *The Emergence of a Scientific Culture: Science and the Shaping of Modernity, 1210–1685* (Oxford, 2006).

Gaukroger, Stephen. *The Collapse of Mechanism and the Rise of Sensibility: Science and the Shaping of Modernity, 1680–1760* (Oxford, 2010).

Gaukroger, Stephen. *Objectivity: A Very Short Introduction* (Oxford, 2012).

Gaukroger, Stephen. 'Kant and the Nature of Matter: Mechanics, Chemistry, and the Life Sciences', *Studies in History and Philosophy of Science* 58 (2016), 108–14.

Gaukroger, Stephen. *The Natural and the Human: Science and the Shaping of Modernity, 1739–1841* (Oxford, 2016).

Gaukroger, Stephen. 'Alexandre Koyré and the History of Science as a Species of the History of Philosophy: The Cases of Galileo and Descartes', in Raffaele Pisano, Joseph Agassi, and Daria Drozdova, eds, *Hypotheses and Perspectives in the History and Philosophy of Science. Hommage to Alexandre Koyré 1964–2014* (New York, 2017), 179–87.

Gaultier, Abbé A. E. C. *A Rational and Moral Game* (London, *c*.1805).

Gavroglu, Kostas and Ana Simões. *Neither Physics Nor Chemistry: A History of Quantum Chemistry* (Cambridge, MA, 2012).

Gay-Lussac, Joseph-Louis. 'Mémoire sur la combinaison des substances gazeuses, les unes avec les autres', *Mémoires de physique et de chimie de la Société d'Arcueil*, 2 (1809), 207–35.

Gehler, Johann Samuel. *Physikalisches Wörterbuch: oder Versuch einer Erklärung der vornehmsten Begriffe und Kunstwörter der Naturlehre, mit kurzen Nachrichten von der Geschichte der Erfindungen und Beschreibungen der Werkzeuge begleitet* (5 vols, Leipzig, 1787–92).

Geoffroy Saint-Hilaire, Étienne. 'Histoire des Makis, ou singes de Madagascar', *Magasin encyclopédique* 1 (1796), 1–48.

Geoffroy Saint-Hilaire, Étienne. *Philosophie anatomique* (2 vols, Paris, 1818–22).

Gibbs-Smith, Charles. *The Aeroplane: An Historical Survey of Its Origins and Development* (London, 1960).

Gillham, Nicholas Wright. *A Life of Sir Francis Galton: From African Exploration to the Birth of Eugenics* (Oxford, 2001).

Gispen, Kees. *New Profession, Old Order: Engineers and German Society, 1815–1914* (Cambridge, 1989).

Glauert, Hermann. 'Theoretical Relationships for the Lift and Drag of an Aerofoil Structure', *Journal of the Royal Aeronautical Society* 27 (1923), 512–18.

Gliboff, Sander. *H. G. Bron, Ernst Haeckel, and the Origins of German Darwinism: A Study in Translation and Transformation* (Cambridge, MA, 2008).

Gmelin, Johann Friedrich. *Grundriß der allgemeinen Chemie zum Gebrauch bei Vorlesunge* (Göttingen, 1804).

Godart, G. Clinton. 'Spencerism in Japan: Boom and Bust of a Theory', in B. Lightman, ed., *Global Spencerism: The Communication and Appropriation of a British Evolutionist* (Leiden, 2015), 56–77.

Goethe, Johann Wolfgang. *Zur Naturwissenschaft überhaupt, besonders zur Morphologie* (Stuttgart, 1817).

Goldman, Lawrence. 'The Origins of British "Social Science": Political Economy, Natural Science and Statistics, 1830–1835', *The Historical Journal* 26 (1983), 587–616.

Golinski, Jan. '"The Nicety of Experiment": Precision of Measurement and Precision of Reasoning in Late Eighteenth-Century Chemistry', in M. Norton Wise, ed., *The Values of Precision* (Princeton, NJ, 1995), 72–91.

Gooday, Graeme. '"Vague and Artificial": The Historically Elusive Distinction Between Pure and Applied Science', *Isis* 103 (2012), 546–54.

Goodey, Christopher F. *A History of Intelligence and 'Intellectual Disability': The Shaping of Psychology in Early Modern Europe* (Farnham, 2011).

Gordin, Michael D. *Scientific Babel: How Science was Done Before and After Global English* (Chicago, 2015).

Gordon, Peter E. *Continental Divide: Heidegger, Cassirer, Davos* (Cambridge, MA, 2010).

Gosse, Philip Henry. *The Aquarium: An Unveiling of the Wonders of the Deep Sea* (London, 1856).

Gosse, Philip Henry. *Actinologia Britannica: A History of British Sea-Anemones and Corals* (London, 1860).

Gougher, Ronald L. 'Comparison of English and American Views of the German University, 1840–1865: A Bibliography', *History of Education Quarterly* 9 (1969), 477–91.

Gould, Stephen Jay. *Hen's Teeth and Horse Toes: Further Reflections on Natural History* (New York, 1983).

Gould, Stephen Jay. *The Structure of Evolutionary Theory* (Cambridge, MA, 2002).

Graham, Loren R. *Science in Russia and the Soviet Union* (Cambridge, 1993).

Grattan-Guiness, Ivor. 'Work for the Workers: Advances in Engineering Mechanics and Instruction in France, 1800–1830', *Annals of Science* 41 (1984), 1–33.

Gray, John. *Enlightenment's Wake: Politics and Culture at the Close of the Modern Age* (London, 1995).

Gray, John. Review of Alex Callinicos, *Equality*, *Times Literary Supplement* 5116 (20 April 2001), 3.

Greenaway, Frank ed. *Archives of the Royal Institution* (15 vols, London, 1971–6).

Gregory, Frederick. *Scientific Materialism in Nineteenth Century Germany* (Dordrecht, 1977).

Gregory, Richard. *Discovery; or, The Spirit and Service of Science* (London, 1917).

Gresswell, Albert and George. *The Wonderland of Evolution* (London, 1884).

Grey, Charles. 'Editorial Comment', *Aeroplane* 12 (1917), 1284.

Griffiths, Paul and Karola Stotz. *Genetics and Philosophy: An Introduction* (Cambridge, 2014).

Grmek, Mirko D. *Raisonnement expérimental et recherches toxicologiques chez Claude Bernard* (Geneva, 1972).

Groc, Léon. *Deux mille ans sous la mer* (Paris, 1924).

Groenewegen, Peter. *A Soaring Eagle: Alfred Marshall 1842–1924* (Aldershot, 1995).

Grove, William Robert. *On the Progress of the Physical Sciences* (London, 1842).

Guizot, François. *Histoire de la civilization en Europe: depuis la chute de l'empire romain jusqu'à la révolution française* (Paris, 1828).

Hacking, Ian. 'The Disunities of the Sciences', in Peter Galison and David J. Stump, eds, *The Disunity of Science* (Stanford, 1996), 37–74.

Haeckel, Ernst. 'Die Gastrula und die Eifurchung der Thiere', *Jenische Zeitschrift für Naturwissenschaft* 9 (1875), 402–508.

Haeckel, Ernst Heinrich. *Generelle Morphologie der Organismen: Allgemeine Grundzüge der organischen Former-Wissenschaft, mechanisch begründet durch die von Charles Darwin reformirte Descendez-Theorie* (2 vols, Berlin, 1866).

Haeckel, Ernst Heinrich. *Natürliche Schöpfungsgeschichte* (Berlin, 1868).

Haeckel, Ernst Heinrich. *Anthropogenie, oder Entwickelungsgeschichte des Menschen* (Leipzig, 1874).

Haeckel, Ernst Heinrich. 'Die Gastrea-Theorie, die phylogenetische Classification des Thierreichs und die Homologie der Keimblätter', *Jenaische Zeitschrift für Naturwissenschaft* 8 (1874), 1–55.

Haeckel, Ernst Heinrich. *Ziele und Wege der heutigen Entwickelungsgeschichte* (Jena, 1875).

Haeckel, Ernst Heinrich. *The History of Creation* (London, 1876).

Haeckel, Ernst Heinrich. 'Ueber die Entwicklunstheorie Darwin's', in Ernst Heinrich Haeckel, *Gesammelte populäre Vorträge aus dem Gebiete der Entwicklungslehre Heft 1* (Bonn, 1878), 1–28.

Haines IV, George. *Essays on German Influence upon English Education and Science, 1850–1919* (Hamden, CT, 1969).

Hakfoort, Casper. 'Science Deified: Wilhelm Ostwald's Energeticist World-View and the History of Scientism', *Annals of Science* 49 (1992), 525–44.

Haldane, J. S. 'An Address on the Relation of Physiology to Physics and Chemistry, Delivered before the Physiological Section of the British Association for the Advancement of Science, Dublin, 1908', *The British Medical Journal* (12 September 1908), 693–6.

Haldane, J. S. *The Causes of Evolution* (London, 1932).

Hale, Piers J. 'Monkeys into Men and Men into Monkeys: Chance and Contingency in the Evolution of Man, Mind and Morals in Charles Kingsley's *Water-Babies*', *Journal of the History of Biology* 46 (2013), 551–95.

Hale, Piers J. *Political Descent: Malthus, Mutualism, and the Politics of Evolution in Victorian England* (Chicago, 2014).

Hall, Brian K. 'Balfour, Garstand and de Beer: The First Century of Evolutionary Biology', *American Zoologist* 40 (2000), 718–28.

Haller, Mark H. *Eugenics: Hereditarian Attitudes in American Thought* (New Brunswick, NJ, 2008).

Hansson, Sven Ove. 'What is Technological Science?', *Studies in History and Philosophy of Science* 38 (2007), 523–7.

Harbou, Thea von. *Metropolis* (Berlin, 1926).

Harbou, Thea von. *Die Frau im Mond* (Berlin, 1928).

Harman, Peter M. *Energy, Force, and Matter: The Conceptual Development of Nineteenth-Century Physics* (Cambridge, 1982).

Harman, Peter M. *The Natural Philosophy of James Clerk Maxwell* (Cambridge, 1998).

Harper, Robert S. 'The First Psychological Laboratory', *Isis* 41 (1950), 158–61.

Harris, Henry. *The Birth of the Cell* (New Haven, CT, 1999).

Harrison, Peter. *The Bible, Protestantism, and the Rise of Natural Science* (Cambridge, 1998).

Harrison, Peter. *The Fall of Man and the Foundations of Science* (Cambridge, 2007).

Hartmann, Nicolai. 'Diesseits von Idealismus und Realismus. Ein betrag zur Scheidung des Geschichtlichen und Übergeschichtlichen in der Kantischen Philosophie', *Kant-Studien* 29 (1924), 160–206.

Hase, Johann Matthias. *Historiae universalis politicae quantum ad eius partemn i. ac ii. idea plane nova et legitima tractationem summorumn imperiorum etc.* (Nuremberg, 1743).

Hecht, Jennifer Michael. *The End of the Soul: Scientific Modernity, Atheism, and Anthropology in France* (New York, 2003).

Heffer, Simon. *High Minds: The Victorians and the Birth of Modern Britain* (London, 2013).

Heffernan, William C. 'Percival Lowell and the Debate over Extraterrestrial Life', *Journal of the History of Ideas* 42 (1981), 527–30.

Heidegger, Martin. *Sein und Zeit* (Tübingen, 1927).

Heidegger, Martin. *Kant und das Problem der Metaphysik* (Bonn, 1929).

Heidegger, Martin. *Nietzsche* (2 vols, Pfullingen, 1961).

Heidegger, Martin. *Kant and the Problem of Metaphysics* (Bloomington, 1962).

Heilbron John L. and Thomas S. Kuhn. 'The Genesis of the Bohr Atom', *Historical Studies in the Physical Sciences* 1 (1969), 211–90.

Helmholtz, Hermann von. 'Ueber den Stoffverbrauch in der Muskelaktion', *Archiv für Anatomie, Physiologie und Wissenschaftliche Medicin* (1845), 72–83.

Helmholtz, Hermann von. *Über die Erhaltung der Kraft: eine physikaliscke Abhandlung* (Berlin, 1847).

Helmholtz, Hermann von. 'Über discontinuirliche Flüssigkeits-Bewegungen', *Monatsbericht der königlich preussischen Akademie des Wissenschaften zu Berlin* (1868), 215–28.

Helmholtz, Hermann von. *Vorträge und Reden* (3rd edition, 2 vols, Berlin, 1884).

Helmholtz, Hermann von. *Science and Culture*, ed. David Cahan (Chicago, 1995).

Henderson, James. *Early Mathematical Economics* (Boulder, CO, 1996).

Henslow, George. *Plants of the Bible* (London, 1896).

Hepp, Noémi. *Homère en France au XVIIe siècle* (Paris, 1968).

Herren, Madeleine, Martin Rüesch, and Christiane Sibille. *Transcultural History: Theories, Methods, Sources* (Berlin, 2012).

Herschel, John. *A Preliminary Discourse on the Study of Natural Philosophy* (London, 1830).

[Herschel, John]. 'Mechanism of the Heavens', *Quarterly Review* 47 (1832), 537–5.

Hesketh, Ian. *The Science of History in Victorian Britain: Making the Past Speak* (London, 2011).

Hesse, Mary. *Models and Analogies in Science* (Notre Dame, IN, 1966).

Hilton, Boyd. *The Age of Atonement: The Influence of Evangelicalism on Social and Economic Thought, 1795–1865* (Oxford, 1988).

Hilts, Victor L. 'Aliis Exterendum, or, the Origins of the Statistical Society of London', *Isis* 69 (1978), 21–43.

Hindle, Brooke. *Emulation and Invention* (New York, 1981).

Hindle, Brooke and Steven Lubar. *Engineers of Change: The American Industrial Revolution 1790–1860* (Washington, DC, 1986).

His, Wilhelm. *Unsere Körperform und das physiologische Problem ihrer Entstehung* (Leipzig, 1874).

His, Wilhelm. 'On the Principles of Animal Morphology', *Proceedings of the Royal Society of Edinburgh* 15 (1888), 287–98.

Hofmann, Johann Valentin. *Somatologie oder Lehre von den inneren Beschaffenheit der Korper auf Grund einer vergleichenden Betrachtung der chemischen, morphologischen und physikalischen Eigenschaften derselben* (Gottingen, 1863).

Hollinger, David A. 'Science as a Weapon in *Kulturkämpfe* in the United States During and After World War II', *Isis* 86 (1995), 440–54.

Hollinger, David A. *Science, Jews, and Secular Culture: Studies in Mid-Twentieth Century American Intellectual History* (Princeton, NJ, 1996).

Holmes, Frederic Lawrence. *Claude Bernard and Animal Chemistry: The Emergence of a Scientist* (Cambridge, MA, 1974).

Holmes, Frederic Lawrence. 'The Complementarity of Teaching and Research in Leibig's Laboratory', *Osiris* (1989), 121–64.

Holzhey, Helmut. *Cohen und Natorp, Band 1: Ursprung und Einheit* (Basle, 1986).

Holzhey, Helmut. *Cohen und Natorp, Band 2: Der Marburger Neukantianismus in Quellen* (Basle, 1986).

Hopwood, Nick. *Haeckel's Embryos: Images, Evolution, Fraud* (Chicago, 2015).

Hossenfelder, Sabine. *Lost in Math: How Beauty leads Physics Astray* (New York, 2018).

Houghton, William. *The Microscope and Some of the Wonders it Reveals* (London, [1871]).

Houghton, William. *Sketches of British Insects: A Handbook for Beginners in the Study of Entomology* (London, 1875).

Houghton, William. *British Fresh-Water Fishes* (London, 1879).

Hovelacque, Abel. *La linguistique* (Paris, 1892).

Hovelacque, Abel, Charles Issaurat, André Lefevre, Charles Letourneau, Gabriel de Mortillet, Henri Thulié, and Eugene Véron, *Dictionnaire des sciences anthropologique* (Paris, 1889).

Hsia, Florence. *Sojourners in a Strange Land: Jesuits and Their Scientific Missions in Late Imperial China* (Chicago, 2009).

Hubert, René. *Les Sciences sociales dans l'Encyclopédie* (Paris, 1923).

Hufeland, Christoph Wilhelm. *Ideen über Pathologie und Einfluß der Lebenskraft auf Entstehung und Form der Krankheit* (Jena, 1795).

Hughes, Jeff. 'Unity through Experiment? Reductionism, Rhetoric and the Politics of Nuclear Science, 1918–40', in H. Kamminga and G. Somsen, eds, *Pursuing the Unity of Science: Ideology and Scientific Practice from the Great War to the Cold War* (London, 2016), 50–81.

Hughes, Thomas P. *Networks of Power: Electrification in Western Society 1880–1930* (Baltimore, 1983).

Hughes, Thomas P. *American Genesis: A Century of Invention and Technological Enthusiasm, 1870–1970* (Chicago, 2004).

Hull, David. *Darwin and his Critics: The Reception of Darwin's Theory of Evolution By The Scientific Community* (Cambridge, MA, 1983).

Humboldt, Alexander von. *Aphorismen aus der chemischen Physiologie der Pflanzen* (Leipzig, 1794).

Humboldt, Alexander von. *Versuche über die gereizte Muskel- und Nervenfaser* (2 vols, Posen and Berlin, 1797–9).

Humboldt, Alexander von. *Essai géognostique sur le gisement des roches dans les deux hémisphères* (Paris, 1823).

Humboldt, Wilhelm von. *Über die Kawi-Sprache auf der Insel Java, nebst einer Einleitung über die Verschiedenheit des menschlichen Sprachbaues und ihren Einfluss auf die geistige Entwickelung des Menschengeschlechts* (3 vols, Berlin, 1836–9).

Humboldt, Wilhelm von. *Gesammelte Schriften* (17 vols, Berlin, 1903–36).

Hume, David. *Essays and Treatises on Several Subjects* (2 vols, Edinburgh, 1793).

Hundert, Edward. *The Enlightenment's Fable: Bernard Mandeville and the Discovery of Society* (Cambridge, 1994).

Husserl, Edmund. 'Philosophie als strenge Wissenschaft', *Logos* 1 (1910), 289–314.

Husserl, Edmund. *The Crisis of European Sciences and Transcendental Phenomenology: An Introduction to Phenomenological Philosophy* (Evanston, IL, 1970).

Hutchinson, Henry Neville. *Extinct Monsters: A Popular Account of Some of the Larger Forms of Ancient Animal Life* (London, 1897).

Hutton, James. 'Theory of the Earth: or an investigation of the laws observable in the composition, dissolution, and restoration of land upon the Globe', *Transactions of the Royal Society of Edinburgh* vol. 1, part 2 (1788), 209–304.

Huxley, Julian. *Evolution, The Modern Synthesis* (London, 1942).

Huxley, Julian. *Memories* (2 vols, New York, 1970–3).

Huxley, Thomas Henry. *Man's Place in Nature* (London, 1863).

Huxley, Thomas Henry. *Lessons in Elementary Physiology* (3rd edition, London, 1872).

Huxley, Thomas Henry. *Lay Sermons, Addresses and Reviews* (London, 1877).

Huxley, Thomas Henry. *Science and Culture, and Other Essays* (London, 1881).

Huxley, Thomas Henry. *Collected Essays* (9 vols, New York, 1893–4).

Huxley, Thomas Henry. 'The Natural History of Creation—by Dr. Ernst Haeckel', *Academy* 1 (1896), 566–80.

Huxley, Thomas Henry. *Discourses Biological and Geological* (New York, 1897).

Huxley, Thomas Henry. *Science and Education* (New York, 1898).

Iggers, Georg G. 'The Idea of Progress in Historiography and Social Thought since the Enlightenment', in G. Almond, M. Chodorow, and R. Pearce, eds, *Progress and its Discontents* (Berkeley, 1982), 41–66.

Igo, Sarah E. *The Averaged American: Surveys, Citizens, and the Making of a Mass Public* (Cambridge, MA, 2007).

Inkster, Ian. 'The Social Context of an Educational Movement: A Revisionist Approach to the English Mechanics' Institutes, 1820–1850', *Oxford Review of Education* 2 (1976), 277–307.

Inkster, Ian. 'London Science and the Seditious Meetings Act of 1817', *The British Journal for the History of Science* 12 (1979), 192–6.

Intorcetta, Prospero. *Chum Yum: Sinarum scientia politico-moralis* (Canton and Goa, 1667–9).

Intorcetta, Prospero. *La Science des chinois* (Paris, 1673).

Isherwood, Benjamin Franklin. *Experimental Researches in Steam Engineering* (2 vols, Philadelphia, 1863).

Israel, Paul B. *Edison: A Life of Invention* (New York, 1998).

[Jacson, Maria]. *Botanical Dialogues, between Hortensia and her four Children . . .* (London, 1799).

Jaeger, Siegfried. 'Origins of Child Psychology: William Preyer', in William R. Woodward and Mitchell G. Ash, eds, *The Problematic Science: Psychology in Nineteenth-Century Thought* (New York, 1982), 300–21.

James, Frank. 'Running the Royal Institution: Faraday as an Administrator', in Frank James, ed., '*The Common Purposes of Life': Science and Society at the Royal Institution of Great Britain* (Aldershot, 2002), 119–46.

James, William. *Pragmatism: A New Name for an Old Way of Thinking* (Buffalo, NY, 1991).

Janich, Peter. *Zweck und Methode der Physik aus philosophischer Sicht* (Konstanz, 1973).

Janich, Peter. 'Physics—Natural Science or Technology', in W. Krohn, E. Layton Jr, and P. Weingart, eds, *The Dynamics of Science and Technology* (Dordrecht, 1978), 3–27.

Jaspers, Karl. *Die geistige Situation der Zeit* (Berlin, 1931).

Jeans, James. *The Mysterious Universe* (Cambridge, 1930).

Jeremy, David. *Artisans, Entrepreneurs and Machines: Essays on the Early Anglo-American Textile Industries* (Aldershot, 1998).

Jevons, W. Stanley. *The Theory of Political Economy* (London, 1871).

Johns, Charles Alexander. *Flora Sacra; or, The Knowledge of the Works of Nature Conducive to the Knowledge of the God of Nature* (London, 1840).

Johns, Charles Alexander. *First Steps to Botany* (London, 1853).

Johns, Charles Alexander. *Birds' Nests* (London, 1854).

Jones, R. T. *Classical Aerodynamic Theory*, NASA Reference Publication 1050, 1979.

Jones, Richard. *An Essay on the Distribution of Wealth and on the Sources of Taxation* (London, 1831).

Jones, Richard. *Literary Remains, consisting of Lectures and Tracts on Political Economy* (London, 1859).

Jones, William. 'On the Gods of Greece, Italy and India', *Asiatick Researches* 1 (1788), 221–75.

Jørgensen, Bent S. 'Berzelius und die Lebenskraft', *Centaurus* 19 (1964), 258–81.

Joule, James Prescott. 'Description of an Electro-Magnetic Engine', *Annals of Electricity, Magnetism, and Chemistry; and Guardian of Experimental Science* 3 (1838), 122–3.

Joule, James Prescott. 'On the Calorific Effects of Magneto-Electricity, and on the Mechanical Value of Heat', *Philosophical Magazine* 23 (1843), 263–76, 347–55, 435–43.

Joule, James Prescott. 'On the Changes of Temperature Produced by the Rarefaction and Condensation of Air', *Philosophical Magazine* 26 (1844), 369–83.

Joyce, Richard. *The Myth of Morality* (Cambridge, 2001).

Jungnickel, Christa and Russell McCormmach, *Intellectual Mastery of Nature: Theoretical Physics from Ohm to Einstein* (2 vols, Chicago, 1986).

Kahn, Hermann. *On Thermonuclear War* (Princeton, NJ, 1960).

Kahneman, Daniel and Amos Tversky. 'On the Psychology of Prediction', *Psychological Review* 80.4 (1973), 237–51.

Kahneman, Daniel and Amos Tversky. 'Prospect Theory: An Analysis of Decision under Risk', *Econometrica* 47.2 (1979), 263–91.

Kaiser, David. *Drawing Theories Apart: The Dispersion of Feynman Diagrams in Postwar Physics* (Chicago, 2005).

Kamminga, Harmke and Geert Somsen. 'Introduction', in Harmke Kamminga and Geert Somsen, eds, *Pursuing the Unity of Science: Ideology and Scientific Practice from the Great War to the Cold War* (London, 2016), 1–11.

Kant, Immanuel. *Gesammelte Schriften* ('Akademie' edition, 29 vols, Berlin, 1900 onwards).

Kant, Immanuel. *Critique of the Power of Judgement* (Cambridge, 2000).

Kapila, Shruti. 'The Enchantment of Science in India', *Isis* 101 (2010), 120–32.

Kay, Lily E. *The Molecular Vision of Life: Caltech, the Rockefeller Foundation, and the Rise of the New Biology* (Oxford, 1993).

Keene, Melanie. *Science in Wonderland: The Scientific Fairy Tales of Victorian Britain* (Oxford, 2015).

Keilin, David. *History of Cell Respiration and Cytochrome* (Cambridge, 1966).

Keith, Arthur. *Darwinism and What it Implies* (London, 1928).

Kekulé, Friedrich August. 'Ueber die Constitution und die Metamorphosen der chemische Verbindungen und über die chemische Natur des Kohlenstoffs', *Annalen der Chemie und Pharmacie* 106 (1858), 129–59.

Kent, Christopher. *Brains and Numbers: Elitism, Comtism, and Democracy in Mid-Victorian England* (Toronto, 1978).

Kepler, Johannes. *Somnium, seu Opus Postvmvm de Astronomia Lvnari, Divulgatum à M. Ludovici Kepplero Filio* (Frankfurt, 1634).

Kerr, J. Graham. 'Biology and the Training of the Citizen', *Nature* 118 (1926), 102–12.

Kertzer, David I. *The Pope Who Would be King: The Exile of Pius IX and the Emergence of Modern Europe* (Oxford, 2018).

Kevles, Daniel J. *The Physicists: The History of a Scientific Community in Modern America* (Cambridge, MA, 1971).

Kevles, Daniel J. 'The Physics, Mathematics, and Chemistry Communities: A Comparative Analysis', in Alexandra Oleson and John Voss, eds, *The Organization of Knowledge in America* (Baltimore, 1979), 139–72.

Kielmeyer, Carl Friedrich. *Über die Verhältniße der organischen Kräfte unter einander in der Reihe der verschiedenen Organisationen, die Gesetze und Folgen dieser Verhältniße* (Stuttgart, 1793).

Kim, Jaegwon. 'Making Sense of Emergence', *Philosophical Studies* 95 (1999), 3–36.

Kim, Mi Gyung. *Affinity, That Elusive Dream: A Genealogy of the Chemical Revolution* (Cambridge, MA, 2003).

Kim, Sangkeun. *Strange Names of God: The Missionary Translation of the Divine Name and the Chinese Responses in Late Ming China, 1583–1644* (New York, 2004).

Kingsley, Charles. *Water Babies* (London, 1863).

Kingsley, Charles. *Scientific Lectures and Essays* (London, 1890).

Kingsley, Charles. *The Water-Babies and Glaucus* (London, 1908).

Kipnis, Naum. 'Luigi Galvani and the Debates on Animal Electricity, 1791–1800', *Annals of Science* 44 (1987), 107–42.

Kirchhoff, Gustav. *Über das Ziel der Naturwissenschaften* (Heidelberg, 1865).

Kirchhoff, Gustav. 'Zur Theorie freier Flüssigkeitsstrahlen', *Crelle's Journal für reine und angewandte Mathematik* 70 (1869), 289–98.

Kirchhoff, Gustav. *Gesammelte Abhandlungen* (Leipzig, 1882).

Klein, Ursula. 'Techniques of Modelling and Paper-Tools in Classical Chemistry', in Mary S. Morgan and Margaret Morrison, eds, *Models as Mediators: Perspectives on Natural and Social Science* (Cambridge, 1999), 146–67.

Klein, Ursula. *Experiments, Models, Paper Tools: Cultures of Organic Chemistry in the Nineteenth Century* (Stanford, 2003).

Klein, Ursula. *Humboldt's Preußen: Wissenschaft und Technik in Aufbruch* (Darmstadt, 2015).

Klein, Ursula. *Nützliches Wissen: Der Erfindung der Technikwissenschaften* (Göttingen, 2017).

Klein, Ursula and Wolfgang Lefèvre, *Materials in Eighteenth-Century Science: A Historical Ontology* (Cambridge, MA, 2007).

Kline, Ronald. 'Construing "Technology" as "Applied Science": Public Rhetoric of Scientists and Engineers in the United States, 1880–1945', *Isis* 86 (1995), 194–221.

Klineberg, Otto. *An Experimental Study of Speed and Other Factors in 'Racial' Differences* (New York, 1928).

Knight, David. 'Getting Science Across', *British Journal of the History of Science* 29 (1997), 129–38.

Knight, David. 'Scientists and Their Publics: Popularization of Science in the Nineteenth Century', in Mary Jo Nye, ed., *The Cambridge History of Nineteenth and Twentieth Century Science* (Cambridge, 2003), 72–90.

Knight, David. *The Making of Modern Science* (Cambridge, 2009).

Kochmann, Wilhelm. 'Das Taylorsystem und seine volkswirtschaftliche Bedeutung', *Archiv für Sozialwissenschaft und Sozialpolitik* 38 (1914), 391–424.

Koerner, Lisbet. *Linnaeus: Nature and Nation* (Harvard, MA, 1999).

Kohler, Robert E. *All Creatures: Naturalists, Collectors, and Biodiversity, 1850–1950* (Princeton, NJ, 2006).

Köhnke, Klaus. *The Rise of Neo-Kantianism: German Academic Philosophy between Idealism and Positivism* (Cambridge, 1991).

Koyré, Alexandre. *Essai sur l'idée de Dieu et les preuves de son existence chez Descartes* (Paris, 1922).

Koyré, Alexandre. *Études galiléennes* (Paris, 1939).

Koyré, Alexandre. 'A Documentary History of the Problem of Fall from Kepler to Newton', *Transactions of the American Philosophical Society* 45 (1955), 329–95.

Kracauer, Siegfried. *The Mass Ornament: Weimar Essays* (Cambridge, MA, 1995).

Kragh, Helge. *Niels Bohr and the Quantum Atom: The Bohr Model of Atomic Structure, 1913–1925* (Oxford, 2012).

Kranakis, Eda Fowlks. 'The French Connection: Giffard's Injector and the Nature of Heat', *Technology and Culture* 23 (1982), 3–38.

Krementsov, Nikolai. *With and Without Galton: Vasilii Florinskii and the Fate of Eugenics in Russia* (Cambridge, 2018).

Kremer, Richard L. 'Physiology', in Peter Bowler and John Pickstone, eds, *The Cambridge History of Science, Volume 6: The Modern Biological and Earth Sciences* (Cambridge, 2009), 342–66.

Krige, John and Dominique Pestre. 'Some Thoughts on the Early History of CERN', in Peter Galison and Bruce Helvy, eds, *Big Science: The Growth of Large-Scale Research* (Stanford, CA, 1992), 78–99.

Kuhn, Thomas. 'Energy Conservation as an Example of Simultaneous Discovery', in M. Clagett, ed., *Critical Problems in the History of Science* (Madison, 1959), 321–56.

Kuhn, Thomas. *Black-Body Radiation and the Quantum Discontinuity, 1894–1912* (Chicago, 1978).

Lachapelle, Sophie. *Conjuring Science: A History of Scientific Education and Stage Magic in Modern France* (London, 2015).

Lagrange, Joseph Louis de. *Mécanique analytique* (Paris, 1788).

Lagrange, Joseph Louis de. *Mécanique analytique, nouvelle édition* (Paris, 1811).

Lahy, Jean-Marie. *Le Système Taylor et la physiologie du travail professionnel* (Paris, 1916).

Laks, André. *The Concept of Presocratic Philosophy: Its Origin, Development, and Significance* (Princeton, NJ, 2018).

Lalande, Jerôme. *Article pour les cahiers dont les 36 rédacteurs sont prier instament et requis expressément de faire usage* ([Paris], 1789).

Lamarck, Jean-Baptiste-Pierre-Antoine de Monet de. *Recherches sur les causes des principaux faits physiques* (2 vols, Paris, 1794).

Lamarck, Jean-Baptiste-Pierre-Antoine de Monet de. *Recherches sur l'organisation des corps vivans et particulièrement sur leur origine, sur la cause de ses développemens et des progrès de sa composition, et sur celle qui, tendant continuellement à la détruire dans chaque individu, amène nécessairement sa mort; précéde du discours d'ouverture du cours de zoologie, donné dans le Muséum national d'Histoire Naturelle* (Paris, 1802).

Lamarck, Jean-Baptiste-Pierre-Antoine de Monet de. *Philosophie zoologique, ou exposition des considérations relatives à l'histoire naturelle des animaux; à la diversité de leur organisation et des facultés qu'ils en obtiennent; aux causes physiques qui maintiennent en eux la vie et donnent lieu aux mouvemens qu'ils exécutent; enfin, à celles qui produisent les unes le sentiment et les autres l'intelligence de ceux qui en sont doués* (2 vols, Paris, 1809).

Lamb, Horace. *A Treatise on the Mathematical Theory of the Motion of Fluids* (Cambridge, 1879).

Lamb, Horace. *Hydrodynamics* (5th edition, Cambridge, 1924).

Lambert, Claude-François. *Histoire générale, civile, naturelle, politique et religieuse de tous les peuples du monde, Avec des observations sur les mœurs, les coutumes, les usages, les caracteres, les differentes langues, le gouvernement . . . les arts & les sciences des différents peuples de l'Europe, de l'Asie, de l'Afrique & de l'Amerique* (15 vols, Paris, 1750).

Lanchester, Frederick. *Aerodynamics, constituting the First Volume of a Complete Work on Aerial Flight* (London, 1907).

Lange, Friedrich Albert. *Die Arbeitfrage: Ihre Bedeutung für Gegenwart und Zukunft* (Leipzig, 1910).

Langlès, Louis Mathieu. *Fables et contes indiens: Nouvelles traduits, avec un discours préliminaire et des notes sur la religion, la littérature, les moeurs, &c. des Hindoux* (Paris, 1790).

Lankester, E. Ray. *Science from an Easy Chair* (London, 1910).

Laplace, Pierre-Simon. *Exposition du système du monde* (2 vols, Paris, 1796).

Laplace, Pierre-Simon. *Traité de mécanique céleste* (5 vols, Paris, 1799–1825).

Largent, Mark A. *Breeding Contempt: The History of Coerced Sterilization in the United States* (New Brunswick, NJ, 2008).

Larmore, Joseph. 'Lord Kelvin', *Proceedings of the Royal Society* 81 (1908), iii–lxxvi.

Larrère, Catherine. 'Mirabeau et les physiocrates: l'origine agrarienne de la civilisation', in Bertrand Binoche, ed., *Les Équivoques de la civilisation* (Seyssel, 2005), 83–105.

Lasswitz, Kurd. *Auf zwei Planeten* (2 vols, Leipzig, 1897).

Latour, Bruno. *Science in Action: How to Follow Scientists and Engineers through Society* (Cambridge, MA, 1987).

Laubichler, Manfred D. 'Does History Recapitulate Itself? Epistemological Reflections on the Origins of Evolutionary Developmental Biology', M. D. Laubichler and J. Maienschein, eds, *From Embryology to Evo-Devo: A History of Developmental Evolution* (Cambridge, MA, 2007), 13–33.

Laudan, Larry. *Science and Hypothesis: Historical Essays on Scientific Methodology* (Dordrecht, 1981).

Laughlin, Harry H. *Eugenical Sterilization in the United States* (Chicago, 1922).

Lavoisier, Antoine-Laurent de. *Oeuvres de Lavoisier publiées par les soins du Ministère de l'Instruction Publique* (6 vols, Paris, 1864–93).

Le Roy, Loys. *De la vicissitude ou variété des choses en l'univers, et concurrence des armes et des lettres par les premieres et plus illustres nations du monde, depuis le temps où a commencé la civilité, et memoire humain jusques à presente* (Paris, 1575).

Leary, David E. 'The Fate and Influence of John Stuart Mill's Proposed Science of Ethology', *Journal of the History of Ideas* 43 (1982), 153–62.

Ledeboer, John Henry. 'The Function of Literature', *Aeronautics* 11 (1916), 33.

Lefèvre, André. *La renaissance de matérialisme* (Paris, 1881).

Legendre, Adrien Marie. *Nouvelles méthodes pour la détermination des orbites des comètes* (Paris, 1805).

Leibniz, Gottfried Wilhelm. *Leibniz Korrespondiert mit China*, ed. R. Widmaier (Frankfurt, 1990).

Lenoir, Timothy. *The Strategy of Life: Teleology and Mechanics in Nineteenth-Century German Biology* (Chicago, 1982).

Lepenies, Wolf. *Das Ende der Naturgeschichte: Wandel kultureller Selbstverständlichkeiten in den Wissenschaft des 18. und 19. Jahrhunderts* (Munich, 1976).

Lepenies, Wolf. *The Seduction of Culture in German History* (Princeton, NJ, 2006).

Lesch, John E. *Science and Medicine in France: The Emergence of Experimental Physiology, 1790–1855* (Cambridge, MA, 1984).

Lessing, Gotthold Ephraim. *Werke* ed. H. G. Gölpert et al. (8 vols, Berlin, 1978).

Letourneau, Charles. *La biologie* (Paris, 1877).

Letourneau, Charles. *La sociologie d'après l'ethnographie* (Paris, 1884).

Leuckart, Rudolph. *Über die Morphologie und Verwandtschaftsverhältnisse der wirbellosen Thiere: Ein Beitrag zur Charakteristik und Classification der thierischen Formen* (Braunschweig, 1848).

Levine, Philippa and Alison Bashford, 'Introduction: Eugenics and the Modern World', in Alison Bashford and Philippa Levine, eds, *The Oxford Handbook of the History of Eugenics* (Oxford, 2010), 5–24.

Lewes, George Henry. *Studies in Animal Life* (London, 1862).

Liebert, Arthur. *Die geistige Krisis der Gegenwart* (Berlin, 1924).

Liebig, Justus. *Die organische Chemie in ihrer Anwendung auf Agricultur und Physiologie* (Braunschweig, 1840).

Liebig, Justus. *Die organische Chimie in ihrer Anwendung auf Physiologie und Pathologie* (Braunschweig, 1842).

Liebig, Justus. 'Das Verhältniß der Physiologie und Pathologie zur Chemie und Physik, und die Methode der Forschung in diesen Wissenschaften', *Deutsche Vierteljahrs Schrift* 3 (1846), 169–243.

Lightman, Bernard. *Victorian Popularizers of Science: Designing Nature for a New Audience* (Chicago, 2007).

Lindley, David. *Boltzmann's Atom: The Great Debate that Launched a Revolution in Physics* (New York, 2001).

Lindsay, David. *Voyage to Arcturus* (London, 1920).

Link, Jürgen. *Versuch über den Normalismus: Wie Normalität produziert wird* (Göttingen, 2013).

Linstrum, Eric. *Ruling Minds: Psychology in the British Empire* (Cambridge, MA, 2016).

Lloyd, G. E. R. *Adversaries and Authorities: Investigations into Ancient Greek and Chinese Science* (Cambridge, 1996).

Lodge, Oliver. *Man and the Universe: A Study of the Influence of the Advance in Scientific Knowledge upon our Understanding of Christianity* (London, 1908).

Lombroso, Cesare. *L'uomo delinquente in rapporto all'antropologia, alla giurisprudenza ed alle discipline carcerarie* (Turin, 1889).

Loschmidt, Johann Josef. 'Zur Grösse der Luftmoleküle', *Sitzungsberichte der kaiserlichen Akademie der Wissenschaften Wien* 52 (1865), 393–413.

Loudon, Jane. *The Mummy: A Tale of the Twenty-Second Century* (London, 1827).

Loudon, Jane. *The Young Naturalist's Journey; or, The Travels of Agnes Merton and Her Mama* (London, 1840).

Loudon, Jane. *The First Book of Botany: Being a Plain and Brief Introduction to That Science, for Students and Young Persons* (London, 1841).

Lowell, Percival. *Mars* (New York, 1895).

Lowell, Percival. *Mars and its Canals* (New York, 1906).

Lowell, Percival. *Mars as an Abode of Life* (New York, 1908).

Löwith, Karl. *Von Hegel zu Nietzsche: Der revolutionäre Bruch im Denken des 19. Jahrhunderts* (Zurich, 1941).

Lubbock, John W. *On Currency* (London, 1840).

Lucier, Paul. 'Geological Industries', in Peter Bowler and John Pickstone, eds, *The Cambridge History of Science, Volume 6: The Modern Biological and Earth Sciences* (Cambridge, 2009), 108–25.

Ludwig, Carl. *Lehrbuch der Physiologie des Menschen* (2 vols, Leipzig, 1852–6).

Ludwig, Carl. 'Zur Ablehnung der Anmuthungen in Herrn R. Wagner in Göttingen', *Zeitschrift für rationelle Medizin* 5 (1854), 269–74.

Luft, Sebastian. *The Space of Culture: Towards a Neo-Kantian Philosophy of Culture (Cohen, Natorp, and Cassirer)* (Oxford, 2015).

Lyell, Charles. *Principles of Geology: being an Inquiry how far the former changes of the Earth's surface are referable to causes now in operation* (3 vols, London, 1830–3).

McCabe, Joseph. *Evolution: A General Sketch from Nebula to Man* (London, 1910).

McCabe, Joseph. *The Existence of God* (London, 1934).

McClelland, Charles E. *State, Society and University in Germany, 1700–1914* (Cambridge, 1980).

McConnell, Anita. 'Instruments and Instrument-Makers, 1700–1850', in Jed Z. Buchwald and Robert Fox, eds, *The Oxford Handbook of the History of Physics* (Oxford, 2013), 326–57.

McCormmach, Russell. 'H. A. Lorenz and the Electromagnetic View of Nature', *Isis* 61 (1970), 459–97.

McGrayne, Sharon Bertsch. *Prometheans in the Lab: Chemistry and the Making of the Modern World* (New York, 2001).

McNeill, William H. *The Pursuit of Power: Technology, Armed Force, and Society Since A. D. 1000* (Chicago, 1982).

Mach, Ernst. *Beträge zur Analyse der Empfindungen* (Jena, 1886).

Mach, Ernst. *Die Principien der Wärmelehre* (Leipzig, 1896).

Mach, Ernst. *The Analysis of Sensations: And the Relation of the Physical to the Psychical* (New York, 1959).

Mach, Ernst. *Knowledge and Error: Sketches on the Psychology of Enquiry* (Dordrecht, 1976).

Mach, Ernst. *Principles of the Theory of Heat: Historically and Critically Elucidated* (Dordrecht, 1986).

Mackintosh, James. *Dissertation on the Progress of Ethical Philosophy, Chiefly During the Seventeenth and Eighteenth Centuries* (Edinburgh, 1830).

MacLeod, Roy. 'Evolutionism, Internationalism, and Commercial Enterprise in Science: The International Science Series, 1871–1910', in A. J. Matthews, ed., *Development of Science Publishing in Europe* (Amsterdam, 1980), 63–93.

Macleod, Roy. 'The "Bankruptcy of Science" Debate: The Creed of Science and its Critics', *Science, Technology, and Human Values* 7:4 (1982), 2–15.

MacLeod, Roy. 'Whigs and Savants: Reflections on the Reform Movement in the Royal Society, 1830–1848', in I. Inkster and J. Morell, eds, *Metropolis and Province: Science in British Culture, 1780–1850* (London, 1983), 55–90.

MacLeod, Roy. 'Der wissenschaftliche Internationalismus in der Krise: Die Akademien der Alliierten und ihre Reaktion auf den Ersten Weltkrieg', in Notker Hammerstein et al., *Die Preussische Akademie der Wissenschaften zu Berlin, 1914–1945* (Berlin, 2000), 317–49.

Magendie, François. *Précis élémentaire de physiologie* (2 vols, Paris, 1836).

Magnus, Gustav. 'Ueber die im Blute enthaltenen Gase, Sauerstoff, Stickstoff und Kohlensäure', *Annalen der Physik und Chemie* 40.4 (1837), 583–605.

Maier, Charles S. 'Between Taylorism and Technocracy: European Ideology and the Vision of Industrial Productivity in the 1920s', *Journal of Contemporary History* 5.2 (1970), 27–61.

Mairan, Jean-Baptiste Dortous de. *Lettres de M. de Mairan, au R. P. Parrenin, Missionaire de la Compagnie de Jesus, à Pekin. Contenant diverses Questions sur la Chine* (Paris, 1769).

Malter, Rudolph. 'Main Currents of the German Interpretation of the *Critique of Pure Reason* since the Beginnings of Neo-Kantianism', *Journal of the History of Ideas* 42 (1981), 531–51.

Malthus, Thomas. *On the Principle of Population, as it affects the Future Improvement of Society. With Remarks on the Speculations of Mr Godwin, M. Condorcet, and other writers* (London, 1798).

Mandeville, Bernard. *The Fable of the Bees* (London, 1714).

Maquet, Louis. 'Condorcet et la création du système métrique décimal', in Pierre Crèpel and Christian Gilian, eds, *Condorcet, mathématicien, économiste, philosophe, homme politique* (Paris, 1989), 52–62.

Marchionni, Caterina. 'Explanatory Pluralism and Complementarity: From Autonomy to Integration', *Philosophy of the Social Sciences* 38 (2008), 314–33.

Marsden, Ben. 'Ranking Rankine: W. J. M. Rankine (1820–72) and the Making of "Engineering Science" Revisited', *History of Science* 51 (2013), 434–56.

[Martin, William]. *Popular Introduction to the Study of Quadrupeds; with a particular notice of those mentioned in Scripture* (London, 1833).

[Martin, William]. *Introduction to the Study of Birds, with a Particular Notice of the Birds Mentioned in Scripture* (London, 1835).

[Martin, William]. *Popular History of Reptiles, or Introduction to the Study of Class Reptilia, on Scientific Principles* (London, 1842).

Martini, Martino. *Sinicae historiae decas prima* (Amsterdam, 1658).

Marx, Leo. *The Machine in the Garden: Technology and the Pastoral Ideal in America* (New York, 1964).

Maxwell, James Clerk. 'On Faraday's Lines of Force', *Transactions of the Cambridge Philosophical Society* 10 (1856), 27–83.

Maxwell, James Clerk. *On the Stability of the Motion of Saturn's Rings* (London, 1859).

Maxwell, James Clerk. 'Illustrations of the Dynamical Theory of Gases. Part I. On the Motions and Collisions of Perfectly Elastic Spheres', *Philosophical Magazine* 19 (1860), 19–32.

Maxwell, James Clerk. 'Illustrations of the Dynamical Theory of Gases. Part II. On the Process of Diffusion of Two or More Kinds of Moving Particles Among One Another', *Philosophical Magazine* 20 (1860), 21–37.

Maxwell, James Clerk. 'On Physical Lines of Force', *Philosophical Magazine and Journal of Science* 21 (1861), 161–75, 281–91, 338–48; 22 (1861), 12–24, 85–95.

Maxwell, James Clerk. 'Grove's "Correlation of Forces"', *Nature* 10 (20 August 1874), 302–4.

Maxwell, James Clerk. 'Are There Real Analogies in Nature?', in Lewis Campbell and William Garnett, *The Life of James Clerk Maxwell, with a Selection from his Correspondence and Occasional Writings and a Sketch of his Contributions to Science* (London, 1882), 235–44.

Maxwell, James Clerk. *The Scientific Papers of James Clerk Maxwell*, ed. W. D. Niven (New York, 1965).

Mayer, Julius Robert. *Die organische Bewegung in ihrem Zusammenhange mit dem Stoffwechsel. Ein Beitrag zur Naturkunde* (Heilbronn, 1845).

Mayr, Ernst. *Systematics and the Origin of Species* (New York, 1942).

Mayr, Ernst. *The Growth of Biological Thought: Diversity, Evolution, and Inheritance* (Cambridge, MA, 1992).

Mayr, Ernst. 'What Was the Evolutionary Synthesis?', *Trends in Ecology and Evolution* 8 (1993), 31–4.

Meckel, Johann. *System der vergleichenden Anatomie* (5 vols, Halle, 1821).

Mendelsohn, Everett. *Heat and Life: The Development of the Theory of Animal Heat* (Cambridge, MA, 1964).

Menzies, Peter, and Huw Price, 'Causation as a Secondary Quality', *British Journal for the Philosophy of Science* 42 (1991), 157–76.

Merrill, Lynn L. *The Romance of Victorian Natural History* (Oxford, 1989).

Merz, John Theodore. *A History of European Thought in the Nineteenth Century* (4 vols, Edinburgh and London, 1904–12).

Meyer, Lucy Ryder. *Real Fairy Folks, Or, the Fairyland of Chemistry: Explorations of the World of Atoms* (Boston, 1887).

Milgate, Murray and Shannon C. Stimson. *Ricardian Politics* (Princeton, NJ, 1991).

Mill, John Stuart. *Collected Works of John Stuart Mill* (33 vols, Toronto, 1963–91).

Miller, Arthur. *Imagery in Scientific Thought* (Cambridge, MA, 1986).

Millikan, Robert. *The Autobiography of Robert A. Millikan* (New York, 1950).

Milne-Edwards, Henri. *Outlines of Anatomy and Physiology* (Boston, 1841).

Milne-Edwards, Henri. 'Considerations sur quelques principes relatifs à la classification naturelle des animaux', *Annales des sciences*, 3rd series, 1 (1844), 65–99.

Mirabeau, Victor Riqueti, marquis de. *L'Ami des hommes, ou Traité de la population* (2 vols, Avignon, 1756–8).

Mirowski, Philip. *More Heat than Light: Economics as Social Physics, Physics as Nature's Economics* (Cambridge, 1989).

Mitchell, Sandra D. *Biological Complexity and Integrative Pluralism* (Cambridge, 2003).

Mitchell, Sandra D. *Unsimple Truths: Science, Complexity, and Policy* (Chicago, 2009).

Mokyr, Joel. *The Lever of Riches: Technological Creativity and Economic Progress* (New York, 1990).

Mokyr, Joel. 'The Second Industrial Revolution, 1870–1914', in V. Castronovo, ed., *Storia dell'Economia Mondiale* (Rome, 1999), 219–45 (http://www.faculty.ecn.northwestern.edu/faculty/mokyr/castronovo.pdf).

Mokyr, Joel. *A Culture of Growth: The Origins of the Modern Economy* (Princeton, NJ, 2017).

Moleschott, Jacob. *Die Physiologie der Nahrungsmittel* (Giessen 1850).

Moleschott, Jacob. *Der Kreislauf des Lebens: Physiologische Antworten auf Liebig's 'Chemische Briefe'* (Mainz, 1852).

Montesquieu, Charles de Secondat, Baron de. *De l'esprit des lois: ou Du rapport que les loix doivent avoir avec la constitution de chaque gouvernement, les moeurs, le climat, la religion, le commerce, &c.* (Geneva, 1748).

Montesquieu, Charles de Secondat, Baron de. *Oeuvres complètes*, ed. R. Caillois (2 vols, Paris, 1949).

Moras, Joachim. *Ursprung und Entwicklung des Begriffs der Zivilisation in Frankreich (1756–1830)* (Hamburg, 1930).

Morel, Charlotte. 'La querelle du matérialisme: présentation historique', in Charlotte Morel, ed., *L'Allemagne et la querelle du matérialisme (1848–1866)* (Paris, 2017), 9–43.

Morel, Charlotte. 'Métaphysique et science chez Lotze: Enjeu d'une position médiane dans la querelle du materialisme', in Charlotte Morel, ed., *L'Allemagne et la querelle du matérialisme (1848–1866)* (Paris, 2017), 129–54.

Morrell, Jack and Arnold Thackray. *Gentlemen of Science: Early Years of the British Association for the Advancement of Science* (Oxford, 1981).

Morrison, Margaret. 'Models as Autonomous Agents', Mary S. Morgan and Margaret Morrison, eds, *Models as Mediators: Perspectives on Natural and Social Science* (Cambridge, 1999), 38–65.

Mortillet, Gabriel de. *La préhistorique* (Paris, 1883).

Morton, Peter. *'The Busiest Man in England': Grant Allen and the Writing Trade, 1875–1900* (New York, 2005).

Morus, Iwan Rhys. *Shocking Bodies: Life, Death and Electricity in Victorian England* (London, 2011).

Morus, Iwan Rhys. *Frankenstein's Children: Electricity, Exhibition, and Experiment in Early-Nineteenth-Century London* (Princeton, 2014).

Moses, A. Dirk and Dan Stone. 'Eugenics and Genocide', in Alison Bashford and Philippa Levine, eds, *The Oxford Handbook of the History of Eugenics* (Oxford, 2010), 192–209.

Moynahan, Gregory B. *Ernst Cassirer and the Critical Science of Germany, 1899–1919* (London, 2013).

Müller, Johannes. *Handbuch der Physiologie des Menschen für Vorlesungen*, vol. 1, pt. 2 (4th edition, Coblenz, 1841).

Munson, Richard. *From Edison to Enron: The Business of Power and what it Means for the Future of Electricity* (Westport, CT, 2005).

Myers, J. L. 'The Beginnings of Science', in F. S. Marvin, ed., *Science and Civilization* (Oxford, 1923), 7–42.

Natorp, Paul. 'Kant und die Marburger Schule', *Kant-Studien* 17 (1912), 192–221.

Natorp, Paul. *Der Tage des Deutschen: Vier Kriegsaufsätze* (Hagen, 1915).

Natorp, Paul. *Philosophie—ihr Problem und ihre Probleme. Einführung in den kritischen Idealismus* (Göttingen, 2008).

Navarrete, Domingo. *The Travels and Controversies of Friar Domingo Navarrete, 1618–1686*, ed. J. S. Cummins (2 vols, Cambridge, 1962).

Needham, Joseph. *Science and Civilisation in China* (7 vols, Cambridge, 1954 onwards).

Needham, Paul. 'Has Daltonian Atomism Provided Chemistry with any Explanations?', *Philosophy of Science* 71 (2004), 1038–48.

Nelson, Daniel. *Frederick W. Taylor and the Rise of Scientific Management* (Madison, WI, 1980).

Neurath, Otto. 'Soziologie im Physikalismus', *Ekenntnis* 2 (1931/2), 393–431.

Neurath, Otto. *International Picture Language: The First Rules of Isotype* (London, 1936).

Neurath, Otto. 'Orchestration of the Sciences by the Encyclopedism of Logical Empiricism', *Philosophy and Phenomenological Research* 6 (1946), 496–508.

Neurath, Otto. 'Unified Science as Encyclopedic Integration', in O. Neurath, R. Carnap, and C. Morris eds, *Foundations of the Unity of Science: Toward an International Encyclopedia of Unified Science* (2 vols, Chicago, 1955–70), 1–27.

Niekamp, Johann Lukas. *Kurtzgefasste Mißions-Geschichte, oder historische Auszug der evangelischen Mißions-Berichte aus Ost-Indien von dem Jahr 1705 bis zu Ende des Jahres 1736* (Halle, 1740).

Nietzsche, Friedrich. *Die Geburt der Tragödie aus dem Geiste der Musik* (Leipzig, 1872).

Nietzsche, Friedrich. *Unzeitgemässe Betrachtungen* (2 vols, Leipzig, 1899).

Nietzsche, Friedrich. *Basic Writings of Nietzsche*, trans. and ed. Walter Kaufman (New York, 1968).

Nipperdey, Thomas. *Deutsche Geschichte, 1866–1918: Erste Band: Arbeitswelt und Bürgergeist* (Munich, 1998).

Nongbri, Brent. *Before Religion: A History of a Modern Concept* (New Haven, CT, 2013).

Nordau, Max. *Entartung* (2 vols, Berlin, 1892–3).

Nordenskiöld, Erik. *The History of Biology* (New York, 1928).

Norton Wise, Matthew. 'Mediations: Enlightenment Balancing Acts, or the Technologies of Rationalism', in Paul Horwich, ed., *World Changes, Thomas Kuhn and the Nature of Science* (Cambridge, MA, 1993), 207–56.

Norton Wise, Matthew. 'Precision: Agent of Unity and Product of Agreement', in M. Norton Wise, ed., *The Values of Precision* (Princeton, NJ, 1995), 92–100.

Nye, Mary Jo. *Molecular Reality: A Perspective on the Scientific Work of Jean Perrin* (London, 1972).

Nye, Mary Jo. *From Chemical Philosophy to Theoretical Chemistry: Dynamics of Matter and Dynamics of Disciplines, 1800–1950* (Berkeley, 1993).

Nye, Mary Jo. *Before Big Science: The Pursuit of Modern Chemistry and Physics, 1800–1910* (Cambridge, MA, 1996).

Nyhart, Lynn K. *Biology Takes Form: Animal Morphology and the German Universities, 1800–1900* (Chicago, 1995).

Nyhart, Lynn K. 'Natural History and the "New" Biology', in N. Jardine, J. A. Secord, and E. C. Spary, eds, *Cultures of Natural History* (Cambridge, 1996), 426–43.

Nyhart, Lynn K. 'Embryology and Morphology', in Michael Ruse and Robert J. Richards, eds, *The Cambridge Companion to the 'Origin of Species'* (Cambridge, 2009), 194–215.

Oberth, Hermann. *Die Rakete zu den Planetenräumen* (Munich, 1923).

O'Boyle, Leonore. 'Learning for its Own Sake: The German University as Nineteenth-Century Model', *Comparative Studies in Society and History* 25 (1983), 3–25.

O'Connor, Ralph. *The Earth on Show: Fossils and the Poetics of Popular Science, 1802–1856* (Chicago, 2007).

Oken, Lorenz. *Die Zeugnung* (Bamberg, 1805).

Oken, Lorenz. *Lehrbuch der Naturphilosophie* (3 vols, Jena, 1809–11).

Oldham, Kalil Swain. 'The Doctrine of Description: Gustav Kirchoff, Classical Physics, and the "Purpose of All Science" in 19th-Century Germany', unpublished PhD dissertation, University of California Berkeley, 2008.

[Ollier, Edmund]. Review of '*Yule-Tide Stories . . .* and *Household Stories*', *The Athenæum* 1322 (25 February 1863), 247–8.

Oppenheim, Janet. *The Other World: Spiritualism and Psychical Research in England, 1850–1914* (Cambridge, 1985).

Orange, A. D. 'The Idols of the Theatre: The British Association and its Early Critics', *Annals of Science* 32 (1975), 277–94.

Osterhammel, Jürgen. *Die Entzauberung Asiens: Europa und die asiatischen Reiche im 18. Jahrhundert* (Munich, 1988).

Osterhammel, Jürgen. *China und die Weltgesellschaft. Vom 18. Jahrhundert bis in unsere Zeit* (Munich, 1989)

Osterhammel, Jürgen. *The Transformation of the World: A Global History of the Nineteenth Century* (Princeton, NJ, 2014).

Ostwald, Wilhelm. 'Theorie des Glückes', *Annalen der Naturphilosophie* 4 (1905), 459–74.

Ostwald, Wilhelm. *Monistische Sonntagspredigten* (5 vols, Leipzig, 1911–16).

Ostwald, Wilhelm. *Die Energie* (Leipzig, 1912).

Ostwald, Wilhelm. 'Elements and Compounds', in C. S. Gibson and A. J. Greenaway, eds, *Faraday Lectures 1869–1928* (London, 1928), 185–201.

Owen, Richard. *Lectures on the Comparative Anatomy and Physiology of the Invertebrate Animals* (2nd edition, London, 1855).

Page, David. *The Past and Present Life of the Globe: Being a Sketch in Outline of the World's Life-System* (London, 1861).

Page, David. *The Earth's Crust: A Handy Outline of Geology* (Edinburgh, 1868).

Paley, William. *Natural Theology: or, Evidences of the Existence and Attributes of the Deity, collected from the Appearances of Nature* (London, 1802).

Pannwitz, Rudolph. *Die Krisis der europäischen Kultur* (Nuremberg, 1917).

Parsons, Charles. *The Steam Turbine: The Rede Lecture 1911* (Cambridge, 1911).

Patrin, Eugène-Louis-Melchior. *Histoire naturelle des minéraux, contenant leur description, celle de leur gîte, la théorie de leur formation, leurs rapports avec la géologie ou l'histoire de la terre, le détail de leurs propriétés et de leurs usages, leur analyse chimique* (5 vols, Paris, 1801).

Patton, Lydia. 'Hermann Cohen's History and Philosophy of Science', unpublished PhD dissertation, McGill University, 2004.

Patton, Lydia. 'Methodology of the Sciences', in Michael Forster and Kristin Gjesdal, eds, *The Oxford Handbook of German Philosophy in the Nineteenth Century* (Oxford, 2015), 595–606.

Paul, Diane B. and Ben Day. 'John Stuart Mill, Innate Differences, and the Regulation of Reproduction', *Studies in History and Philosophy of the Biological and Biomedical Sciences* 39 (2008), 222–31.

Paul, Diane B. and James Moore, 'The Darwinian Context: Evolution and Inheritance', in Alison Bashford and Philippa Levine, eds, *The Oxford Handbook of the History of Eugenics* (Oxford, 2010), 27–42.

Paulsen, Friedrich. *Immanuel Kant, sein Leben und seine Lehre* (Stuttgart, 1898).

Peach, Terry. *Interpreting Ricardo* (Cambridge, 1993).

Peacock, Thomas Love. 'Four Ages of Poetry', in F. H. B. Brett-Smith, ed., *Peacock's Four Ages of Poetry, Shelley's Defence of Poetry, Browning's Essay on Shelley* (Boston, 1921), 1–19.

Pearson, Karl. *The Grammar of Science* (London, 1892).

Pearson, Karl. *Speeches Delivered at a Dinner Held in University College, London, in Honour of Professor Karl Pearson, 23 April, 1934* (Cambridge, 1934).

Pepe, Luigi. 'Volta, the Istituto Nazionale and Scientific Communication in Early Nineteenth-Century Italy', in *Nuova Voltiana: Studies on Volta and his Time* 4 (Milan, 2002), 101–16.

Pepper, John Henry. *The Boy's Playbook of Science* (London, 1860).

Pepper, John Henry. *Playbook of Metals* (London, 1861).

Perkins, Franklin. *Leibniz and China: A Commerce of Light* (Cambridge, 2004).

Pérochon, Ernest. *Les Hommes frénétiques* (Paris, 1925).

Perrault, Charles. *Paralelle des anciens et des modernes, en ce qui regarde les arts et les sciences. Dialogves. Avec le poëme du Siecle de Louis le Grand, et une epistre en vers sur le genie* (Paris, 1688).

Perrin, Jean. 'L'agitation moléculaire et le mouvement brownien', *Comtes rendus de l'Académie des Sciences* 146 (1908), 967–70.

Perrin, Jean. 'La loi de Stokes et le mouvement brownien', *Comtes rendus de l'Académie des Sciences* 147 (1908), 475–6.

Perrin, Jean. 'L'origine de mouvement brownien', *Comtes rendus de l'Académie des Sciences* 147 (1908), 530.

Perrin, Jean. *Les Atoms* (Paris, 1913).

Petzold, Joseph. 'Zur Krisis des Kausalitätsbegriffs', *Naturwissenschaften* 19 (1922), 693–95.

Pezron, Paul. *L'Antiquité des tems rétablié et defenduë contre les Juifs & les nouveaux chronologists* (Paris, 1687).

Pfaff, Christoph Heinrich. *Über thierische Elektricität und Reizbarkeit. Ein Beytrag zu den neuesten Entdeckungen über diese Gegenstände* (Leipzig, 1795).

Phillips, Denise. *Acolytes of Nature: Defining Natural Science in Germany 1770–1850* (Chicago, 2012).

Pickering, Mary. *Auguste Comte: An Intellectual Biography* (3 vols, Cambridge, 1993–2009).

Pickstone, John V. 'Vital Actions and Organic Physics: Henri Dutrochet and French Physiology During the 1820s', *Bulletin of the History of Medicine* 50 (1976), 191–212.

Piszkiewicz, Dennis. *The Nazi Rocketeers: Dreams of Space and Crimes of War* (Mechanicsberg, PA, 2007).

Plumb, John H. 'The New World of Children in Eighteenth-Century England', *Past and Present* 67 (1975), 64–95.

Pocock, John G. A. *Barbarism and Religion* (6 vols, Cambridge, 1999–2015).

Pocock, John G. A. 'Perceptions of Modernity in Early Modern Historical Thinking', *Intellectual History Review* 17 (2007), 55–63.

Poincaré, Henri. *Thermodynamique: Cours de physique mathématique de la faculté des sciences de Paris* (2nd edition, Paris, 1908).

Poliquin, Rachel. *The Breathless Zoo: Taxidermy and the Cultures of Longing* (State University, PA, 2012).

Pool, Robert. *Beyond Engineering: How Society Shapes Technology* (Oxford, 1997).

Popper, Karl R. *The Poverty of Historicism* (London, 1957).

Popper, Karl R. *The Logic of Scientific Discovery* (revised edition, London, 1968).

Porter, Theodore M. *The Rise of Statistical Thinking 1820–1900* (Princeton, NJ, 1986).

Porter, Theodore M. *Karl Pearson: The Scientific Life in a Statistical Age* (Princeton, NJ, 2004).

Porter, Theodore M. 'How Science Became Technical', *Isis* 100 (2009), 292–309.

Poynting, John Henry. 'On the Transfer of Energy in the Electromagnetic Field', *Philosophical Transactions* 175 (1884), 343–61.

Prandtl, Ludwig. 'Über Flüssigkeitbewegung bei sehr kleiner Reibung', in A. Krazer, ed., *Verhandlungen des dritten Internationalen Mathematiker-Kongresses* (Leipzig, 1905).

Prandtl, Ludwig. 'Applications of Modern Hydrodynamics to Aeronautics', *National Advisory Committee for Aeronautics*, Report No. 116 (1923).

Prandtl, Ludwig. 'The Generation of Vortices in Fluids of Small Viscosity', *Journal of the Royal Aeronautical Society* 31 (1927), 720–41.

Prétot, Etienne André Philippe de. *Analyse chronologique de l'histoire universelle, depuis le commencement du monde, jusqu'a l'empire de Charlemagne inclusivement* (Paris, 1753).

Preyer, William. *Darwin. Sein Leben und Wirken* (Berlin, 1869).

Preyer, William. *Naturwissenschaftliche Tatsachen und Probleme* (Berlin, 1880).

Preyer, William. *Die Seele des Kindes, Beobachtungen über die geistige Entwicklung des Menschen in den ersten Lebensjahren* (Leipzig, 1882).

Preyer, William. *Naturforschung und Schule* (Stuttgart, 1887).

Preyer, William. *Die geistige Entwicklung in der ersten Kindheit nebst Anweisungen für die Eltern, dieselbe zu beobachten* (Stuttgart, 1893).

Provine, William B. 'Adaptation and Mechanisms of Evolution after Darwin: A Study in Persistent Controversies', in David Kohn, ed., *The Darwinian Heritage* (Princeton, NJ, 1985), 825–66.

Quetelet, Lambert Adolphe Jacques. *Sur l'homme et le développment de ses facultés, ou essai de physique sociale* (2 vols, Paris, 1835).

Quine, Williard van Orman. 'Mr Stawson on Logical Theory', *Mind* 62 (1953), 433–51.

Rabinbach, Anson. *The Human Motor: Energy, Fatigue, and the Origins of Modernity* (Berkeley, CA, 1992).

Rackstrow, Benjamin. *Miscellaneous Observations, together with a Collection of Experiments on Electricity* (London, 1748).

Ramberg, Peter J. 'The Death of Vitalism and the Birth of Organic Chemistry: Wöhler's Urea Synthesis and the Disciplinary Identity of Organic Chemistry', *Ambix* 47 (2000), 170–95.

Ramsay, Andrew Michael. *The Philosophical Principles of Natural and Revealed Religion Unfolded in a Geometrical Order* (2 vols, Glasgow, 1748–9).

Rankine, William John Macquorn. *A Manual of Applied Mechanics* (London, 1858).

Rankine, William John Macquorn. *A Manual of the Steam Engine and other Prime Movers* (London, 1859).

Rapoport, Anatol. 'Scientific Approach to Ethics', *Science* 150 (1957), 796–9.

Rapoport, Anatol. *Strategy and Conscience* (New York, 1964).

Rasmussen, Anne. 'Science and Technology', in John Home, ed., *A Companion to World War I* (Chichester, 2012), 307–22.

Rasmussen, Nicolas. 'The Decline of Recapitulationism in Early Twentieth-Century Biology: Disciplinary Conflict and Consensus on the Battleground of Theory', *Journal of the History of Biology* 24 (1991), 51–89.

Raspail, François-Vincent. 'Observations and Experiments Tending to Demonstrate that the Granules which are discharged in the Explosion of a Grain of Pollen, Instead of being Analogous to Spermatic Animalcules, are not even Organised Bodies', *Edinburgh Journal of Science* 10 (1829), 96–106.

Raspail, François-Vincent. *Nouveau système de physiologie végétale et de botanique fondé sur les méthodes d'observation, qui ont été dévelopées dans le nouveau système de chimie organique* (Paris, 1833).

Rayleigh, Lord. 'On the Resistance of Fluids', *Philosophical Magazine* 2 (1876), 430–41.

Rayleigh, Lord. 'On the Irregular Flight of a Tennis Ball', *Messenger of Mathematics* 7 (1877), 14–16.

Reddy, William M. 'The Eurasian Origins of Empty Time and Space: Modernity as Temporality Reconsidered', *History and Theory* 55 (2016), 325–56.

Rees, Martin. *Just Six Numbers: The Deep Forces that Shape the Universe* (London, 1999).

Reichenbach, Hans. *The Rise of Scientific Philosophy* (Berkeley, 1951).

Reill, Johann Christian, 'Von der Lebenskraft', *Archiv für die Physiologie* 1 (1795), 8–162.

Reill, Peter Hanns. 'Science and the Construction of the Cultural Sciences in Late Enlightenment Germany: The Case of Wilhelm von Humboldt', *History and Theory* 33 (1994), 345–66.

Reill, Peter Hanns. *Vitalizing Nature in the Enlightenment* (Berkeley, 2005).

Reingold, Nathan. 'Joseph Henry on the Scientific Life: An AAAS Presidential Address of 1850', in Nathan Reingold, ed., *Science, American Style* (New Brunswick, NJ, 1991), 156–68.

Reisch, George. *How the Cold War Transformed Philosophy of Science: To the Icy Slopes of Logic* (Cambridge, 2005).

Renault, Emmanuel. 'Le concept hégélien de civilisation et ses signifiants: *Bildung, Kulture* (et *Zivilisation?*)', in Bertrand Binoche, ed., *Les Équivoques de la civilisation* (Seyssel, 2005), 225–39.

Rensch, Bernhard. *Neuere Probleme der Abstammungslehre* (Stuttgart, 1947).

Ribot, Théodule. *La philosophie de Schopenhauer* (Paris, 1874).

Ricardo, David. *Principles of Political Economy and Taxation* (London, 1817).

Ricardo, David. *The Works and Correspondence of David Ricardo*, ed. P. Sraffa and M. Dobb (11 vols, Cambridge, 1951).

Ricci, Matteo and Nicolas Trigault. *China in the Sixteenth Century: The Journals of Matthew Ricci, 1583–1610* (New York, 1953).

Richards, Evelleen. 'Huxley and Woman's Place in Science', in J. Moore, ed., *History, Humanity, and Evolution* (Cambridge, 1989), 253–84.

Richards, Joan L. *Mathematical Visions: The Pursuit of Geometry in Victorian England* (San Diego, 1988).

Richards, Joan L. 'Rigor and Clarity: Foundations of Mathematics in France and England, 1800–1840', *Science in Context* 4 (1991), 297–319.

Richards, Joan L. 'The Geometrical Tradition: Mathematics, Space, and Reason in the Nineteenth Century', in Mary Jo Nye, ed., *The Cambridge History of Nineteenth and Twentieth Century Science* (Cambridge, 2003), 449–67.

Richards, Robert J. *Darwin and the Emergence of Evolutionary Theories of Mind and Behavior* (Chicago, 1987).

Richards, Robert J. *The Meaning of Evolution: The Morphological Construction and Ideological Reconstruction of Darwin's Theory* (Chicago, 1992).

Richards, Robert J. 'Kant and Blumenbach on *Bildungstrieb*: A Historical Misunderstanding', *Studies in the History and Philosophy of Biological and Biomedical Sciences* 31 (2000), 11–32.

Richards, Robert J. *The Tragic Sense of Life: Ernst Haeckel and the Struggle over Evolutionary Thought* (Chicago, 2008).

Rickert, Heinrich. *Kants al Philosoph der moderner Kultur. Ein geschichtsphilosophischer Versuch* (Heidelberg, 1924).

Ridley, Matt. *Francis Crick: Discoverer of the Genetic Code* (New York, 2009).

Riehl, Alois. *Der philosophische Kritizismus, Geschichte und System* (Leipzig, 1878).

Riehl, Alois. *Über wissenschaftliche und nichtwissenschaftliche Philosophie: eine akademische Antritte* (Freiberg, 1883).

Ring, Katy. 'The Popularisation of Elementary Science through Popular Science Books, c. 1870–c. 1939', unpublished PhD dissertation, University of Kent at Canterbury, 1988.

Ringer, Fritz. *The Decline of the German Mandarins: The German Academic Community, 1890–1933* (Cambridge, MA, 1969).

Ritter, Christopher. 'An Early History of Alexander Crum Brown's Graphical Formulas', in U. Klein, ed., *Tools and Modes of Representation in Laboratory Sciences* (Dordrecht, 2001), 35–46.

Ritter, Joachim. 'Ernst Cassirers Philosophie der symbolischen Formen', *Neue Jahrbücher für Wissenschaft und Jugendbildung* 6 (1930), 593–605.

Ritter, Johann Wilhelm. *Beweis, dass ein beständiger Galvanismus den Lebensprozess in Therreich begleite* (Weimar, 1798).

Ritter, Johann Wilhelm. *Das elektonrische System der Körper* (Leipzig, 1805).

Ritter von Ettingshausen, Andreas. *Anfangsgründe der Physik* (Vienna, 1844).

Ritvo, Harriet. *The Platypus and the Mermaid and Other Figments of the Classifying Imagination* (Cambridge, MA, 1997).

Robida, Albert. *Le Vingtième Siècle* (Paris, 1883).

Robida, Albert. *La Guerre au vingtième siècle* (Paris, 1887).

Robida, Albert. *Le Vingtième Siècle. La vie électrique* (Paris, 1890).

Robinson, David K. 'Reaction-Time Experiments in Wundt's Institute and Beyond', in R. W. Rieber and D. K. Robinson, eds, *Wilhem Wundt in History* (Boston, 2001), 160–204.

Rocke, Alan J. *Chemical Atomism in the Nineteenth Century: From Dalton to Cannizzaro* (Columbus, Ohio, 1984).

Rocke, Alan J. *Image and Reality: Kekulé, Kopp, and the Scientific Imagination* (Chicago, 2010).

Roger, A. and F. Chernoviz. *Lettres Apostoliques de Pie IX, Grégoire XVI, Pie VIII* (Paris, 1901).

Roger, Jacques. *The Life Sciences in Eighteenth-Century French Thought* (Stanford, 1997).

Roland, Alex. 'Science, Technology, and War', in Mary Jo Nye, ed., *The Cambridge History of Science, volume 5: The Modern Physical and Mathematical Sciences* (Cambridge, 2003), 561–78.

Roll-Hansen, Nils. 'Eugenics and the Science of Genetics', in Alison Bashford and Philippa Levine, eds, *The Oxford Handbook of the History of Eugenics* (Oxford, 2010), 80–97.

Roscoe, Henry E. and Arthur Harden. *A New View of the Origins of Dalton's Atomic Theory* (London, 1896).

Rosenberg, Alexander. 'Reductionism (and Antireductionism) in Biology', in David Hull and Michael Ruse, eds, *The Cambridge Companion to the Philosophy of Biology* (Cambridge, 2007), 120–38.

Rosny, J. H. *Les Navigateurs de l'infini* (Paris, 1925).

Rothfels, Nigel. *Savages and Beasts: The Birth of the Modern Zoo* (Baltimore, 2002).

Routh, Edward John. *A Treatise on the Stability of a Given State of Motion, Particularly Steady Motion* (London, 1877).

Rubner, Max. *Kraft und Stoff im Haushalte der Natur* (Leipzig, 1909).

Rudwick, Martin J. S. 'The Glacial Theory', *History of Science* 8 (1969), 136–57.

Rudwick, Martin J. S. *Bursting the Limits of Time: The Reconstruction of Geohistory in the Age of Revolution* (Chicago, 2005).

Rudwick, Martin J. S. *Worlds before Adam: The Reconstruction of Geohistory in the Age of Reform* (Chicago, 2008).

Ruse, Michael. 'Darwin's Debt to Philosophy: An Examination of the Influence of the Philosophical Ideas of John F. W. Herschel and William Whewell on the Development of Charles Darwin's Theory of Evolution', *Studies in History and Philosophy of Science* 6 (1975), 159–81.

Russell, Colin A. *The History of Valency* (Leicester, 1971).

Russell, Colin A. *The Structure of Chemistry* (Milton Keynes, 1976).

Rylance, Rick. *Victorian Psychology and British Culture 1850–1880* (Oxford, 2000).

St. Sure, Donald F., Ray Robert Noll, and Edward Malatesta. *100 Roman Documents Concerning the Chinese Rites Controversy (1645–1941)* (San Francisco, 1992).

Sale, George, et al. *An Universal History, from the Earliest Account of Time* (68 vols, London, 1747–68).

Saler, Michael. 'Modernity and Enchantment: A Historiographical Review', *American Historical Review* 111 (2006), 692–716.

Sapp, Jan. *Evolution by Association: A History of Symbiosis* (Oxford, 1994).

Sapp, Jan. *The New Foundations of Evolution: On the Tree of Life* (Oxford, 2009).

Sarton, George. *The History of Science and the New Humanism* (New York, 1956).

Saussure, Horace-Benédict de. 'Agenda, ou tableau général des observations et des recherches dont les résultats doivent servir de base à la théorie de la terre', *Journal des mines* 4.20 (1795), 1–70: 32–3.

Sayers, W. H. 'The Arrest of Aerodynamic Development', *Aeroplane* 22 (1922) 138.

Schabas, Margaret. *The Natural Origins of Economics* (Chicago, 2005).

Schäfer, Wolf. 'Global Civilization and Local Cultures', *International Sociology* 16 (2001), 301–19.

Scheele, Carl Wilhelm. *Chemische Abhandlung von der Luft und dem Feuer* (Uppsala and Leipzig, 1777).

Scheele, Carl Wilhelm. *Chemical Observations and Experiments on Air and Fire . . . with a Prefatory Introduction by Torbern Bergman, translated from the German by J. R. Forster* (London, 1780).

Schelling, Friedrich W. J. *Von der Weltseele: Eine Hypothese der höheren Physik zur Erklärung des allgemeinen Organismus* (Hamburg, 1798).

Schelling, Friedrich W. J. *Einleitung zu dem Entwurf eines Systems der Naturphilosophie. Oder über den Begriff der Speculativen Physik und die innere Organisation eines Systems dieser Wissenschaft* (Jena and Leipzig, 1799).

Schermer, Michael. *In Darwin's Shadow: The Life and Science of Alfred Russel Wallace* (New York, 2002).

Scherren, Henry. *The Zoological Society of London: A Sketch of its Foundations and Development* (London, 1905).

Schiaparelli, Giovanni. *La vita sul pianeta Marte* (Milan, 1893).

Schiller, Joseph. *Claude Bernard et les problèmes scientifiques de son temps* (Paris, 1967).

Schipperges, Heinrich. *Weltbild und Wissenschaft: Eröffnungsreden zu den Naturforscherversammlungen, 1822–1972* (Hildesheim, 1976).

Schmidt, Oscar. *Entwicklung der verleichenden Anatomie: Ein Betrag zur Geschichte der Wissenschaft* (Jena, 1855).

Schopenhauer, Arthur. *Die Welt als Wille und Vorstellung* (Leipzig, 1819).

Schopenhauer, Arthur. *Die Welt als Wille und Vorstellung* (2nd edition, 2 vols, Leipzig, 1844).

Schroeder-Gudehus, Brigitte. *Deutsche Wissenschaft und internationale Zusammenarbeit: Ein Beitrag zum Studium kultureller Beziehungen in politischen Krisenzeiten* (Geneva, 1966).

Schroeder-Gudehus, Brigitte. 'Challenges to Transnational Loyalties: International Scientific Organizations after the First World War', *Science Studies* 3 (1973), 93–118.

Schubring, Gert. 'The Conception of Pure Mathematics as an Instrument in the Professionalisation of Mathematics', in H. Mehrtens, H. Bos, and T. Schneider, eds, *Social History of Nineteenth Century Mathematics* (Basel, 1981), 111–34.

Schubring, Gert. 'Die deutsche mathematische Gemeinde', in J. Fauvel, R. Flood, and R. Wilson, eds, *Möbius und sein Band. Der Aufstieg von Mathematik und Astronomie im Deutschland des 19. Jahrhunderts* (Basle, 1994), 394–406.

Schultz, Bart. *Henry Sidgwick, Eye of the Universe: An Intellectual Biography* (Cambridge, 2004).

Schwab, Raymond. *The Oriental Renaissance: Europe's Rediscovery of India and the East, 1680–1880* (New York, 1984).

Schwann, Theodor. *Mikroskopische Untersuchungen über die Uebereinstimmung in der Struktur und den Wachstum der Thiere und Pflanzen* (Berlin, 1839).

Schwann, Theodor. *Microscopical Researches into the Accordance in the Structure and Growth of Animals and Plants* (London, 1847).

Schweber, Silvan S. 'Scientists as Intellectuals: The Early Victorians', in J. Paradis and T. Postlewait, eds, *Victorian Science and Victorian Values: Literary Perspectives* (New Brunswick, 1985), 1–38.

Secord, James A. *Victorian Sensation: The Extraordinary Publication, Reception, and Secret Authorship of Vestiges of the Natural History of Creation* (Chicago, 2000).

Secord, James. 'Monsters at Crystal Palace', in Soraya de Chadarevian and Nick Hopwood, eds, *Models: The Third Dimension of Science* (Stanford, CA, 2004), 138–69.

Secord, James A. *Visions of Science: Books and Readers at the Dawn of the Victorian Age* (Oxford, 2014).

Sedgwick, Adam. 'Address of the President', *Proceedings of the Geological Society of London* 20 (1831), 281–316.

Serres, Étienne R. A. *Anatomie comparée du cerveau* (2 vols, Paris, 1824–6).

Seth, Suman. *Crafting the Quantum: Arnold Sommerfield and the Practice of Theory, 1890–1926* (Cambridge, MA, 2010).

Settle, Thomas B. 'An Experiment in the History of Science', Science 133 (1961), 19–32.

Seward, A. C., ed. *Darwin and Modern Science* (Cambridge, 1909).

Sewell, William. *A Second Letter to a Dissenter on the Opposition of the University of Oxford to the Charter of the London College* (Oxford, 1834).

Shapin, Steven. 'Phrenological Knowledge and the Social Structure of Early Nineteenth-Century Edinburgh', *Annals of Science* 32 (1975), 219–43.

Shapin, Steven. '"Nibbling at the Teats of Science": Edinburgh and the Diffusion of Science in the 1830s', in I. Inkster and J. Morell, eds, *Metropolis and Province: Science in British Culture, 1780–1850* (London, 1983), 151–78.

Shapin, Steven. *The Scientific Life: A Moral History of a Late Modern Vocation* (Chicago, 2008).

Shiach, Morag. *Discourse on Popular Culture: Class, Gender and History in Cultural Analysis, 1730 to the Present* (Cambridge, 1989).

Shils, Edward and John Roberts, 'The Diffusion of European Models Outside Europe', in Walter Rüegg, ed., *A History of the University in Europe, Volume III: Universities in the Nineteenth and Early Twentieth Centuries (1800–1945)* (Cambridge, 2004), 163–230.

Shrefin, Jill. ' "Make it a Pleasure not a Task": Educational Games for Children in Georgian England', *Princeton University Library Chronicle* 60 (1999), 251–75.

Sibum, Otto. 'Reworking the Mechanical Value of Heat: Instruments of Precision and Gestures of Accuracy in Early Victorian England', *Studies in History and Philosophy of Science* 26 (1994), 73–106.

Siddiqi, Asif. A. *The Red Rockets' Glare: Spaceflight and the Soviet Imagination, 1857–1957* (Cambridge, 2010).

Sieg, Ulrich. *Aufstieg und Niedergang des Marburger Neukantianismus: Die Geschichte einer philosophischen Schulgemeinschaft* (Würzburg, 1994).

Siemens, Werner von. 'Das Naturwissenschaftliche Zeitalter', *Tageblatt der 59. Versammlung deutscher Naturforscher uns Ärzte zu Berlin* (Berlin, 1886), 92–6.

Simon, Herbert A. *Models of Man, Social and Rational: Mathematical Essays on Rational Human Behaviour in a Social Setting* (Oxford, 1957).

Simpson, George Gaylord. *Tempo and Mode in Evolution* (New York, 1944).

Simpson, Thomas K. *Maxwell on the Electromagnetic Field: A Guided Study* (New Brunswick, 1997).

Sinclair, Steven V. 'J. J. Thomson and the Chemical Atom: From Ether Vortex to Atomic Decay', *Ambix* 34 (1987), 89–116.

Skelton, Matthew. 'The Paratext of Everything: Constructing and Marketing H. G. Wells's *The Outline of History*', *Book History* 4 (2001), 237–75.

Skidelsky, Edward. *Ernst Cassirer: The Last Philosopher of Culture* (Princeton, NJ, 2008).

Slotten, Ross A. *The Heretic in Darwin's Court: The Life of Alfred Russel Wallace* (New York, 2004).

Smart, J. J. *Philosophy and Scientific Realism* (London, 1963).

Smil, Vaclav. *Creating the Twentieth Century: Technical Innovations of 1867–1914 and Their Lasting Impact* (Oxford, 2005).

Smith, Crosbie. ' "Mechanical Philosophy" and the Emergence of Physics in Britain: 1800–1850', *Annals of Science* 33 (1976), 3–29.

Smith, Crosbie. 'William Hopkins and the Shaping of Dynamical Geology: 1830–1860', *British Journal for the History of Science* 22 (1989), 27–52.

Smith, Crosbie. *The Science of Energy: A Cultural History of Energy Physics in Victorian Britain* (London, 1998).

Smith, Crosbie and M. Norton Wise. *Energy and Empire: A Biographical Study of Lord Kelvin* (2 vols, Cambridge, 1989).

Smocovitis, Vassiliki Betty. *Unifying Biology: The Evolutionary Synthesis and Evolutionary Biology* (Princeton, NJ, 1996).

Smolin, Lee. *The Trouble with Physics: The Rise of String Theory, The Fall of Science, and What Comes Next* (New York, 2006).

Snow, C. P. 'Chemistry', in H. Wright, ed., *University Studies: Cambridge 1933* (London, 1933), 97–121.

Snow, C. P. *Variety of Men* (London, 1939).

Snyder, Laura. *Reforming Philosophy: A Victorian Debate on Science and Society* (Chicago, 2006).

Sokal, Michael M. 'James McKeen Cattell and the Failure of Anthropometric Mental Testing, 1890–1901', in William R. Woodward and Mitchell G. Ash, eds, *The Problematic Science: Psychology in Nineteenth-Century Thought* (New York, 1982), 322–45.

Somerville, Mary. *Mechanism of the Heavens* (London, 1831).

Somerville, Mary. *On the Connexion of the Physical Sciences* (9th edition, London, 1858).

Spallanzani, Lazzaro. *Mémoires sur la respiration* (Geneva, 1803).

Spencer, Herbert. *Social Statics: Or, the Conditions Essential to Human Happiness Specified, and the First of them Developed* (London, 1851).

Spencer, Herbert. 'A Theory of Population, Deduced from the General Law of Animal Fertility', *Westminster Review* 57 (1852), 468–501.

Spencer, Herbert. *An Autobiography* (2 vols, New York, 1904).

Spengler, Oswald. *Der Untergang des Abendlands* (2 vols, Munich, 1918–28).

Sraffa, Piero. *Production of Commodities by Means of Commodities* (Cambridge, 1960).

Stapledon, Olaf. *Last and First Men: A Story of the Near and Far Future* (London, 1930).

Stapledon, Olaf. *Star Maker* (London, 1937).

Stark, Johannes. *Die gegenwärtige Krise in der deutschen Physik* (Leipzig, 1922).

Staudenmaier, John. *Technology's Storytellers: Reweaving the Human Fabric* (Cambridge, MA, 1985).

Stearns, Peter N. *The Industrial Turn in World History* (London, 2017).

Stebbins, G. Ledyard. *Variation and Evolution in Plants* (New York, 1950).

Steffens, Henrich. *Beyträge zur inneren Naturgeschichte der Erde* (Freiberg, 1801).

Steffens, Henrich. *Grundzüge der philosophischen Naturwissenschaft* (Berlin, 1806).

Stephen, Leslie. 'An Attempted Philosophy of History', *Fortnightly Review* 27 (1880), 672–95.

Stichweh, Rudolph. *Zur Entstehung des modernen Systems wissenschaftlicher Disziplinen: Physik in Deutschland, 1740–1890* (Frankfurt, 1984).

Stillingfleet, Edward. *Origines Sacræ, or a Rational Account of the Grounds of Christian Faith, as to the Truth and Divine Authority of the Scriptures, And the matter therein contained* (London, 1662).

Stites, Richard. *Revolutionary Dreams: Utopian Vision and Experimental Life in the Russian Revolution* (Oxford, 1989).

Stokes, Donald E. *Pasteur's Quadrant: Basic Science and Technological Innovation* (Washington, DC, 1997).

Sturgeon, William. 'Historical Sketch of the Rise and Progress of Electro-Magnetic Engines for Propelling Machinery', *Annals of Electricity, Magnetism, and Chemistry; and Guardian of Experimental Science* 3 (1838/9), 429–37.

Sudhoff, Karl. *Rudolf Virchow und die deutschen Naturforscherversammlungen* (Leipzig, 1922).

Sullivan, William Francis. *The Young Liar!!! A Tale of Truth and Caution; for the Benefit of the Rising Generations* (London, 1817).

Sulloway, Frank J. 'Freud and Biology: The Hidden Legacy', in William R. Woodward and Mitchell G. Ash, eds, *The Problematic Science: Psychology in Nineteenth-Century Thought* (New York, 1982), 198–227.

Suvin, Darko. 'The Utopian Tradition in Russian Science Fiction', *Modern Languages Review* 66 (1971), 139–59.

Sysling, Fenneke. 'Science and Self-Assessment: Phrenological Charts 1840–1940', *British Journal for the History of Science* 51 (2018), 261–80.

Talairach-Vielmas, Laurence. *Fairy-Tales, Natural History and Victorian Culture* (Basingstoke, 2014).

Taylor, Frederick W. *The Principles of Scientific Management* (New York, 1911).

Taylor, Geoffrey Ingram. 'Pressure Distribution Around a Cylinder', *Reports and Memoranda of the Advisory Committee for Aeronautics* 191 (1916).

Taylor, Geoffrey Ingram. 'Pressure Distribution over the Wings of an Aeroplane in Flight', *Reports and Memoranda of the Advisory Committee for Aeronautics* 287 (1916).

Thackray, Arnold. *Atoms and Powers: An Essay on Newtonian Matter-Theory and the Development of Chemistry* (Cambridge, MA, 1970).

Thackray, Arnold. 'Natural Knowledge in Cultural Context: The Manchester Model', *American Historical Review* 79 (1974), 672–709.

Thaler, Richard H. *Misbehaving: The Making of Behavioural Economics* (London, 2016).

Theunissen, Bert. 'Darwin and his Pigeons: The Analogy between Artificial and Natural Selection Revisited', *Journal of the History of Biology* 45 (2012), 1–34.

Theunissen, Bert. 'Unifying Science and Human Culture: The Promotion of the History of Science by George Sarton and Frans Verdoorn', in H. Kamminga and G. Somsen, eds, *Pursuing the Unity of Science: Ideology and Scientific Practice from the Great War to the Cold War* (London, 2016), 157–82.

Thiel, Hugo. *Über eigene Formen der landwirtschaftliche Genossenschaften* ([Poppelsdorf], 1868).

Thomson, J. Arthur. *Science and Religion* (London, 1924).

Thomson, S. P. *The Life of William Thomson, Baron Kelvin of Largs* (2 vols, London, 1919).

Thomson, William. 'On the Dynamical Theory of Heat, with Numerical Results Deduced from Mr Joule's Equivalent of a Thermal Unit, and M. Regnault's Observations on Steam', *Transactions of the Royal Society of Edinburgh* 20, Part II (1851), 261–8, 289–98.

Thomson, William. 'On the Secular Cooling of the Earth', *Philosophical Magazine* 25 (1863), 1–14.

Thomson, William. *Mathematical and Physical Papers* (6 vols, Cambridge, 1882–1911).

Thomson, William. *Notes of Lectures on Molecular Dynamics and the Wave Theory of Light* (Baltimore, 1884).

Thomson, William. *Popular Lectures and Addresses* (3 vols, London, 1891–4).

Thomson, William, and Peter Guthrie Tait. *Treatise on Natural Philosophy* (Oxford, 1867).

Thorpe, Thomas E. *Essays in Historical Chemistry* (London, 1894).

Thurston, R. H. 'On the Economy Resulting from Expansion of Steam', *Journal of the Franklin Institute* 71 (1861), 193–5.

Tiedemann, Friedrich. *Physiologie des Menschen. Erster Band* (Darmstadt, 1830).

Todes, Daniel P. 'Pavlov's Physiology Factory', *Isis* 88 (1997), 205–46.

Todhunter, Isaac. *William Whewell, DD. Master of Trinity College Cambridge: An Account of his Writings with Selections from his Literary and Scientific Correspondence* (2 vols, London, 1976).

Toews, John E. *Hegelianism: The Path Towards Dialectical Humanism, 1805–1841* (Cambridge, 1980).

Tolstoy, Aleksey. *Aelita* (Moscow, 1923).

'Tom Telescope' [John Newbery]. *The Newtonian System of Philosophy Adapted to the Capacities of Young Gentlemen and Ladies* (London, 1770).

'Tom Telescope' [John Newbery]. *The Newtonian Philosophy, and Natural Philosophy in General; Explained and Illustrated by Familiar Objects in a Series of Entertaining Lectures* (London, 1838).

Topham, Jonathan. 'Science and Popular Education in the 1830s: The Role of the *Bridgewater Treatises*', *British Journal for the History of Science* 25 (1992), 397–430.

Topham, Jonathan. 'Beyond the Common Context: The Production and Reading of the *Bridgewater Treatises*', *Isis* 89 (1998), 233–62.

Töpner, Kurt. *Gelehrte Politiker und politisierende Gelehrte: die Revolution von 1918 im Urteil deutscher Hochhschullehrer* (Göttingen, 1970).

Trendelenburg, Adolf. *Kuno Fischer und sein Kant: Ein Entgegnung* (Leipzig, 1870).

Trenn, Taddeus J. *The Self-Splitting Atom: A History of the Rutherford–Soddy Collaboration* (London, 1977).

Treviranus, Gottfried Reinhold. *Biologie, oder Philosophie der lebenden Natur* (6 vols, Göttingen, 1802–22).

Trimmer, Sarah. *An Easy Introduction to the Knowledge of Nature and the Holy Scriptures* (London, 1780).

Troeltsch, Ernst. 'Die Krisis des Historismus', *Die neue Rundschau* 33 (1922), 572–90.

Trumpler, Marie Jean. 'Questioning Nature: Experimental Investigations of Animal Electricity in Germany, 1791–1810', unpublished PhD dissertation, Yale University, 1992.

Tsiolkovskii, Konstantin. *Na Lune* (Moscow, 1893).

Turgot, Anne Robert Jacques. *Oeuvres de Turgot*, ed. Eugène Daire (2 vols, Paris, 1844).

Turner, Frank M. 'Public Science in Britain, 1880–1919', *Isis* 71 (1980), 589–608.

Turner, Frank M. *Contesting Cultural Authority: Essays in Victorian Intellectual Life* (Cambridge, 1993).

Turner, R. Steven. 'The Growth of Professorial Research in Prussia, 1818 to 1848', *Historical Studies in the Physical Sciences* 3 (1971), 137–82.

Turner, R. Steven. 'University Reformers and Professorial Scholarship in Germany, 1760–1806', in Lawrence Stone, ed., *The University in Society, volume 2* (Oxford, 1974), 495–531.

Turner, R. Steven. 'The *Bildungsbürgertum* and the Learned Professions in Prussia, 1770–1830', *Histoire Social—Social History* 13 (1980), 105–35.

Turner, R. Steven. 'The Great Transition and the Social Patterns of German Science', *Minerva* 25 (1987), 56–76.

Tyler, Edward. *Primitive Culture* (London, 1871).

Tyndall, John. 'Physics and Metaphysics', *Saturday Review* 10 (4 August 1860), 141.

Valentin, Gabriel Gustav. *Lehrbuch der Physiologie des Menschen* (2 vols, Braunschweig, 1844).

Van Dongen, Jeroen. *Einstein's Unification* (Cambridge, 2010).

Van Kley, Edwin J. 'Europe's "Discovery" of China and the Writing of World History', *The American Historical Review* 76 (1971), 358–85.

Van Wyhe, John. *Phrenology and the Origins of Victorian Naturalism* (Aldershot, 2004).

Vasold, Manfred. *Rudolf Virchow: Der große Arst und Politiker* (Stuttgart, 1988).

Veblen, Thorstein. 'The Place of Science in Modern Civilization', *American Journal of Sociology* 11 (1906), 585–609.

Verne, Jules. *De la terre à la lune* (Paris, 1865).

Verne, Jules. *Vingt mille lieues sous les meres* (Paris, 1870).

Véron, Eugène. *L'esthétique* (Paris, 1878).

Vierhaus, Rudolph. 'Bildung', in O. Brunner, W. Conze, and R. Koselleck, eds, *Geschichtliche Grundbegriffe* (5 vols, Stuttgart, 1972–89), i. 508–51.

Vierordt, Carl. *Physiologie des Athmens, mit besonderer Rücksicht auf die Außcheidung der Kohlensäure* (Karlsruhe, 1845).

Vierordt, Carl. 'Über die gegenwärtigen Standpunkt und die Aufgabe der Physiologie', *Archiv für physiologische Heilkunde* 8 (1849), 297–316.

Vincent, David. *Literacy and Popular Culture: England 1750–1914* (Cambridge, 1989).

Vincenti, Walter. *What Engineers Know and How They Know it: Analytical Studies from Aeronautical History* (Baltimore, 1990).

Virchow, Rudolph. *Die Cellularpathologie in ihrer Begründung auf physiologische und pathologische Gewebelehre* (Berlin, 1858).

Virchow, Rudolph. 'Die Naturwissenschaften in ihrer Bedeutung für die sittliche Erziehung der Menschheit', in Karl Sudhoff, ed., *Rudolf Virchow und die deutschen Naturforscherversammlungen* (Leipzig, 1922), 122–47.

Virchow, Rudolph. 'Ueber die Aufgaben der Naturwissenschaften im neuen nationalen Leben Deutschlands', in Karl Sudhoff, ed., *Rudolf Virchow und die deutschen Naturforscherversammlungen* (Leipzig, 1922), 108–18.

Virchow, Rudolph. *Disease, Life, and Man: Selected Essays by Rudolph Virchow*, ed. and trans. L. J. Rather (Stanford, 1958).

Vogt, Karl. *Physiologische Briefe für Gebildete alle Stände* (3 vols, Stuttgart, 1847).

Vogt, Karl. *Ocean und Mittelmeet: Reisebriefe* (2 vols, Frankfurt am Main, 1848).

Vogt, Karl. *Bilder aus dem Thierleben* (Frankfurt am Main, 1852).

Vogt, Karl. *Köhlerglaube und Wissenschaft: Ein Streitschrift gegen Hofrath R. Wagner in Göttingen* (Giessen, 1855).

Volney, Constantin-François Chasebeuf de Boisgiray de. *Les Ruines, ou méditation sur les révolutions des émpires* (Paris, 1791).

Voltaire, François Marie Arouet de. *Le Siècle de Louis XIV* (Berlin, 1751).

Voltaire, François Marie Arouet de. *Essai sur les mœurs, et l'esprit des nations et sur les principaux faits de l'histoire depuis Charlemagne jusqu'à Louis XIII* (Geneva, 1756).

Voltaire, François Marie Arouet de. *Oeuvres complètes de Voltaire* (72 vols, Gotha, 1784–90).

Von Mises, Richard. 'Über die gegenwärtige Krise der Mechanik', *Zeitschrift für angewandte Mathematik und Mechanik* 1 (1921), 425–31.

Von Neumann, John. 'Zur Theorie der Gesellschaftsspiele', *Mathematische Annalen* 100 (1928), 295–320.

Von Neumann, John and Oskar Morgenstern. *Theory of Games and Economic Behaviour* (Princeton, NJ, 1944).

Von Wright, Georg Hendrik. *Explanation and Understanding* (Ithaca, NY, 1971).

Vries, Hugo de. *Die Mutationstheorie. Versuche und Beobachtungen über die Entstehung von Artem im Pflanzenreich* (2 vols, Leipzig, 1901–3).

Wagner, Günter P. 'The Current State and the Future of Developmental Evolution', in M. D. Laubichler and J. Maienschein, eds, *From Embryology to Evo-Devo: A History of Developmental Evolution* (Cambridge, MA, 2007), 525–45.

Wagner, Rudolph. *Menschenschöpfung und Seelensubstanz* (Göttingen, 1854).

Wakefield, Priscilla. *An Introduction to Botany, in a Series of Familiar Letters with Illustrative Engravings* (London, 1796).

Wakefield, Priscilla. *Mental Improvement; or the Beauties and Wonders of Nature and Art. In a Series of Instructive Conversations* (New Bedford, MA, 1799).

Walch, Johann Georg. *Philosophisches Lexicon* (Leipzig, 1726).

Wallace, Alfred Russel. 'On the Law Which has Regulated the Introduction of New Species', *Annals and Magazine of Natural History*, 2nd series, 16.93 (1855), 184–96.

Wallace, Alfred Russel. 'On the Habits of the Orang-Utan in Borneo', *Annals and Magazine of Natural History*, 2nd series, 17.103 (1856), 26–32.

Wallace, Alfred Russel. *The Malay Archipelago: The Land of the Orang-utan and the Bird of Paradise, A Narrative of Travel with Studies of Man and Nature* (London, 1869).

Wallace, Alfred Russel. *My Life: A Record of Events and Opinions* (2 vols, New York, 1905).

Walter, Ryan. 'The Enthusiasm of David Ricardo', *Modern Intellectual History* 15 (2018), 353–80.

Walters, Helen B. *Hermann Oberth: Father of Space Travel* (New York, 1962).

Ward, Mary and Lady Jane Mahon. *Entomology in Sport, and Entomology in Earnest* (London, [1859]).

Warwick, Andrew. *Masters of Theory: Cambridge and the Rise of Mathematical Physics* (Chicago, 2003).

Weber, Wilhelm. *Elektrodynamische Maassbestimmungen*, (Leipzig, 1846).

Webster, Charles. *The Great Instauration: Science, Medicine and Reform (1626–1660)* (London, 1975).

Weightman, Gavin. *The Industrial Revolutionaries: The Creation of the Modern World 1776–1914* (London, 2007).

Weil, Hans. *Die Entstehung des deutsches Bildungsprinzips* (Bonn, 1930).

Weininger, Stephen. 'Contemplating the Finger: Visuality and the Semiotics of Chemistry', *Hyle* 4 (1998), 3–27.

Weintraub, E. Roy. *How Economics Became a Mathematical Science* (Durham, NC, 2002).

Weismann, August. 'Zur Frage nach der Verebung erworbener Eigenschaften', *Biologisches Centralblatt* 6 (1886), 33–48.

Weiss, John Hubbel. *The Making of Technological Man: Social Origins of French Engineering Education* (Cambridge, MA, 1982).

Weisz, George. *The Emergence of Modern Universities in France* (Princeton, NJ, 1983).

Wells, H. G. 'Popularising Science', *Nature* 50 (26 July 1894), 300–1.

Wells, H. G. *The Outline of History: Being a Plain History of Life and Mankind* (2 vols, London, 1920).

Wells, H. G. *The Work, Wealth and Happiness of Mankind* (London, 1932).

Wells, H. G. *The Shape of Things to Come* (London, 1933).

Werner, Alfred. *Neuere Anschauungen auf der Gebiete der anorganischen Chemie* (Braunschweig, 1905).

Wernick, Andrew. *Auguste Comte and the Religion of Humanity: The Post-Theistic Program of French Social Theory* (Cambridge, 2001).

Weyl, Herman. 'Über die neue Grundlagenkrise der Mathematik', *Mathematische Zeitschrift* 10 (1921), 39–79.

Whately, Richard. *Introductory Lectures on Political Economy* (London, 1832).

Wheeler, John A. *Geometrodynamics* (New York, 1963).

Whewell, William. 'Jones—on the Distribution of Wealth and the Sources of Taxation', *British Critics* 10 (1831), 41–61.

Whewell, William. 'Mathematical Exposition of Some Leading Doctrines in Mr. Ricardo's *Principles of Political Economy and Taxation*', *Transactions of the Cambridge Philosophical Society* 3 (1831), 192–230.

Whewell, William. 'Modern Science—Inductive Philosophy', *Quarterly Review* 45 (1831), 374–407.

Whewell, William. Review of *A Preliminary Discourse on the Study of Natural Philosophy* by J. F. W. Herschel, *Quarterly Review* 45 (1831), 374–407.

Whewell, William. *Astronomy and General Physics Considered with Reference to Theology* (London, 1833).

Whewell, William. 'Mrs Somerville on the Connexion of the Sciences', *Quarterly Review* 51 (1834), 54–68.

Whewell, William. *The Elements of Morality, Including Polity* (2 vols, London, 1845).

Whewell, William. *Lectures on Systematic Morality* (London, 1846).

Whewell, William. *The Philosophy of the Inductive Sciences, A New Edition* (3 vols, London 1847).

Whewell, William. *Of Induction* (London, 1849).

Whewell, William. *Lectures on the History of Moral Philosophy in England* (London, 1852).

Whewell, William. *On the Philosophy of Discovery* (London, 1860).

White, Andrew Dickson. *A History of the Warfare of Science and Theology in Christendom* (2 vols, New York, 1896).

White, Paul. *Thomas Huxley: Making the 'Man of Science'* (Cambridge, 2003).

White, Sheldon. 'Conceptual Foundations of IQ Testing', *Psychology, Public Policy, and Law* 6 (2000), 33–43.

Whitely, Richard. 'Knowledge Producers and Knowledge Acquirers: Popularisation as a Relation between Scientific Fields and Their Publics', in T. Shinn and R. Whitley, eds, *Expository Science: Forms and Functions of Popularisation* (Dordrecht, 1985), 3–28.

Willey, Thomas E. *Back to Kant: The Revival of Kantianism in German Social and Historical Thought, 1860–1914* (Detroit, 1978).

Williams, Elizabeth. 'Sciences of Appetite in the Enlightenment, 1750–1800', *Studies in History and Philosophy of Biological and Biomedical Sciences* 43 (2012), 392–404.

Williams, L. Pearce. 'Science, Education and the French Revolution', *Isis* 44 (1953), 311–30.

Williams, L. Pearce. 'Science, Education and Napoleon I', *Isis* 47 (1956), 369–82.

Williamson, Alexander. 'Theory of Etherification', *Journal of the Chemical Society* 4 (1852), 229–39.

Wills, Ian. 'Edison and Science: A Curious Result', *Studies in History and Philosophy of Science* 40 (2009), 157–66.

Wilson, Catherine. *The Invisible World: Early Modern Philosophy and the Invention of the Microscope* (Princeton, NJ, 1995).

Wilson, E. O. *Sociobiology: The New Synthesis* (Cambridge, MA, 1975).

Wimsatt, William C. 'Echoes of Haeckel? Reentrenching Development in Evolution', in M. D. Laubichler and J. Maienschein, eds, *From Embryology to Evo-Devo: A History of Developmental Evolution* (Cambridge, MA, 2007), 309–55.

Windelband, Wilhelm. *Geschichte und Naturwissenschaft: Rede zum Antritt des Rectorats der Kaiser-Wilhelms-Universität, Strassburg, geh. am 1. Mai 1894* (Strassburg, 1894).

Windelband, Wilhelm. *Präludien: Aufsätze und Reden zur Philosophie und ihrer Geschichte Band I/II* (Tübingen, 1924).

Wise, Matthew Norton. 'Mediations: Enlightenment Balancing Acts, or the Technologies of Rationalism', in Paul Horwich, ed., *World Changes, Thomas Kuhn and the Nature of Science* (Cambridge, MA, 1993), 207–56.

Wise, Matthew Norton. 'Precision: Agent of Unity and Product of Agreement', in M. Norton Wise, ed., *The Values of Precision* (Princeton, NJ, 1995), 92–100.

Wolfe, Charles. '"Cabinet d'histoire naturelle", or: The Interplay of Nature and Artifice in Diderot's Naturalism', *Perspectives on Science* 17 (2009), 58–77.

Wolfe, Charles. 'On the Role of Newtonian Analogies in Eighteenth-Century Life Science: Vitalism and Provisionally Inexplicative Devices', in Z. Biener and E. Schliesser, eds, *Newton and Empiricism* (Oxford, 2014), 223–61.

Wolin, Richard. '"Modernity": The Peregrinations of a Contested Historiographical Concept', *American Historical Review* 111 (2011), 741–51.

Wood, John George. *Common Objects of the Country* (London, 1858).

Wood, John George. *Boy's Own Book of Natural History* (London, 1861).

Wood, John George. 'The Dulness of Museums', *Nineteenth Century* 21 (1887), 384–96.

Woodger, Joseph H. *Biological Principles: A Critical Study* (London, 1929).

Woodward, James. *Making Things Happen: A Theory of Causal Explanation* (Oxford, 2003).

Woodward, William A. 'Wundt's Program for the New Psychology: Vicissitudes of Experiment, Theory, and System', in William R. Woodward and Mitchell G. Ash, eds, *The Problematic Science: Psychology in Nineteenth-Century Thought* (New York, 1982), 167–97.

Wright, Gavin. 'The Truth about Scientific Management', *Reviews in American History* 9 (1981), 88–92.

Wright, Howard Theophilus. 'Aeroplanes from an Engineer's Point of View', *Aero* 6 (1912), 374–80.

Wright, Sewall. 'Evolution in Mendelian Populations', *Genetics* 16 (1931), 97–159.

Wright, Terence R. *The Religion of Humanity: The Impact of Comtean Positivism on Victorian Britain* (Cambridge, 1986).

Wundt, Wilhelm. *Grundzüge der physiologischen Psychologie* (Leipzig, 1874).

Yan Haiyan. 'Knowledge Across Borders: The Early Communication of Evolution in China', in B. Lightman, G. McOuat, and L. Stewart, eds, *The Circulation of Knowledge Between Britain, India, and China* (Leiden, 2013), 181–208.

Yeo, Richard. 'William Whewell, Natural Theology and the Philosophy of Science in Mid-Nineteenth Century Britain', *Annals of Science* 36 (1979), 493–512.

Yeo, Richard. 'Science and Intellectual Authority in Mid-Nineteenth Century Britain: Robert Chambers and *Vestiges of the Natural History of Creation*', *Victorian Studies* 28 (1984), 5–31.

Yeo, Richard. *Defining Science: William Whewell, Natural Knowledge and the Public Debate in Early Victorian Britain* (Cambridge, 1993).

Zammito, John H. *Kant, Herder, and the Birth of Anthropology* (Chicago, 2002).

Zammito, John H. 'The Lenoir Thesis Revisited: Blumenbach and Kant', *Studies in History and Philosophy of Biological and Biomedical Sciences* 43 (2012), 120–32.

Zammito, John H. *The Gestation of German Biology: Philosophy and Physiology from Stahl to Schelling* (Chicago, 2018).

Zeller, Eduard. *Über Bedeutung und Aufgabe der Erkenntniss-Theorie. Ein akademischer Vortrag* (Heidelberg, 1862).

'Zeta'. 'The Scientific Amusements of London', *London Polytechnic Magazine and Journal* 1 (1844), 225–34.

Zloczower, Avraham. *Career Opportunities and the Growth of Scientific Discovery in Nineteenth Century Germany, with Special Reference to the Development of Physiology* (New York, 1981).

Zourabichvili, François. 'Leibniz et la barbarie', in Bertrand Binoche, ed., *Les Équivoques de la civilisation* (Seyssel, 2005), 33–53.

General Index

This is a combined index to all four volumes in the 'Science and the Shaping of Modernity' series. The volumes are designated as follows:

I   *The Emergence of a Scientific Culture: Science and the Shaping of Modernity, 1210–1685*
II   *The Collapse of Mechanism and the Rise of Sensibility: Science and the Shaping of Modernity, 1680–1760*
III   *The Natural and the Human: Science and the Shaping of Modernity, 1739–1841*
IV   *Civilization and the Culture of Science: Science and the Shaping of Modernity, 1795–1935*

Newton, Isaac (1642–1727) **I**. v, 6, 16, 25, 282, 397–9, 414, 434–8, 463–4, 504; alchemy 467–8; 'Certain Philosophical Questions' 446; *De gravitatione* 446–7, 463–4, 466–7; *De motu* 447–9; *Opticks* 463n.17; optics of the spectrum 379–80, 390–7, 456, 466, 471, 508; *Principia* 20, 170, 257, 290, 319, 322, 352. 449–51, 463, 467, 507; **II**. 11–16, 35–37, 38, 55–92, 119, 157–9, 174, 184–6, 225, 234n.10, 245, 263–5, 280, 284, 296, 299, 305, 314, 325–6, 330–1, 338, 363, 367, 368, 376–7, 409, 437–40, 450; alchemy/chemistry 84–6, 89–90; calculus 125–6, 135–40, 146, 243, 255, 305; *Chronology of Ancient Kingdoms* 373–4; *De aere et aethere* 88; *De gravitatione* 55, 88–9, 113, 346; *De motu corporum* 56, 64–6, 309; *De natura acidorum* 84n.64, 90; dynamics 61–83, 107, 113–15, 117, 119–20, 145, 147, 148, 248, 253–6, 305–11, 315–16, 371; experiments on the spectrum 6, 94, 153–5, 173, 181, 197–8, 205, 209, 217, 300, 301; magnetism and electricity 200–1, 337, 344; *Opticks* and *Optice*, 16, 91–4, 184, 210–12, 251–2, 329, 331, 333–5, 340, 450–1; *Principia*, 7, 13, 16, 31, 37, 55–92, 97, 115–16, 125–6, 146, 184–6, 219, 247, 250, 251, 294, 304–5, 309, 318–20, 332–3, 345, 376, 437n.44, 450–1; *Questiones quaedem philosophicae* 84–5; Newtonianism 3–4, 7, 184–6, 229, 261, 267, 274, 334–5, 345–6, 351; in France: 97, 229, 247–56, 257–8, 304–17, 344, 360–1, 437n.44; **III**. 20, 23, 32, 36–8, 53, 54, 61, 63, 86, 88–90, 111, 116, 157, 168–9, 203–4, 223; on chronology 337; **IV**. 52, 54, 60, 85–6, 89, 91, 105–6, 120, 121, 123n.6, 129, 133, 177, 241, 243,273, 278, 296, 318, 328–9, 350

Nicaea, Council of (325) **I**. 53, 61

Nicéron, Jean–François (1613–46) **II**. 101

Nicholas I, Pope (r. 856–67) **I**. 60n.50

Nicholas of Autrecourt (c.1300 to post–1350) **I**. 69, 74, 200n.15

Nicholas of Cusa (1401–64) **I**. 88

Nicholson, William (1753–1815) **III**. 95, 110

Nicolai, Christoph Friedrich (1733–1811) **III**. 22, 40

Nicole, Pierre (1625–95) **II**. 169

Nicomachus of Gerasa (fl. 100) **I**. 247

Nietzsche, Friedrich (1844–1900) **I**. 11; **IV**. 19, 43, 287

Nieuwentijdt, Bernard (1654–1718) **II**. 135n.102, 141, 243

Nifo, Agostino (c.1496–1538) **I**. 109, 160–4

nitrous oxide **III**. 166–7

Nitsch, Friedrich August (d. *c.* 1813) **III**. 60

Noah **I**. 497, 500–1; **II**. 277–8; **III**. 326; **IV**. 33, 35

Nolan, Frederick (1784–1864) **IV**. 84n.24.

Nollet, Jean Antoine (1700–70) **II**. 206, 332, 339–44, 349–50, 356–7

Nordau, Max (1849–1923) **IV**. 43

Northumberland, Earl of, *see* Percy, Henry

Norman, Robert (fl. 1590) **I**. 223

norms **IV**. 14, 391–420

North British Review **IV**. 5

Nouvelles de la République des Lettres **II**. 240, 273, 431n.25

Novalis (Georg Friedrich Philipp von Hardenberg) (1772–1801) **III**. 313

nuclear physics **I**. 15; **IV** 98–101

Oberth, Hermann (1894–1989) **IV**. 382–3

objective being **I**. 242, 328

objectivity **I**. 1–2, 196, 222, 228–49, 378; **II**. 27, 152

occasionalism **II**. 26–7, 89, 159, 172

occult qualities **II**. 82, 93, 116, 198, 247, 248, 250–1, 265, 437

Ocellus Lucanus (2nd century BCE) **I**. 498

Ockham, William of (c.1285–c.1349) **I**. 74n.98, 81–2, 210n.53; **II**. 122

Odoevskii, Vladimir F. (1804–69) **IV**. 384

oil **IV**. 69–71.

Oldenburg, Henry (c.1617–77) **I**. 37, 380, 503n.106; **IV**. 2

Olivi, Pierre (c.1248–98) **I**. 84

ontology **IV**. 11, 80, 120, 142, 156, 183–7, 265

optics and the theory of light **I**. 5, 6, 21, 33, 47, 95–6, 213, 229, 257, 268–9, 311, 331, 337, 371–2, 402, 413; **II**. 6, 8–19, 34, 197–8, 206; chromatic aberration 325–6; reflection 68; refraction 197–8, 251–2, 325–6; **III**. 3, 32, 35, 77, 159, 179; **IV**. 2, 46, 90–1, 111, 129n.21, 145, 222, 241, 244, 260, 296, 298, 312, 322, 428, 434

orangutan **III**. 233–4, 237–42

orbits, planetary **I**. 6–7, chaotic 439n.80; eccentric 120, 122, 169, 182

Oresme, Nicholas (c.1320–82) **I**. 74–5, 88

organic realm **I**. 259–60

original sin **I**. 52

Origen (c.185–c.254) **I**. 55, 62, 84, 90, 135–6, 199n.10; **II**. 31

Ørsted, Hans Christian (1777–1851) **III**. 114; **IV**. 106, 122

Osterhammel, Jürgen **IV**. 20, 26–7, 34

Ostwald, Friedrich Wilhelm (1853–1932) **I**. 25n.56; **IV**. 93, 96–8, 120, 144, 148–9

Otto III, Holy Roman Emperor (980–1002) **I**. 142n.38

Ottoman Empire, **IV**. 34–5

Overton, Richard (fl. 1646) **I**. 109